Nanoscale Science and Engineering Forum 2015

Core Programming Area at the 2015 AIChE Annual Meeting

Salt Lake City, Utah, USA
8-13 November 2015

ISBN: 978-1-5108-1866-8

Printed from e-media with permission by:

Curran Associates, Inc.
57 Morehouse Lane
Red Hook, NY 12571

Printed by Curran Associates, Inc. (2016)

AIChE
120 Wall Street, FL 23
New York, NY 10005-4020

Phone: (800) 242-4363
Fax: (203) 775-5177

www.aiche.org

Additional copies of this publication are available from:

Curran Associates, Inc.
57 Morehouse Lane
Red Hook, NY 12571 USA
Phone: 845-758-0400
Fax: 845-758-2634
Email: curran@proceedings.com
Web: www.proceedings.com

TABLE OF CONTENTS

432868 Low-Temperature Approaches to Inorganic Photovoltaic Thin Films

Abstract:

Solution-based precursors are promising route to synthesizing thin film photovoltaic semiconductors which can be printed using low cost, high-throughput techniques such as roll-to-roll processing. A major limitation is the processing of these precursors to semiconductors; in order to achieve electronic and optoelectronic properties which are suitable for application in photovoltaics, high temperatures and harsh processing atmospheres are often required, which limits the benefits of solution-based precursors. In this presentation, we discuss some low-temperature approaches to synthesizing thin films at low temperature with excellent electronic properties. For example, we find that metastable wurtzite CdTe tetrapods undergo a phase transformation at 350°C and undergo sintering to produce grain sizes of several hundred nanometers. Additionally, thin films passivated with compositionally-matched chalcogenidomettalate clusters can have electron carrier mobilities in excess of 200 $cm^2 V^{-1} s^{-1}$. When integrated into photovoltaic devices, we find that the power conversion efficiencies greater than 12% along with evidence that much higher efficiencies could be achieved with optimization of contact layers.

429232 Symmetry-Breaking in Light-Trapping Nanostructures on Silicon for Solar Photovoltaics

Abstract:

In thin-film photovoltaics, highly absorptive materials are conventionally used. However, these materials have achieved efficiencies that are not comparable to those of thick crystalline silicon (c-Si) photovoltaics and, in some cases, suffer from their toxicity and low supply. A viable solution to these problems would be to use c-Si for thin-film photovoltaics. However, thin c-Si films absorb sunlight weakly because of its indirect band gap and strong light-trapping should be provided to achieve high efficiency. For thin-film photovoltaics, nanoscale structures are typically involved for light trapping because the film thickness becomes comparable to the wavelength of sun light. While diverse nanostructures have been studied to break the light-trapping limit of geometric optics, known as the Lambertian limit, highly efficient nanostructures that can be easily manufactured have not been demonstrated. We have previously predicted that symmetry-breaking in light-trapping periodic nanostructures on thin films can approach the Lambertian limit very closely. Herein, we will present how the systematic symmetry-lowering increases light-trapping in c-Si thin-film photovoltaics. We will demonstrate the experimental realization of such low-symmetry structures using simple wet etching methods on c-Si(100) wafers without any off-cut, tilt angle. Further, we will discuss the optical characterization of our fabricated structures on thin c-Si films.

425940 Controlling Morphology of Photovoltaic Thin Films By Phase Transformation of Metastable Colloidal Nanocrystals

Abstract:

Colloidal nanocrystals are promising solution processible precursors to thin films of complex semiconductors, including $CuZnSnS_4$ and $CuInSe_2$. Detrimental impurity phases and spatial fluctuations of the composition can be minimized by controlling stoichiometry during colloidal synthesis. However, the preparation of large-grained, electronic quality thin films from such nanocrystals has been challenging. We synthesize nanocrystals of these materials in a metastable hexagonal crystal phase and leverage their thermally driven conversion to the stable phase to engineer grain size and morphology in the resulting polycrystalline films. The activation energy for the phase transformation, determined by variable heating rate *in situ* X-ray diffraction, is found to depend strongly on nanocrystal shape. Using nanorods around 40 nm in length, we apply classical nucleation and growth principles to tune the thermal processing conditions to achieve large, micron scale grains in a matter of minutes. To exclude carbon impurities that can negatively impact optoelectronic properties, the same nanocrystals can be processed using inorganic chalcogenidometallates (ChaMs) as ligands. Upon heating, the ChaMs form a glassy matrix around each nanocrystal that inhibits nucleation of the stable phase (observed as a higher activation energy for nucleation), yet facilitates grain growth by coarsening. Interdiffusion ultimately results in alloyed thin films with grains that span their full thickness. Optical and electronic characterization of the resulting thin films shows promise for their application in thin film photovoltaic devices.

427513 Plasmonic Enhancement of Mesoporous Solar Cells with Shape Controlled Nanostructures

Abstract:

Plasmonic Enhancement of Mesoporous Solar Cells with Shape Controlled Nanostructures

Rizia Bardhan

Department of Chemical and Biomolecular Engineering,

Vanderbilt University, Nashville, TN 37235

Next generation mesoporous solar cells (MSCs) including dye-sensitized solar cells (DSSCs) and perovskite-sensitized solar cells (PSSCs) have rapidly emerged with efficiencies reported up to 13% for DSSCs and 20% for PSSCs. Current efforts to improve the performance of MSCs are focused on altering the photoanode morphology, and manipulating the sensitizer composition. In contrast to these approaches that often lead to incremental improvement tailored to each MSC system, plasmonic enhancement provides a universal route applicable to the whole family of MSCs that can significantly boost the optical absorption and carrier generation. Through small additions of metal nanoparticles to MSCs (< 2%), the amount of sensitizer required to achieve high efficiency can be drastically reduced enabling thin film architectures compatible with scalable nanomanufacturing routes including roll-to-roll processing and inkjet printing. Here we demonstrate enhanced light harvesting in both DSSCs and PSSCs by embedding shape-controlled Au nanostructures, and Au/Ag core/shell bimetallic nanostructures in the photoanodes. Plasmonic enhancement resulted in 25 – 35 % enhancement in DSSCs and >15% enhancement in PSSCs. Photocurrent and IPCE spectra revealed that device performance in MSCs is controlled by particle density of plasmonic nanostructures and monotonically decreases at high concentrations. Transient absorption pump-probe spectroscopy revealed that presence of metal nanostructures leads to enhanced exciton generation, and faster electron injection into TiO_2 conduction band before recombination can occur in the bulk, resulting in shorter exciton lifetime. This work shows that shape-controlled plasmonic nanostructures can be employed as a universal platform for enhanced light-trapping in a range of solar devices.

420359 Effect of Uniquely Assembled Nanostructures on Photovoltaic Properties

Abstract:

This study focuses on the energy transfer and electron transport in the photovoltaic device with uniquely designed nanostructures of an acceptor/donor energy system and an electron carrier delivery system. Furthermore, the effect of both the assembled nanostructures of acceptor/donor energy delivery and the different morphology of nanomaterials on photovoltaic performance are studied.

Förster resonance energy transfer between the xanthene dye (donor of energy) and a new polymethine dye (acceptor of energy) was studied on the surface of TiO_2 films in DSSCs. The sensitization of semiconductor films by the donor acceptor compound leads to a doubling of energy conversion efficiency. Measurements show that this is due to the widening of photo sensitivity of the cell in the blue region of the spectrum. Increasing the total energy absorbed leads to increasing the number of generated charge carriers at the interface of TiO_2 and the dyes.

Zinc Oxide nanoarrays with rod and sheet morphologies were fabricated and assembled for electron transport study of the dye sensitized solar cells (DSSCs). A photoluminescence spectra study indicated that nanosheets had more defect density compared to nanorods, which resulted in a lower open circuit photovoltage. Electrochemical impedance spectroscopy (EIS) and dye absorption amount analysis further explained the high short-circuit current density (J_{sc}) of the DSSC with ZnO nanosheet structure, based on the dye-loading amount, effective electron lifetime, and electron density.

418019 Applying the Chemistry of Amine-Thiol Mixtures for Solution-Processed CdTe Thin Films

Abstract:

Solution-processing promises to revolutionize the electronic industry by providing cost-effective and rapid fabrication of flexible electronic devices. Whereas traditional electronic devices are processed on hard substrates using slow and costly vacuum-based batch processing, solution-processing enables rapid, roll-to-roll manufacturing on plastic films and foils. Electronic devices on such substrates are desirable for wearable electronics and to apply conformally to any surface. Important applications of solution-processed electronics include photovoltaics, thermoelectrics, thin film transistors, batteries, and sensors.

Ultimately, it is desirable to combine the superior performance of inorganic electronic materials with the facile processing of organic solutions. Unfortunately, most inorganic salts that would be useful precursors are insoluble in many organic solvents compatible with roll-to-roll, solution-based techniques. We have overcome this challenge by developing a mixture of commonly available thiols and amines to dissolve a host of materials that are otherwise insoluble in either solvent by itself. The solvent system not only takes the form as pure amine-thiol mixtures, but also as mixtures such as thioglycolic acid-ethanolamine,[1] ethanolamine-thioacetamide,[2] and ammonium thioglycolate solutions[3] in water. These mixtures have been used with great success for nanoparticle synthesis,[4,5] photovoltaic absorber layers,[1,6,7] luminescent quantum dot films,[1] and other thin films[8]making it a very general solvent system for processing of inorganic thin films.

Here we present for the first time the fabrication of CdTe thin films via a solution-processed molecular precursor approach. We further elucidate the chemistry of the amine-thiol system using advanced tandem mass spectrometry techniques such as ESI and APCI to identify species in solution and verify their structure using quantum chemical calculations and modeling. In a typical preparation $CdCl_2$and Te are each dissolved in 1-propanethiol and ethylenediamine in concentrations ranging from 0.15 M to 0.22 M. ITO-coated or Mo-coated glass are then spin-coated with 375 μL of the solution. To obtain a film thickness of about 1 μm, eight to twenty-five coatings are applied, depending on the concentration and spin coating speed employed. Between each coating, the film is annealed on a hot plate at 500 °C to grow the CdTe phase. Grazing incidence x-ray diffraction reveals that pure CdTe can be recovered under optimized annealing conditions. A plane view and cross section of a CdTe film are given in Figure 1. However, we have found that very short annealing times (less than 2 mins) have the potential to produce a small amount of a segregated Te phase. Scanning electron microscopy coupled with energy dispersive x-ray spectroscopy (EDS) linescans has confirmed the presence of this phase and found that it is located at the surface of the film.

Modest efficiencies of about 0.2% are currently obtained from these films. One limitation we have found is significant chlorine residue remaining in the film, often in between grain boundaries. Cadmium acetylacetonate can be used to either partially or entirely replace $CdCl_2$ as a precursor to reduce the presence of chlorine, but films prepared in this manner have much smaller grains than films prepared with $CdCl_2$and have more voids. Further insight into the solution chemistry and grain growth will enable the removal of chlorine contamination and improvement of device performance.

References

(1) Tian, Q.; Wang, G.; Zhao, W.; Chen, Y.; Yang, Y.; Huang, L.; Pan, D. *Chem. Mater.* **2014**, *26*(10), 3098.

(2) Sun, Y.; Zhang, Y.; Wang, H.; Xie, M.; Zong, K.; Zheng, H.; Shu, Y.; Liu, J.; Yan, H.; Zhu, M.; Lau, W. *J. Mater. Chem. A* **2013**, *1*(23), 6880.

(3) Tian, Q.; Huang, L.; Zhao, W.; Yang, Y.; Wang, G.; Pan, D. *Green Chem.* **2015**, *17*(2), 1269.

(4) Yang, W.-C.; Miskin, C. K.; Hages, C. J.; Hanley, E. C.; Handwerker, C.; Stach, E. A.; Agrawal, R. *Chem. Mater.* **2014**, *26*(11), 3530.

(5) Walker, B. C.; Agrawal, R. *Chem. Commun. (Camb).* **2014**, *50*(61), 8331.

(6) Yang, Y.; Wang, G.; Zhao, W.; Tian, Q.; Huang, L.; Pan, D. *ACS Appl. Mater. Interfaces* **2014**, *7*(1), 460.

(7) Zhang, R.; Szczepaniak, S. M.; Carter, N. J.; Handwerker, C. A.; Agrawal, R. *Chem. Mater.* **2015**, *27*(6), 2114.

(8) Antunez, P. D.; Torelli, D. A.; Yang, F.; Rabu, F. A.; Lewis, N. S.; Brutchey, R. L. *Chem. Mater.* **2014**, *26* (19), 5444.

Production of Cu(InGa)(SeS)$_2$ absorber layers via rapid thermal processing of selenium-capped metal precursors for thin film solar cells

R.J. Lovelett*†, W.N. Shafarman†, R.W. Birkmire†, B.A. Ogunnaike*

University of Delaware
*Department of Chemical and Biomolecular Engineering
†Institute of Energy Conversion

November 8, 2015

Introduction

Solar energy is experiencing rapid growth

- 36% of all newly installed US electric capacity in 2014 is solar-based

Thin films offer advantages over dominant technology, crystalline silicon

- Simple, low-cost manufacturing
- CuInGaSe$_2$ (CIGS) record cell efficiency is 21.7%, best among thin films and greater than mc-Si

There are multiple viable manufacturing routes to produce CIGS modules

- Elemental coevaporation
- Precursor reaction/selenization

Introduction

Solar energy is experiencing rapid growth

- 36% of all newly installed US electric capacity in 2014 is solar-based

Thin films offer advantages over dominant technology, crystalline silicon

- Simple, low-cost manufacturing
- $CuInGaSe_2$ (CIGS) record cell efficiency is 21.7%, best among thin films and greater than mc-Si

There are multiple viable manufacturing routes to produce CIGS modules

- Elemental coevaporation
- **Precursor reaction/selenization**

CIGS Properties

CIGS is an inorganic, p-type, polycrystalline semiconductor

- Chalcopyrite crystal structure
- Direct band gap with high optical absorption
- Forms a **continuous solid solution** of $CuInSe_2$ and $CuGaSe_2$
- Band gap can be adjusted by In:Ga ratio or alloying with Ag, Al, S or other elements
  - **Band gap can be tuned** to match the solar spectrum
  - **Through-film gradient** in material affects electrical properties

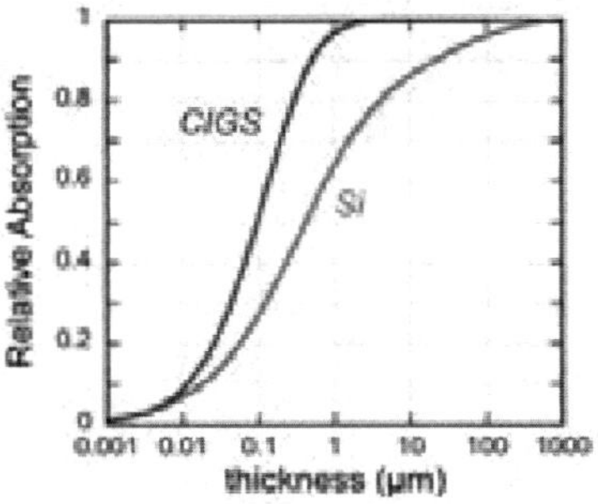

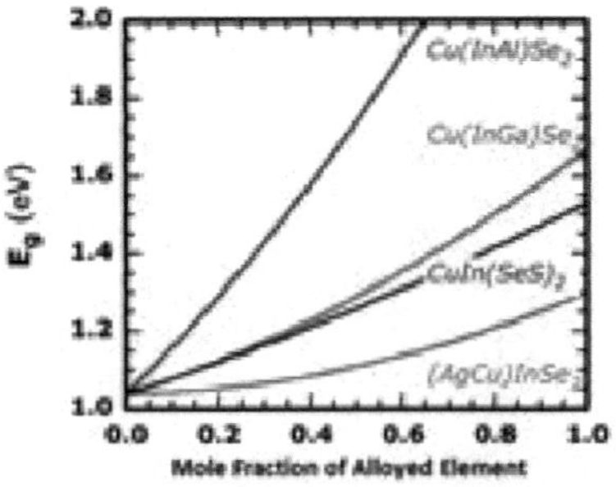

CIGS Solar Cells

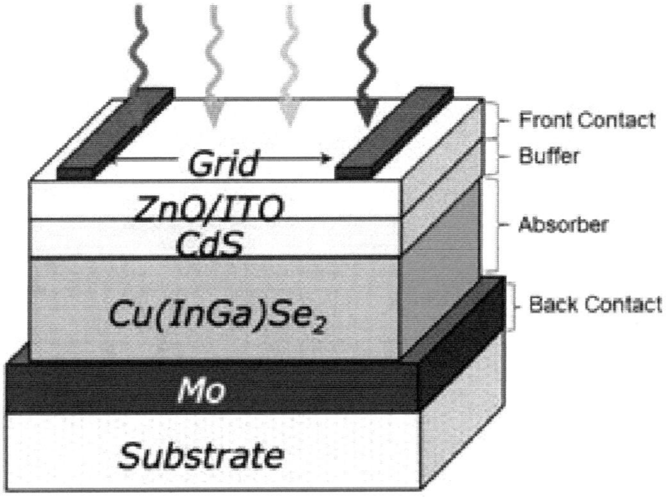

Device structure of a typical CIGS cell[1]

[1]Shafarman, Siebentritt, and Stolt, *Handbook of Photovoltaics*, 2011

CIGS Production by Precursor Reaction/Selenization

- Consists of two step process:
  1. Deposit Cu-In-Ga precursor
  2. React in H_2Se and/or H_2S gas
- Alternative approach: include selenium in precursor; react in H_2S
- Challenges:
  - Slow kinetics mean **low process throughput** (residence time ~1 hour)
  - **Asymmetric reactions cause gradients** in through-film compositions: accumulation of Ga often seen at back contact

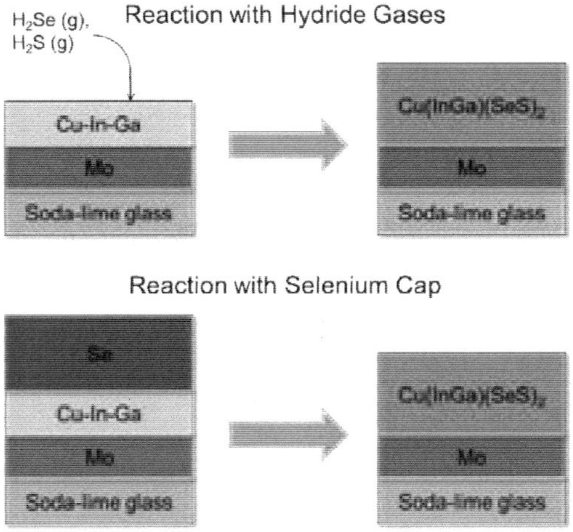

CIGS Production by Precursor Reaction/Selenization

- Consists of two step process:
  1. Deposit Cu-In-Ga precursor
  2. React in H_2Se and/or H_2S gas
- Alternative approach: include selenium in precursor; react in H_2S
- Challenges:
  - Slow kinetics mean **low process throughput** (residence time ∼1 hour)
  - **Asymmetric reactions cause gradients** in through-film compositions: accumulation of Ga often seen at back contact

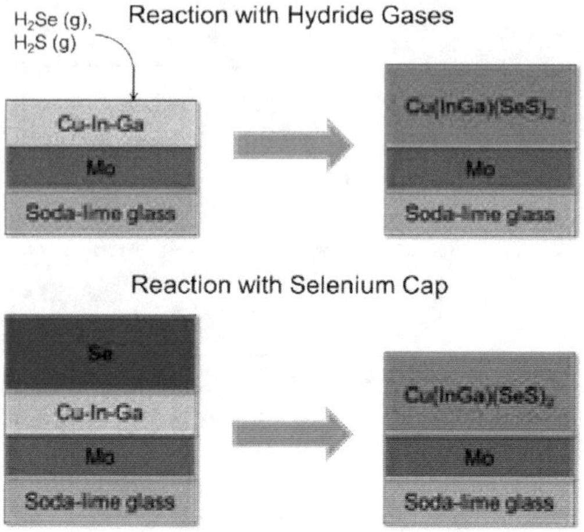

Approach: study reactions of Se-capped precursors in custom-built rapid thermal processing (RTP) reactor to develop fast process for absorber synthesis.

Motivation and Goals

- Selenization process promising method to produce high-efficiency modules with low cost and high throughput
- Slow kinetics and toxic gas handling limit throughput and technology deployment

Motivation and Goals

- Selenization process promising method to produce high-efficiency modules with low cost and high throughput
- Slow kinetics and toxic gas handling limit throughput and technology deployment

High temperature process using **Se-capped precursors** is viable alternative

- Process to be investigated with:
  1. Experimental studies: film and device properties
  2. Growth modeling and simulation

Motivation and Goals

- Selenization process promising method to produce high-efficiency modules with low cost and high throughput
- Slow kinetics and toxic gas handling limit throughput and technology deployment

High temperature process using **Se-capped precursors** is viable alternative

- Process to be investigated with:
  1. **Experimental studies: film** and device **properties**
  2. Growth modeling and simulation

Reaction with Se-Capped Precursor

CIGS Deposition Process:

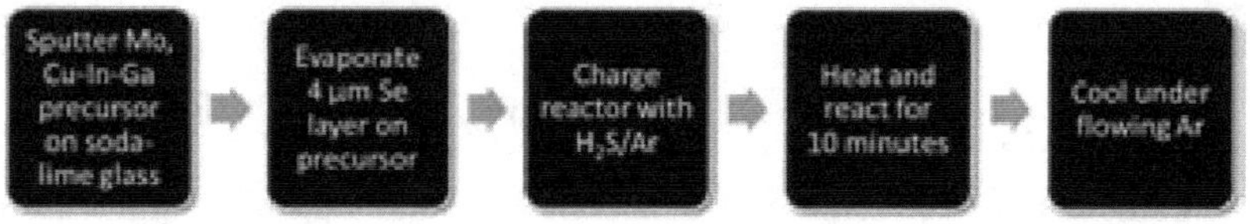

Approach: 2^3 Factorial Design
Factors:

- Temperature ramp rate (2–6 K/s)
- Dwell temperature (500–600 °C)
- H_2S concentration (1–10%)

Material Characterization/Responses:

- Elemental composition (EDX and XRF)
- Phase composition and analysis (XRD)

Film Properties

Process effects from EDX (~ 1/2 depth of film sampled) results:

- S/(S+Se) and Cu/(In+Ga) increase with H_2S concentration
- Ga/(In+Ga) increases with temperature
- No effect of ramp rate within the range studied

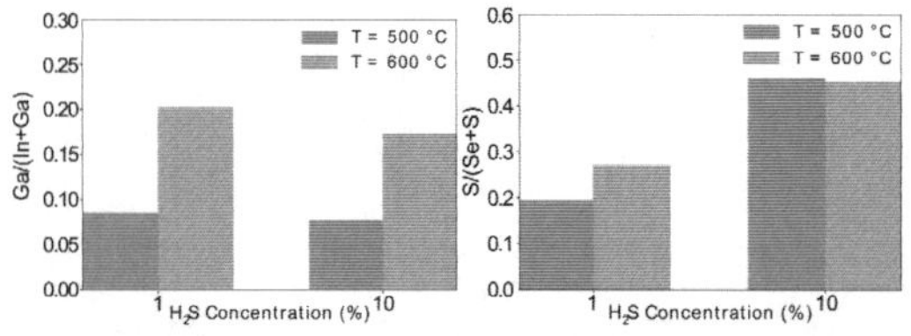

Film Properties

Process effects from EDX (~ 1/2 depth of film sampled) results:

- $S/(S+Se)$ and $Cu/(In+Ga)$ increase with H_2S concentration
- $Ga/(In+Ga)$ increases with temperature
- No effect of ramp rate within the range studied

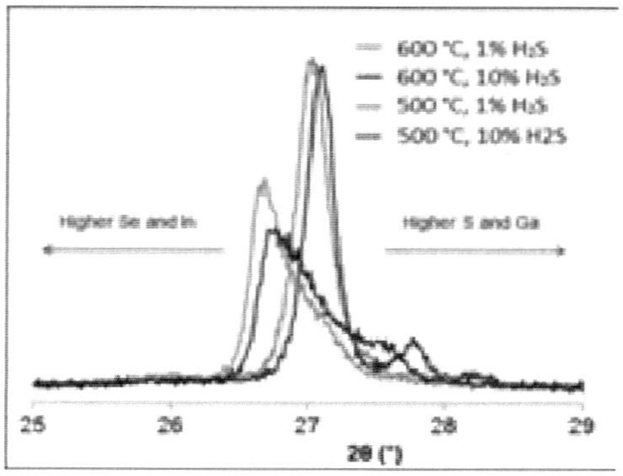

XRD patterns of the CIGS 112 peak of films produced with varying process conditions.

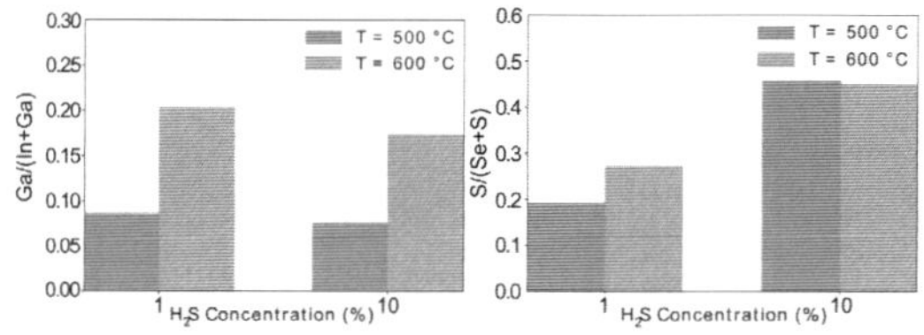

Time Evolution

Film reaches steady states within 5 minute reaction:

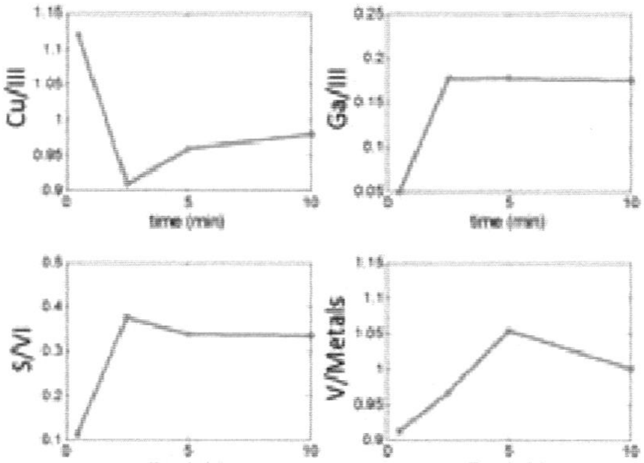

Time evolution of film composition

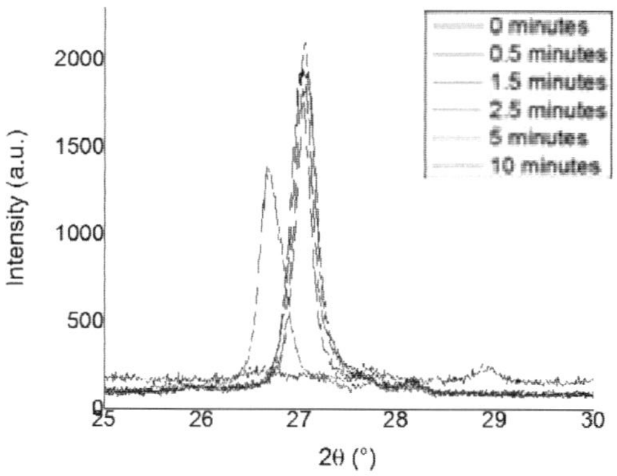

Time evolution of 112 peak

Motivation and Goals

- Selenization process promising method to produce high-efficiency modules with low cost and high throughput
- Slow kinetics and toxic gas handling limit throughput and technology deployment

High temperature process using **Se-capped precursors** is viable alternative

- Process to be investigated with:
  1. **Experimental studies:** film and **device properties**
  2. Growth modeling and simulation

Reaction with Se-Capped Precursor

CIGS Deposition Process:

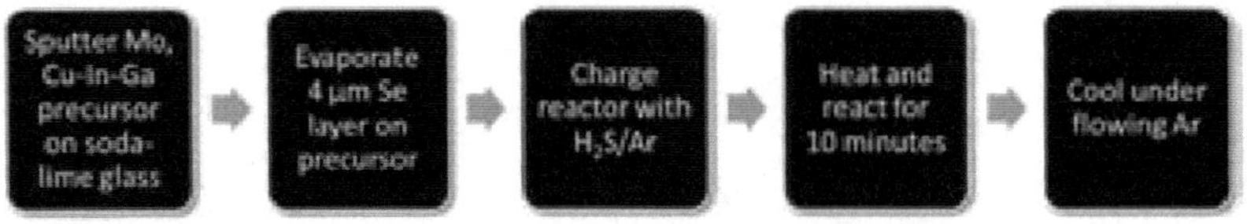

- Complete device synthesis: CdS CBD; sputter ZnO/ITO; Evaporate grids; mechanically scribe cells
- Measure JV and QE

Device Results: JV Curves

- Devices produced from CIGS from Se-capped process at varying H_2S and temperature.
- Devices from low temperature process shunted ($\eta \approx 0$). Small concentration unreacted phases likely.

Introduction Thin Film Characterization **Device Results** Stochastic Modeling Conclusions and Future Work

Device Results: JV Curves

- Devices produced from CIGS from Se-capped process at varying H_2S and temperature.
- Devices from low temperature process shunted ($\eta \approx 0$). Small concentration unreacted phases likely.

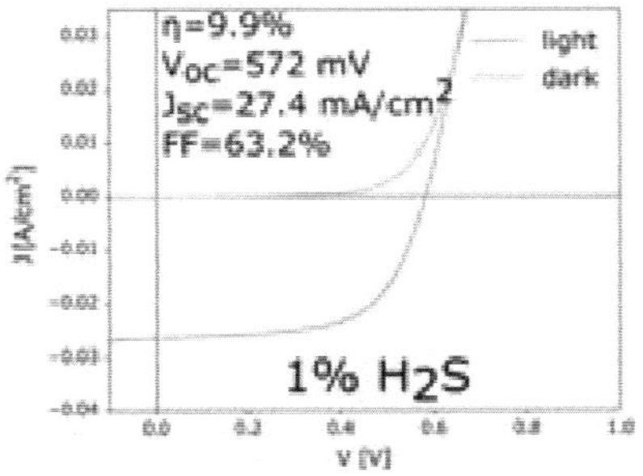

R.J. Lovelett, W.N. Shafarman, R.W. Birkmire, B.A. Ogunnaike University of Delaware
CIGS Production via Rapid Thermal Processing of Se-capped Precursors 13 of 22

Device Results: JV Curves

- Devices produced from CIGS from Se-capped process at varying H_2S and temperature.
- Devices from low temperature process shunted ($\eta \approx 0$). Small concentration unreacted phases likely.

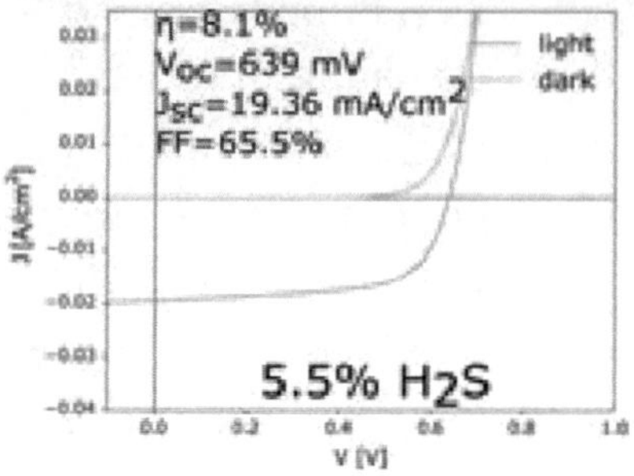

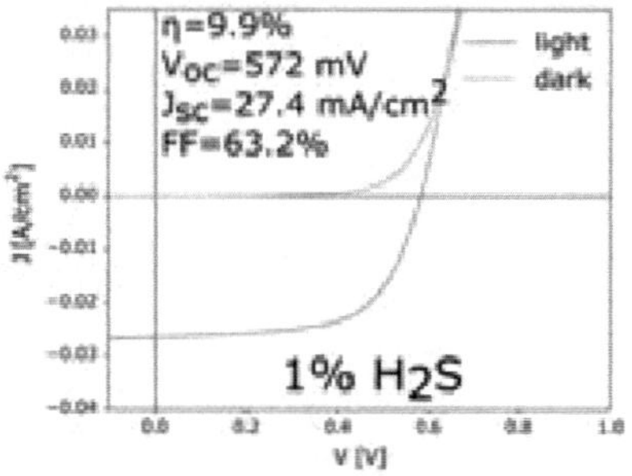

Introduction | Thin Film Characterization | Device Results | Stochastic Modeling | Conclusions and Future Work

Device Results: JV Curves

- Devices produced from CIGS from Se-capped process at varying H_2S and temperature.
- Devices from low temperature process shunted ($\eta \approx 0$). Small concentration unreacted phases likely.

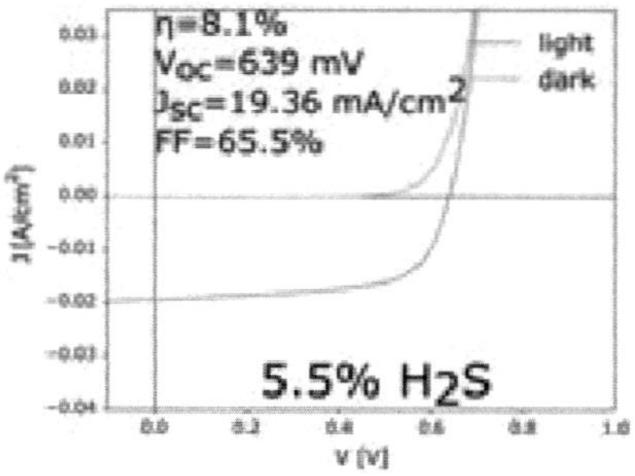

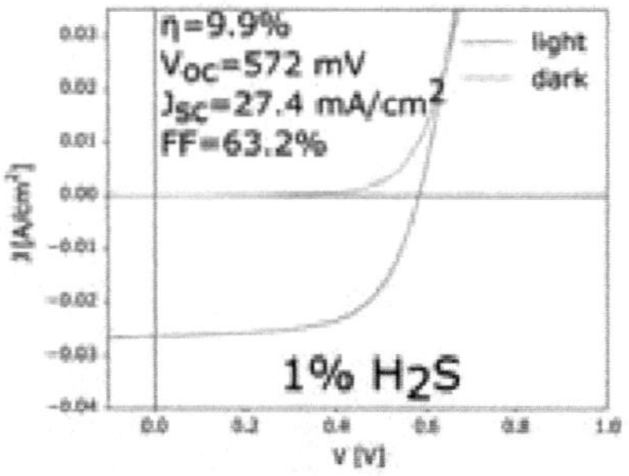

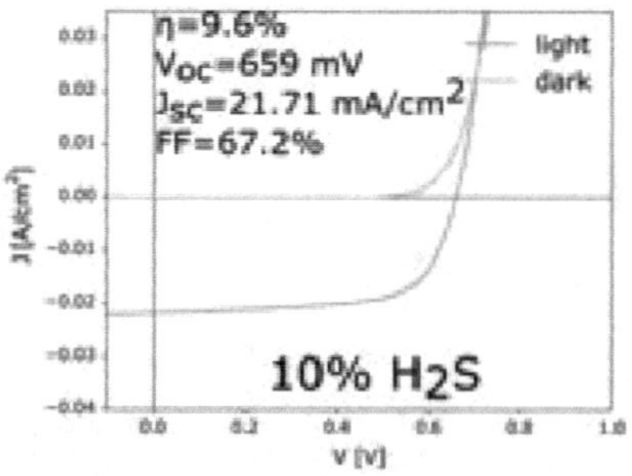

R.J. Lovelett, W.N. Shafarman, R.W. Birkmire, B.A. Ogunnaike | University of Delaware
CIGS Production via Rapid Thermal Processing of Se-capped Precursors | 13 of 22

Device Results: QE Curves

- Quantum efficiency (QE) gives fraction of radiation converted to electricity at each wavelength
- Optical bandgap of device determined from inflection point in long-wavelength region

Introduction Thin Film Characterization Device Results Stochastic Modeling Conclusions and Future Work

Device Results: QE Curves

- Quantum efficiency (QE) gives fraction of radiation converted to electricity at each wavelength
- Optical bandgap of device determined from inflection point in long-wavelength region

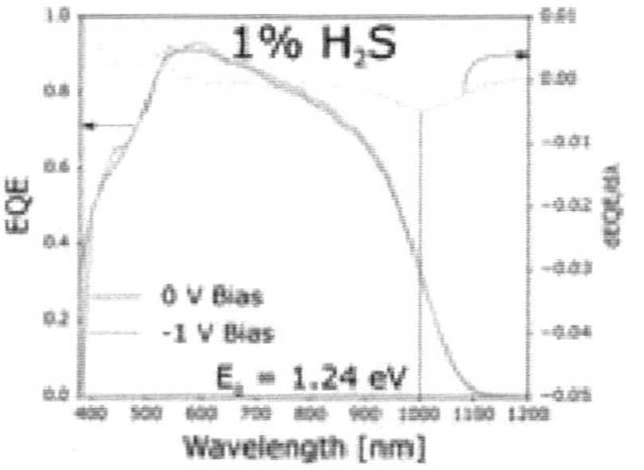

R.J. Lovelett, W.N. Shafarman, R.W. Birkmire, B.A. Ogunnaike University of Delaware
CIGS Production via Rapid Thermal Processing of Se-capped Precursors 14 of 22

Device Results: QE Curves

- Quantum efficiency (QE) gives fraction of radiation converted to electricity at each wavelength
- Optical bandgap of device determined from inflection point in long-wavelength region

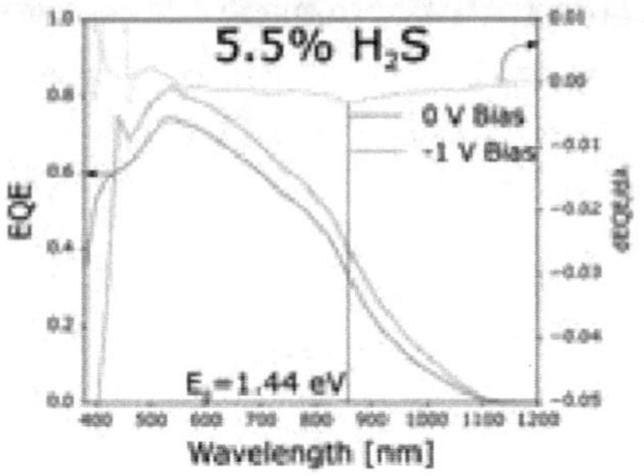

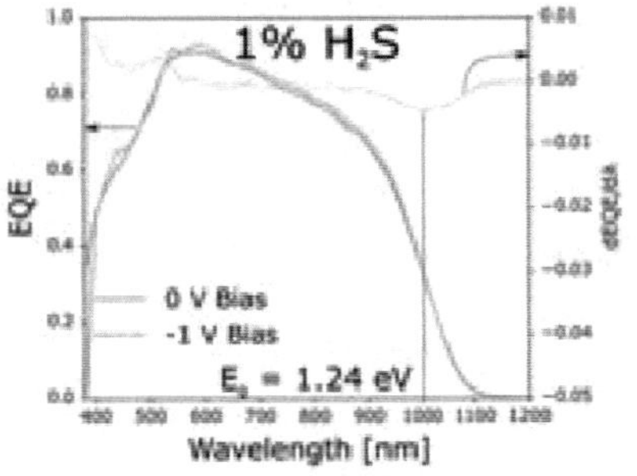

Introduction Thin Film Characterization **Device Results** Stochastic Modeling Conclusions and Future Work

Device Results: QE Curves

- Quantum efficiency (QE) gives fraction of radiation converted to electricity at each wavelength
- Optical bandgap of device determined from inflection point in long-wavelength region

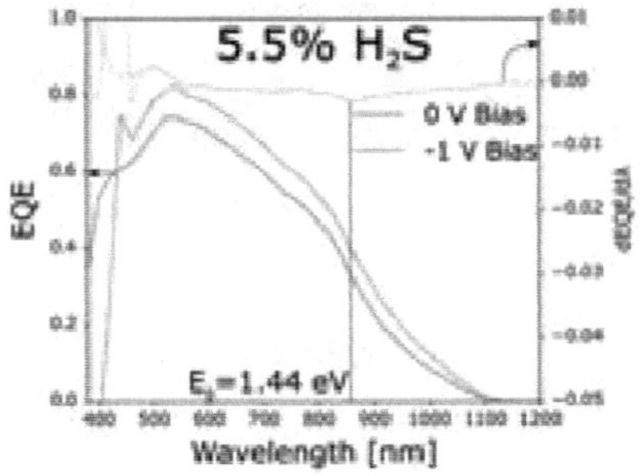

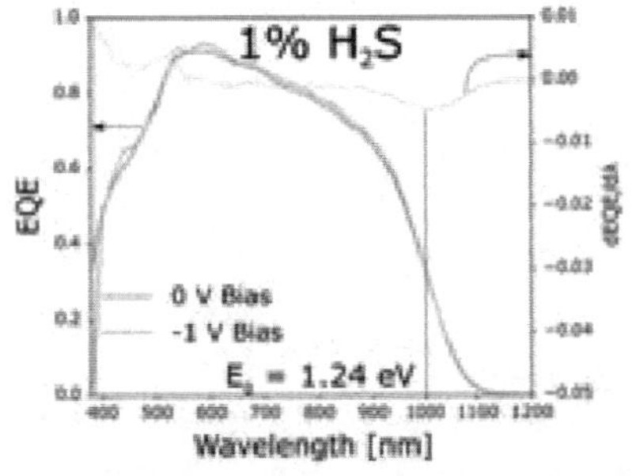

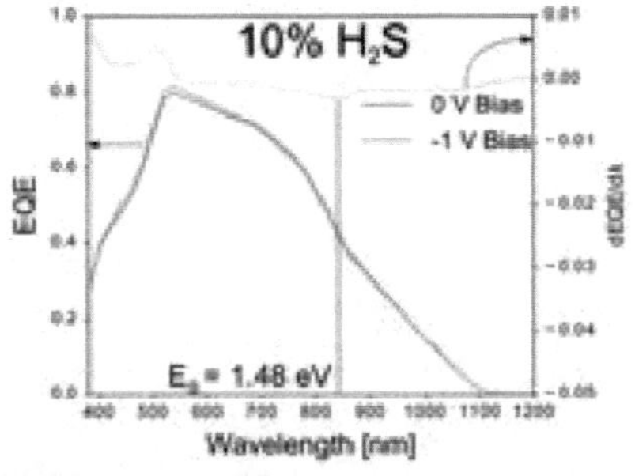

R.J. Lovelett, W.N. Shafarman, R.W. Birkmire, B.A. Ogunnaike
University of Delaware
CIGS Production via Rapid Thermal Processing of Se-capped Precursors
14 of 22

Experimental Results: Summary and Analysis

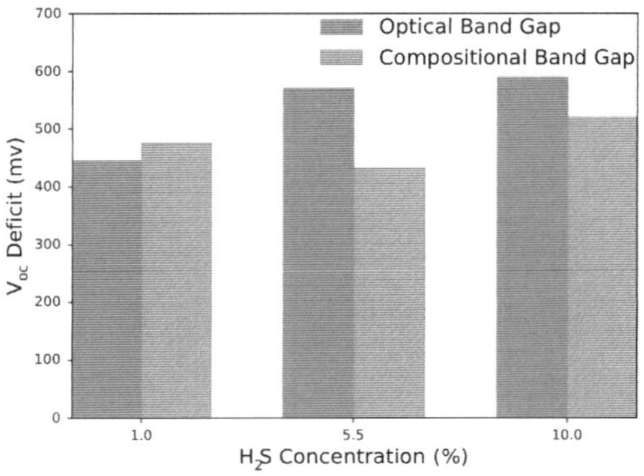

Voltage deficit vs. radiative limit using QE and compositional[1] band gaps

[1] Bär et al., *J. Appl. Phys.*, 2004

- Devices have well-behaved JV curves, but substantial losses compared to state-of-the-art.
- Losses in V_{OC} and FF; possibly high series resistance.
- Future Work: Identify source of V_{OC} deficit and voltage dependent QE with more thorough device analysis: JVT, capacitance profiling

Motivation and Goals

- Selenization process promising method to produce high-efficiency modules at low cost
- Slow kinetics and toxic gas handling limit throughput and technology deployment

High temperature process using **Se-capped precursors** is viable alternative

- Process to be investigated with:
  1. Experimental studies: film and device properties
  2. **Growth modeling and simulation**

Stochastic Model

Motivation: Explain experimental results and to assist with process improvements

Stochastic simulation provides a method to:

- Predict through-film composition
- Decouple and estimate kinetic and transport properties
- Access spatially-confined film properties (e.g., agglomerations)

Model used previously to simulate reactions in H_2Se and H_2S[2]

[2]R.J. Lovelett et al., *IEEE PVSC*, 2015

Introduction | Thin Film Characterization | Device Results | Stochastic Modeling | Conclusions and Future Work

Stochastic Model

Motivation: Explain experimental results and to assist with process improvements

Stochastic simulation provides a method to:

- **Predict through-film composition**
- Decouple and estimate kinetic and transport properties
- Access spatially-confined film properties (e.g., agglomerations)

Model used previously to simulate reactions in H_2Se and H_2S[2]

[2]R.J. Lovelett et al., *IEEE PVSC*, 2015

R.J. Lovelett, W.N. Shafarman, R.W. Birkmire, B.A. Ogunnaike | University of Delaware
CIGS Production via Rapid Thermal Processing of Se-capped Precursors | 17 of 22

Stochastic Simulation of Reaction

Reaction Mechanism/Lattice Events	
Adsorption/ Desorption	Selenium Adsorption = 0 Sulfur Adsorption Selenium Desorption Sulfur Desorption

Introduction | Thin Film Characterization | Device Results | Stochastic Modeling | Conclusions and Future Work

Stochastic Simulation of Reaction

Reaction Mechanism/Lattice Events	
Adsorption/ Desorption	Selenium Adsorption = 0 Sulfur Adsorption Selenium Desorption Sulfur Desorption
Binary Selenide/ Sulfide Formation	$CuIn + 2\,Se \longrightarrow CuSe + InSe$ $CuGa + 2\,Se \longrightarrow CuSe + GaSe$ $CuIn + 2\,S \longrightarrow CuS + InS$ $CuGa + 2\,S \longrightarrow CuS + GaS$

R.J. Lovelett, W.N. Shafarman, R.W. Birkmire, B.A. Ogunnaike | University of Delaware
CIGS Production via Rapid Thermal Processing of Se-capped Precursors | 18 of 22

Stochastic Simulation of Reaction

	Reaction Mechanism/Lattice Events
Adsorption/ Desorption	Selenium Adsorption = 0
	Sulfur Adsorption
	Selenium Desorption
	Sulfur Desorption
Binary Selenide/ Sulfide Formation	$CuIn + 2\,Se \longrightarrow CuSe + InSe$
	$CuGa + 2\,Se \longrightarrow CuSe + GaSe$
	$CuIn + 2\,S \longrightarrow CuS + InS$
	$CuGa + 2\,S \longrightarrow CuS + GaS$
Chalcopyrite Formation	$CuSe + InSe \longrightarrow CuInSe_2$
	$CuSe + GaSe \longrightarrow CuGaSe_2$
	$CuS + InS \longrightarrow CuInS_2$
	$CuS + GaS \longrightarrow CuGaS_2$
	$CuS + InSe \longrightarrow 0.5\,CuInSe_2 + 0.5\,CuInS_2$
	$CuS + GaSe \longrightarrow 0.5\,CuGaSe_2 + 0.5\,CuGaS_2$
	$CuSe + InS \longrightarrow 0.5\,CuInSe_2 + 0.5\,CuInS_2$
	$CuSe + GaS \longrightarrow 0.5\,CuGaSe_2 + 0.5\,CuGaS_2$

Stochastic Simulation of Reaction

	Reaction Mechanism/Lattice Events
Adsorption/ Desorption	Selenium Adsorption = 0
	Sulfur Adsorption
	Selenium Desorption
	Sulfur Desorption
Binary Selenide/ Sulfide Formation	$CuIn + 2\,Se \longrightarrow CuSe + InSe$
	$CuGa + 2\,Se \longrightarrow CuSe + GaSe$
	$CuIn + 2\,S \longrightarrow CuS + InS$
	$CuGa + 2\,S \longrightarrow CuS + GaS$
Chalcopyrite Formation	$CuSe + InSe \longrightarrow CuInSe_2$
	$CuSe + GaSe \longrightarrow CuGaSe_2$
	$CuS + InS \longrightarrow CuInS_2$
	$CuS + GaS \longrightarrow CuGaS_2$
	$CuS + InSe \longrightarrow 0.5\,CuInSe_2 + 0.5\,CuInS_2$
	$CuS + GaSe \longrightarrow 0.5\,CuGaSe_2 + 0.5\,CuGaS_2$
	$CuSe + InS \longrightarrow 0.5\,CuInSe_2 + 0.5\,CuInS_2$
	$CuSe + GaS \longrightarrow 0.5\,CuGaSe_2 + 0.5\,CuGaS_2$
Diffusion	Any pair of species

Stochastic Simulation of Reaction

Reaction Mechanism/Lattice Events	
Adsorption/ Desorption	Selenium Adsorption = 0
	Sulfur Adsorption
	Selenium Desorption
	Sulfur Desorption
Binary Selenide/ Sulfide Formation	$CuIn + 2\,Se \longrightarrow CuSe + InSe$
	$CuGa + 2\,Se \longrightarrow CuSe + GaSe$
	$CuIn + 2\,S \longrightarrow CuS + InS$
	$CuGa + 2\,S \longrightarrow CuS + GaS$
Chalcopyrite Formation	$CuSe + InSe \longrightarrow CuInSe_2$
	$CuSe + GaSe \longrightarrow CuGaSe_2$
	$CuS + InS \longrightarrow CuInS_2$
	$CuS + GaS \longrightarrow CuGaS_2$
	$CuS + InSe \longrightarrow 0.5\,CuInSe_2 + 0.5\,CuInS_2$
	$CuS + GaSe \longrightarrow 0.5\,CuGaSe_2 + 0.5\,CuGaS_2$
	$CuSe + InS \longrightarrow 0.5\,CuInSe_2 + 0.5\,CuInS_2$
	$CuSe + GaS \longrightarrow 0.5\,CuGaSe_2 + 0.5\,CuGaS_2$
Diffusion	Any pair of species

- Discretize film into a square lattice of finite volume elements
- Each lattice element is phase pure
- Time evolution of film takes place in discrete lattice events

Introduction Thin Film Characterization Device Results Stochastic Modeling Conclusions and Future Work

Stochastic Simulation of Reaction

Reaction Mechanism/Lattice Events	
Adsorption/ Desorption	Selenium Adsorption = 0
	Sulfur Adsorption
	Selenium Desorption
	Sulfur Desorption
Binary Selenide/ Sulfide Formation	$CuIn + 2\,Se \longrightarrow CuSe + InSe$
	$CuGa + 2\,Se \longrightarrow CuSe + GaSe$
	$CuIn + 2\,S \longrightarrow CuS + InS$
	$CuGa + 2\,S \longrightarrow CuS + GaS$
Chalcopyrite Formation	$CuSe + InSe \longrightarrow CuInSe_2$
	$CuSe + GaSe \longrightarrow CuGaSe_2$
	$CuS + InS \longrightarrow CuInS_2$
	$CuS + GaS \longrightarrow CuGaS_2$
	$CuS + InSe \longrightarrow 0.5\,CuInSe_2 + 0.5\,CuInS_2$
	$CuS + GaSe \longrightarrow 0.5\,CuGaSe_2 + 0.5\,CuGaS_2$
	$CuSe + InS \longrightarrow 0.5\,CuInSe_2 + 0.5\,CuInS_2$
	$CuSe + GaS \longrightarrow 0.5\,CuGaSe_2 + 0.5\,CuGaS_2$
Diffusion	Any pair of species

- Discretize film into a square lattice of finite volume elements
- Each lattice element is phase pure
- Time evolution of film takes place in discrete lattice events
- Two random variables, that are functions of lattice state and **propensity constants**, are chosen at each time step
  - Event Propensity: $a_i = p_i N_i$
  - Event Probability: $P_i = \frac{a_i}{\Sigma a_j}$
  - Event Time PDF: $f(\tau) = \frac{1}{\Sigma a_i} \exp\left(-\frac{\tau}{\Sigma a_i}\right), \tau \geq 0$

Model Results

Higher sulfur concentration moves band gap gradient closer to front of film

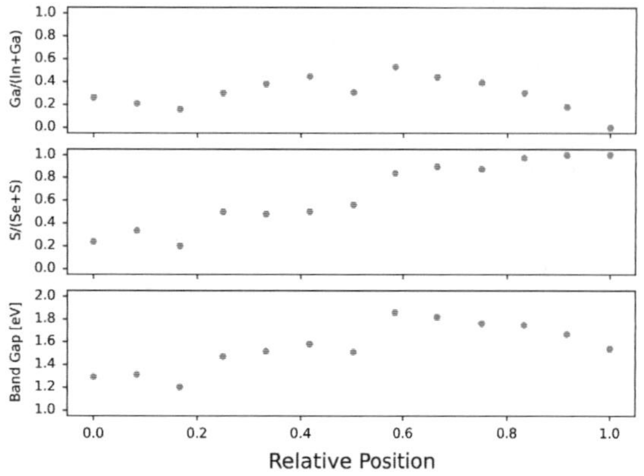

Profiles for low sulfur.

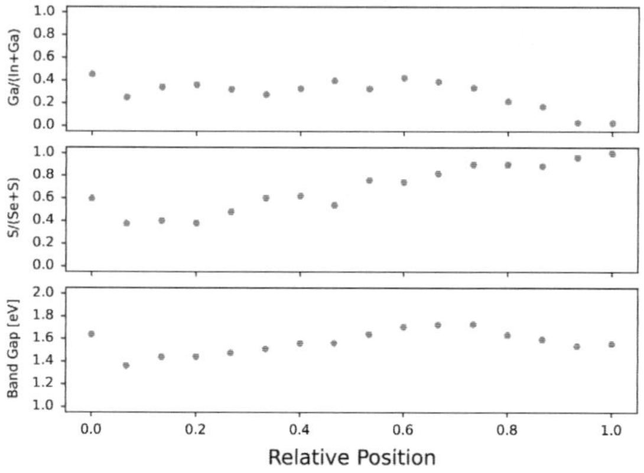

Profiles for high sulfur.

Conclusions and Future Work

Experimental Results:

- Films produced in <10 minutes at 600 °C

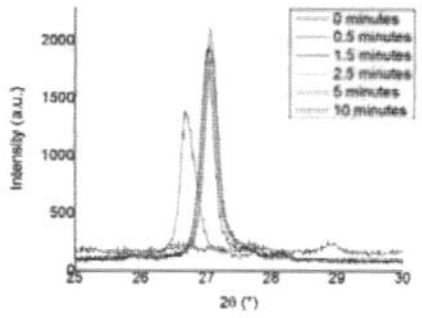

Conclusions and Future Work

Experimental Results:

- Films produced in <10 minutes at 600 °C
- $\eta \approx 10\%$ devices obtained with range of band gaps (varied S incorporation)
- Processing conditions affect the band gap and band gap gradient
- Next step: more thorough electrical characterization

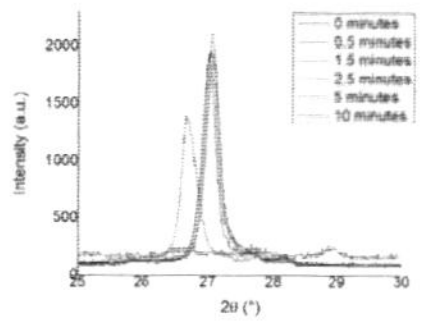

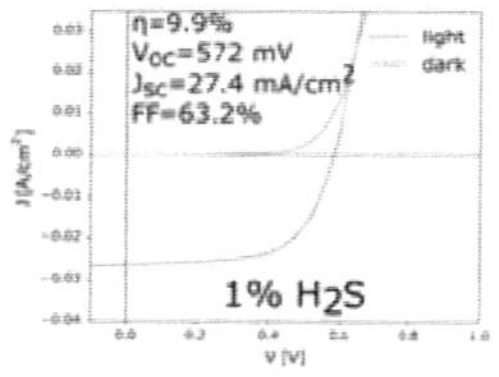

Introduction Thin Film Characterization Device Results Stochastic Modeling **Conclusions and Future Work**

Conclusions and Future Work

Experimental Results:

- Films produced in <10 minutes at 600 °C
- $\eta \approx 10\%$ devices obtained with range of band gaps (varied S incorporation)
- Processing conditions affect the band gap and band gap gradient
- Next step: more thorough electrical characterization

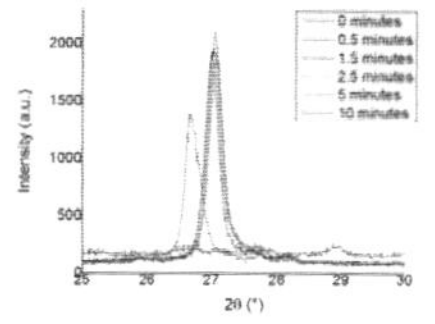

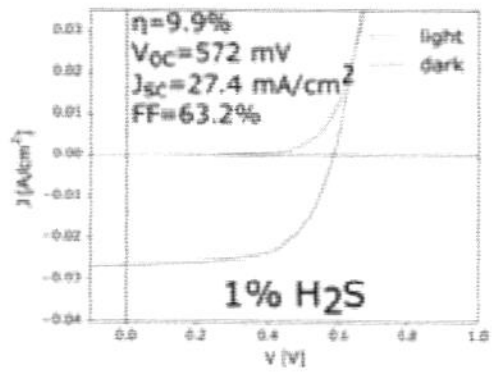

Modeling Results:

- Simulation results suggest mechanism for band gap grading

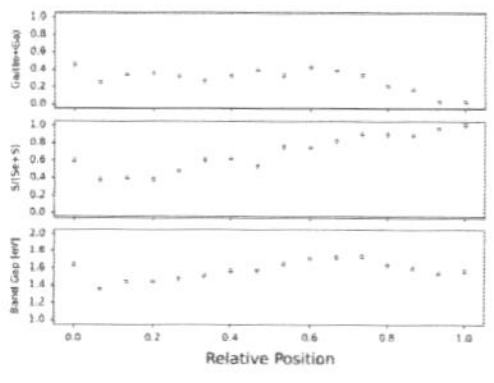

R.J. Lovelett, W.N. Shafarman, R.W. Birkmire, B.A. Ogunnaike University of Delaware
CIGS Production via Rapid Thermal Processing of Se-capped Precursors 20 of 22

Acknowledgments

- Ogunnaike Research Group
  - Daniel Cook
  - Dr. Qian Gou
  - Dr. Jacob McGill
  - Dr. Lijia Luo
  - James Park
  - Xin Peng
  - Devesh Radhakrishnan
  - Dr. Melissa St. Amand
  - Chia-Hung Tsai
  - Dr. Zachary Whiteman
- Institute of Energy Conversion
  - Dr. Dominik Berg
  - Wayne Buchanan
  - John Elliot
  - Shannon Fields
  - Kevin Hart
  - Dr. Gregory Hanket
  - Brian McCandless
  - Sina Soltonmohammad

This material is based upon work primarily supported by the National Science Foundation (NSF) and the Department of Energy (DOE) under NSF CA No. EEC-1041895. Any opinions, findings and conclusions or recommendations expressed in this material are those of the author(s) and do not necessarily reflect those of NSF or DOE.

Introduction Thin Film Characterization Device Results Stochastic Modeling Conclusions and Future Work

Production of $Cu(InGa)(SeS)_2$ absorber layers via rapid thermal processing of selenium-capped metal precursors for thin film solar cells

R.J. Lovelett*†, W.N. Shafarman†, R.W. Birkmire†, B.A. Ogunnaike*

University of Delaware
*Department of Chemical and Biomolecular Engineering
†Institute of Energy Conversion

November 8, 2015

Enhanced Liquid-Phase Synthesis and Interfacial Self-Assembly of Amphipathic Polypeptides

Matthew B. Kubilius & Raymond Tu, The City College of New York

Abstract

Synthesizing useful, self-assembling, periodically-sequenced amphipathic polypeptides is challenging due to the increasing polydispersity index in high molecular weight polymers. To offset this, we designed synthetic amino acid dimers that polymerize into an amphipathic chain with alternating hydrophilic/hydrophobic side groups. This periodicity is typical of peptide sequences prone to self-assembly into β-sheets. In this way, we influence polydispersity in the growing polypeptide chains: controlling the kinetics of growth through transport-limited chain elongation. Our experiments show that in the absence of a micellar interface, standard bulk-phase condensation polymerization occurs. The amphipathic character of the peptide chain increases with increasing molecular weight, resulting in a polypeptide that partitions into surfactant micelles as a function of molecular weight. This type of kinetically-limited growth serves to narrow the polydispersity of our periodically-sequenced polypeptide. We quantify the dynamics of chain elongation and interfacial assembly using multi-angle light scattering and define the evolving sheet-like secondary structure using circular dichroism for various peptides of differing amino acid pairs. Our results show that the peptides grown in the presence of micelles show significantly enhanced self-assembly and a narrowed polydispersity index. These self-assemblies are characterized further with Langmuir-Blodgett films and imaged using Brewster angle microscopy. We conclude that the transport-limited chain elongation polymerization method shows great promise in the manufacture of low-cost, interfacially-assembling polypeptides.

Introduction

Amino acids and peptides are readily available chemical resources. Despite this, synthesized or biologically expressed peptides are typically valued at hundreds of dollars per milligram, compared to the dollars per gram cost of the amino acids that form them. Typical synthesis and purification processes give rise to a six-orders-of-magnitude increase in cost between reagents and products. In this work, we have designed a low-cost, polymerization-based peptide synthesis method for growing narrow polydispersity, amphipathic peptides for use in biomimetic materials.

Our periodically-sequenced peptides are designed to self-assemble their peptide backbones into β-sheets in the plane of the interface[1]. This intermolecular folding represents overcoming entropic forces in favor of hydrogen bonding and hydrophobic interactions. In low-molecular weight peptide molecules, the self-assembly-driving forces are insufficient to overcome entropic effects[2-7]. Additionally, low molecular weight peptides have greater numbers of terminal acids and amines (unreacted ends). These ends correspond to gaps in the self-assembled state, lacking the alignment of hydrogens and oxygens on the peptide backbone conducive to the formation of hydrogen bond network formation.

Predicting that self-assembly is a function of molecular weight, we use Langmuir adsorption kinetics to predict increasing partitioning of peptides to interfaces as they grow. Early evidence justifying this hypothesis has been shown[1, 8]. Polypeptide chain growth only takes place in the liquid phase when chain ends are exposed to water-soluble activating reagents. If a polypeptide is out of solution on an interface, our kinetic model predicts that its interfacially-adsorbed ends becomes unreactive. This effect can be bundled into the kinetic model by redefining k_r as a function of molecular weight:

$$\boldsymbol{k_{eff} = k_r e^{-\xi * N}}$$

Here, k_r is the reaction coefficient associated with standard condensation polymerization. Equal reactivity of all terminal amines and acids is assumed for k_r. ξ is a constant defining the change in reactivity as a function of molecular weight, and N is the number of dimers defining the chain length. Assuming an adsorption-limited reaction to define ξ, we predict a significant drop in our product's polydispersity index, from 2.0 in the standard condensation reaction case, to approximately 1.1 in the transport-limited system[1]. A similar model was previously proposed to explain smaller, observed polydispersity deviations [9].

Experimental Section

Dimethyl lysine-valine, an amino acid dimer, was synthesized by liquid-phase peptide synthesis and analyzed for purity via mass spectrometry (Figure 1). The hydrophilic amino acid Fmoc-Lys(Me)2-OH HCl (AnaSpec) was reacted with hydrophobic L-valine tert-butyl ester hydrochloride (Sigma-Aldrich). The Fmoc group was removed with 20% piperidine in dimethylformamide, followed by the t-butyl ester deprotection in 95% trifluoroacetic acid in water. The product was then additionally purified by distillation, filtration, and drying, yielding a di-amino acid monomer for polymerization.

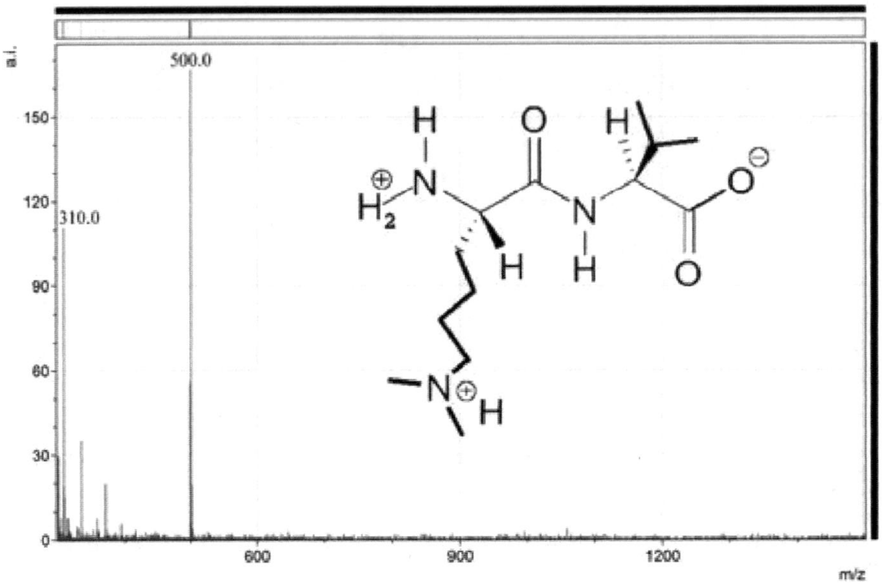

Figure 1. MALDI mass spectrometry scan of pure di-amino-acid dimethyl lysine-valine. The peak at m/z = 310.0 is the potassium salt of the dimer, and the peak at 500.0 is a complex of four dimers, two sodium and two hydrogen ions.

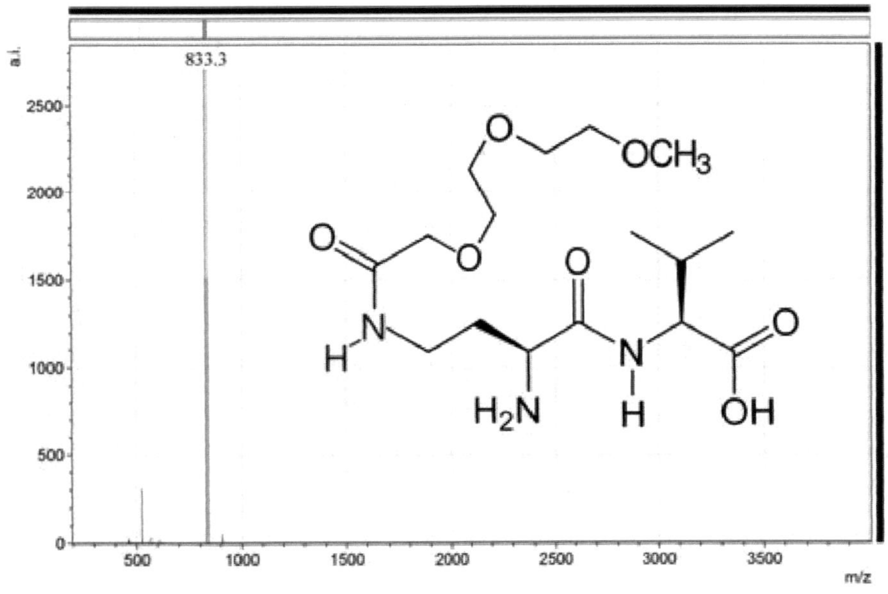

Figure 2. MALDI mass spectrometry scan of pure di-amino-acid lysine((PEG)$_3$Me)-valine. The peak at m/z = 833.3 corresponds to a complex of two dimer molecules, one sodium and one hydrogen ion.

A second amino acid dimer, lysine((PEG)$_3$Me)-valine was synthesized by liquid-phase peptide synthesis and analyzed for purity via mass spectrometry (Figure 2). Methoxy(ethoxy-ethoxy) carboxylic acid (Aldrich) was reacted to an NHS ester intermediate with dicyclohexylcarbodiimide. The NHS ester was purified by filtration, then reacted with Fmoc-Lys-OH HCl (AnaSpec). The resulting PEGylated lysine was then reacted with hydrophobic L-valine tert-butyl ester hydrochloride (Sigma-Aldrich), yielding a fully-protected amino acid dimer. The product was then additionally purified by distillation, filtration, and drying, yielding a di-amino acid monomer for polymerization.

We grew poly-dimethyl-lysine-valine peptides under two conditions. First, the monomer was polymerized in water. Polymerization was activated by a 1:1 molar ratio of dimer to 1-Ethyl-3-(3-dimethylaminopropyl)carbodiimide (EDC) and Sulfo-NHS. Second, polymerization was activated as in the first case, but after the addition of 5% Triton-X 100 surfactant (by volume). Spherical micelles are prevalent under these conditions, as Triton-X 100 has a critical micelle concentration of 0.6 g/L[10]. As such, this addition modified the base case to include plentiful micellar surface area for the growing peptides to partition into as they polymerized.

Both systems were left to react for eight hours, then polymerization was quenched by addition of 5% acetic acid (by volume). Following the quench, Triton-X 100 was added to the first (control) case, in an amount equivalent to that of the control.

The two samples were dried, then purified by three precipitations in t-butyl methyl ether to remove the surfactant. Peptide size and polydispersity was analyzed via size-exclusion chromatography multi-angle light scattering.

Poly-lysine((PEG)$_3$Me)-valine was polymerized in acetone using dicyclohexylcarbodiimide as a linking reagent. Under these conditions, the linking spent linking reagent (dicyclohexylurea) precipitates out of solution and was removed by cold filtration.

Results & Discussion

MALDI mass spectrometry was used to verify and characterize polymerization in the dimethyl lysine valine system (Figure 3.) Molecular weights fifty times larger than those seen for the dimer dominate the scan, and polypeptides with molecular weights as high as 150 kD can be seen in the sample.

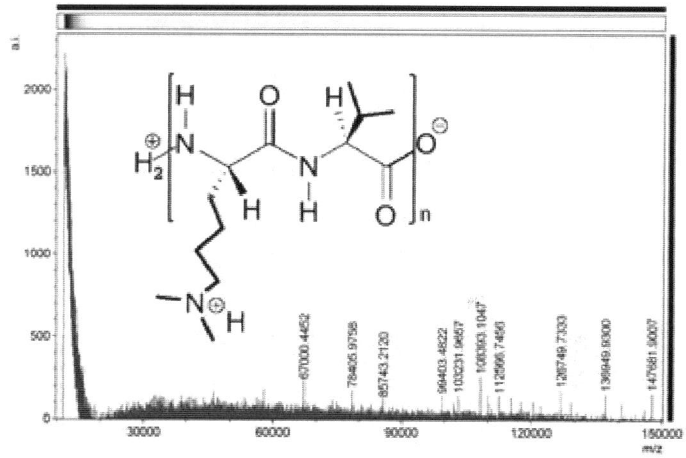

Figure 3. MALDI mass spectrometry scan of poly-dimethyl lysine valine. In contrast to the scan of the dimer (Figure 1.), polypeptide molecular weights of 150 kD are seen, indicating extensive polymerization.

Polydispersities of the dimethyl lysine valine polypeptides were analyzed using multi-angle light scattering. As expected, the polydispersity index for the control case of bulk condensation polymerization was 2.0 and showed a distribution of peptide lengths ranging from high molecular weight to continued existence of dimers. This broad distribution matched the hypothesis for the control case. Polypeptides grown with access to Triton-X 100 micelles showed increased average molecular weights. Additionally, dimeric species are no longer visible, showing these reaction conditions enhance reactivity of the smallest polypeptide species. In water, these polypeptides formed assemblies with polydispersities of 1.1.

Larger self assemblies are possible in the case of poly-lysine((PEG)$_3$Me)-valine. Here, the polypeptide is not soluble in water and will form β-sheets on an air/water interface. Previous work has shown that hysteresis in this system exists[1], but did not investigate how the system responds to repeated compression and expansion cycles. To characterize this behavior, Langmuir isotherms were generated for the water-insoluble poly-lysine((PEG)$_3$Me)-valine (Figure 4). During the first compression, there is a linear relationship between sample area and surface pressure up to pressures of 40 mN/m. At that point, formation of β-sheets becomes visible as a non-linear region in the curve. Hysteresis is observed on re-expansion, and subsequent cycles are essentially linear. This, paired with small pressure drops in those same regions, show that material is being gradually lost to the sub-phase with each cycle.

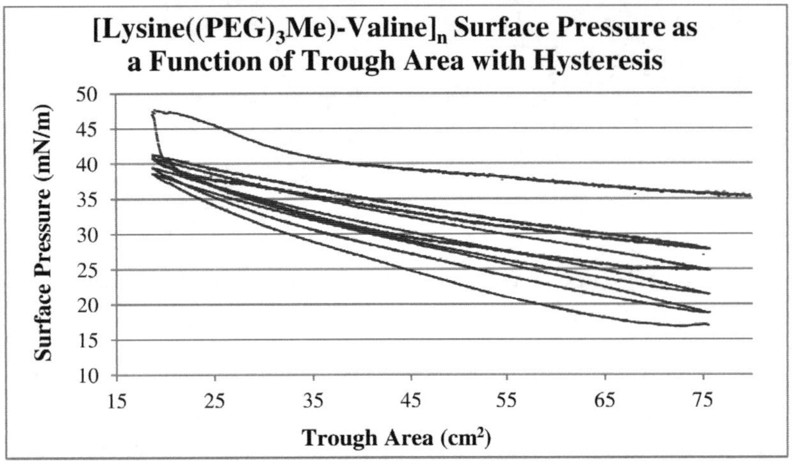

Figure 4. Poly-lysine((PEG)$_3$Me)-valine self-assembles into β-sheets at an air/water interface at 40 mN/m. The first compression cycle (top, starting at a trough area of 80 cm^2 and decreasing) shows this behavior with the change in slope of the curve above that surface pressure. The system showing a steep decrease in surface pressure when the barriers first reverse (hysteresis), corresponding to the large degree of self-assembly associated with the formation of these sheets. Subsequent cycles show slight, similar drops in surface pressures for the complete run, corresponding to small losses of material to the subphase over time. Reversible behavior, either in the form of dissociation from sheets to fibrils or return to the interface from the sub-phase, is not seen.

Surface pressure data for the poly-lysine((PEG)$_3$Me)-valine assemblies can be further understood and visualized using Brewster angle microscopy. Fibrils first become visible at surface pressures of 12 mN/m (Figure 5). At 25 mN/m, fibrils are still seen as the dominant phase, but are visibly thicker and more pronounced (Figure 6). At 42 mN/m, β-sheets are visible as a new, dominant phase. At 44mN/m, this self-assembled phase has become continuous. Re-expansion of these compressed assemblies back to moderate surface pressure (35 mN/m, Figure 8) continues to show patches of these thick self-assemblies, but black regions corresponding to a clean air/water interface now separate them. Fibril reformation is not seen, corresponding to the observed hysteresis effects.

Figure 5. Low surface pressure (12mN/m) Brewser angle microscopy image showing fibrils of Poly-lysine((PEG)$_3$Me)-valine at the air/water interface.

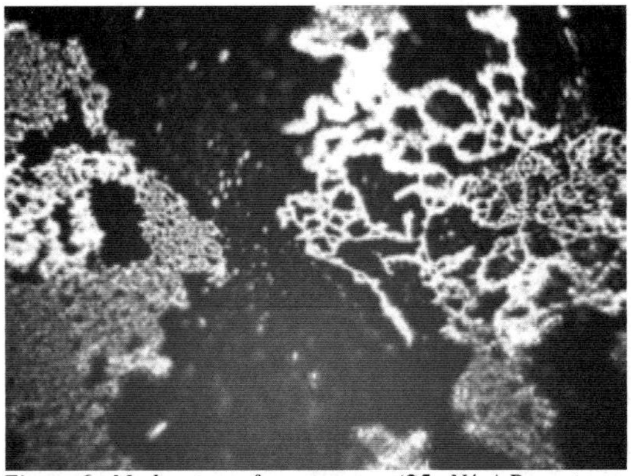

Figure 6. Moderate surface pressure (25mN/m) Brewster angle microscope image of Poly-lysine((PEG)$_3$Me)-valine fibrils at the air/water interface. At this pressure, fibrils are thicker and easily seen.

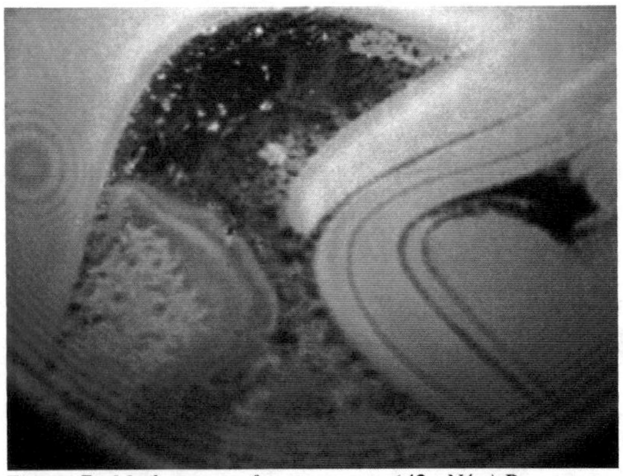

Figure 7. Moderate surface pressure (42mN/m) Brewster angle microscopy image of Poly-lysine((PEG)$_3$Me)-valine (white) assemblies at the air/water interface. Large regions of β-sheet structures are visible at this pressure, separated by the fibrils seen at lower pressures and clean water (dark regions).

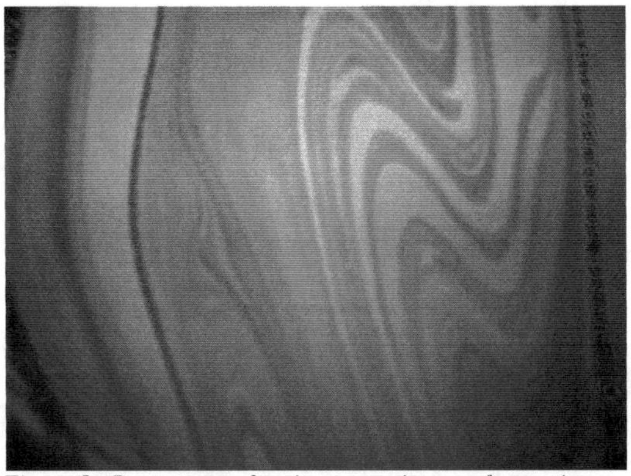

Figure 8. Brewster angle microscopy image of a continuous, 44 mN/m β-sheet of Poly-lysine((PEG)$_3$Me)-valine at the air/water interface. At this pressure, the fibrils have been completely taken up by the β-sheet assemblies.

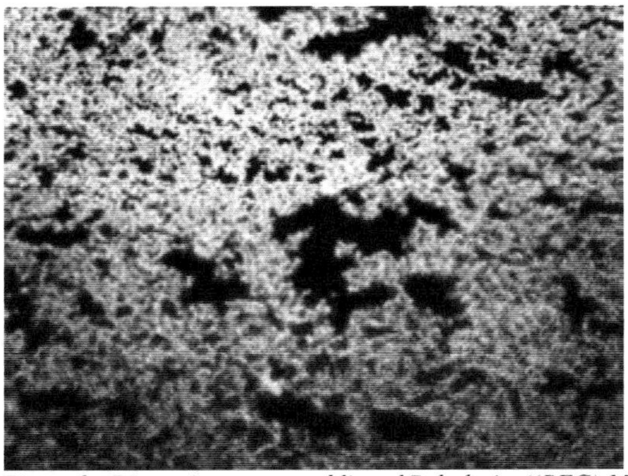

Figure 9. Discontinuous assemblies of Poly-lysine((PEG)$_3$Me)-valine at the air/water interface (35 mN/m) imaged with Brewster angle microscopy after initiating a re-expansion cycle. White regions of assembled β-sheets are still visible, but black regions of naked air/water interface now separate them. The sharp contrast between dense material and clean water highlights the irreversible assembly of the compressed β-sheets.

Conclusions

This work demonstrates a low-cost method of synthesizing self-assembling, amphipathic polypeptides. By designing self-assembling, interfacially-active polypeptides, the polydispersity index of the growing chains can be attenuated by transport of the growing chains into surfactant micelles. These polypeptides demonstrate a strong affinity for patterning macroscopic interfacial areas with β-sheet structures when compressed at high (~45mN/m) surface pressures.

Acknowledgements

This work was funded by the National Science Foundation, DMR grant #1006407 and NSF-CBET 1067501.

References

1. Kubilius, M.B. and R.S. Tu, *Surfactant-Mediated Polydispersities in the Synthesis of Self-Assembling, Amphipathic Polypeptides.* AIChE - Annual Meeting Extended Abstract (Proceedings Paper), 2014.
2. Harrison, P., et al., *Probing structure-function relations in ferritin and bacterioferritin.* Adv. Inorg. Chem, 1991. **36**: p. 449-486.
3. Klimov, D.K. and D. Thirumalai, *Mechanisms and kinetics of β-hairpin formation.* Proceedings of the National Academy of Sciences, 2000. **97**(6): p. 2544-2549.
4. Dill, K.A., *Polymer principles and protein folding.* Protein Science, 1999. **8**(06): p. 1166-1180.
5. Carlsson, U. and B.-H. Jonsson, *Folding of β-sheet proteins.* Current Opinion in Structural Biology, 1995. **5**(4): p. 482-487.
6. Dill, K.A., K.M. Fiebig, and H.S. Chan, *Cooperativity in protein-folding kinetics.* Proceedings of the National Academy of Sciences, 1993. **90**(5): p. 1942-1946.
7. Xu, Y., et al., *Probing the Kinetic Cooperativity of β-Sheet Folding Perpendicular to the Strand Direction.* Biochemistry, 2008. **47**(7): p. 2064-2070.
8. Guo, C., et al., *Mechanisms of cooperativity underlying sequence-independent beta-sheet formation.* The Journal of Chemical Physics, 2002. **116**(10): p. 4353-4365.
9. Nanda, V. and S. Jain, *Statistical study of the effect of variation of bimolecular rate constant in condensation polymerization.* Journal of Polymer Science Part A 1: Polymer Chemistry, 1970. **8**(7): p. 1871-1885.
10. Gonick, E. and J.W. McBain, *Cryoscopic Evidence for Micellar Association in Aqueous Solutions of Non-ionic Detergents.* Journal of the American Chemical Society, 1947. **69**(2): p. 334-336.

422045 DNA Origami Self-Assembly into Superstructures Via Explicit Characterization and Optimization of the Thermodynamics and Kinetics of the Assembly Process

Abstract:

The hierarchical self-assembly of different types of DNA origami nanostructure components into well-defined complexes would make it possible to build larger DNA nanostructures where locations across the structure (up to a micron in scale) can be addressed with nanometer scale precision. These structures could serve as templates for novel biomolecular devices for sensing cellular interactions for the design of metamaterials.

However, such self-assembly processes have typically been plagued by low yields that drop rapidly as complex size increases. To understand more about how DNA origami structures self-assemble into complexes, we constructed origami monomers designed to self-assemble into heterotetrameric ring structures via specific Watson-Crick base pairing interactions. We used fluorescence spectroscopy to characterize how origami interface structure and design control the kinetics and thermodynamics of dimerization, and correlation fluorescence microscopy to characterize the thermodynamics of all of the reactions in the assembly. We used this information to optimize the assembly process of both origami complexes with well-formed structure and the assembly of 2-dimensional origami lattices.

430483 Engineering Surface Hydrophobicity of Zein Films Via Self-Assembly

Abstract:

Zein, a corn prolamin, has been widely studied as a biopolymer due to its film-forming abilities. Zein protein has almost equal amount of hydrophobic and hydrophilic amino acids, which allows for self-assembly. In this research, we utilized this property to modify the surface properties of films obtained via soft-lithography method. Zein films were prepared by casting on polydimethylsilohexane (PDMS) substrate, which was either exposed (PDMS O+) or non-exposed to oxygen plasma (PDMS O-). Oxygen plasma is known to make PDMS hydrophilic by inducing –OH groups on the surface. Zein solution was prepared by dissolving in aqueous ethanol and mixing with a plasticizer, oleic acid, and emulsifier. After zein films were formed in desiccator, they were peeled off of PDMS with ease. It was found that zein's surface that was in contact had different water contact angle (WCA) values depending on the plasticizer content and whether PDMS was exposed to oxygen plasma. WCA of zein films were decreased when it was cast on oxygen plasma exposed PDMS compared to regular PDMS. In addition, increase in oleic acid content resulted in a larger change in the difference between the cases of PDMS O- and PDMS O+. The change in the hydrophilicity, which was measured by WCA method, were further investigated by Atomic Force Microscopy experiments. Both topography and force measurements were conducted on zein films with varying oleic acid and PDMS substrates. We also demonstrated that by using a mask with a pattern, it is possible to obtain zein films with varying hydrophilicity. This enables us to develop microfluidic, sensor, tissue engineering platforms based on the need at hand. We believe that zein's ability to self-assemble can be exploited by using this facile technique to obtain free-standing films with different water affinities.

434876 Preventing Antibiotic-Resistant Bacterial Infections Using an Antimicrobial Coating Composed of "Janus" Nanoparticles

Abstract:

Antibiotic-resistant bacteria are one of the largest problems in the field of medicine today. According to the Center for Disease Control (CDC), at least two million people become infected annually with bacteria resistant to one or more of our current antibiotics that would normally be used to kill the infection. Of those two million people, at least 23,000 die from these infections annually. The CDC also estimates that between healthcare costs and the cost to society for lost productivity, antibiotic-resistant bacteria are costing the United States as much as $55 billion dollars per year. One of the major areas where infection occurs is from contamination from improperly sterilized equipment and surfaces (hospitals, medical clinics, etc.). Nanomedicine provides some potentialfor preventing the spread of antibiotic-resistant bacteria.

"Smart" nanoparticles have been developed that self-assembly on surfaces and adhere to and destroy antibiotic resistant bacteria. The particles consisted of two distinct hemispheres, so called "Janus" nanoparticles, with one side being hydrophobic and the other containing antimicrobial properties.

The nanoparticles were characterized using TEM imaging, DLS analysis, FTIR analysis, while ongoing studies will confirm its antimicrobial properties. This work has the potential to revolutionize antimicrobial coatings, as well as save thousands of lives every year.

437495 Nanogel Formation By Self-Assembly of Polyethylene Glycol Macromers Sequentially Chain-Extended with Short Lactide and Glycolide Segments

Sunday, November 8, 2015: 4:50 PM
253A (Salt Palace Convention Center)
Danial Barati, Seyedsina Moeinzadeh and Esmaiel Jabbari, University of South Carolina, Columbia, SC

Introduction: poly(lactide-co-glycolide) (PLGA) nanoparticles due to their wide range of resorption times are used extensively for immobilization and timed-release of proteins in drug delivery and regenerative medicine. However, protein denaturation by surface adsorption and acidic degradation products of PLGA significantly reduces protein bioactivity in vitro and in vivo. Due to chain flexibility and electrical neutrality of polyethylene glycol (PEG) macromers, PEGylation is used to increase stability of aqueous nanoparticle suspensions. The objective of this work was to investigate the effect of sequential chain-extension of PEG with short segments of lactide (L) and glycolide (G) monomers by ring-opening polymerization on self-assembly and stability in aqueous solution, and bioactivity of proteins grafted to the self-assembled nanogels (NGs).

Materials and method: The PEG-LG block copolymers were synthesized by melt ring-opening polymerization with sequential addition of L and G monomers to the reaction using PEG as the initiator. The synthesized copolymer was functionalized with succinimide groups by reacting hydroxyl end-groups of the copolymer with N,N/-disuccinimidyl carbonate. The maromers were dissolved in DMSO and self-assembled to form NGs by dialysis (3.5 kDa MW cutoff) against PBS for 8 h. For protein grafting, NGs were suspended in 0.5 mL PBS by sonication for 1 min. Next, 0.5 mL of the protein in PBS (800 ng/mL) was added to the NG suspension, and the amine group of the protein was allowed to react with succinimide end-groups of the chain extended macromer in NGs under ambient conditions for 12 h. Size distribution of the NGs was measured by dynamic light scattering. Zeta potential of the NGs was measured with a ZetaPlus analyzer. NGs degradation kinetic was measured by incubation in PBS at 37°C as we described previously. Release kinetics of the grafted protein was measured by incubating 1 mg protein-loaded NGs in 1 mL PBS at 37°C as we previously described. At each time point, the suspension was centrifuged at 18350 rcf for 10 min, the supernatant was removed, the NGs were resuspended in 1 mL fresh PBS and incubated until the next time point. The amount of bovine serum albumin (BSA) in the supernatant was measured with the ninhydrin reagent. For BMP2-grafted NGs, the protein concentration in the supernatant was measured by ELISA. The abbreviation P4 is for 4.6 kDa PEG MW and 10 LG:PEG molar ratio; P8 is for 8 kDa PEG MW and 20 LG:PEG molar ratio; P12 is for 12 kDa PEG MW and 30 LG:PEG molar ratio. Abbreviation I, II, and III are for L/G ratio of 75/25, 60/40, and 50/50, respectively.

Results: Figure 1a to 1c show the effect of PEG molecular weight, LG segment length, and L/G ratio in the chain-extended macromer on NG size distribution. The morphology of the dried NGs is shown in the inset SEM images in (a-c). NGs size was <200 nm. The average size ranged from 120±6 to 140±13 nm for P4 NGs, 150±11 to 160±17 nm for P8 NGs, and 170±12 to 190±15 nm for P12 NGs, respectively. The NGs were relatively monodisperse with PD<0.2 and the mean NG size increased with decreasing lactide fraction. All NGs were negatively charged with zeta potential between -23 and -15 mV. The absolute value of the zeta potential increased with increasing PEG molecular weight while the lactide fraction did not significantly affect zeta potential of the NGs. The effect of concurrent (random, c) and sequential (block, s) addition of L and G monomers to the polymerization reaction on colloidal stability of the NGs was also investigated. The NGs generated from P4-I macromers with concurrent addition of LG monomers (P4-Ic) had a significantly wider distribution (PD=0.31±0.05) than the sequential addition (0.07±0.01). P4-Is NGs was more stable in aqueous solution than P4-Ic NGs because their mean size did not change with incubation time and their precipitation from solution occurred over a narrower centrifugation speeds. Grafting BMP2 and VEGF to the NGs from sequential chain-extension led to only 30% and 20% denaturation of the proteins, respectively. Results of this work indicate that nanogels self-assembled from PEG macromers chain-extended with sequential polymerization of lactide and glycolide are a more efficient platform for delivery of proteins compared to PEGylated PLGA nanoparticles.

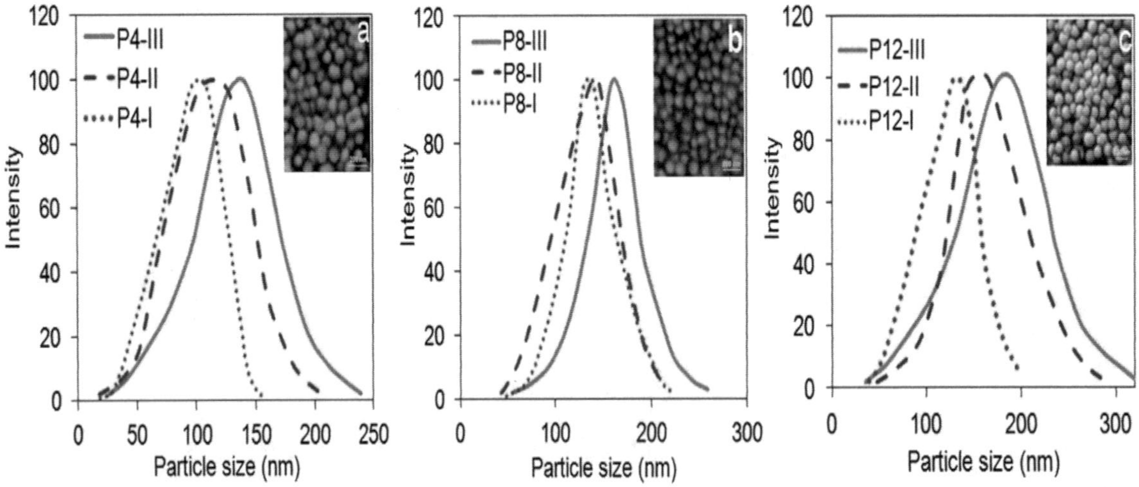

Figure 1. Effect of PEG molecular weight, LG segment length, and L/G ratio in PEG macromers chain-extended with sequential polymerization of L and G on NG assembly and size distribution; Morphology of the dried NGs is shown in the inset SEM images in (a-c).

441498 Dynamic Covalent Assembly of Peptoid-Based Ladder Oligomers

Abstract:

Dynamic covalent chemistry, in conjunction with template-directed assembly, enables the fabrication of extended nanostructures that are both precise and tough. Here, we demonstrate the dynamic covalent assembly of peptoid-based molecular ladders with up to 12 rungs via scandium(III)-catalyzed imine metathesis by employing the principle of Vernier templating, where small precursor units with mismatched numbers of complementary functional groups are co-reacted to yield larger structures with sizes determined by the respective precursor functionalities. Owing to their monomer diversity and synthetic accessibility, sequence-specific oligopeptoids bearing dynamic covalent pendant groups were employed as precursors for molecular ladder fabrication.

439679 NSEF Young Investigator Award Presentation: Nanostructured Materials and Bionanotechnology to Engineer Cells from the inside out and the Outside in

Abstract:

The Green lab synthesizes biodegradable polymers to construct nanostructured polymer/DNA polyplexes that are safe and effective for intracellular delivery of DNA and siRNA and that can be used to engineer target cells from the "inside out." Additionally, they develop biomimetic nanoparticles that act as artificial antigen presenting cells and that can program biological immune cells from the "outside in." Both of these nanobiotechnologies will be discussed. For intracellular delivery, the effect of differential polymer structure on multiple steps of intracellular delivery of nucleic acids will be discussed. Modification of biomaterial structure enables cell-type specificity of nanostructured polyplexes in a range of cell types, including enabling cancer targeting. The role of particle size and morphology in constructing artificial antigen presenting cells and how these biomimetic particles can be used for immunotherapy will be also be presented.

439681 NSEF Forum Award Presentation: Functionalization of Fine Particles By Atomic/Molecular Layer Deposition (ALD/MLD)

Abstract:

The functionalization of fine primary particles, including nanoparticles, is easily carried out using sequential self-limiting surface reactions in a fluidized bed reactor. The self-limiting reactions result in the deposition of atomic or molecular layers, i.e. ALD or MLD. This functionalization process, referred to as Particle ALD/MLD, can be used to deposit conformal and pinhole-free films of refractory oxides, non-oxides, metals, and hybrid polymer-based materials, among others. It is also possible to deposit conformal porous or semi-continuous films where partial coverage is preferred instead. Fluidized bed reactors are well suited for large scale operations.

In this process, the particles are normally fluidized under reduced pressure conditions using an inert gas, although atmospheric pressure operation can also be carried out. Precursor doses can be delivered to the bed of particles sequentially and, in most cases, can be utilized at nearly 100% efficiency without precursor breakthrough and loss. The progress of the coating process can be monitored continuously using an in-line downstream mass spectrometer. The ability to use precursors with efficiencies approaching 100% opens the door for a unique opportunity to utilize precursors that previously might have been considered to be too expensive. Fluidized beds containing particles comprising hundreds of thousands, even millions of m^2 of surface area can be coated efficiently. The dose times depend on the amount of surface area required to be coated and the flow rate of the precursor into the fluidized bed. Although dose times could potentially be very long for coating millions of m^2 of particles, i.e. minutes to hours, the efficient use of the precursor to place nearly perfect films on primary particles makes Particle ALD/MLD the low cost and potentially only process that can cost-effectively functionalize high surface area ultra-fine dry particles.

Particle ALD/MLD has been demonstrated to place nanofilms on primary nanoparticles as small as 10 nm as well as on nanotubes having surface areas approaching 1000 m^2/g and within the porous structure of polymeric materials having porosity near 95%. Physical, optical, electrical, and magnetic properties of the particles can be controlled in order to passivate, activate, or is some manner functionalize the particles. Current applications of interest include passivation of phosphors and nanocrystals used in LEDs, improved cycling of Li-ion batteries, and high-activity sintering resistant catalysts among others.

Functionalized Plasmonic Nanostructure Arrays for Direct and Accurate Mapping Extracellular pH of Living Cells Using SERS

Fang Sun, Shaoyi Jiang and Qiuming Yu*
Department of Chemical Engineering
University of Washington
2015 AIChE Annual Meeting
Salt Lake City

UNIVERSITY *of* WASHINGTON

Outline

- ❑ Introduction
- ❑ Quasi-3D Plasmonic Nano Arrays
- ❑ pH Detection using 4-Mercaptobenzoic Acid (4-MBA)
- ❑ Reproducibility and Interaction with Cations
- ❑ pH_e Distribution: Normal Cell vs. Cancer Cell
- ❑ Conclusions

1

Mapping extracellular pH of Living Cells

Significance

pH_e Influences

- Cycle progression
- Proliferation
- Differentiation

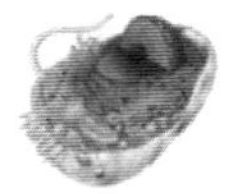

Tumor Cell - Dysregulated pH

- Transformation
- Adhesion
- Migration
- Drug uptake

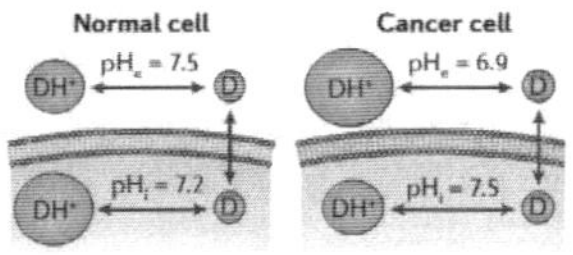

Conventional Methods

- pH microelectrodes
- MRI

Limitations

- Invasive method
- Low spatial resolution

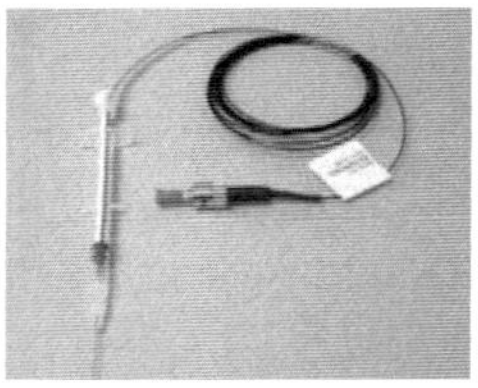

Microelectrode

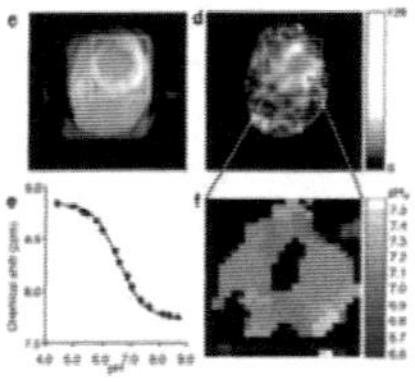

MRI

BA, Webb, et al. *Nature Reviews Cancer* . 2011, 9, 671-677

W 2

Surface Enhanced Raman Scattering

Enhancement Mechanisms:

- Electromagnetic enhancement (EF: 10^6-10^7)
- Chemical enhancement (EF: 10-10^2)

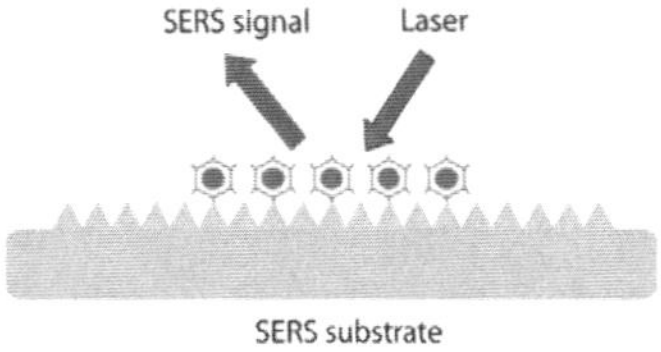

Advantages

- Ultra-sensitivity: single molecule detection
- Rich information: fingerprint spectra (molecular identity)

Applications

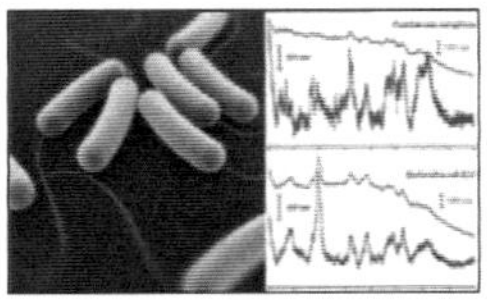

Bacteria

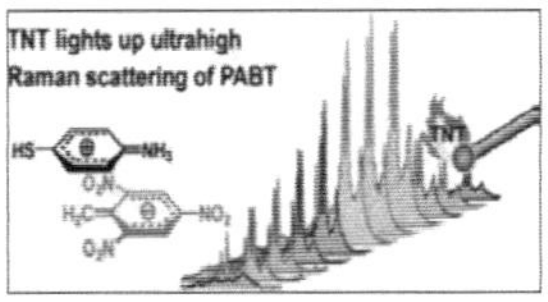

Chemical threats

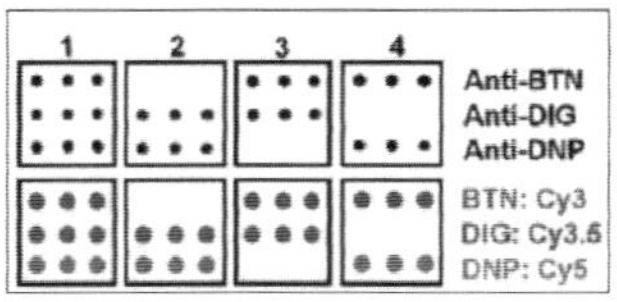

Diagnostics

J. Xu, et. al. *Small*. 2011, 3, 371
H. Zhou, et. al. *Anal. Chem.* 2011, 83, 6913
Y. Cao, et. al. *J.Am.Chem.Soc.* 2003, 125, 14676

UNIVERSITY *of* WASHINGTON 3

SERS Substrate: Quasi-3D Plasmonic Nano Arrays

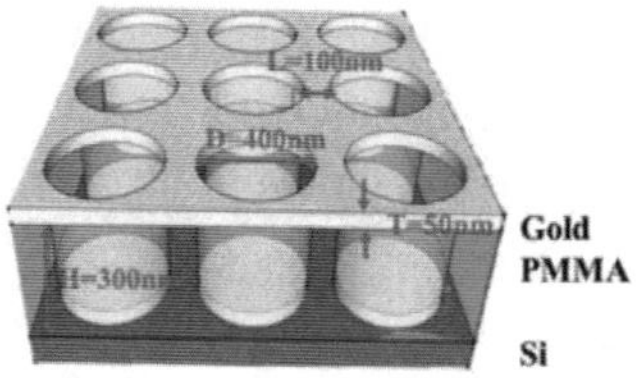

Q3D SERS substrate

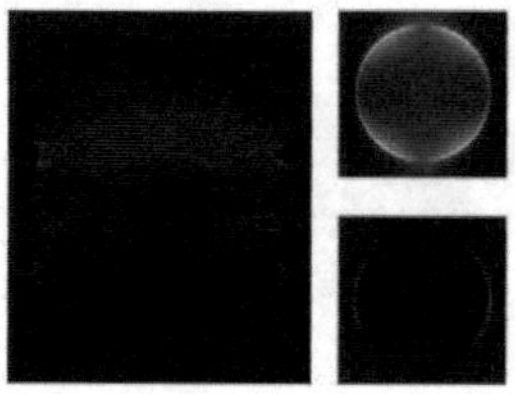

FDTD simulation

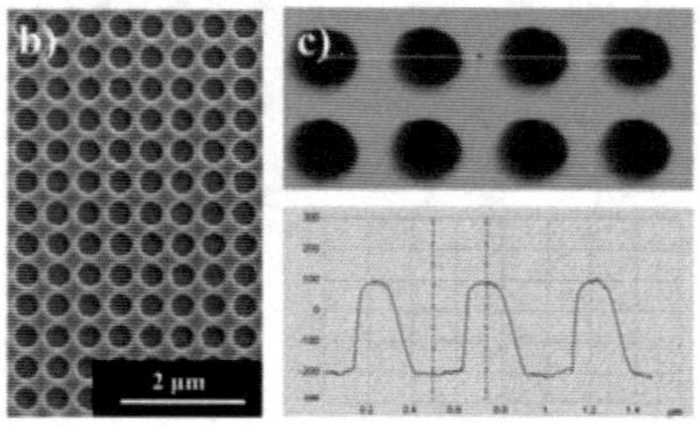

SEM and AFM

- ✓ High sensitivity
- ✓ High reproducibility

F. Sun, et al. *Anal. Chem.* 2014, 86, 2387-2394

UNIVERSITY *of* WASHINGTON 4

pH Detection using 4-Mercaptobenzoic Acid (4-MBA)

4-MBA modified SERS substrate

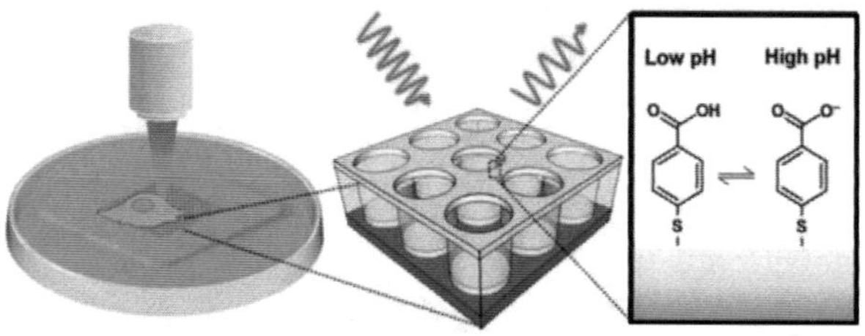

SERS spectra with increasing pH

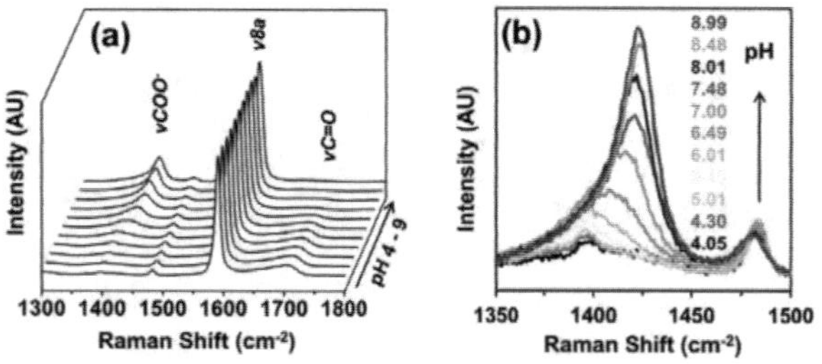

✓ pH detection is achieved by monitoring the SERS spectral changes of the peak at 1420 cm^{-1} ($vCOO^{-}$) of 4-MBA.

F. Sun, et al. *Biosensors and Bioelectronics.* 2015, 73, 202-207

UNIVERSITY *of* WASHINGTON 5

Reproducibility and Interaction with Cations

SERS contour of 4-MBA from 32 spots

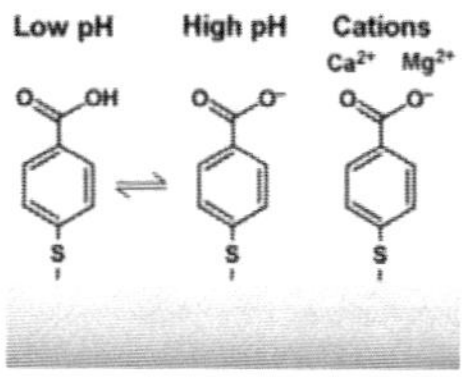

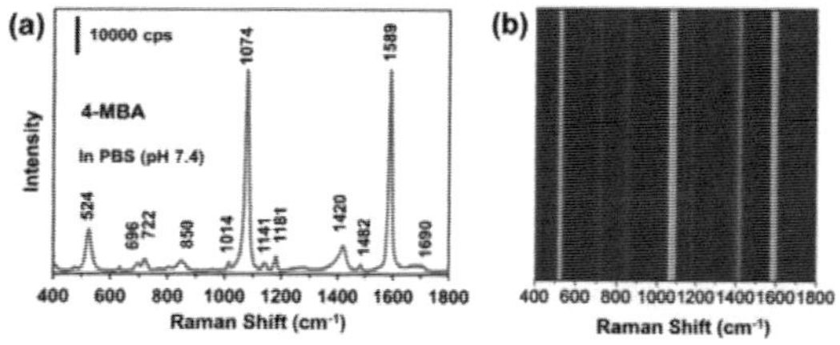

Influences from the cations and calibrations

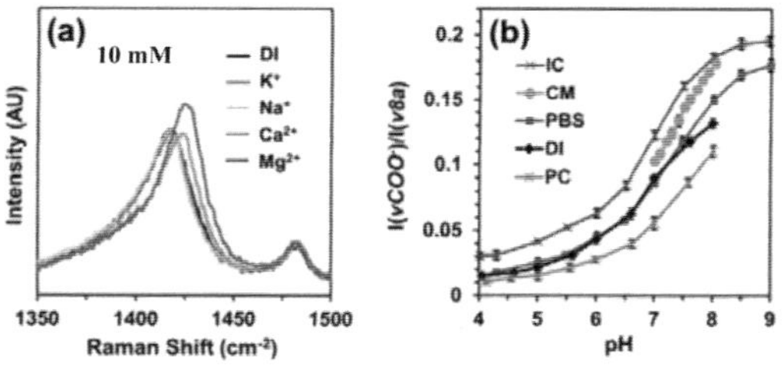

- Intracellular mimic (IC)
- Cell culture medium (CM)
- PBS
- DI water
- Phosphate citrate (PC)

✓ Reproducibility
✓ Reliability

F. Sun, et al. *Biosensors and Bioelectronics*. 2015, 73, 202-207

pH_e Distribution: Normal Cell vs. Cancer Cell

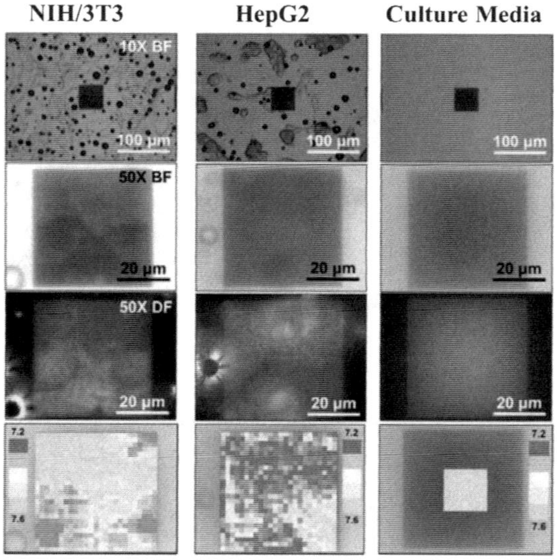

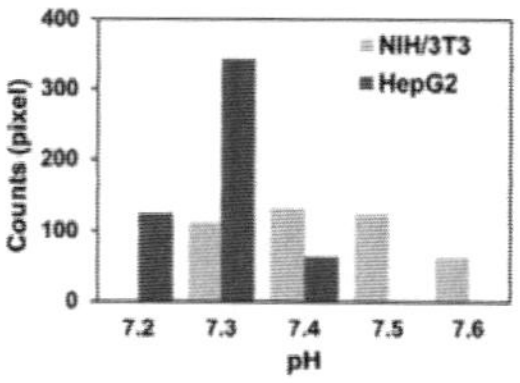

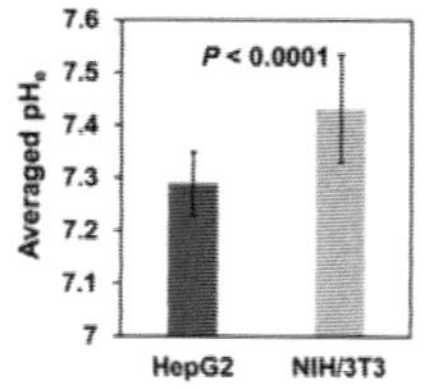

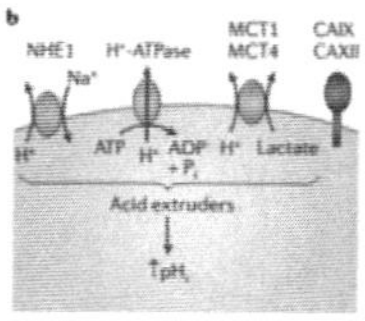

✓ pH_e domains are observed
✓ Tumor cells show more acid averaged pH_e

F. Sun, et al. *Biosensors and Bioelectronics*. 2015, 73, 202-207

Conclusions

- Q3D-PNAs SERS substrate demonstrated excellent reproducibility.
- 4-MBA modified Q3D-PNAs SERS substrate can detect the pH variation. Resolution: 0.1 pH unit.
- Divalent cations influenced the pH detection by 4-MBA.
- pH_e distributions of normal and tumor cells were mapped and imaged.
- Tumor cells demonstrated more acid averaged pH_e.

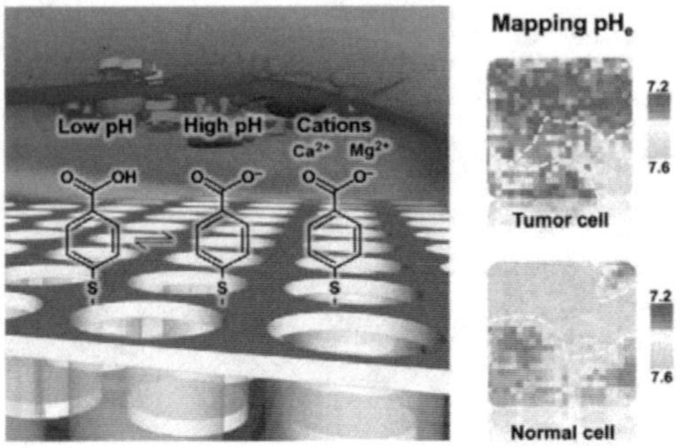

F. Sun, et al. *Anal. Chem.* 2014, 86, 2387-2394
F. Sun, et al. *Biosensors and Bioelectronics.* 2015, 73, 202-207

W 8

Acknowledgements

Prof. Shaoyi Jiang Research Group

Prof. Qiuming Yu Research Group

Funding Sources

DARPA NSF NNIN Nanoscale Science, Engineering & Technology

W 9

Questions?

416957 Conductivity, pH, and Oxidation Reduction Potential Sensing with a Simple Three-Electrode Microfabricated System

Abstract:

Detection of pH and oxidation reduction potential (ORP) in aqueous solutions typically requires a reference electrode that supplies a constant and known voltage regardless of solution conditions. These electrodes are challenging to construct for benchtop operation and are even more difficult to microfabricate. Current microfabricated reference electrodes and ion-selective electrodes use relatively difficult processing steps and have relatively short user lifetimes. For instance, silver/silver-chloride devices require construction of a constant concentration silver-chloride reservoir, and ISFET fabrication requires silicon doping.

We have constructed a three-electrode platinum sensor that can detect the conductivity, pH, and ORP of an aqueous solution. The three-electrodes system measures conductivity and ORP sequentially using an AC and a DC signal, respectively, passed through two electrodes. To measure pH, a controlled current signal generates both a constant voltage and a pH-varying voltage on the three-electrode system. This novel method, unlike other micro-fabricated reference electrodes, uses platinum electrodes and, hence, has an almost unlimited lifetime. Further, the fabrication procedure is comparatively easy and cheap: only single layer of platinum deposition on a glass wafer was necessary. The novel pH and ORP sensor can measure conductivity, pH, and ORP in non-oxidizing (below 1 ppm chlorite) solutions. Measurements in oxidizing solutions (above 12 ppm chlorite) are also possible although modification of one electrode with a surface coating is necessary.

435200 A Low-Noise Glass Nanopore: Comparison of Nanopores in Glass and Silicon Nitride

Abstract:

Synthetic nanopores provide benefits over biological nanopores in sensing applications; they are stable over a broader range of environmental conditions and offer greater flexibility with regard to pore size and surface chemistry.[1] Micromachining processes have been well developed over the past five decades with the growth of IC technology, which can be employed to fabricate synthetic nanopores. For example, nanopores may be milled straightforwardly in thin, solid-state membranes using a focused ion or electron beam. Since nanopore-based sensing devices rely on changes in conductance as analytes block the nanopore or are translocated through it, there have been many attempts to reduce signal noise to improve sensitivity. For example, insulating material (*e.g*., PDMS, silicon dioxide or silicon nitride) is commonly deposited on the membrane surface to decrease noise. Although most micromachined nanopores are silicon based, it has been reported that glass capillary nanopores show lower noise and a higher signal-to-noise ratio for DNA translocation compared to nanopores in silicon nitride membranes.[2] However, the pulled capillary glass nanopore exhibits a smaller current due to its long conical shape and is not amenable to high-throughput manufacturing. Recently, a low-noise, solid-state nanopore based on an insulating substrate has been introduced in which a thin silicon nitride membrane is attached to a quartz substrate to decrease low frequency noise by several orders of magnitude.[3]

In this work, the creation and characterization of a glass nanopore is presented. Both the membrane and the substrate are derived from a single glass wafer using batch micromachining. By thinning the substrate with a dilute mixture of hydrofluoric and hydrochloric acid,[4] submicron-thick membranes are successfully produced. In addition, a comparison of the glass nanopore versus the silicon-based nanopore is investigated based on the stability of the base current, the measured noise-level (*e.g*., RMS ionic current noise and current power spectral density in frequency domain), and the presence of undesired peaks upon changing applied voltage polarity. A relation between noise level and pore sizes is also established. To test the nanopore as a sensor, the glass chip with membrane and nanopore was sandwiched between two buffer-filled chambers. Carboxylate-terminated polystyrene beads were placed in one of the buffer-filled chambers, and a constant electric field was applied across the pore. Due to its negative charge, the bead migrates to the pore mouth and blocks it giving rise to a decrease in conductance. Using this test, the performance of the nanopores in glass and those in silicon nitride are compared.

[1] Dekker C. Solid-state nanopores. Nat. Nanotechnol. 2007; 2: 209-215.

[2] Steinbock LJ, Bulushev RD, Krishnan S, Raillon C, Radenovic A. DNA Translocation through Low-Noise Glass Nanopores. ACS Nano. 2013; 12: 11255-11262.

[3] Lee M-H, Kumar A, Park K-B, Cho S-Y, Kim H-M, Lim M-C, Kim Y-R, Kim K-B. A Low-Noise Solid-State Nanopore Platform Based on a Highly Insulating Substrate. Sci. Rep. 2014; 4: 1-7.

[4] Iliescu C, Chen B, Miao J. On the wet etching of Pyrex glass. Sens. Act. A. 2008; 143: 154-161.

430572 Detection of Acetone in Exhaled Breath By Gold Nanoparticle Gas Sensors

Abstract:

Detection of acetone in exhaled breath by gold nanoparticle gas sensors

Zhenzhen Xie[a], Xiao-an Fu[a*]

a Department of Chemical Engineering, University of Louisville, Louisville, KY 40292

*Corresponding author: xiaoan.fu@louisville.edu

Abstract

With the worldwide prevalence of diabetes, the methods for diagnosing have received considerable interest. The conventional method of blood tests is painful and has some limitations. Acetone in the human breath has been identified as an important biomarker for diagnosis of diabetes. The average concentration of exhaled acetone in the healthy human breath is found to be lower than 0.8 ppm, while that is higher than 1.8 ppm in diabetic patients. Gas sensors with sub-ppm acetone detection capacity play a significant role in the development of noninvasive monitors for diabetic patients. Gold nanoparticle-based sensors offer several advantages over conventional metal oxide-based sensors, including low power consumption, flexibility, ambient temperature operation and good chemical selectivity. They can be operated at room temperature and have shown limits of detection in the parts-per-million (ppm) or even parts-per-billion (ppb) level.

The main objective of this work is to increase the sensitivity of the sensors between different gold-thiolate nanoparticles for the detection of volatile organic compounds (VOCs) in human breath and dry air, and also the effect of the relative humidity. Gold nanoparticles with small size (< 5 nm) were synthesized. We found that the response follows a linear regression with increasing concentration of VOC in the range from 0.1 ppb to 5 ppm. Our results show that gas sensors based on gold nanoparticles could be used to distinguish healthy human from diabetic patients.

Keywords: diabetes, acetone, gold nanoparticle, sensors

429421 Carbon Nanofiber Mats for Surface Enhanced Raman Spectroscopy

Abstract:

Carbon Nanofiber Mats for Surface Enhanced Raman Spectroscopy

by

Juan F. Yee-De-León[1], Victor H. Perez-Gonzalez[1], Roberto C. Gallo-Villanueva[1], Sergio O. Martinez-Chapa[1,*]

[1] Sensors and Devices Research Group, School of Engineering and Sciences, Tecnológico de Monterrey, México

*Correspondence should be addressed to smart@itesm.mx

Abstract

Surface Enhanced Raman Spectroscopy (SERS) is a powerful readout technique that yields qualitative and quantitative information of chemical species and biomolecules at low concentrations. The effectiveness of the technique is closely related to the SERS-active substrate employed. Currently, nano-fabrication techniques have enabled the development of highly active and reproducible SERS-substrates, however, very specialized personnel and equipment are needed to construct these substrates, increasing the related cost of sample analysis using SERS as the readout technique. This situation has hampered the use of SERS in real life situations, which has motivated to the quest of novel nanomaterials and methodologies for producing homogenous, highly sensitive, scalable and low cost SERS-active substrates for chemical trace detection and biological applications.

In this work, polyacrylonitrile (PAN) nanofiber mats were fabricated via far-field electrospinning. These mats were subjected to a carbonization step to obtain nanofiber mats composed of carbon. The carbon nanofiber mats obtained were oxidized using a solution composed of 3:1 (H_2SO_4:HNO_3) v/v, and later, the surface of the nanofibers was functionalized with aminosilanes, in order to attach gold nanoparticles to the carbon nanofiber mats. Here, we present a first approximation to the use of carbon nanofiber mats as structural three-dimensional scaffolds for supporting metallic nanoparticles for its use as SERS-active substrates. The structure's high surface area and porosity, derivatives from the three-dimensional entanglement of the nanofibers, allows easy access and interaction of more analyte molecules with the metallic nanostructures located onto the nanofiber's surface. A higher number of metallic nanoparticles adsorbed increases the sensitivity and saturation levels of these substrates with regard to 2D SERS-active substrates.

430056 A Comparison of Cobalt and Gold Functionalized TiO2 Nanotubes Based Sensing Platform for Enhanced Electrochemical Detection of Volatile Organic Biomarkers

Abstract:

One class of biomarkers that is known to have associations with certain diseases, and yet has found limited use as a diagnostic tool is volatile organic biomarkers (VOBs). From a healthcare standpoint, sensing and detection of VOBs from the breath is important for early diagnosis of several pulmonary diseases, including tuberculosis, whose distinct biomarkers have been successfully identified by researchers using mass spectroscopy. Time of detection, cost, equipment portability, and performance have been identified as some of the major roadblocks in the development of a suitable device for point-of-care (POC) detection of tuberculosis (TB). In this study, we demonstrate the use of metal functionalized TiO_2 nanotubes array (TNA) based robust sensing platform for rapid electrochemical detection of the four prominent TB biomarkers viz. methyl nicotinate, methyl p-anisate, methyl Phenylacetate, and o-phenylanisole.

Solid-state TiO_2 nanotube array has demonstrated to be a promising sensing platform for VOCs detection. Highly ordered TNA were synthesized through electrochemical anodization, and functionalized separately with cobalt and gold, which were determined as leading frontrunners for efficient binding with the VOBs. Cobalt functionalization was first achieved using the incipient wetting impregnation method (aka dip coating method) by sonicating the TNA in 0.1M $CoCl_2$ solution for 30 minutes. The sensor response was recorded through electrochemical detection of the four VOBs based on the observable change in current at fixed low bias upon exposure of the sensor to the vapors of the biomarker. In an effort to increase the sensitivity of the sensor, cobalt was incorporated into the titania nanotube lattice through a one step in-situ anodization-co-functionalization process in a CoF_2 based ethylene glycol solution. As expected, enhanced sensor response was achieved with reduced time of detection. Further, TNA was functionalized with gold using an electroless deposition technique. Results indicate that sensor response from Au-TNA increased manifold over its Co-TNA counterparts.

The sensor substrates were characterized using SEM, TEM, and XPS. Indirect band gap was calculated based on UV-Vis absorbance. Further, a mechanism has been proposed to describe the attachment of predominate biomarkers to the sensor surface, based on band theory. Overall, the robust synthesis and approach of the sensor demonstrates promise for developing and commercializing a hand-held device for point-of-care diagnosis of pulmonary diseases.

430207 Biocompatibility Evaluation of Skin Wearable Silver Nanowire Biosensor Systems

Abstract:

Recent advances *in vitro* diagnostics on human skin or organ surfaces rely on the use of sensors. Wearable diagnostic and therapeutic devices are ubiquitous to health care, where technology can improve patient's life. Due to the high aspect ratio and excellent electrical conductivity, silver nanowires (AgNWs) have been widely explored for flexible/stretchable conductors that can accommodate large deformations, make intimate contact with curvilinear surfaces (such as human skin) and maintain high conductivity under large strain. A plethora of wearable devices using AgNWs have been demonstrated in the past few years, such as strain (motion) sensors, pressure sensors, touch sensors, and electrophysiological electrodes. However, biocompatibility of silver nanowires to human skin is largely unknown, which is critical for these wearable devices as many of them are in contact with skin for long time for long-term, continuous health and wellness monitoring. Lack of biocompatibility in a skin-attachable sensor, can result in complications including: cell cytotoxicity, skin irritation, skin sensitization, skin chronic inflammation and dermatitis.

We fabricated a highly conductive and stretchable conductor with AgNWs embedded in the surface layer of poly(dimethylsiloxane) (PDMS). Assessment of the materials biocompatibilities on skin is currently performed with the use of animals. Alternative non-animal related *in vitro* methods are demanded by Animal welfare and EU legislation. We have developed novel *in vitro* biocompatibility test methodology for screening of skin-wearable AgNW/PDMS sensors. These silver sensor components were assessed *in vitro* using two skin cell lines and 3-dimensional tissues. Three characterized in vitro assays were used to measure the cytotoxicity, irritation and sensitization of the skin-attachable sensor materials. These *in vitro* assays could meet the FDA requirement, as well as minimize the usage of live animals. We also characterized these AgNW/PDMS sensors to ensure that their electrical conductivities were performed functional for different applications, such as ECG and hydration sensors, without eliciting any undesirable local or systemic effects, including cytotoxicity and skin irritation.

426934 Metallization of DNA Origami to Form Thin, Electrically Conductive Nanowires

Abstract:

Bottom-up nanotechnology, which enables construction of complex structures from molecular building blocks, is seen as a promising technology to develop future generations of electronic devices. The bottom-up approach takes advantage of molecular self-assembly to create useful architectures of nanometer dimensions. DNA, with its small size, functional groups and complementary base pairs, is a powerful template for nanodevice fabrication via bottom-up assembly. Recent advances in the field of "DNA origami" have enabled fabrication of nanostructures by folding single-stranded DNA into a variety of 2- and 3-D shapes. However, electrically conductive structures are required for electronic applications.

In this study, we report the fabrication of electrically conductive DNA origami-templated nanowires with the use of gold nanorods (Au NR's) as a step towards DNA nanodevices. Our process begins with the seeding step where Au NR's electrostatically bind to the DNA origami backbone. The seeds are then connected by electroless deposition of gold to make continuous wires. The use of nanorods as seeds provides two key advantages over previous work with particulate seeds. First, the surfactant-enabled anisotropic growth used to initially fabricate the nanorods appears to persist through the electroless plating process on the DNA template. This permits improved control of the width of the final metallized structure since growth in the axial direction is preferred. The average final width of our Au-metallized DNA wires was 11.5 nm (n=30) with a standard deviation of 2.1 nm. We were able to obtain a 60% yield of continuous ~400 nm long DNA wires. In doing so, gaps between seed particles ranging from 10.6 nm on average to as wide as 25 nm were successfully filled. The second advantage of nanorods is the reduction in the number of connection points between seed particles by a factor of three relative to that observed for 5 nm seed particles. The reduced number of connection points is expected to increase the conductivity of the wires. Electrical characterization of the DNA wires is underway and complete current/voltage measurements will be presented.

In summary, we have demonstrated a metallization process to fabricate thin, high aspect ratio, conductive wires on DNA-origami templates with the use of Au NR's and electroless plating. Continuing efforts seek to use local functionalization to enable site-specific metallization of DNA-origami templates with Au NR's. The thin nanowires produced in this study will be interfaced with appropriate semiconducting elements to help enable fabrication of self-assembled nanoscale electronic circuits.

References:

1. Uprety, B.; Gates, E. P.; Geng, Y.; Woolley, A. T.; Harb, J. N. Site-Specific Metallization of Multiple Metals on a Single DNA Origami Template. *Langmuir* **2014**, 30, 1134-1141.
2. Gates, E.P.; Jensen, J.K.; Harb, J.N.; Woolley, A.T. Optimizing Gold Nanoparticle Seeding Density on DNA Origami. *RSC Advances* **2015**, 5, 8134-8141.

429915 Novel Light-Activated Therapy for Mutli-Drug Resistant Pathogens

Abstract:

Current design of therapeutics (drugs and nanoparticle-based therapies) relies heavily on a spatially selective drug-delivery system (colloids and vesicles coated with desired antibodies or proteins) due to non-specificity of the drugs or metal nanoparticles. Typically, these non-specific drugs once spatially delivered kill the targeted cells (e.g. tumor or pathogen) as well as healthy cells in close proximity (including metal nanopartciles used for cancer therapy). Further complexities in non-specific binding and incorrect delivery can cause undesirable side-effects. Here we present a novel nanoparticle-based therapy where the (non-metallic) nanoparticles specifically target selective cell phenotypes, based on their biochemical potentials. Using a specific light induced targeting mechanism, these light-activated reactive species (LARS) have the capability to target and kill specific-pathogens on illumination with visible light, without affecting the growth of other cells. While the growth of selected cells and the targeted pathogens is unaffected in dark, we selectively designed different LARS species which can either kill or proliferate the growth of selected target cells. This is demonstrated when targeting Escherichia coli cells, HEK293 mammalian cells, as well as, a co-culture of E. coli and HEK293 cells. Furthermore, within the same strain of E.coli, we find that while lab strain DH5Î± is susceptible to LARS, clinically isolated multi-drug resistant clinical strain of E. coli is not; likely due to presence of a non-specific antibiotic resistance mechanism. The ability to tune the biochemical potential of nanoparticles allows for selective cell phenotypes over a range of organisms, and has potential applications in antimicrobials and targeted organism-specific therapeutics.

422802 Rapid Synthesis of Unilamellar Liposomes

Abstract:

Liposomes present as a versatile drug carrying vehicle, capable of loading hydrophobic and hydrophilic drugs, bio-compatibility, long-circulation, wide mono-disperse size range (from 10 nm to 10s of microns), and targeting specific tissues. Several methods of liposome synthesis are being employed that result in production of unilamellar vesicles in a laboratory time scale of minutes. These fast methods employ external driving to achieve a monodisperse population. We present an alternate method of mono-disperse unilamellar liposome synthesis that results from spontaneous self-assembly of lipids in a solution (and therefore does not require external driving), and which is rapid. The lipids self-assemble when the solvent in which they are dissolved is rapidly replaced by a non-solvent (water) in a stationary phase interdiffusion. There has also been no attempt to resonably compare the synthesis times for various methods of liposome production. We provide estimates of the time of synthesis of a single liposome for various methods known in the literature, and show that the present method is among the fastest of all the methods, and is much faster than earlier methods of spontaneous assembly. The method is presently carried out in a batch mode, and requires device and process design to make it semi-batch or continuous.

426317 Encapsulation and Release of Hydrophilic Drug Molecules Via a Poly(lactic acid)-Montmorillorite Composite Micro/Nano-Particle System

Abstract:

Poly (lactic acid) (PLA) is one of the most widely used materials to fabricate biodegradable micro/nanoparticle drug delivery carriers. However, its hydrophobicity limits its application in the encapsulation of hydrophilic drug molecules, like nisin. In our group, a novel dynamic encapsulation process via a single emulsion method has been developed to encapsulate hydrophilic molecules within hydrophobic PLA micro/nano-particles. The uniqueness of this process is the emulsification of PLA solution in the water-glycerol medium under high viscous turbulent shear flow, where water-in-oil-water (W/O/W) droplets formed via one single emulsion step with the high temperature (HT) (60) condition, and oil-in-water (O/W) droplets under the room temperature (RT) (25) condition. The formation of W/O/W droplets and O/W droplets is reversible between the HT and RT condition, which can be controlled by a cooling circulating system. SEM results showed that open hollow microparticles were obtained under HT condition while nanospheres were prepared under RT condition. By increasing the repeating times of reversible droplets formation, more hydrophilic drug molecules can be encapsulated into the PLA particle matrix. The encapsulation efficiency and release profile will be tested on a series of samples with different repeating times. The addition of montmorillorite (MMT) to PLA is to improve the strength and the biocompatibility of polymer particles. The SEM results showed that more closed hollow microparticles were obtained compared with the samples without MMT. The anti-crack property of the PLA/MMT composite particle may lead to higher encapsulation efficiency because it prevents the breaking leakage of water phase, which contains the hydrophilic drug molecules.

432431 Biocompatibility of Lipid Coated Nanocomposites

Abstract:

There is significant interest in the use of a variety of nanoparticles for drug delivery applications. Solid nanoparticles are unique in that they enable incorporation of drug on the surface of nonporous particles as well as within mesoporous systems. As such, nanoparticles have been examined for use in many routes of delivery such as pulmonary, transdermal, oral and intravenous. Despite the breadth of research examining their use as drug delivery platforms, there is a lack of information regarding the health and safety of solid nanomaterials. In most studies investigating cell toxicity little regard is given to the characterization of the nanomaterials being used. Important aspects often left unconsidered include size, size distribution, surface charge, aspect ratio, morphology, and preparation method.

In addition to traditional solid nanoparticles, liposomes have been highly researched and represent one of the few nanocarriers to successfully transition to clinical applications. Studies of liposome toxicity have shown a strong correlation to their composition. Surface charge, size and concentration have been shown to be significant factors affecting toxicity of liposomes.

Recent literature has focused on combining nanomaterials to develop novel nanocarriers such as liposome encapsulated solid nanoparticles. However, little concern has been given to the health and safety of these newly formed nanocomposites. This study aims to elucidate the missing cellular toxicity effects associated with these new materials and provide a direct comparison to traditional solid nanoparticles. In addition to lipid encapsulation, the factors investigated included size-dependence, dose-dependence, and surface modification effects.

433024 Biomimetic Amphiphilic Polymers for Intracellular Therapeutic Delivery and Theranostic Applications

Abstract:

Biomacromolecules represent a powerful new class of medicine with potential for treatment of a wide variety of previously intractable human diseases. However, it remains a big challenge to effectively deliver them through extracellular and intracellular barriers. There is a need to better understand the mechanisms of entry into the cell cytoplasm and nucleus in order to design optimal delivery systems for biological molecules. On the one hand, this would open up significant opportunities to deliver potent drug payloads against intracellular targets to positively impact human health. In addition this would enable us to develop a more general understanding of the rules governing the uptake of biological molecules into cells. This presentation will cover our recent efforts to design, synthesis and in-vitro/in-vivo evaluation of novel bio-functional polymers. The pH-responsive, metabolite-derived, amphiphilic polymers are designed to mimic factors that enable efficient viral transfection. Strict control over the size, structure, hydrophobicity-hydrophilicity balance and aromaticity of the polymers can effectively manipulate their conformational characteristics and dynamic behaviour in aqueous solution and their interactions with lipid membrane, cell and tissue models. It has been demonstrated that the biomimetic polymers can traverse the extracellular matrix in 3-D multicellular spheroids, reach individual cells in the tumour models, and enable efficient cellular entry and substantial endosomal release of payloads (e.g. siRNA and therapeutic proteins) and imaging moieties into the cytoplasm or the nucleus. This could represent a promising therapeutic delivery platform, suitable for research and theranostic applications in the treatment of various diseases including cancer.

427890 Evolution of the Catalyst Nanoparticle Structure and Composition during Single Walled Carbon Nanotube Nucleation and Growth

Abstract:

The incorporation of single-walled carbon nanotubes (SWCNTs) into commercial technologies that exploit their remarkable properties is to a great extent dependent on the ability to synthesize them with specific structures and chiralities. The structural features of the metal catalyst nanoparticle during synthesis conditions by catalytic chemical vapor deposition (CCVD) are believed to play a fundamental role in determining the structure of the nascent nanotube cap throughout nucleation and subsequent growth. With the aid of computational tools such as reactive and *ab initio* molecular dynamics, and density functional theory together with real-time atomic-resolution transmission electron microscope observation of SWCNT growth on supported Co nanoparticles, we are able to look at the structure of the nanoparticle at different stages of the process. The evolution of the nanoparticle in terms of structural parameters and distribution of C atoms in the nanoparticle is monitored during the C dissolution, nanotube nucleation and growth stages. The interactions of the nanoparticle with the substrate and with the nucleating nanotube are found to play an important role in determining the nanoparticle shape and the distribution of C atoms inside the nanoparticle. Moreover, the observation of a C gradient and identification of pure metal and carbide-like size-fluctuating regions inside the nanoparticle and the correlation observed between such phenomena with the rates of incorporation of C atoms to the nanotube shed some light into the transport and reaction mechanisms.

429374 Physicochemical Properties of Engineered Catalyst Supports and Implications for Growth of Carbon Nanotube Carpets

Abstract:

Catalysts supported on alumina have been widely heralded for their unparalleled efficiency in the growth of high-quality carbon nanotube (CNT) carpets via CVD with hydrocarbon precursors. However, some important applications of CNT carpets require carpets to be grown directly on specific substrates. Thus, there is significant interest in understanding the properties of a good catalyst substrate and how to create and maximize these properties in inactive substrates. The recent transformation of an inactive catalyst substrate (c-cut sapphire) to an active substrate for CNT carpet growth via ion beam bombardment provides an excellent model system for this study (*Nano Lett.* **2014**, *14*, 4997). The properties of pristine and ion beam damaged sapphire substrates are probed using a combination of contact angle measurements and X-ray reflectivity (XRR). The contact angle data are analyzed using the van Oss-Good-Chaudhury model and the Young-Dupre equation, enabling the determination of the acid-base properties of the substrates. For the XRR data, a Fourier analysis of the interference pattern allows extraction of layer thickness while the atomic density and interfacial roughness are extracted by analyzing the amplitude of the interference oscillations. The results show interesting correlations between the physicochemical properties of the catalyst substrate and the activity of Fe catalyst during CNT carpet growth. Detailed discussion of the implications of the structural and acid-base properties of the pristine and engineered substrates in CNT carpet growth are presented.

423557 Sulfur-Induced Chirality Changes in Single-Walled Carbon Nanotube Synthesis on Co/SiO2 Catalyst

Abstract:

Selective synthesis of single-walled carbon nanotubes (SWCNTs) with different chiral structures is desirable for their potential applications. We show that introducing sulfur-containing compounds into carbon feedstock may efficiently alter the chiral selectivity of a Co/SiO2 catalyst toward different chiral species. When carbon disulfide (0.0001 wt%) was added to ethanol, the carbon yield increased significantly from 4.8 to 14 wt% without chirality changes. The changes in chiral selectivity are correlated with the significant changes in the carbon yield. With further increase of carbon disulfide concentration, the chirality distribution shifted from smaller diameter chiral nanotubes, such as (7,6) at 0.9 nm, toward larger diameter chiral ones, such as (8,7) at 1 nm and (9,8) at 1.2 nm, in addition to a sharp decrease in the carbon yield. Further, when a different sulfur-containing compound was used, the chiral selectivity changed differently. Thiophene (0.1 wt%) in ethanol led to a good chiral selectivity toward (9,8) nanotubes with a relative abundance of 43.1% among all semiconducting nanotubes. We propose that sulfur may selectively block active sites on Co particles through dynamic interactions among sulfur, hydrogen, carbon and Co metal particles. Sulfur-containing compounds may be used as an efficient additive to tune the chirality of SWCNTs. (Ref: *Journal of Materials Chemistry A*, **2015**, *3, 3310–3319)*

430158 Porous TiO2 Conformal Coating on Carbon Nanotubes As Energy Storage Materials

Abstract:

TiO_2 is a promising material for lithium-ion battery (LIB) and supercapacitor (SC) due to its high specific energy density, low cost, nontoxicity, eco-friendliness, and abundant availability. However, the practical electrochemical performance of anatase TiO_2 is still not optimal due to its low electronic conductivity. Hybridizing with carbon nanotubes (CNTs) or other carbon materials is one of the most facile and effective approaches. CNTs can facilitate faster electron transfer, which would increase the performance of LIB or SC. Here we report a simple sol-gel method for facile synthesis of TiO_2/CNTs hybrid. The porous anatase TiO_2 nanoparticles are uniformly coated on the CNTs conducting network, which leads to remarkably improved electrochemical performances such as exceptional cycling stability, good high rate durability, and reduced resistance. This hybrid exhibits a reversible capacity as high as 200 mA·h g^{-1} at a current density of 0.1 A g^{-1} as an anode in LIB. As a SC, it shows a specific supercapacitance of 145 F g^{-1} in 0.5 M H_2SO_4 electrolyte, higher than that of the previously reported TiO_2 based supercapacitors. Moreover, this hybrid also exhibits excellent durability after 1000 cycles for both LIBs and SCs. The superior performance and cycling durability demonstrate the reinforced synergistic effects between the porous TiO_2 and interweaved CNTs network, indicating a great application potential for such hybrid materials in high power LIBs and SCs.

427096 Fabrication of High-Quality Graphene Oxide Nanoscrolls and Application in Supercapacitor

Abstract:

We reported a simple and effective way of fabricating one-dimensional (1D) graphene oxide nanoscrolls (GONS) from graphene oxide (GO) sheets through shock cooling by liquid nitrogen. The corresponding mechanism of rolling was proposed. One possibility is the formation of ice crystals during the shock cooling process in liquid nitrogen to be the driving force. The other might be due to the uneven stress of the sheets inside or outside ice during the lyophilization. After reducing, graphene nanoscrolls (GNS) exhibited good structural stability, high specific surface area, and high specific capacitance. The capacitance properties were investigated by cyclic voltammetry, galvanostatic charge-discharge, and electrical impedance spectroscopy. A specific capacity of 156 F/g for the GNS at the current density of 1.0 A/g was obtained comparing with the specific capacity of 108 F/g for graphene sheets. Those results indicated that GNS-based rolling structure could be a kind of promising electrode material for supercapacitors and batteries.

432560 Carbon Nano-Tube Addition to Improve the Conductivity of a Molecular Sieve Adsorbent

Abstract:

During resistive heating regeneration, an electric current passed through an adsorbent with sufficiently low resistivity (typically, 0.1 – 10 Ω.m) and heat is generated by the Joule effect. This method allows for rapid adsorbent regeneration, as the heat flow is decoupled from the purge gas flow. Molecular sieves are effective adsorbents because they have tailored and structured pore size distributions. Unlike carbonaceous materials, however, molecular sieves have high electrical resistivity that prevents their regeneration by resistive heating. In this study, carbon nano tube (CNT) was added to zeolite Y (high resistivity, >E+7 Ω.m) to decrease its resistivity. CNT was added by CH4 decomposition on cobalt impregnated zeolite Y. Different cobalt impregnation conditions and decomposition temperatures were used for CNT growth on zeolite Y. Modified samples were characterized using bulk elemental (CHN) analysis, scanning electron microscopy (SEM), X-ray diffraction (XRD), and resistivity measurements. CHN and SEM confirmed the addition of carbon onto zeolite Y in the form of CNT. Resistivity measurements showed that CNT growth decreased the resistivity of zeolite Y by up to 8 orders of magnitude. The results were compared to the resistivity of samples physically mixed with powdered carbon in different ratios. It was found that the resistivity can be decreased more efficiently by adding carbon directly to the structure of zeolite in the form of CNT. Finally, the modified samples were heated resistively, suggesting that CNT addition to Zeolite Y allows faster and more energy efficient heating compared to conductive-convective heating.

439673 Specific DNA Sequences for High Purity Carbon Nanotube Enantiomer Separation

Abstract:

Despite the remarkable electrical, thermal, mechanical, and optical properties of carbon nanotubes (CNTs), polydispersity has been one of the major challenges limiting their fundamental research and technological development. In addition, separating carbon nanotube enantiomers is of great scientific interest since it provides important information into the mode of DNA binding and molecular recognition towards specific nanotube structures. Polymer aqueous two-phase (ATP) extraction has been recently demonstrated as an effective technique to sort carbon nanotubes of different chiralities in high purity and yield. The spontaneous partition of single-stranded DNA-wrapped single-wall carbon nanotubes (ssDNA-SWCNTs) in a given ATP system is strongly sequence-dependent. Additionally, circular dichroism (CD) spectroscopy shows that purified semiconducting and metallic nanotubes are also enantiomer enriched. In this work, we have conducted an exhaustive screening of a 12-mer DNA library using a specific DNA sequence pattern. To our surprise, this new pattern enabled the identification of a super-sequence which can recognize and separate both enantiomers of a single chirality (6,5) species in high purity and yield, as well as the isolation of multiple high purity enantiomers in an ATP system.

428802 Engineering Nanoscale Protein Scaffolds with Modular Functionalities

Abstract:

Proteins are the most versatile among the various biological building blocks and efforts in protein engineering have resulted in many industrial and biomedical applications. However, the strength of proteins - their versatility and specific interactions - also complicates and hinders their systematic design and engineering. One potential approach is to exploit the modular nature of different protein domains in order to design synthetic protein scaffolds that can perform completely new biological functions. In this talk, I will outline several successful examples in connecting exchangeable protein domains for predicative engineering applications in (1) bioenergy applications, (2) Fuel cell applications, and (3) biosensing.

432276 Self-Assembling Polypeptoids for Nanoscale Materials Applications

Abstract:

Self-assembling synthetic polymers offer the stability and robustness necessary for materials applications and can be used to pattern on the nanometer scale, however our ability to directly include functionality in these materials is at its infancy. Biological polymers, such as polypeptides, are responsible for many of life's most sophisticated functions due to precisely evolved hierarchical structures. Protein structures are the result of monodisperse sequences of amino acids that fold into well-defined chain shapes and tertiary structures with a degree of complexity that is very difficult to understand. In order bridge this gap between our ability to direct the self-assembly of synthetic polymers and the sophisticated intertwining of structure and function in biopolymers, we must first develop an understanding of the relationship between monomer sequence, chain shape, and directed self-assembly. Polypeptoids are non-natural, sequence specific polymers that offer the opportunity to probe the effect of sequence on self-assembly with much simpler molecular interactions and more scalable synthesis than traditional polypeptides. In this talk, I will discuss the use of this model system to understand the role of sequence on chain shape and block copolymer self-assembly. I will then discuss potential application as surface active agents for anti-fouling.

434526 Molecular Technology for Improved Treatment of Disease

Abstract:

Our understanding of disease states, both communicable and non-communicable, progresses at a remarkable pace in this age of large-scale, data-rich biology. Unfortunately, our current ability to make use of this knowledge on a routine basis in clinical settings is limited by a number of factors. The focus of this presentation will be three examples of how chemical engineering design and biomolecular engineering can be used to further the goals of precision medicine and improved access to the benefits of knowledge of disease at the molecular level.

425800 Nanosheet Conformation Dynamics during Flow

Abstract:

We develop a general coarse-grained Brownian Dynamics (BD) methodology for simulating conformation dynamics of two-dimensional nanosheets (such as graphene, boron nitride, and metal dichalcogenides). Such nanosheets are promising materials for a variety of materials and electronics applications. Flow processing (shear, extensional) are frequently to cast liquid nanosheet dispersions and mixtures, and little has been reported in regard to the effects of such processing techniques on the shape of the nanosheets. Our BD simulations show the effects of both shear and extension on nanosheets over time. The bending moduli play a major role in these dynamics. Nanosheets with low bending moduli are prone to show a repeated coil-stretch transition under shear and an extended "washrag" conformation under elongational flow, whereas nanosheets with high moduli tend to show a tumbling response under shear and a flat extended conformation under elongational flow. Most importantly, these simulations can actually shed light on recent experimental studies of the relationship between hydrodynamic radius (measured using dynamic light scattering) and nanosheet dimensions.

430514 Characterizing the Differences in Adsorbed Surfactant and Hydration Layers Around Single Species Metallic and Semiconducting Single-Wall Carbon Nanotubes with Analytical Ultracentrifugation

Abstract:

Over the past decade, numerous methods have emerged for the dispersion and liquid phase purification of single-wall carbon nanotubes (SWCNTs). The success and efficiency of the dispersion and purification methods, however, are believed to be strongly dependent on the characteristics of the bound layer of dispersant (such as small molecule surfactant(s) or DNA) both in the amount of bound dispersant as well as the specific structure it forms. However, only limited direct measurements of the differences in the bound layers on well-resolved SWCNT populations have been reported. Thus, it is important to elucidate the structure of the adsorbed surfactant and hydration shells around different single species SWCNTs and surfactant-SWCNT complexes, respectively, to quantitatively address the differences in structure that drive differential separation. To study the structure of the adsorbed surfactant layer, we utilize analytical ultracentrifugation (AUC) to resolve the density and radial distribution of adsorbed bile salt surfactants (sodium deoxycholate, sodium cholate) and sodium dodecyl sulfate on different species of SWCNTs at various solution conditions. In our AUC experiments, we use nanotubes which have been both chirality and length sorted, giving us a population of particles which not only have a narrow length and diameter distribution, but also unique optical transitions and, in theory, unique interactions with surfactants. In this work, we measure and compare the anhydrous and buoyant radii around multiple semiconducting and metallic SWCNT species dispersed in sodium deoxycholate and provide direct measurements of the effects of adding co-surfactant to the observed sedimentation.

424606 Electronic and Mechanical Properties of Hydrogenated Irradiated and Amorphous Graphene

Abstract:

Electronic and Mechanical Properties of

Hydrogenated Irradiated and Amorphous Graphene

Asanka Weerasinghe,[1] Ashwin Ramasubramaniam,[2] and Dimitrios Maroudas[3]

[1]Department of Physics, University of Massachusetts, Amherst

[2]Department of Mechanical and Industrial Engineering, University of Massachusetts, Amherst

[3]Department of Chemical Engineering, University of Massachusetts, Amherst

Defect engineering and chemical functionalization of graphene are promising routes for fabrication of carbon nanostructures and two-dimensional metamaterials with unique properties and function. Toward this end, we have developed computer models of irradiated graphene through insertion of random distributions of vacancies, which have generated configurations with structural and morphological characteristics in very good agreement with experimental observations. We have predicted the onset of an amorphization transition at an inserted vacancy concentration between 5% and 10%, with the transition becoming less abrupt with defect concentration as the temperature increases. We have found that the electronic density of states of vacancy-amorphized graphene is characterized by introduction of localized states near the Fermi level. We have also found that the vacancy-induced crystalline-to-amorphous transition is accompanied by a brittle-to-ductile transition in the failure response of irradiated graphene sheets. While point defects and larger voids appreciably degrade the strength of irradiated graphene compared to that of pristine graphene, even heavily damaged samples exhibit tensile strengths in significant excess of those typical of engineering materials.

In this presentation, we use hydrogenation of irradiated and irradiation-induced amorphized graphene to study the effects of chemical functionalization on the properties of graphene damaged by irradiation. Our study is focused on the impact of hydrogenation on the electronic structure and properties and on the mechanical response of irradiated and irradiation-induced amorphized graphene. We use molecular-dynamics simulations according to a reliable bond-order potential to prepare the hydrogenated configurations and to carry out mechanical tests of dynamic deformation of the prepared configurations at constant strain rate and temperature. Computations of the electronic band structure and electronic density of states of fully relaxed hydrogenated configurations are performed based on a density functional tight binding approach. We report results for the electronic properties, as well as the ultimate tensile strength, fracture strain, and toughness of a broad class of irradiated graphene structures as a function of the hydrogen concentration/coverage used in the hydrogenation process.

435219 Stability of Carbon Nanotube Dispersions with a Mixture of Nonionic and Ionic Surfactants in DPD Simulation

Abstract:

Carbon nanotubes (CNTs) have provided a wide range of potential applications over the past few years. However, pristine CNTs tend to agglomerate in solvents because of strong Van der Waals attraction [1]. Investigating ways to disperse CNTs in solutions has been an active research area. In experiments at the University of Oklahoma, the mixture of ionic (Alfoterra) and non-ionic (Tergitol) surfactants have been used to successfully disperse CNTs in water. Based on this finding, Dissipative Particle Dynamics (DPD) simulations are used to figure out the assembly and the surfactant adsorption on the CNT surface. First of all, the DPD interaction parameters for surfactants were validated by determining CMC of surfactants from DPD calculations in equilibrium. The micelle shape of each surfactant and the mixture of two surfactants in water were also determined. After that, a CNT (i.e., a hollow carbon cylinder in the simulation) was inserted into the system of water and surfactants. The competition of forming micelles and adsorption on the CNT surface was observed in DPD simulation. Surfactants on CNT make the CNT surface to be more hydrophilic and repel other surfactant molecules at high concentration. Besides, the effect of CNT on surfactant CMC and adsorption density of surfactant was also investigated. Our work can give a more detailed understanding of the configuration of surfactants on the CNT surface and the conditions necessary for the development of stable CNT dispersions.

ACKNOWLEDGEMENTS

The financial support of the Advanced Energy Consortium (AEC BEG08-022) and the computational support of XSEDE (CTS090017) are acknowledged.

REFERENCES

1. L. A. Girifalco, M. Hodak, R. S. Lee, Phys. Rev. B, 2000, 62, 13101-13110.

435022 A Reversible-Growth Cluster Aggregation Model for Graphene Particle Dispersions

Abstract:

Graphene is widely used as a filler for materials serving in a variety of applications such as polymer composites and the anodes of lithium ion batteries. In these applications, the colloidal dispersions of graphene are used to control the composite structures and the resulting properties of the material. To provide a fundamental basis for the aggregation mechanisms of graphene during consolidation, we have developed a reversible aggregation model based on diffusion-limited aggregation and diffusion-limited cluster aggregation models. The structures produced by this model are in excellent agreement with the experimentally observed structures produced in a system of functionalized graphene sheets dispersed in an aqueous SDS solution. The generalization of this model to predict the structures of fillers in composite materials is expected to be of use in the calculation of the resulting properties of graphene-filled materials.

426948 Characterization of Viscoelastic and Cured Composite Properties of Styrene Functionalized Single-Walled Carbon Nanotubes in Unsaturated Polyester Resin

Abstract:

Single-walled carbon nanotube (SWNT) functionalization has been predominantly used to improve dispersion state in polymer composites. Unsaturated polyester resin (UPR)-SWNT composites have huge potential to improve UPR properties. This could enable UPR, which already has the largest market share of any thermoset resin, to be used in higher performance applications. Previously, we have investigated the dispersion state of different commercial SWNT products in UPR. However, within the range SWNT concentrations tested, cured composites have not shown significant mechanical property improvements. Urena-Benavides et al showed styrene functionalized MWNT in UPR had less aggregation and significantly different viscoelastic properties than pristine MWNT dispersions. However, they did not investigate curing kinetics or the final mechanical properties. We have hypothesized that styrene functionalized SWNT in UPR will show better cured composite properties due to the potential for intra-SWNT and SWNT- polyester resin network crosslinking. This talk will show the differences between SWNT and styrene functionalized SWNT in terms of their 1) dispersion state and viscoelastic properties of in UPR, 2) curing kinetics, and 3) final cured composite mechanical and thermal properties.

419025 Single-Walled Carbon Nanotube Dispersion in a Complex Biological Medium

Abstract:

This research aims to examine the ability of Tryptic Soy Broth (TSB), a bacterial culture medium, to disperse single-walled carbon nanotubes (SWNT). There has been notable research into how specific proteins and biological macromolecules interact with SWNT; however, relatively little is known about SWNT's interactions with complex biological media frequently used in bacteriology. The dispersions examined in this study were prepared via probe ultrasonication of SWNT and TSB mixtures in the 0.05 wt% to 0.1 wt% range at 60W for 30 minutes or settings that delivered equivalent energy. These dispersions were centrifuged for 3 hours at 17,000 x *g*. The resulting supernatants were analyzed via various methods including ultraviolet-visible spectroscopy and circular dichroism spectroscopy in order to characterize the ability of TSB to disperse SWNT. Dried samples were analyzed with atomic force microscopy, scanning electron microscopy, optical microscopy and infrared spectroscopy in order gain additional insight into TSB-SWNT interactions. These results combined with earlier work on SWNT-Lysozyme interactions suggest that π-π stacking interactions between SWNT and the amino acid tryptophan are a key factor in polypeptide based dispersion. The ability to disperse SWNT in biologically relevant media provides a starting point for further research to assess possible SWNT interactions with pathogenic bacteria without concerns about antimicrobial properties of the dispersion medium.

429688 Suppression of Infrared Absorption in Nanostructured Metals By Controlling Faraday Inductance and Electron Path Length

Abstract:

While metals are widely used for electrodes in many optoelectronic devices, the device efficiency is limited by optical losses due to metals. As thinner devices are desired for reduced cost without sacrificing performance, the metal electrodes can be nanostructured to concentrate light in the photoactive material. A popular approach for the light concentration is to excite surface plasmon polaritons (SPPs). However, in such SPP-based devices, strong metal absorption is a major source of losses. The problem of this parasitic absorption becomes critical when a high device efficiency is desired. Moreover, if the device operates over a broad band as in solar cells, metal absorption needs to be controlled over the whole spectrum of interest and this poses a significant scientific and engineering challenge. Here, we investigate the loss suppression in metal nanocoil structures in the infrared range. Surprisingly, we find that the nanostructured metals can behave almost like a vacuum with a negligible optical loss in the infrared region. The loss is reduced in comparison to flat films by more than an order of magnitude over most of the very broad spectrum between short and long wavelength infrared. Our detailed analysis shows that this suppression of optical losses is due to both a large Faraday inductance and a long electron path. These two factors render the conduction electron in the metal effectively heavier than a hydrogen molecule. To investigate the usefulness of this effect in infrared detectors, we include a photoactive material in the coiled metal structures. In this case, the fraction of absorption in the active material increases by two orders of magnitude compared to non-coiled structures. Moreover, we demonstrate that the coiling of the infrared detector nanostrips increases the photoresponsivity by 15 times without additional electrical losses. These findings could benefit many metal-based applications that require low loss such as photovoltaics, photoconductive detectors, solar selective surfaces, infrared-transparent defrosting windows, and other metamaterials.

415043 Plasmon-Enhanced Energy Transfer and Other Photophysical Effects in Doped-Lanthanide Nanocrystals

Abstract:

Upconversion of infrared radiation into visible light has been investigated for applications in biological imaging and photovoltaics. However, low conversion efficiency due to small absorption cross-section for infrared light (Yb^{3+}), and slow rate of energy transfer (to Er^{3+} states) has prevented application of upconversion photoluminescence (UPL) for diffuse sunlight or imaging tissue samples. Here, we utilize resonant surface plasmon polaritons (SPP) waves to enhance UPL in doped-lanthanide nanocrystals. Our analysis indicates that SPP waves not only enhance the electromagnetic field, and hence weak Purcell effect, but also increases the rate of resonant energy transfer from Yb^{3+} to Er^{3+} ions by 6 fold. While we do observe strong metal mediated quenching (14 fold) of green fluorescence on flat metal surfaces, the nanostructured metal is resonant in the infrared, and hence enhances the nanocrystal UPL. This strong columbic effect on energy transfer can have important implications for other fluorescent and excitonic systems too. We also investigated the effect of surface plasmons on color tuning and other optoelectronic properties in upconverting nanoparticles. SPP strongly affected the emission color in different-codoped lanthanide UCNPs by tuning the energy transfer between lanthanide ions (Er, Yb and Tm here), depending on the energy overlap between resonant SPP modes and multiple energy states in respective ions. This selective enhancement of resonant energy transfer provides valuable physical insights and useful design principles for electronically coupling different upconverting nanoparticles with matched semiconductor devices and plasmonic absorbers. To test the role of upconversion enhancement in optoelectronics, upconversion enhanced photocurrent was measured on 2D semiconductor nanosheets coupled with UCNPs. The semiconductor 2D nanosheets were nominally transparent to infrared radiation, and no photocurrent is observed on illumination of sub-bandgap radiation. However, strongly enhanced photocurrent was observed in these 2D nanosheets using 980 nm light irradiation, which can enable new applications in development of ultrathin room-temperature infrared detectors and high-resolution imaging systems.

Quantum Dynamical Simulations of the Photoinduced Charge Transfer Process in Donor-Bridge-Acceptor

Ma. Belen Oviedo and Brya M. Wong

1. Department of Chemical Engineering, University of California, Riverside, Riverside, California, United States
2. Materials Science & Engineering Program, University of California, Riverside, Riverside, California, United States

One of the major challenges in science and technology is the efficient conversion of solar energy into readily usable forms. In particular, the enormous size and time scales in solar energy harvesting requires the convergence of multiple scientific disciplines to both improve and optimize these photoconversion processes [1]. Quantum-based simulations play a crucial role in this area since they provide accurate predictions of optical properties of new complex materials. Although very accurate methods exist for predicting optical absorption spectra such as linear-response time-dependent density functional theory [2] or other wave-function based methods, these methods are computationally expensive and can only be applied to relatively small molecules. Consequently, large systems relevant to realistic experiments (such as molecular self-assembled nanostructures and nanomaterials) pose a serious challenge since they lie outside the computational capabilities of these conventional methods.

To this end, we have developed a new real-time time-dependent density functional tight binding (RT-TDDFTB) [3] approach to calculate the electron dynamics of donor-bridge-acceptor complexes. Our approach significantly differs from previous linear-response TD-DFT methods in that we directly propagate the one-electron density matrix in the presence of a non-perturbative external field. Furthermore, this RT-TDDFTB approach permits the calculation of large systems composed of thousands of atoms, enabling new quantum simulations that are beyond the reach of traditional Kohn-Sham approaches in conventional TDDFT. We have successfully used this method to described the photoinduced charge-transfer processes in light-harvesting systems within a fully quantum dynamic framework. In this work Furthermore, we show the charge-transfer state can be modulated by the molecular configuration of the bridge subunit and the charge-transfer processe is enhanced and controlled by the molecular architecture of this subunit.

[1] Nozik, A. J., Miller, J. Chem. Rev., 110, 6443-5 (2010).
[2] Petersilka, M., Gossmann, U. J., Gross, E. K. U. Phys. Rev. Lett., 76, 1212-1215 (1996).
[3] Oviedo, M. B., Zarate, X., Negre, C. F. A., Schott, E., et al. J. Phys. Chem. Lett. 3, 2548-2555 (2012).

415019 Multiple Energy "Exciton-Shelves" in Quantum-Dot-DNA Nanobioelectronic Materials

Abstract:

Quantum dots (QDs) are semiconductor nanocrystallites with size-dependent multiple quantum-confined states which are being explored for utilizing broadband radiation. While DNA has been used for the self-assembly of nanocrystals, it has not been investigated for the formation of simultaneous conduction pathways for transporting multiple energy charges or excitons. These exciton shelves can be formed by coupling the conduction band, valence band, and hot-carriers states in QDs with different HOMO-LUMO levels of DNA nucleobases, resulting from varying degrees of conjugation in the nucleobases. Here, we present studies on the electronic density of states (DOS) in naturally occurring nucleobases (guanine, thymine, cytosine, and adenine), which energetically couple with quantized states in semiconductor QDs. Using scanning tunneling spectroscopy of single nanoparticle-DNA constructs, we demonstrate composite DOS of chemically-coupled DNA oligonucleotides and cadmium chalcogenide QDs (CdS, CdSe, CdTe). While perfectly aligned QD-DNA states lead to exciton-shelves for multiple energy charge transport and energy transfer, mismatched energy levels in QD-DNA hybrids introduce intra-bandgap states which can lead to charge trapping and recombination. Using single nanoparticle and ensemble device measurements we also show successful extraction and transport of both bandedge and high-energy charge carriers, and energy transport using excitons. These results can have important implications for the development of a new class of nano-bio electronics and biological transducers.

419731 Integration of Photosystem I Proteins within Conductive Polymer Matrices Using Vapor Phase Techniques

Abstract:

Organic Solvent-Free Integration of Photosystem I Proteins within Conductive Polymer Matrices Using Vapor Phase Techniques

Maxwell Robinson, Evan Gizzie, David Cliffel, and G. Kane Jennings

Photosystem I (PSI) is a 500 kDa photocatalytic protein complex residing within chloroplast granum of photosynthetic organisms, converting incident solar radiation to electrons with near-unity internal quantum efficiency. Extracted PSI has been employed successfully as a molecular photosensitizer within a host of low-cost electrochemical and solid-state photovoltaic architectures. This presentation explores this group's recent efforts to assimilate PSI within a chemically and energetically tuned conductive polymer matrix. Composite assembly follows a vapor phase Friedel-Crafts grafting procedure, whereby dispersed Fe(III) catalyzes rapid polymer growth from aromatic moieties inherent to PSI. The technique avoids temperature and solvent exposure requirements standard to bulk heterojunction fabrication in organic photovoltaics and can be employed using a wide library of high vapor pressure conjugated monomers. Our composite assemblies enhance charge shuttling processes from individual proteins within multilayer assemblies—greatly reducing charge transfer resistances and improving overall efficiency of photocells.

438527 Controlled Release from Polyelectrolyte Complex Drug Carriers

Abstract:

Comparison studies of release of a model drug, Rhodamine B, from two different types of polyelectrolyte complex beads (PECs) made of poly(styrenesulfonate, PSS)-poly(allylamine, PAA) and chitosan-alginate were investigated. The release was carried out by using a modified Franz diffusion cell into deionized water at 37°C. The long term sustained release (> 7 days) of Rhodamine B was obtained with PSS-PAA PECs due to their tighter networks. By tuning the compositions of PSS-PAA, the time needed to completely release the drug could be varied from two days to 14 days. For the case of chitosan-alginate PECs, 100% accumulated release occurred within 24 hours. With the incorporation of a thermo-responsive polymer, e.g., poly(N-isopropylacrylamine), the release could be extended to two or more days for the chitosan-alginate PECs. By drying the beads, the release of Rhodamine B was found to be further extended.

438450 3D Printed Microfluidic Device for Dynamic Investigation of the Blood Brain Barrier

Abstract:

The “Blood Brain Barrier” (BBB), surrounding the capillaries in the Central Nervous System, is a tissue composed of endothelial cells joined to each other by tight junctions. It serves to impede the passage of any impurities from the blood to the brain and is essential to brain homeostasis, disruption of which can result in a myriad of brain disorders like Alzheimer’s, Multiple Sclerosis, and Parkinson’s disease. Current in-vitro models of the BBB include a seeded Transwell ® Membrane under static conditions, where the apical and basal side fluids do not exhibit dynamic flow and apply shear stress on the membrane. Being a simplified model of the tissue, it does not adequately model the tissue as it is in-vivo, where there is dynamic flow of both basal (cerebrospinal fluid) and apical (blood) fluids. Consequently, we have designed a microfluidics device, the μTRANS chip, to model the BBB under dynamic conditions, allowing us to study not only the tissue structure of the BBB but also the cellular level dynamics of it under different environmental conditions. Our initial results, revolving around the construction of the device, its characterization, and the biochemical/mechanical modulations of the tissue will be presented at the conference. In terms of construction, the device was built using a seeded Transwell ® membrane (consisting of a layer of human brain endothelial cells on the apical side and a layer of astrocytes on the basal side) inserted between two glass slides. Each of the glass slides served as a distinct chamber, containing two 3D printed micro-channels, interconnected by smaller micro-channels. The smaller channels were designed to allow for fluid exchange between the larger micro-channels. Gold electrodes were also built into the device to analyze the permeability of the BBB using electrochemical impedance spectroscopy (EIS). The channels, glass substrate layers, and Transwell ® Membrane were all placed into a 3D printed vacuum chamber as the final step in construction, in order to maintain contact between the different layers of the device during experimental modulations. The permeability characteristics of the BBB were studied using fluorescence and EIS, specifically in the form of transendothelial electrical resistance tests (TEER), under different biochemical and mechanical conditions, including varying levels of pH, shear stress, and nutrients. We believe that our ability to study the BBB in a dynamic environment allows us to not only accurately model the tissue under physiological conditions but to control its characteristics using biochemical and mechanical means.

438448 Self-Assembly Simulations of Polymer Functionalized Virus Capsids

Abstract:

Viral capsids have been the subject of a significant amount of research in a variety of fields. Their potential relevance to biomedical science arises from their ability to encapsulate therapeutic agents and deliver them to targeted cellular sites. The capsids can be decorated with polymers to target specific sites or promote controlled aggregation for creating novel materials. We used coarse-grained molecular dynamics to simulate the aggregation dynamics of amphiphilic functionalized icosahedral capsids, representing the Cowpea Mosaic Virus capsid, when grafted with diblock copolymers of polyethylene glycol and polylactic acid. Additional variations of the system includes changing the polymer grafting density and capsid volume fraction to show how these changes affect the self-assembled structures. Our observations suggest that (a) the excluded volume of the icosahedral and (b) the polymer grafting density play the largest role on aggregation dynamics and morphologies when compared to the other variants.

438410 Investigation of the Interaction Between a Novel Drug Delivery System and an Epithelial Cell Layer

Abstract:

Surgery and other invasive processes are the current treatment options for patients with diabetic retinopathy, glaucoma, and age-related macular degeneration, which are eye diseases that affect at least nine million Americans. We are developing a targeted drug-delivery system, which would provide a non-invasive option for treatment. The system features a collagen membrane with drug-loaded poly(lactic-co-glycolic acid) (PLGA) nanoparticles incorporated into a thin membrane. This membrane can be worn by a patient in conjunction with a contact lens, and provides for continuous drug release. The purpose of this project is to study the interactions of lidocaine-loaded nanoparticles in a collagen membrane with a layer of corneal epithelial cells. Lidocaine-loaded PLGA nanoparticles were made and characterized by measuring size and drug content. Nanoparticles were mixed into a collagen solution and then gelled to form a thin membrane. Corneal epithelial cells were cultured on semipermeable well inserts to form a confluent layer of cells. The cell layer was exposed to the collagen membrane containing the lidocaine-loaded nanoparticles. Measurements were taken to determine cell viability and permeability, as well as the amount of lidocaine that crosses the cell layer. Results showed that lidocaine was released from the system and delivered across the cell layer. The release profile shows an initial burst release, followed by steady drug release over the testing period. Cell viability decreased with the application of the membranes, compared to a control sample of cells only. Further work is being done to determine the cause of cell death and to test the system for longer periods of time

438407 Quantitative Analysis of Fundus Images for Grading of Vitreous Haze

Abstract:

Purpose: Intraocular inflammatory processes lead to progressive accumulation of cells and protein exudate in the vitreous, which can be detected by ophthalmoloscopic inspection of the eye as decreased visibility of the retinal vasculature, optic nerve head, and other fundus details. The objective of this work was to develop a quantitative method for grading the blurriness of ocular fundus images in order to provide automated grading of the severity of uveitis.

Methods: A computer algorithm was developed to quantify image blurriness through application of entropy filtering and Fourier spatial-frequency analysis. The algorithm was refined using a set of 8 reference images that were created by optically filtering a single standard fundus picture to varying degrees that simulate increasing vitreous haze. It was then applied without modification to a dataset of 120 digital fundus images collected from patients with different grades of uveitis. The algorithm performance was evaluated against the acutance method, which is commonly used to quantify the overall sharpness of edges in an image. Computed scores of vitreous haze were also compared for both methods in a masked fashion with the subjective readings of three expert clinicians.

Results: The Pearson correlation coefficient between computed automated grades of the algorithm and clinician grades of blurriness was 0.81-0.88 (R^2 = 0.66-0.78). Acutance measurements calculated from the raw images and from images with background fluctuations subtracted out did not perform as well. The Pearson correlation coefficient between acutance and clinician grades was found to be -0.05 (R = 0.003) and 0.33 (R=0.109), respectively, which is quite weak.

Conclusions: We have developed a computer algorithm for grading vitreous haze in an unbiased and quantitative manner that correlates strongly with the subjective readings of expert clinicians. The algorithm performance with fundus images is much better than that of standard acutance measures.

McCormick

Northwestern Engineering

Department of Chemical and Biological Engineering

Developing a Strategy for Constructing Modular Biosensors

Andrew K.D. Younger[1], **Neil C. Dalvie[2]**, Austin G. Rottinghaus[3], and Joshua N. Leonard[2,4,5,*]

[1] Interdisciplinary Biological Sciences Graduate Program, Northwestern University, Evanston, IL, 60208, USA
[2] Department of Chemical and Biological Engineering, Northwestern University, Evanston, IL, 60208, USA
[3] Department of Chemical and Biological Engineering, University of Wisconsin-Madison, Madison, WI, 53706, USA
[4] Chemistry of Life Processes Institute, Northwestern University, Evanston, IL, 60208, USA
[5] Member, Robert H. Lurie Comprehensive Cancer Center, Northwestern University, Evanston, IL, 60208, USA

1

McCormick

Northwestern Engineering

Department of Chemical and Biological Engineering

Motivation: Quantifying metabolism

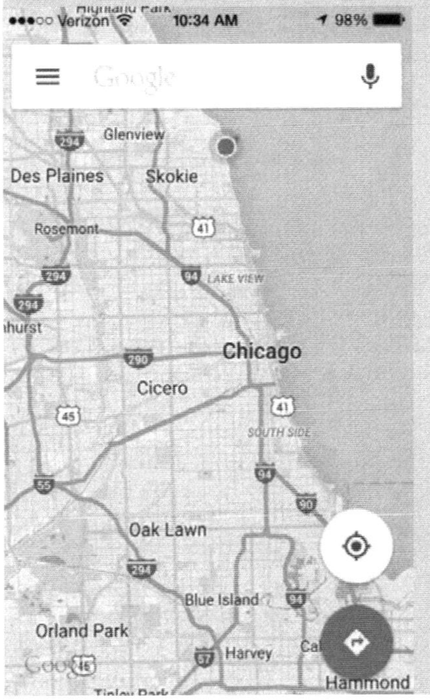

2

McCormick
Northwestern Engineering
Department of Chemical and Biological Engineering

Motivation: Quantifying metabolism

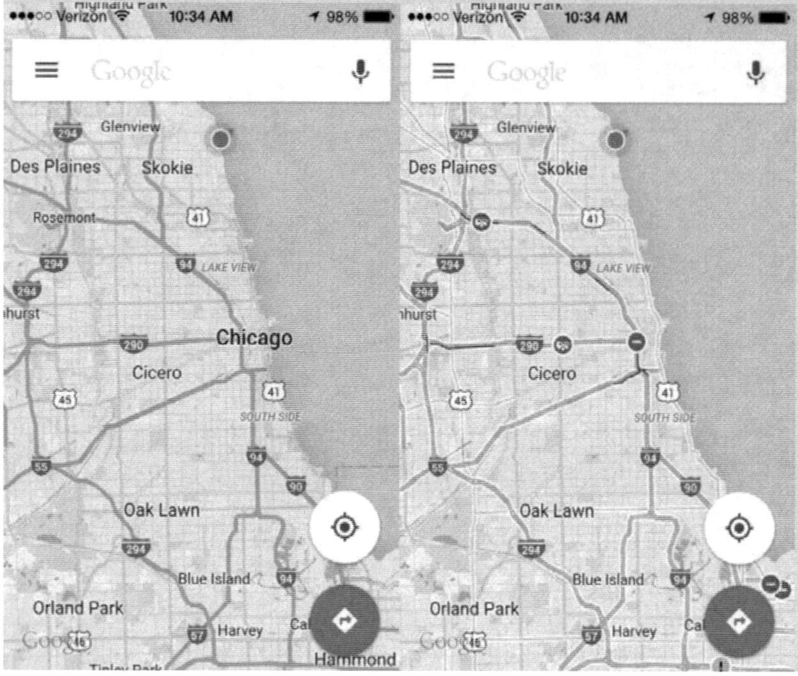

3

McCormick
Northwestern Engineering
Department of Chemical and Biological Engineering

Motivation: Quantifying metabolism

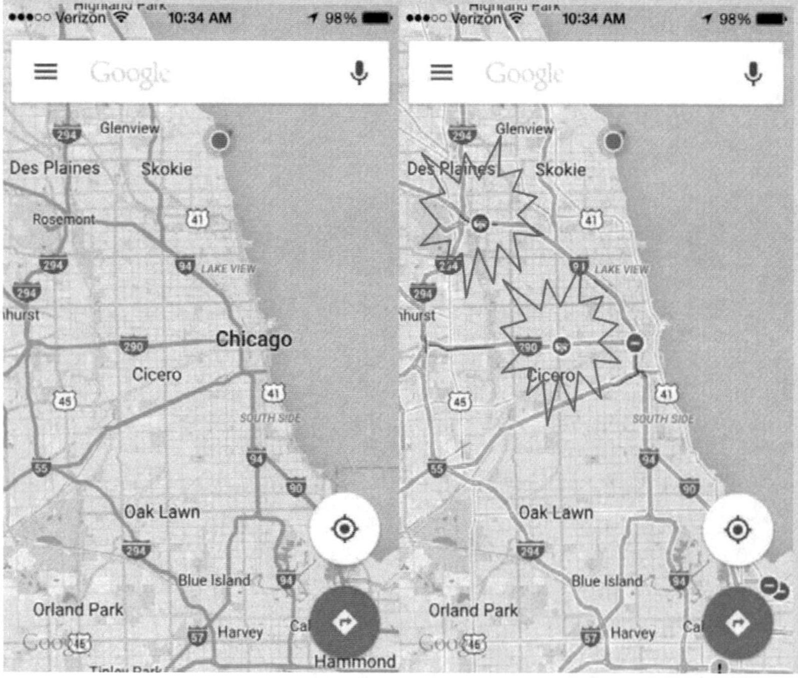

4

McCormick

Northwestern Engineering

Department of Chemical and Biological Engineering

Motivation: Quantifying metabolism

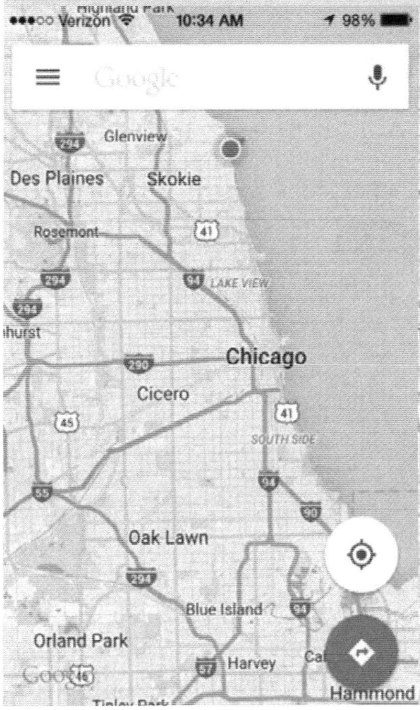

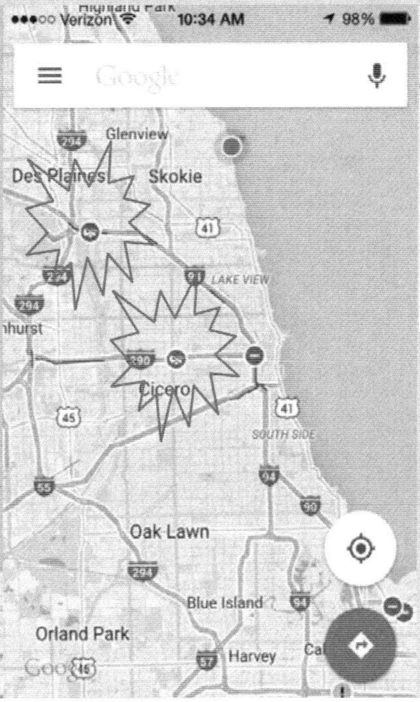

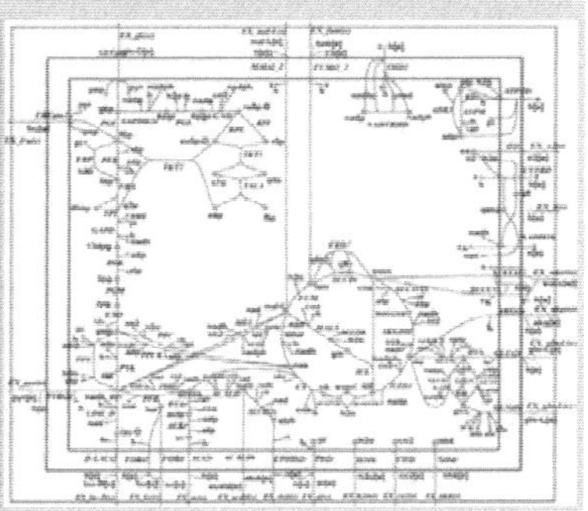

5

Orth *et al*. 2011

McCormick

Northwestern Engineering

Department of Chemical and Biological Engineering

Motivation: Quantifying metabolism

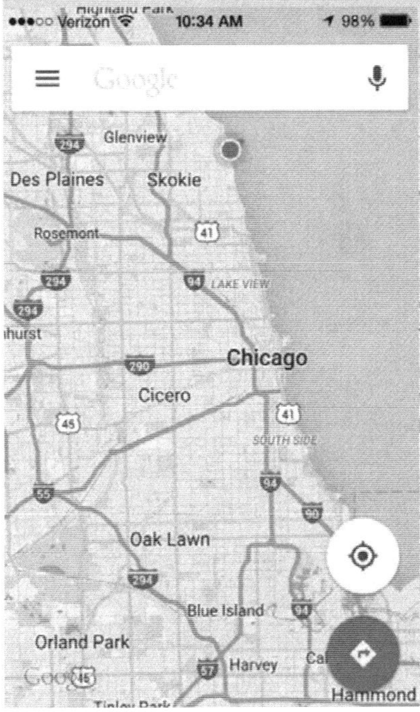

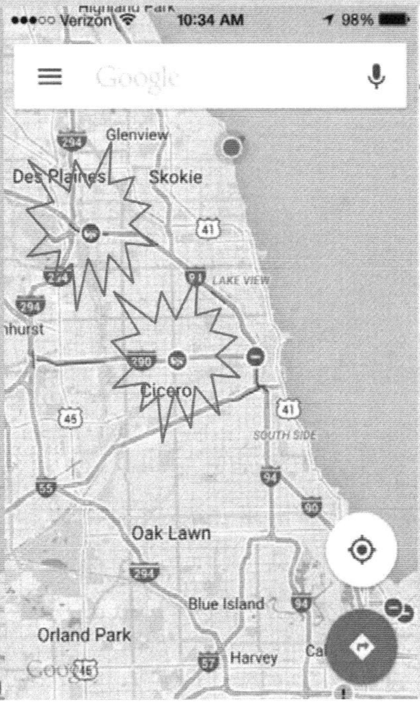

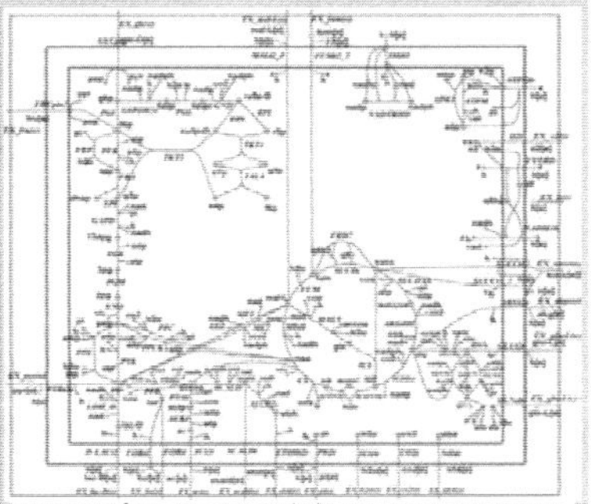

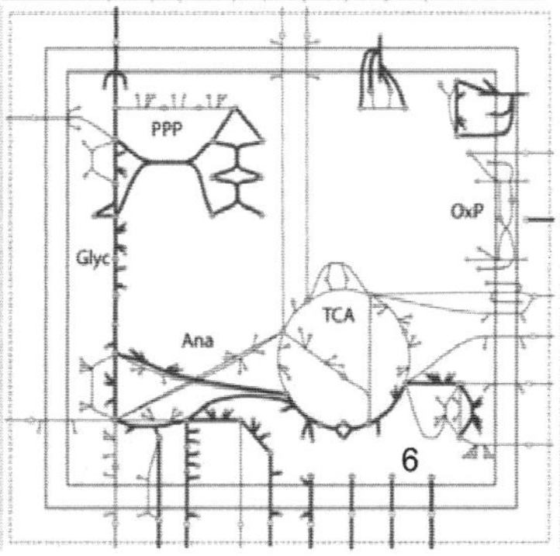

6

McCormick

Northwestern Engineering

Department of Chemical and Biological Engineering

Useful functions of metabolite biosensors

High Throughput Library Screening

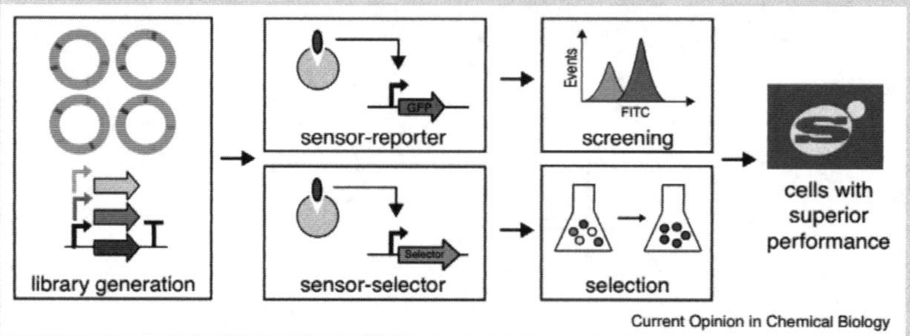

Current Opinion in Chemical Biology

7

Zhang *et al.* 2015

McCormick

Northwestern Engineering

Department of Chemical and Biological Engineering

Useful functions of metabolite biosensors

High Throughput Library Screening

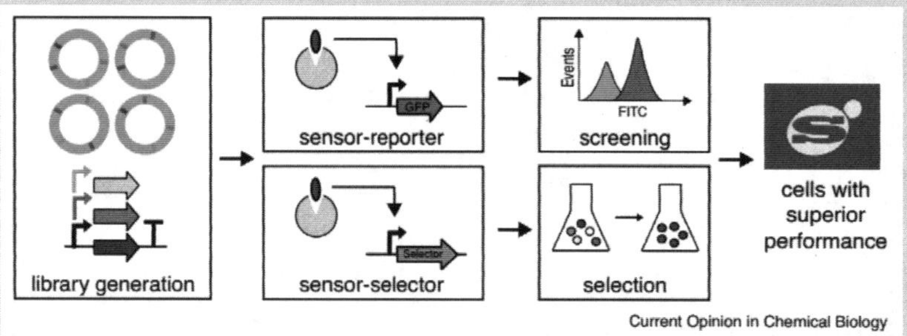

Current Opinion in Chemical Biology

Feedback control

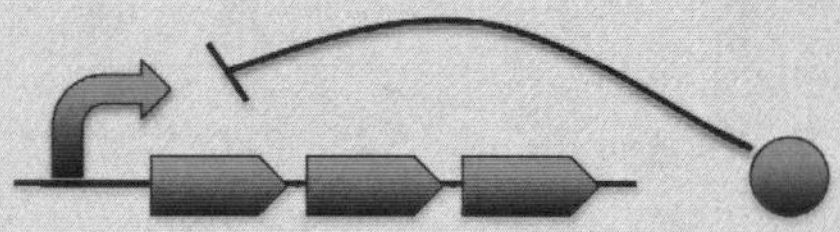

8

Zhang *et al.* 2015

Useful functions of metabolite biosensors

High Throughput Library Screening

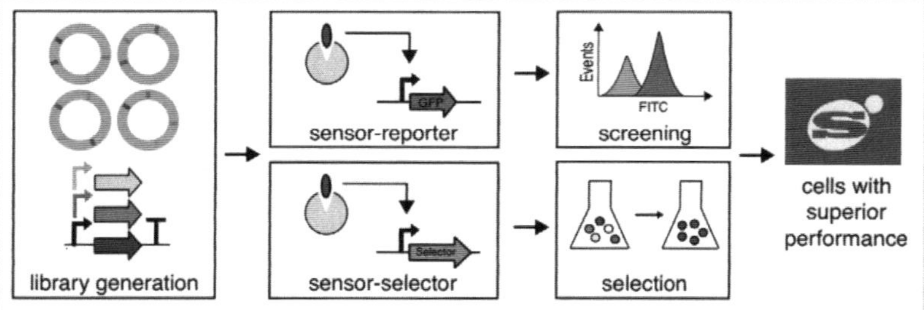

Current Opinion in Chemical Biology

Feedback control

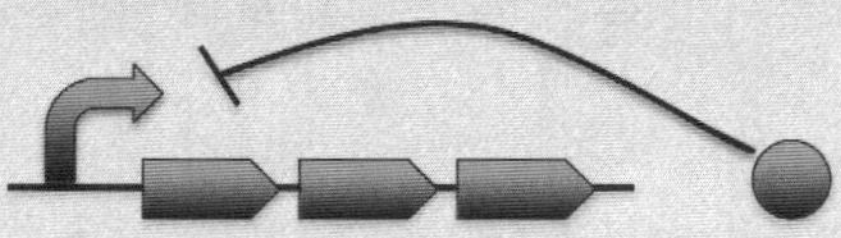

Limited natural options

9

Zhang *et al.* 2015

McCormick
Northwestern Engineering
Department of Chemical and Biological Engineering

Metabolite responsive transcriptional regulators

Goal:

Ligand responsive transcriptional regulator

Approach:

Modular design

-Convert ligand binding protein to biosensor

-DNA binding domain

10

McCormick
Northwestern Engineering
Department of Chemical and Biological Engineering

Metabolite responsive transcriptional regulators

Goal:

Ligand responsive transcriptional regulator

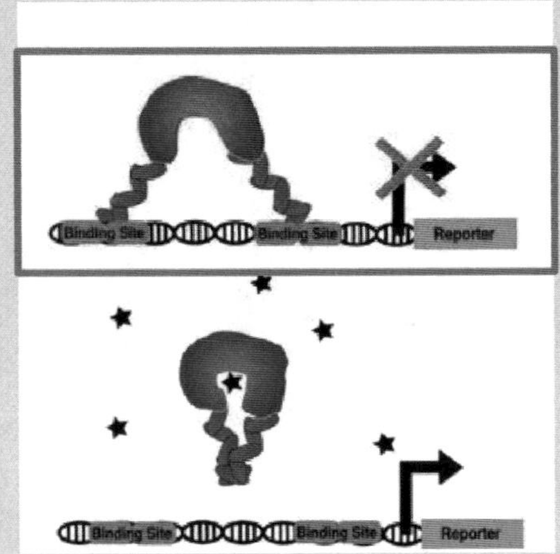

Approach:

Modular design

-Convert ligand binding protein to biosensor

-DNA binding domain

11

McCormick
Northwestern Engineering
Department of Chemical and Biological Engineering

DNA binding domain: BCR1 zinc finger

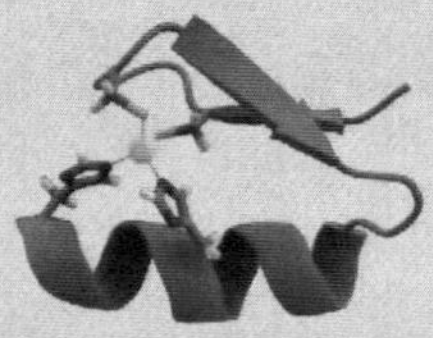

12

PDB #1A1L Thomas Splettstoesser

DNA binding domain: BCR1 zinc finger

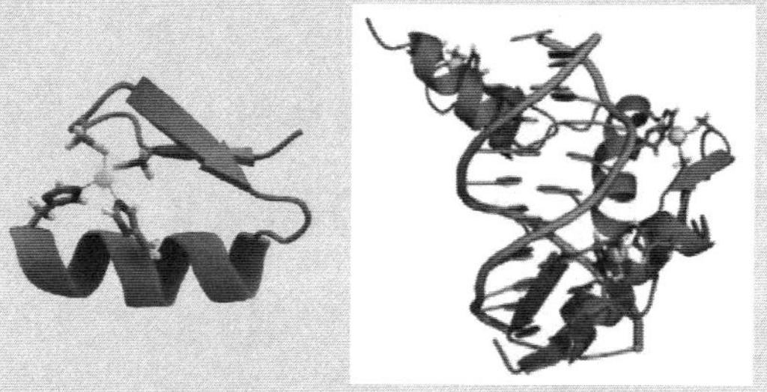

9 base pair binding site

13

PDB #1A1L Thomas Splettstoesser

McCormick
Northwestern Engineering
Department of Chemical and Biological Engineering

DNA binding domain: BCR1 zinc finger

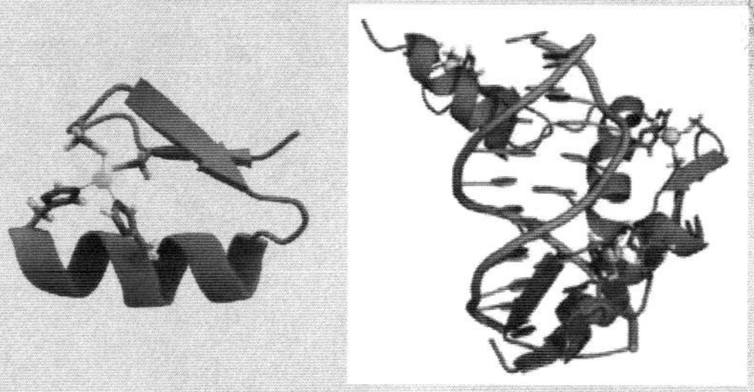

9 base pair binding site

Constitutive GFP reporter

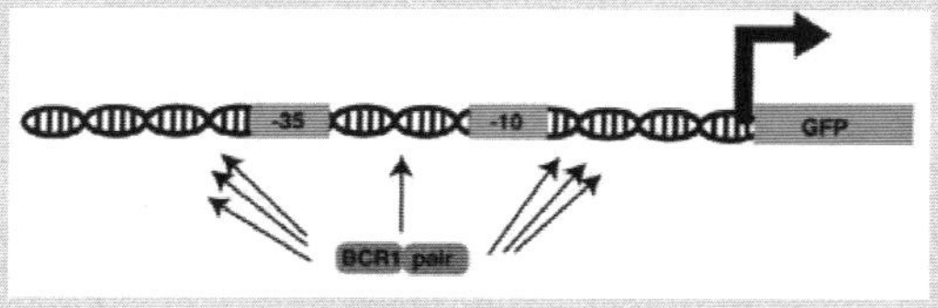

14

PDB #1A1L Thomas Splettstoesser

McCormick

Northwestern Engineering

Department of Chemical and Biological Engineering

DNA binding domain: BCR1 zinc finger

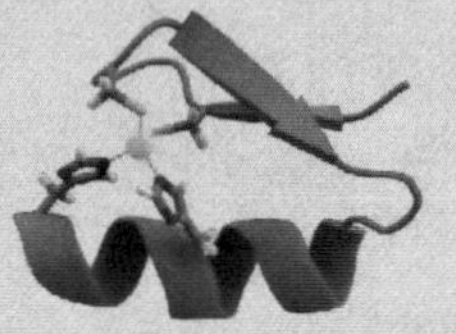

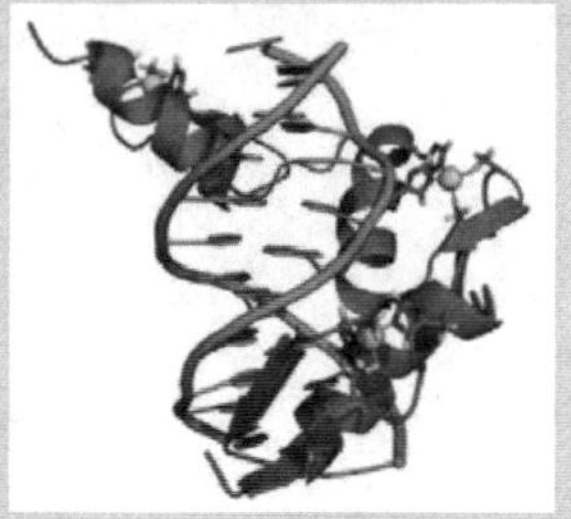

9 base pair binding site

Constitutive GFP reporter

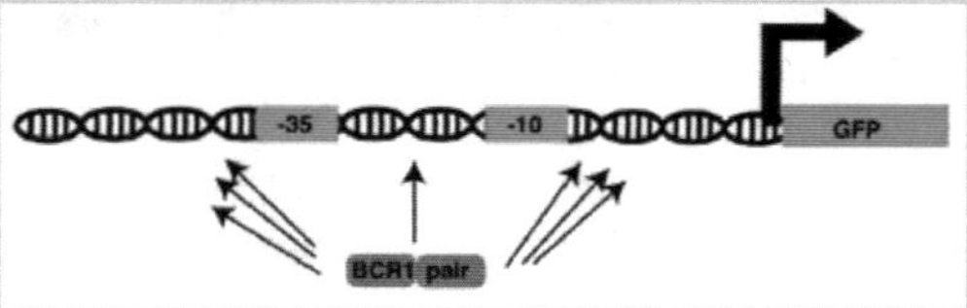

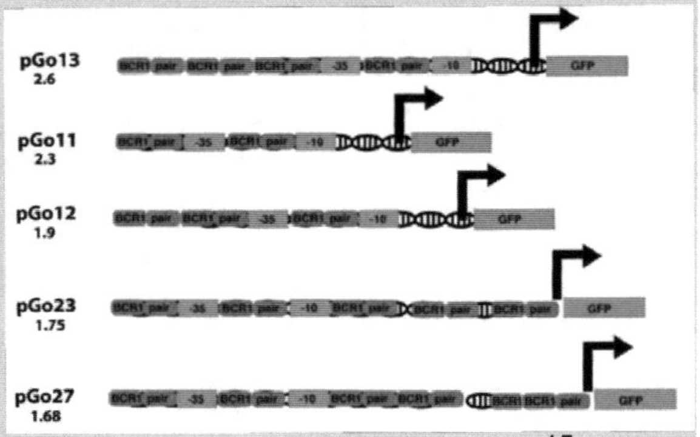

15

PDB #1A1L Thomas Splettstoesser

McCormick

Northwestern Engineering

Department of Chemical and Biological Engineering

Investigate expression of promoter library

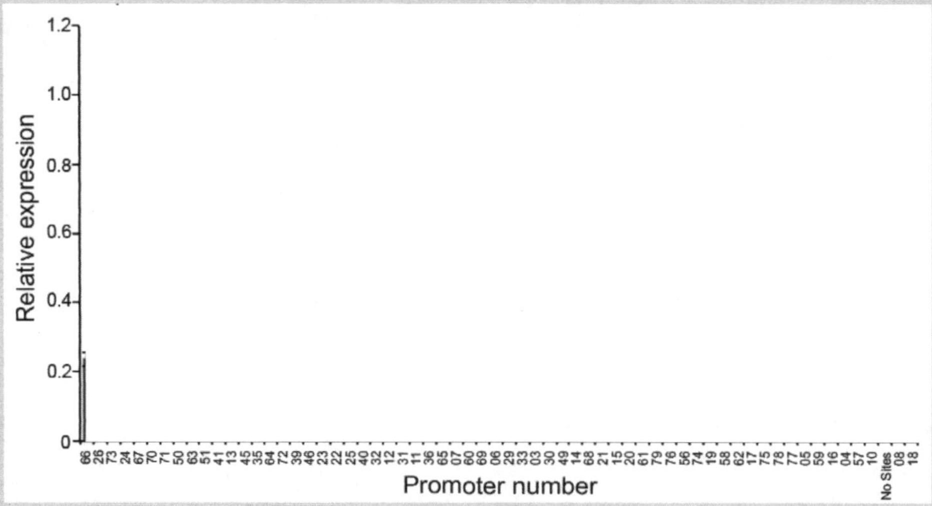

McCormick

Northwestern Engineering

Department of Chemical and Biological Engineering

Investigate expression of promoter library

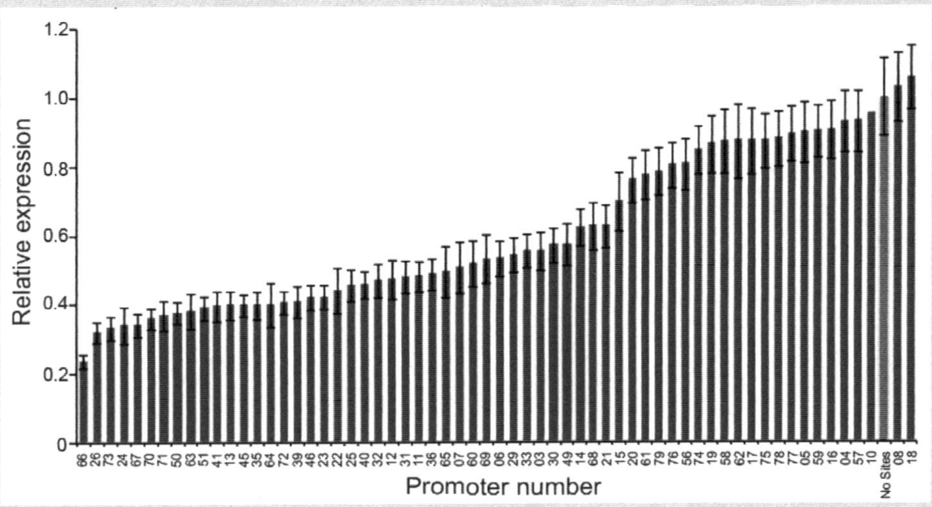

McCormick

Northwestern Engineering

Department of Chemical and Biological Engineering

Investigate expression of promoter library

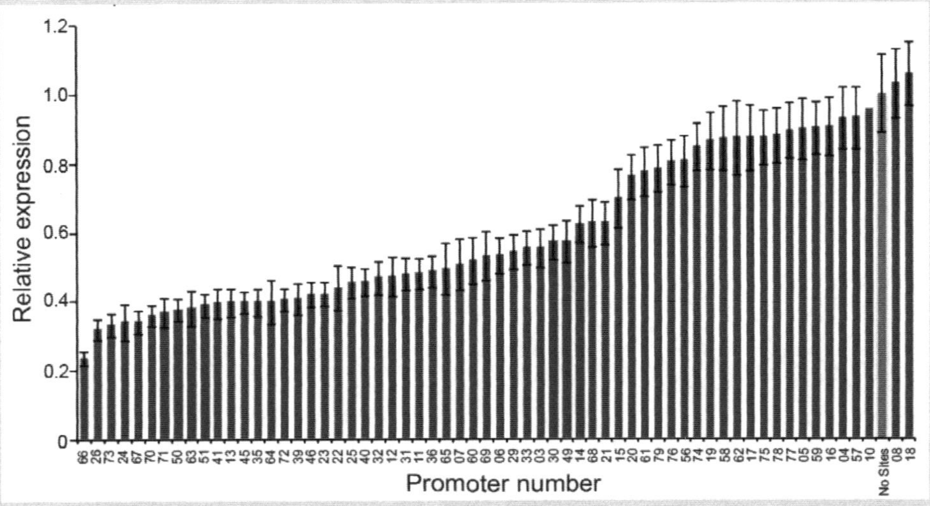

Wide range of repression – Why?

McCormick
Northwestern Engineering
Department of Chemical and Biological Engineering

Method for quantitatively describing the library

Goal: Select a set of descriptors that describe promoters.

Risk: Too many descriptors will over fit the system.

19

McCormick
Northwestern Engineering
Department of Chemical and Biological Engineering

Method for quantitatively describing the library

Goal: Select a set of descriptors that describe promoters.

Risk: Too many descriptors will over fit the system.

Assumption: The only thing that affects reporter expression is the binding of the biosensor

Features only describe the location of zinc finger binding sites

20

McCormick
Northwestern Engineering
Department of Chemical and Biological Engineering

Number	Shorthand	Description of Feature
1 ★	*BS in core*	Number of binding sites in the core
2 ★	*pairs in core*	Number of pairs in the core
3	*BS upstream*	Number of binding sites upstream of the -35 region
4	*pairs upstream*	Number of pairs upstream of the -35 region
5	*BS downstream*	Number of binding sites downstream of the -10 box
6	*pairs downstream*	Number of pairs downstream of the -10 box
7	*length*	Length of the promoter in base pairs
8 ★	*spacer -10*	Base pairs downstream of the -10 box, before the first site
9 ★	*spacer -35*	Base pairs upstream of the -35 region, before the first site
10 ★	*spacer inside -10*	Base pairs upstream of the -10 box, before the first site
11 ★	*spacer insider -35*	Base pairs downstream of the -35 region, before the first site
12	*core middle space*	Non-binding site base pairs in the core
13	*ave downstream spacing*	Average number of base pairs between sites downstream
14	*ave upstream spacing*	Average number of base pairs between sites upstream
15	*down length*	Base pairs downstream of the -10 box
16	*up length*	base pairs upstream of the -35 box
17	*lopsidedness*	Absolute value of (sites upstream) - (sites downstream)

21

McCormick
Northwestern Engineering
Department of Chemical and Biological Engineering

Determining important features

Define output:
"repressibility" = (-1)*expression

Questions:

What is the relative importance of features in determining repressibility?

How many of these features are actually significant?

22

McCormick

Northwestern Engineering

Department of Chemical and Biological Engineering

Approach: Three independent feature selection methods

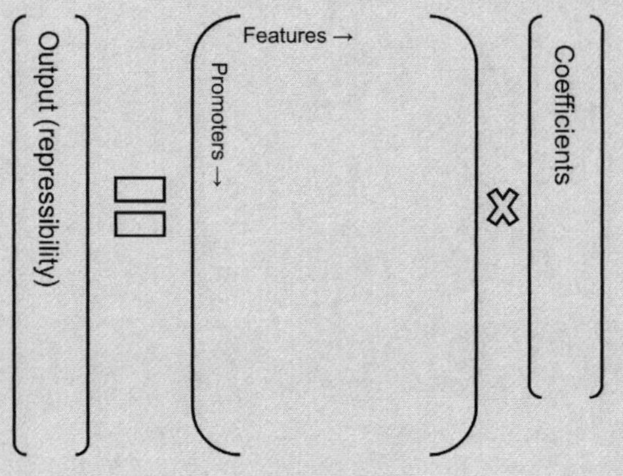

- Partial least squares regression ← Iterative
- Random forest ← Iterative
- Lasso regression ← Regularized

23

McCormick

Northwestern Engineering

Department of Chemical and Biological Engineering

How many features are significant?

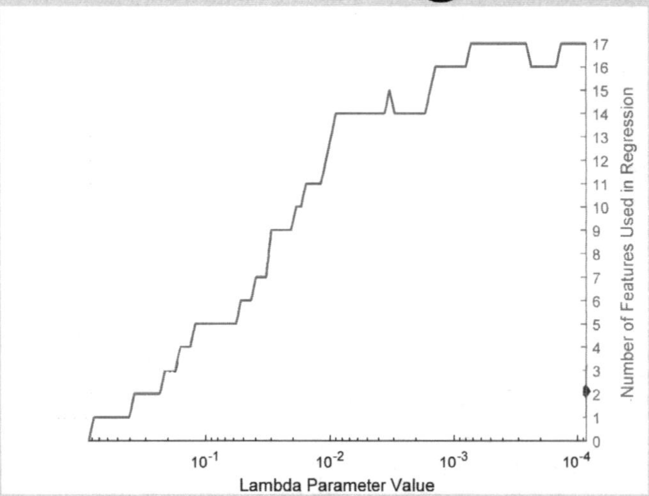

Decreasing penalty on coefficients

Lasso regression shrinks the coefficients of unimportant features to 0

After a certain number of features, the fits stops improving

24

McCormick

Northwestern Engineering

Department of Chemical and Biological Engineering

How many features are significant?

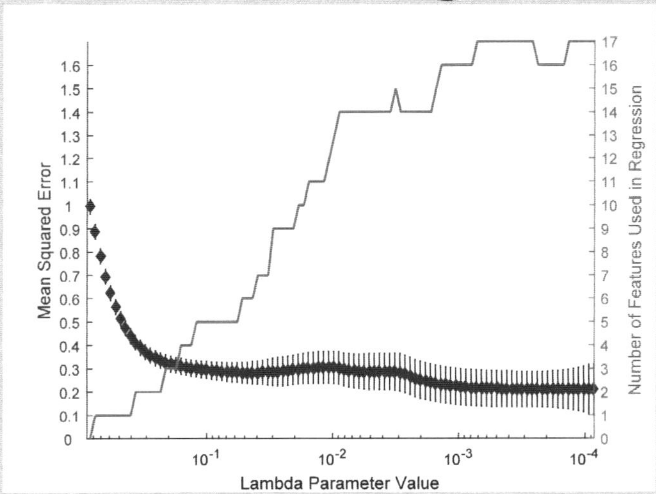

Lasso regression shrinks the coefficients of unimportant features to 0

After a certain number of features, the fits stops improving

Decreasing penalty on coefficients

25

McCormick

Northwestern Engineering

Department of Chemical and Biological Engineering

How many features are significant?

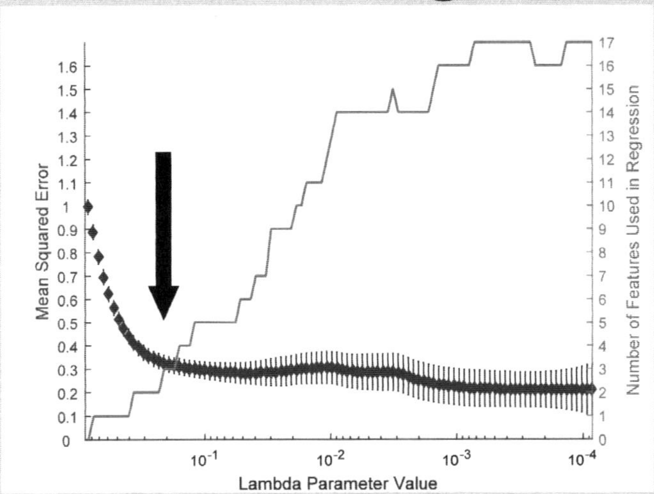

Lasso regression shrinks the coefficients of unimportant features to 0

After a certain number of features, the fits stops improving

Decreasing penalty on coefficients

26

McCormick
Northwestern Engineering
Department of Chemical and Biological Engineering

How many features are significant?

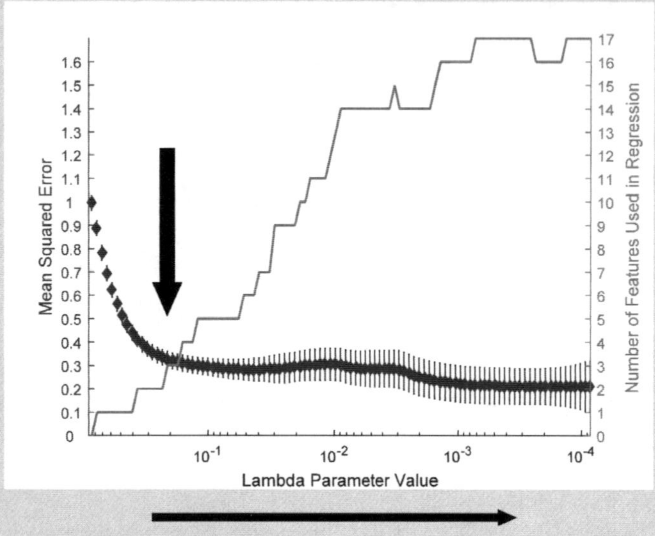

Decreasing penalty on coefficients

Lasso regression shrinks the coefficients of unimportant features to 0

After a certain number of features, the fits stops improving

~3 important features

27

McCormick
Northwestern Engineering
Department of Chemical and Biological Engineering

Comparison of all three methods

Green = positive regression coefficient (positively correlated with repressibility)

Pink = negative regression coefficient (negatively correlated with repressibility)

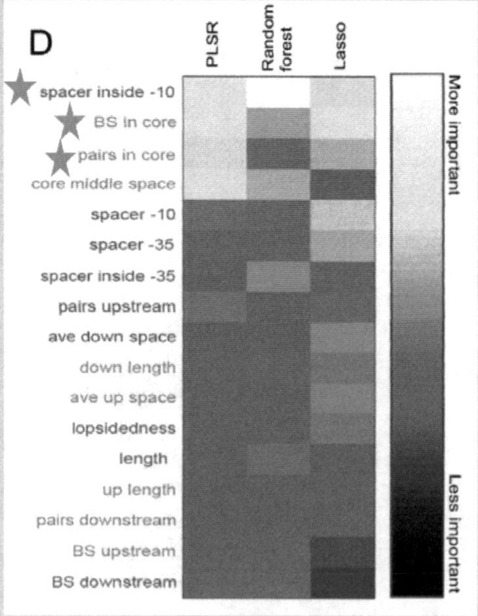

28

McCormick
Northwestern Engineering
Department of Chemical and Biological Engineering

Rules for zinc finger repression

Crowding promoter boxes, especially -10, tends to create strong repression

Binding site quantity does not matter
- No titration effect

Regions away from the boxes are not important (upstream, downstream)

Proposed mechanism: The zinc finger blocks the sigma factor when near the -10 box

29

McCormick
Northwestern Engineering
Department of Chemical and Biological Engineering

Application to biosensors

Promoter library analysis revealed elegant design rules for synthetic promoters

Useful feature selection methods for describing transcription factor binding

Application of this DNA binding domain in biosensors

McCormick
Northwestern Engineering
Department of Chemical and Biological Engineering

Application to biosensors

Promoter library analysis revealed elegant design rules for synthetic promoters

Useful feature selection methods for describing transcription factor binding

Application of this DNA binding domain in biosensors

Development of a small molecule responsive transcriptional regulator platform.
Andrew Younger

Session 734: Emerging Tools for Synthetic Biology

Thursday 3:15, Salt Palace Convention Center 150G

McCormick
Northwestern Engineering
Department of Chemical and Biological Engineering

Acknowledgements

Josh Leonard
Andrew Younger
Austin Rottinghaus
Rachel Dudek
Andy Scarpelli
Michelle Hung
Kelly Schwarz
Taylor Dolberg
Alia Zander
Danny Wells

Collaborators
Declan Bates (U. Warrick)
Paul Freemont (Imperial College London)
Jay Keasling (UC Berkeley)
Anne Osbourn (John Innes Center)
Susan Rosser (U. Edinburgh)

Chemistry of Life Processes Institute

Office of Undergraduate Research UR@NU

NSF

EPSRC Pioneering research and skills

32

McCormick

Northwestern Engineering

Department of Chemical and Biological Engineering

McCormick

Northwestern Engineering

Department of Chemical and Biological Engineering

Decision tree regression: Random forest

Uses decision trees on subsets of features

Quantify how much accuracy is lost when feature is not present

Creates a ranked list of importance

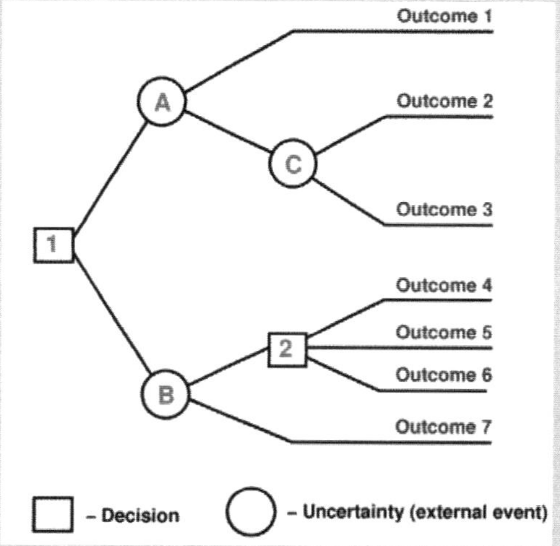

34

McCormick
Northwestern Engineering
Department of Chemical and Biological Engineering

Partial Least Squares Regression: Permutation Test

Shuffle output randomly

Obtain a variance for each feature coefficient

Compare the real output to the random variance

Again, creates a ranked list

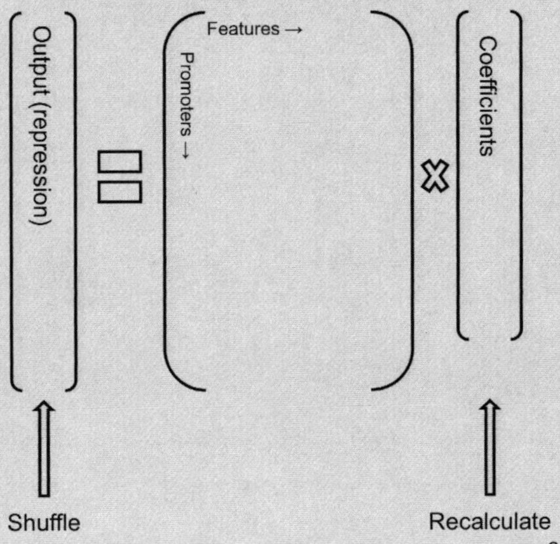

35

McCormick
Northwestern Engineering
Department of Chemical and Biological Engineering

Regularized regression: Lasso

Constrains $\sum|\beta|$
(sum of absolute coefficients)

penalty

$$\sum_{i=1}^{N}(y_i - \sum_{j} \beta_j x_{ij})^2 + \lambda \sum_{j} |\beta_j|$$

Eliminates features by shrinking unimportant coefficients to 0

Creates a ranked list of importance (again)

36

438528 Immunomodulatory Amphiphilic Polyanhydride Microparticles for Peripheral Nerve Regeneration

Abstract:

: Peripheral nerve repair using implanted biomaterial conduits seeded with cells face many challenges, including the host immune response. Tissue engineering approaches often employ the use of drug delivery vehicles to supply sustained release of trophic factors which support the differentiation and continued growth of the implanted cells. However, host immune cells responding to inflammatory cytokines released by the damaged nerve may inhibit this growth. This project focuses on the fabrication of amphiphilic degradable polyanhydride micro particles that can release surface adsorbed Interleukin-4 and encapsulated nerve growth factor simultaneously. Interleukin-4, an anti-inflammatory cytokine, serves to promote an M2-like phenotype in invading macrophages as a method to support wound healing. These M2-like macrophages may reduce the inflammatory response, thereby generating an environment which supports cell growth and tissue regeneration, thus allowing implanted cells to benefit from the sustained release of neurotrophic factors. The release of nerve growth factor helps stimulate the peripheral nerve regeneration. We have focused on the fabrication and release kinetics of the two macromolecules from the polyanhydride microparticles.

435707 At the Interface of Immunology and Engineering: Elucidating the Biophysical and Molecular Interactions of Pathogens and Polymers with the Immune System

Abstract:

This presentation will explore our lab's recent progress in elucidating (i) a novel and exceptionally potent mechanism of mucosal immunity based on transient antibody-mucin interactions, and (ii) the interactions between PEGylated nanoparticles and the immune system. More antibodies are secreted into mucus than either the blood or the lymph, yet the mechanisms by which antibodies can protect against pathogens in mucus are not fully understood. Using high resolution fluorescence microscopy, we reveal that the array of pathogen-bound antibodies can trap individual pathogens in the mucin matrix at sub-neutralizing antibody concentrations through N-glycans on the Fc-domain. The carefully tuned antibody-mucin affinity transforms the mucus gel into an effective adhesive barrier against a diverse array of viral and bacterial pathogens *in vitro* and *in vivo*. We are developing computational models to further reveal the dynamics of this 'molecular shield'. Separately, we are exploring how the physicochemical properties of polymeric coatings on nanoparticles may influence their interactions with immune cells and antibodies. Using nanoparticles with precisely tunable polymer grafting, we show that evading immune cell uptake and clearance in mice requires PEG grafting at densities substantially exceeding the minimum for polymer brush conformation.

414062 Nanomagnets and Polymers: Design of Vehicles for Medical Theranostics

Abstract:

The combination of nano-scale magnetic particles with biocompatible polymers affords opportunities for multifunctional devices for theranostics applications. In particular, further combination with targeting ligands allows therapies and medical diagnostics to be conducted on a cellular or sub-cellular level. In the area of diagnostics, magnetic nanoparticles have proven useful as phase contrast agents for MRI, and although most do not provide contrast enhancement as great as gadolinium, iron oxide-based materials offer additional functionality such as localization or magnetic hyperthermia in addition avoiding the toxicological issues observed with gadolinium. The application of static magnetic fields can collect magnetic nanoparticles in simulated blood flow to focus the therapeutic effect in a region of interest. By applying high frequency (kHz-MHz) magnetic fields, iron oxide and other magnetic nanoparticles generate significant heat, which can be harnessed for magnetothermal therapies, which has been used successfully in cancer treatment. Polymer coatings are used to improve biocompatibility and decrease particle aggregation. Surface chemistry plays an important role in determining hydrodynamic sizes and the stability of nanoparticle dispersions. The polymers can also host drugs and deliver them to targeted cells and tissue. Specifically, the combination of magnetic hyperthermia with thermally responsive polymers, such as copolymers of poly(N-isopropyl acrylamide), which display a lower critical solution temperature between 37-45 ^{o}C, or nanocrystalline polycaprolactone, which melts around 40-45 ^{o}C, have proven useful for magnetically-activated drug release. Optimization of these systems, including nanoparticle and polymer design, will be discussed.

424070 Pollen Grains, Gold Nanoparticles and Microneedles: Delivery Systems for Mucosal Vaccination

Abstract:

Mucosal vaccination, i.e., vaccine administration through the mucosal surfaces such as the oral or nasal routes continues to be an important goal. This is because mucosal vaccination is needle-free, painless, child-friendly and can be self-administered. Furthermore, from a functional perspective, delivery of vaccines through mucosal surfaces can additionally induce mucosal immunity, which can help to combat pathogens on mucosal surfaces before they can cause infection. However, mucosal barriers such as the tightly juxtaposed cells and the harsh vaccine-degrading environment in the stomach, make mucosal vaccination challenging.

This presentation describes our efforts to engineer micro and nano-systems for vaccine delivery across the mucosal surfaces. Engineering and characterization of three different delivery systems will be presented. First, our efforts to engineer and transform pollen grains from the bane of allergies into 'trojan horses' for oral vaccination will be presented. Our results show that native plant matter from lycopodium spores can be removed to create allergen-free durable empty microcapsules that can be filled with vaccine antigens. In vivo evaluation of this novel delivery approach using ovalbumin as a model antigen has shown that high levels of anti-ovalbumin antibodies can be induced in mouse serum and fecal matter, indicative of systemic and mucosal immune responses, respectively. The second system that will be discussed relates to engineering of gold nanoparticles as nanocarriers for intranasal vaccination to help develop a universal influenza vaccine. Our studies show that conjugation of the ectodomain of a conserved membrane protein (M2) of the influenza virus to gold nanoparticles can provide protection against lethal influenza challenge in a mouse model. Lastly, use of coated microneedles to deliver antigens into the rabbit oral-cavity-mucosa will be discussed. Our results show that delivery of antigens into the oral cavity can stimulate production of antigen-specific antibodies in saliva. Overall, these delivery systems show potential for vaccine delivery across mucosal surfaces.

437145 Graphene Oxide Hollow Fibre Membranes: Stability, Modifications, and Challenges

Abstract:

Graphene, a two dimensional single atomic carbon layer, is an attractive material for thin film and membrane applications. Graphene oxide (GO) synthesized from chemical exfoliation is a commonly used precursor for graphene based materials. GO membranes made by assembling small GO flakes have demonstrated great potential in gas separation and liquid filtration. For upscale applications, GO membranes in hollow fibre geometry are of particular interest due to the high surface area per volume and easy assembly features at module level.

In our research group, inorganic hollow fibres have been developed for various membrane processes. Among all types of hollow fibre substrates, ceramic hollow fibre is an excellent candidate for GO membranes due to its low mass transfer resistance, robust mechanical property and inert chemical nature, which enable it to work under extreme conditions. However, our recent study revealed that GO membranes on ceramic hollow fibre are unstable in dry state, mainly due to drying-related shrinkage [1]. A simple way to tackle the stability problem is to reserve GO hollow fibre membranes in water after initial drying to avoid further shrinkage and the formation of fatal defects. Though such a strategy worked well to preserve the microstructure of GO membranes, it limits the applications of GO membranes to wet processes. Modifications of the membranes at dry state will also be difficult. In another approach, GO hollow fibre membranes were stabilized by using a porous poly(methyl methacrylate) (PMMA) sacrificial layer, which created a space between the hollow fibre substrate and the GO membrane that allows stress-free shrinkage. Defect free GO hollow fibre membrane was successfully determined and the membrane was stable in a long term (1200 hours) gastight stability test.

Post treatment of the stabilized GO membranes was also successfully carried out using UV radiation. The UV treatment changed the surface of GO membranes to superhydrophilic, and induced mild reduction that could have led to some controlled defect formations. All these modifications have enhanced the performance of GO hollow fibre membranes in both aqueous and solvent nanofiltration, while still maintaining the molecular sieve property of GO membranes. The pure water and acetone flux increased from 0.05 to 2.8 LMH, and from 0.14 to 7.5 LMH, respectively, with a molecular weight cut-off lower than 250 Da.

[1] Aba, N.F.D., et al., Graphene oxide membranes on ceramic hollow fibers – Microstructural stability and nanofiltration performance. Journal of Membrane Science, 2015. 484(0): p. 87-94

414949 Nanometric Graphene Oxide Framework Membranes with Enhanced Heavy Metal Removal Via Nanofiltration

Abstract:

A novel dual-modification strategy, including (1) the crosslinking and construction of a GO framework by ethylenediamine (EDA) and (2) the amine-enrichment modification by hyperbranched polyethyleneimine (HPEI), has been proposed to design stable and highly charged GO framework membranes with the GO selective layer thickness of 70 nm for effective heave metal removal via nanofiltration (NF). Results from sonication experiments and positron annihilation spectroscopy confirmed that EDA crosslinking not only enhanced structural stability, but also enlarged the nanochannels among the laminated GO nanosheets for higher water permeability. HPEI 60K was found to be the most effective post-treatment agent that resulted in GO framework membranes with a higher surface charge and lower transport resistance. The newly developed membrane exhibited a high pure water permeability of 5.01 L m^{-2} h^{-1} bar^{-1} and comparably high rejections towards Mg^{2+}, Pb^{2+}, Ni^{2+}, Cd^{2+} and Zn^{2+}. These results have demonstrated the great potential of GO framework materials in wastewater treatment and may provide insights for the design and fabrication of the next generation 2D-based NF membranes.

(195c) Freestanding Ultrathin Graphene-Based Membranes for Water Purification

Liu, H., Monash University
Zhang, X., Monash University
Originally presented on: 11/9/2015 16:05:00 - 16:30:00

Graphene-based membranes, consisting of two-dimensional (2D) graphene oxide (GO) or reduced graphene oxide (rGO) nanosheets, have recently attracted increasing attention in separation of liquid mixtures. The nanochannels of these membranes, formed between 2D GO or rGO nanosheets, allow gases and ions with smaller sizes than those of the channels to permeate, while blocking all other larger species. However, hydration of GO in aqueous solution enlarges the channels, making it more challenging to be used in various exciting applications, such as desalination, energy production and hydrofracking water treatment. Although the hydration effect can be inhibited by reduction, a substantial loss in membrane permeability inevitably occurs. In this research, we aim to develop a next generation of high-flux and energy-efficient membranes based on graphene materials, and at the same time maintain their superior selectivity for precise ionic and molecular sieving in aqueous solution. Preliminary work has demonstrated the fabrication of freestanding ultrathin rGO membranes which are more selective and permeable than the state-of-the-art membranes in forward osmosis process. Future work will include further enhancement of the performance of graphene-based membranes in terms of permeability and selectivity, exploration of the uses of graphene-based membranes in other water treatment processes and understanding the mechanisms of mass transport in graphene-based membranes.

426311 Tailoring Inter-Layer Nanospace of Graphene Oxide Laminates for High-Efficient Membrane Gas Separation

Abstract:

As a well-known two-dimensional material, graphene exhibits preeminent electrical, mechanical and thermal properties owing to its unique one-atom-thick structure. Graphene and its derivatives (e.g., graphene oxide) become emerging nano-building blocks for separation membranes featuring distinct laminar structure and tunable physicochemical properties. Extraordinary molecular separation properties for purifying water and gases have been demonstrated by graphene-based membranes, which attract a huge surge of interest during last a few years [1]. The pioneering work of Geim and co-workers found that laminar graphene oxide (GO) membranes are impermeable to gases and liquids but solely allow unimpeded evaporation of water. Thus, GO membranes have been reported for application in water-purification and solvent-dehydration processes [2]. Meanwhile, recent works found that GO membranes can also exhibit gas-separation characteristics if their stacking structure is care-fully controlled [3]. As the prominent characteristic of GO laminates, inter-layer space has been proved to play a significantly important role in molecular transport.

We are trying to tailor the nanospace between graphene oxide sheets for fast and selective transport of gases. Two approaches were demonstrated here. First, membranes with fast and selective CO_2 transport channels of GO laminates were proposed based on the construction of hydrogen bonding between GO and polymer. Our explorations (e.g., TEM, positron annihilation) found that the formed molecular-sieving interlayer spacing and straight diffusion pathways of the GO laminates play critical roles in determining gas transport properties. Second, molecular-intercalated GO membranes were developed by a facile spin-casting assisted assembly technique. With a well-ordered stacking structure and enlarged interlayer spacing, the intercalated GO membranes show excellent molecular sieving properties. Our prepared GO membranes exhibited preferential CO_2 (1st approach) or H_2 (2ndapproach) permeation performance for gas mixtures that transcends the upper bounds of state-of-the-art membranes. Moreover, we firstly realized GO-based membranes having extraordinary operational stability, which is believed to be a notable step toward pushing this membrane into industrialized implementation.

Keywords: Graphene oxide, inter layer space, membrane, gas separation

Acknowledgements

The work was supported by the National Natural Science Foundation of China (Nos. 21406107, 2149580015, 21476107), Innovative Research Team Program by the Ministry of Education of China (No. IRT13070), Natural Science Foundation of Jiangsu Province (No. BK20140930).

References

1. Liu G. P., Jin W. Q. & Xu N. P. Graphene-based membranes. *Chem. Soc. Rev.* DOI: 10.1039/c4cs00423j.
2. Shen J., Liu G. P., Huang K., Jin W. Q., Lee K.-R. & Xu N. P. Membranes with fast and selective gas-transport channels of laminar graphene oxide for efficient CO2 capture. *Angew. Chem. Int. Ed.* **54**, 578-582 (2015).
3. Huang K., Liu G. P., Lou Y. Y., Dong Z. Y., Shen J. & Jin W. Q. A graphene oxide membrane with highly selective molecular separation of aqueous organic solution. *Angew. Chem. Int. Ed.* **53**, 6929-6932 (2014).

431010 Scalable and Ultra-Selective Carbon Molecular Sieve (CMS) Membranes for Challenging Gas Separations

Abstract:

Scalable and Ultra-Selective Carbon Molecular Sieve (CMS) Membranes for Challenging Gas Separations

Chen Zhang, William J. Koros

School of Chemical and Biomolecular Engineering, Georgia Institute of Technology, Atlanta, GA

Carbon molecular sieve (CMS) membranes have received increasing attention in the past years for advanced gas separations. CMS membranes can be formed into asymmetric hollow fibers, by controlled pyrolysis of polymeric precursor hollow fiber membranes, and are capable of delivering simultaneously attractive productivity and separation efficiency without compromising scalability. Pyrolysis temperature is among the parameters that control CMS membrane's ultramicropore size distribution and therefore, gas separation performance. In general, more densely packed sp^2-hybridized graphene-like sheets with lower permeability and higher selectivity are obtained with increasing pyrolysis temperature. For example, previous studies[1] showed that CO_2/CH_4 selectivity of Matrimid®-derived CMS membranes was enhanced by 200% as pyrolysis temperature increased from 650 °C to 800 °C.

Nevertheless, formation of CMS membranes at pyrolysis temperatures above 800 °C has been rarely reported. This was partially due to challenges involved with processing brittle CMS dense films. We have discovered that this challenge can be successfully overcome by using hollow fiber precursors instead of dense film precursors. This talk will present CMS hollow fiber membranes formed at pyrolysis temperatures above 800 °C. Gas permeation measurements suggested that these CMS hollow fiber membranes were capable of delivering superior selectivities that are well-above the polymer upper bound.

References:

1. (a) Bhuwania, N.; Labreche, Y.; Achoundong, C. S. K.; Baltazar, J.; Burgess, S. K.; Karwa, S.; Xu, L.; Henderson, C. L.; Williams, P. J.; Koros, W. J., Engineering substructure morphology of asymmetric carbon molecular sieve hollow fiber membranes. *Carbon* **2014,** *76* (0), 417-434; (b) Ning, X.; Koros, W. J., Carbon molecular sieve membranes derived from Matrimid® polyimide for nitrogen/methane separation. *Carbon* **2014,** *66* (0), 511-522.

433115 Vertically-Aligned Carbon Nanotube Membranes for Breathable and Protective Fabrics

Abstract:

Vertically-aligned Carbon Nanotube Membranes for Breathable and Protective Fabrics

N. Bui, S. Kim, E. Meshot, J. Pena, S. Guo, K. J. Wu, F. Fornasiero*

Lawrence Livermore National Laboratory, Livermore, CA

P. Gibson

U.S. Army Natick Soldier Research, Development and Engineering Center, Natick, MA

Membrane materials with high breathability, high degree of protection from chemical and biological (CB) agents, and rapid response to CB threats are highly desirable for protective fabrics. At LLNL, we recently proposed a CB threat responsive fabric platform based on a carbon nanotube membrane. The membrane consists of an array of a-few-nm-wide vertically-aligned carbon nanotubes (VACNTs) as through pores and a flexible, impermeable parylene-N matrix filling the inter-tube spacing. It promises to provide high water-vapor transport rate and an effective barrier function to CB agents by best leveraging the outstanding mass transport properties of narrow CNT channels. Furthermore, this membrane platform is ideally suited to fundamentally understand the mechanisms of mass transport through CNT pores.

In this work, we empirically investigated the transport properties through the well-aligned, narrow CNT channels at varied length scale. Results showed that these VACNTs membranes sustained an ultrahigh liquid and gas transport rates when compared with conventional theory predictions. Also, they performed outstanding moisture vapor transport rates (> 4000 $g.m^{-2}.day^{-1}$ at all relative humidities), comparable to or exceeding state-of-the-art breathable fabrics. Selectivity study showed that these membranes completely excluded 5-7nm-Au nanoparticles and charged 3-nm dye molecules during dead-end filtration tests. Furthermore, our membranes demonstrated bioprotection capability through size-exclusion mechanism when Dengue virus was used as a surrogate for a biological threat agent. These findings suggest that, with appropriate functionalization at the CNTs pore entrances, VACNT-polymer composite membranes may provide excellent candidates for next generation of breathable fabrics with active protection from CB threats.

This work was performed under the auspices of the U.S. Department of Energy by Lawrence Livermore National Laboratory under Contract DE-AC52-07NA27344.

428835 Comparative Dynamics and Sequence Dependence of DNA and RNA Binding to Single Walled Carbon Nanotubes

Abstract:

Non-covalent polymer- single wall carbon nanotube (SWCNT) conjugates have gained recent interest due to their prevalent use as electrochemical and optical sensors, SWCNT-based therapeutics, and for SWCNT purification methods [1-4]. However, little is known about the effects of polymer-SWCNT molecular interactions on functional properties of these conjugates [2]. In this work, we show that SWCNT complexed with related polynucleotide polymers (DNA, RNA) have dramatically different fluorescence stability. Surprisingly, we find a difference of nearly 2500-fold in fluorescence emission between the most fluorescently stable DNA-SWCNT complex, C_{30} DNA-SWCNT, compared to the least fluorescently stable complex, $(AT)_7A$-$(GU)_7G$ DNA-RNA hybrid-SWCNT. We further reveal the existence of three regimes in which SWCNT fluorescence varies non-monotonically with SWCNT concentration. We utilize molecular dynamics simulations to elucidate the conformation and atomic details of SWCNT-corona phase interactions. Our results show that variations in polynucleotide sequence or sugar backbone can lead to large changes in the conformational stability of the polymer SWCNT corona and the SWCNT optical response. Finally, we demonstrate the effect of the coronae on the response of a recently-developed dopamine nanosensor [3], based on $(GT)_{15}$ DNA- and $(GU)_{15}$ RNA-SWCNT complexes. Our results clarify several features of the sequence dependence of corona phases produced by polynucleotides adsorbed to single walled carbon nanotubes, and the implications for molecular recognition in such phases.

1. Landry, M.P.*, Zhang, J.Q.*, et al., *Molecular recognition using corona phase complexes made of synthetic polymers adsorbed on carbon nanotubes.* Nature Nanotechnology, 2013. **8**(12): p. 959-968.

2. Landry, M.P., et al., *Experimental tools to study molecular recognition within the nanoparticle corona.* Sensors (Basel), 2014. **14**(9): p. 16196-211.

3. Landry, M.P.*, Kruss, S.*, et al., *Neurotransmitter Detection Using Corona Phase Molecular Recognition on Fluorescent Single-Walled Carbon Nanotube Sensors.* Journal of the American Chemical Society, 2014. **136**(2): p. 713-724.

4. Jain, R.M.*, Tvrdy, K.* et al., *Quantitative Theory of Adsorptive Separation for the Electronic Sorting of Single-Walled Carbon Nanotubes.* Acs Nano, 2014. **8**(4): p. 3367-3379.

412661 Selective Ion Transport through Carbon Nanotubes

Abstract:

Carbon nanotubes are promising materials with a vast range of potential applications, such as nanoscale electronic devices, nanoscale sensors, selective molecular filters, gas storage devices, nanofluidic devices, and targeted drug delivery devices, among many others. Water has unique properties and an understanding of the interaction between water and carbon nanotubes is key to the development of many of these devices. The confinement of water in carbon nanotubes as compared to bulk water has been shown to produce some remarkable features, which include but are not limited to a lower number of hydrogen bonds, the increased lifetime of hydrogen bonds, the layered structure of water under confinement, its reduced density and viscosity, and an increased proton mobility in 1-D water chains. One of the more exciting results of water under confinement is the high slip lengths that result at high flow rates, which is conducive to ultrafiltration and nanofiltration. High water flow rates along with selective ion transport is an important feature of ion channels in living organisms; however experimental analysis to study ion channels is a real challenge as ion channels can degrade under experimental conditions, have dimensions of the order of few nanometers, possess non-uniform surfaces, and uneven charge distributions that make them a complex system to study. Carbon nanotubes can be easily embedded into lipid membranes, and hence can be used as biomimetic devices to study water and ion transport at nanoscopic levels.

Flow of water and ions in nanotubes can be achieved by application of an electric field, pressure difference, concentration gradient, uneven charge distribution, and osmotic pressure. The rate and selectivity of ion transport can be controlled by various means such as varying the pore size, the ion concentration, the charge distribution, the density of charge and through chemical functionalization. Because of the large surface to volume ratio as compared with macro and microscopic length scales, surface charge has a pronounced effect on the fluid volume in the hydrophobic cavity at the nanoscopic length scale, leading to a rejection of the co-ions. This phenomenon can thus help improve the selectivity of the ions beyond steric hindrance to include the use of larger diameter nanotubes enabling greater water efflux, which can be exploited to model nano-level filters. Since continuum Navier-Stokes equations for fluid transport break down at nanometer level, molecular dynamics (MD) simulations, which explicitly calculate the motion of all particles in the system described by Newton's equations of motion, is the favored method of study.

Water and ion transport through uncharged and charged carbon nanotube membranes (d = 0.8nm -3.25nm) were studied using long-time MD simulations. The OPLS force field parameters of benzene are employed for carbon atoms. The carbon atoms were held fixed during the course of the simulation. The TIP4P water potential was used to model the water intermolecular interactions. All simulations were carried out in GROMACS 4.6.1. in the NVT ensemble. Various cases were simulated by varying pore size at different electrolyte (NaCl) concentrations and application of electric field for both charged and uncharged CNT membranes at these different pore sizes and concentrations. For uncharged CNTs in the absence of electric field, steric hindrance is the only factor controlling transport of ions. For small pore diameters (<0.8nm) the ion occupancy and passage of ions through the tube is negligible as the hydration radius of the ions is comparable to the pore size. In this case, water diffuses to the chamber containing electrolyte through osmosis. At larger pore sizes the concentration gradient drives ion transport with ion occupancy in the CNT and ion passage progressively increasing. In the case of uncharged CNTs in the presence of electric field, electro-osmosis dominates with increasing field strength. In charged CNTs at low electrolyte concentrations, the thickness of the electric double layer is of the same order as the diameter of the pore and this results in selectivity governed by the Donnan equilibrium. As the concentration strength increases the Debye length becomes smaller, the charges are screened and both ions can pass through. Electrokinetic flow is observed in the case of charged CNT under an electric field, which can be described by the Poisson-Boltzmann equation along with the Nernst-Planck equation. Modulating the strength of the surface charges can be used to tune both ion transport and water flow through the nanotubes.

The structural characteristics of water and ions confined in CNT for the various cases simulated are compared. Measurements of ion occupancy in the nanotube, water flux, ion flux, rejection rates, and other transport properties are obtained using MD simulations. A complete account of ion and water transport under the influence of charge is obtained in this study spanning a large number of CNT diameters. Fast water flow can thus be integrated with ion selectivity through modulation of surface charges and pore size, which not only helps to understand biological processes, but can also help design advanced nanofluidic devices in the future.

419190 Low-Temperature Water Does Not Freeze in Nanotubes

Abstract:

Low-Temperature Water does not Freeze in Nanotubes

Mahdi Khademi and Muhammad Sahimi

Mork Family Department of Chemical Engineering & Materials Science, University of Southern California, Los Angeles, California 90089-1211, USA

There is intense interest in studying the properties of confined water at very low temperatures below its freezing point in the bulk. The challenge is partly motivated by water's critical importance to biological phenomena, and the question of how microorganisms survive at very low temperatures. The cellular structure of such microorganisms contains nanochannels and nanopores and, thus, what happens to water in such nanoscale structures has significant effect on the future of preserving cells and live microorganisms. We have carried out an extensive study of water dynamics inside a nanotube at and below the freezing temperature of water under the bulk conditions. In particular we have evaluated various correlation functions, such as the cage correlation (CC) function, the radial distribution function, and the velocity autocorrelation function. The CC function follows a stretched exponential function, which has important implications for diffusion of water in nanochannels, conformational dynamics of proteins in "crowded" cellular environments, and transport in confined media, as well as the validity of the Stokes-Einstein formula relating the viscosity to self-diffusivity. We show, based on molecular dynamics (MD) simulation, that water molecules inside nanotubes do not follow typical freezing behavior seen in nature under the bulk conditions. Our simulations include both carbon and silicon-carbide nanotubes. We conclude the observed behavior is strongly related to size of the nanotubes, rather than their chemical structure. We argue that due to spatial limitations and steric hindrance inside small enough nanotubes, there is not enough room for the formation of hexagonal ice and, therefore, water inside such nanotubes attains a lower energy at temperatures below the typical freezing point *without ice formation.*

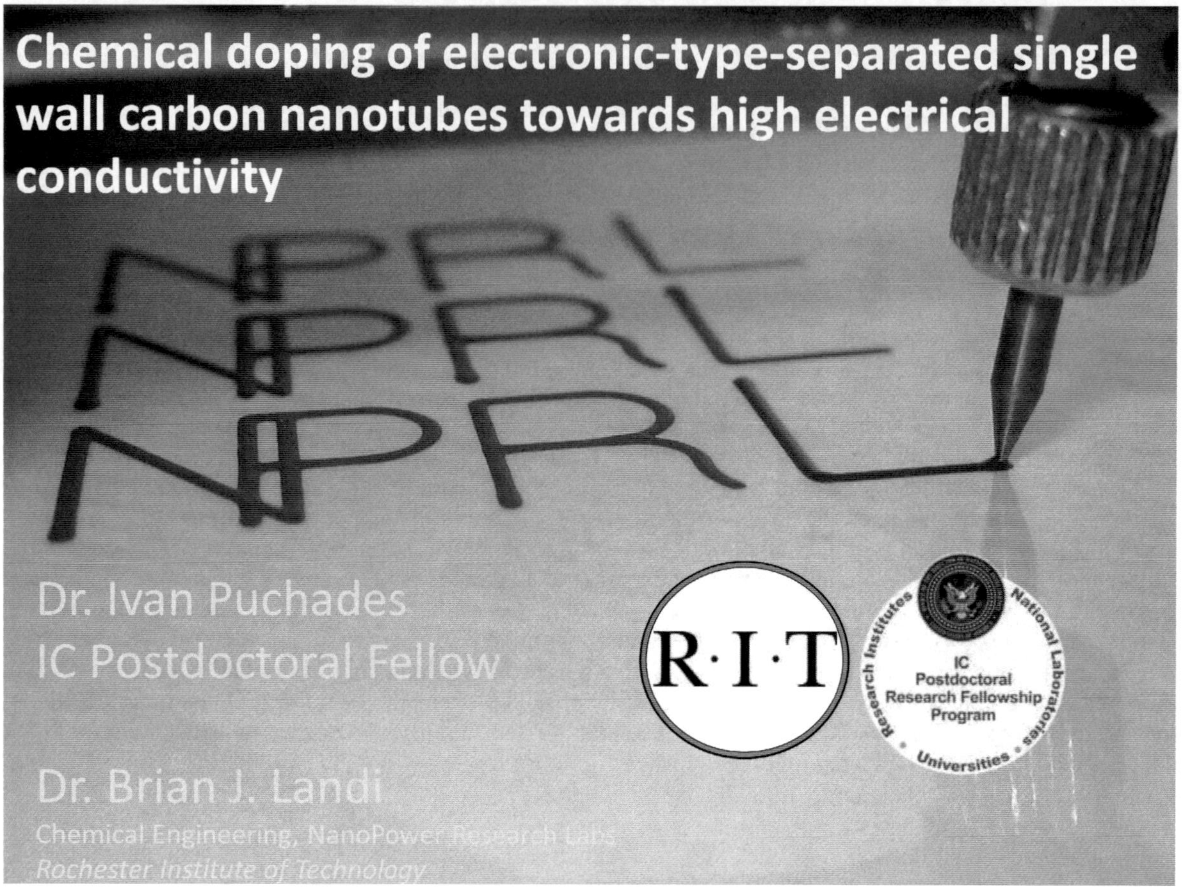

Outline

Chemical doping of electronic-type-separated single wall carbon nanotubes towards high electrical conductivity

- ❑Introduction
  - ❑Carbon nanotubes as conductors
  - ❑Single Wall Carbon Nanotubes – synthesis and characterization
  - ❑Separation techniques
  - ❑Chemical doping
- ❑Experiment
  - ❑Characterization of doped electronic type-separated SWCNTs
- ❑Results
- ❑Conclusions

Puchades et al., J. Mater. Chem. C, 2015,3, 10256-10266

CNT Conductors as wires and cables

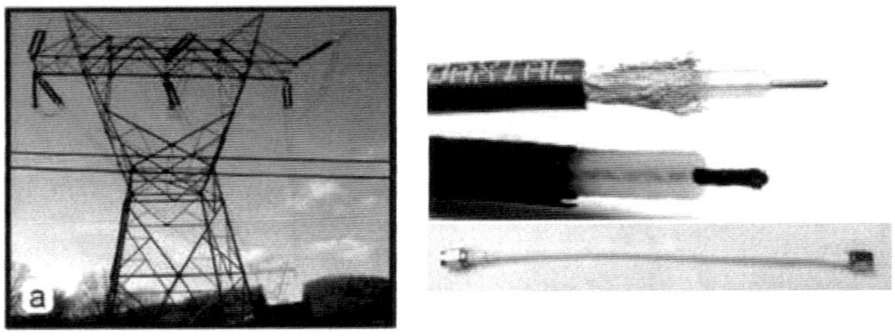

Density:
Copper 8.9 g/cm^3
Densified CNT ~1.0 g/cm^3

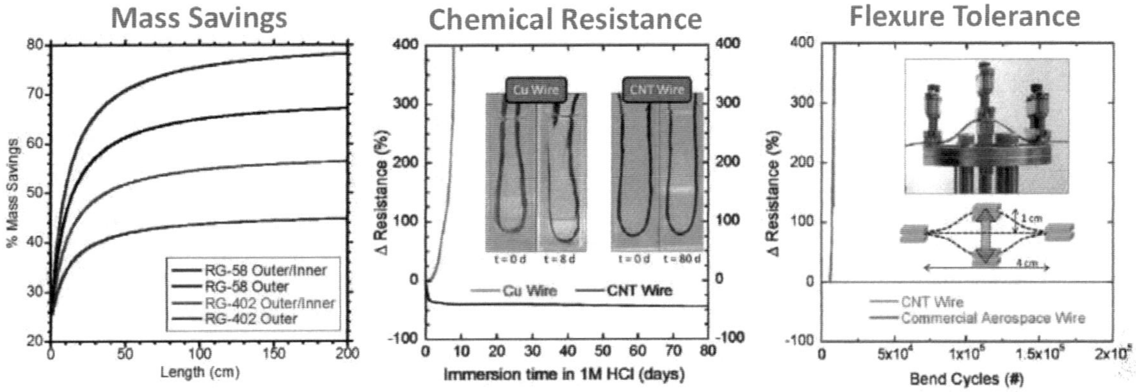

Jarosz, et al., Nanoscale 3.11 (2011): 4542.

3

Electrical conductivity of carbon nanotubes

4

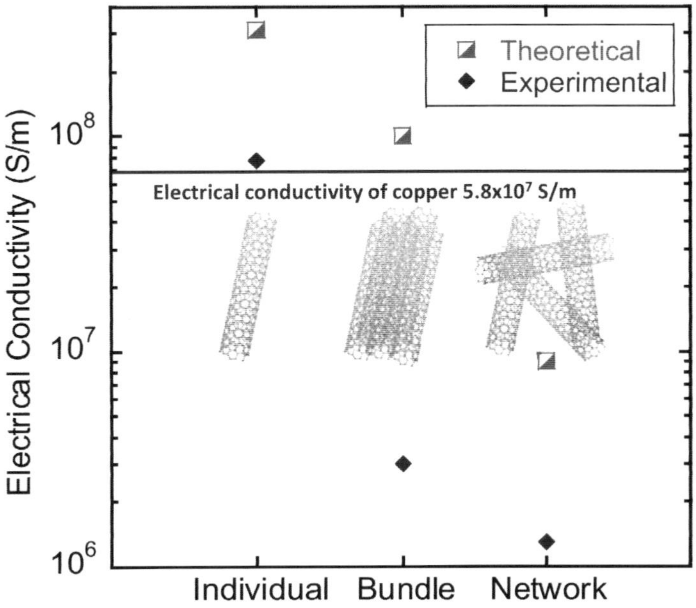

Synthesis of Single Wall Carbon Nanotubes

- SWCNTs are synthesized using a variety of methods including Laser Vaporization, Arc Discharge, CoMoCAT, HiPCo and others.
- As synthesized SWCNTs contain different electronic types, diameter sizes and chirality of nanotubes as well as impurities; all affecting **carrier transfer**.

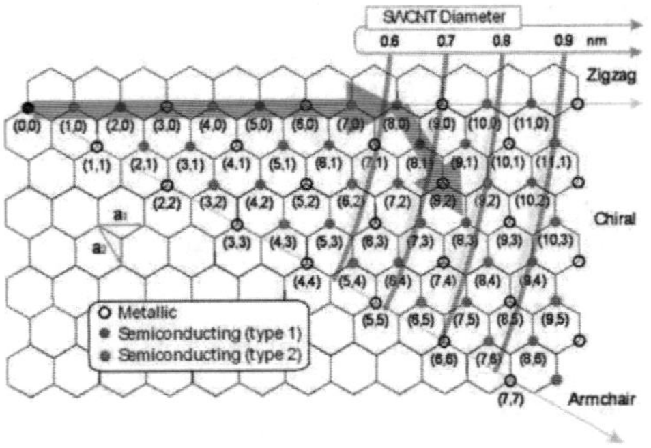

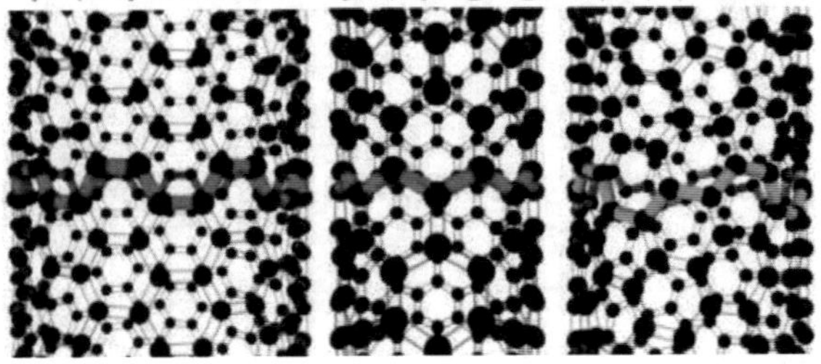

- Typically ~33% of the "as synthesized" tubes are considered metallic, the rest are semiconducting with different diameters and chiralities.
- The presence of the different SWCNTs are commonly characterized using optical absorption, fluorescence or Raman spectroscopy.

Characterization techniques

Optical Absorbance

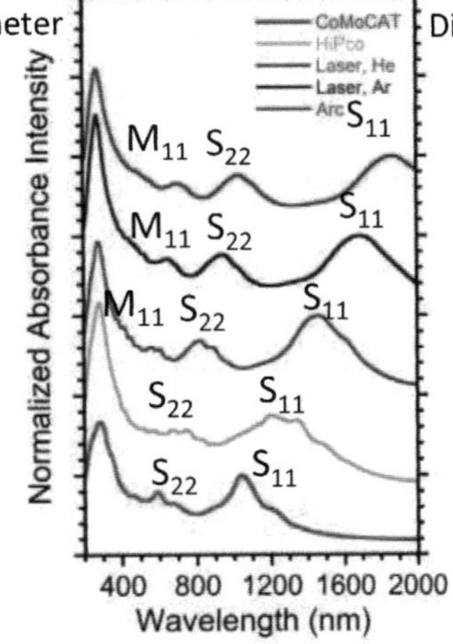

Absorbance at transition energy wavelengths: electronic type and diameter size.

Resonant Raman Spectroscopy

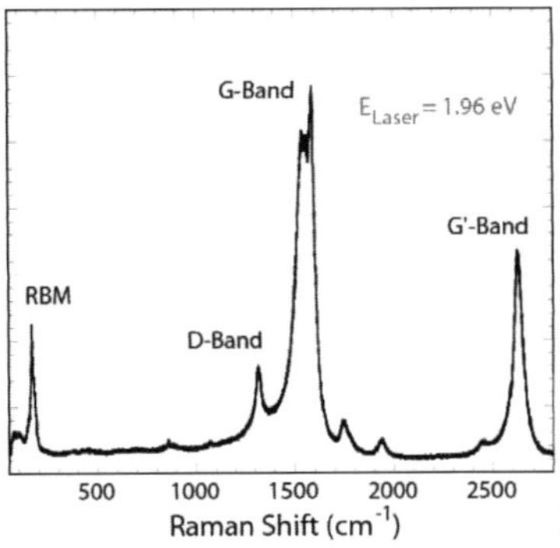

SWCNT resonance determines diameter, defects, electronic configuration, available DOS.

Electrical Conductivity

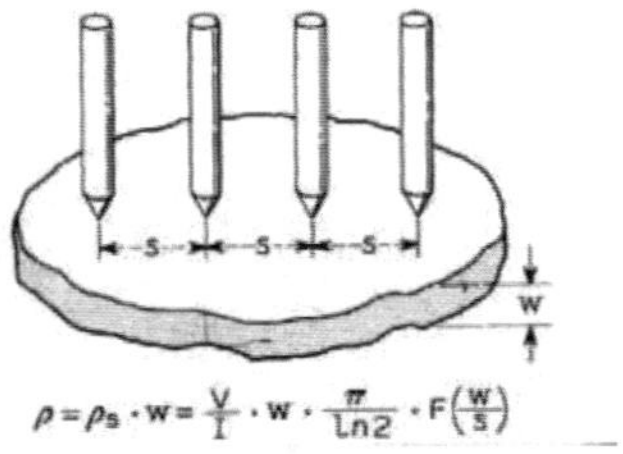

$\rho = \rho_s \cdot w = \frac{V}{I} \cdot w \cdot \frac{\pi}{\ln 2} \cdot F\left(\frac{w}{s}\right)$

Thickness

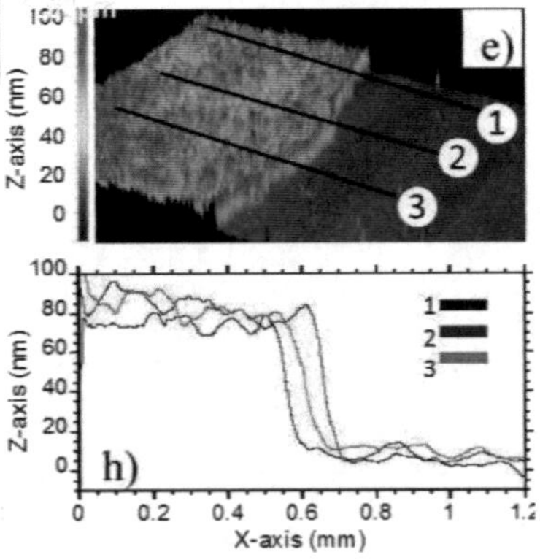

Carbon nanotube wires and cables

- Electrical conductivity of CNTs
  - Individual CNTs– higher than copper
  - Network/bundle dependent on
    - Chirality
    - ➡ Electronic type
    - Alignment
    - Density
    - Defects
    - ➡ Impurities
    - ➡ Contact resistance between CNTs

- Opportunities:
  - Lower contact resistance between same electronic type **(separations)**
  - **Chemical doping** can be used to promote carrier transfer and individual CNT resistance as well as to reduce contact resistance between CNTs.

Electronic-type SWCNT separation techniques

- Different separation techniques are currently being evaluated with respect to yield and commercial scalability.
  - Selective growth
  - Non-covalent functionalization
  - DNA-based Ion Exchange Chromatography
  - Selective Oxidation/Functionalization
  - Density Gradient Ultra centrifugation
  - Column Chromatography
- *Density gradient ultracentrifugation* is so far the only commercially scalable process.

He, M. et al. *Chem. Comm.*, ***2011***, *47*, 1219-1221.

Tu, X. et al. *Nature Lett.*, **2009**, *460*, 250-253.

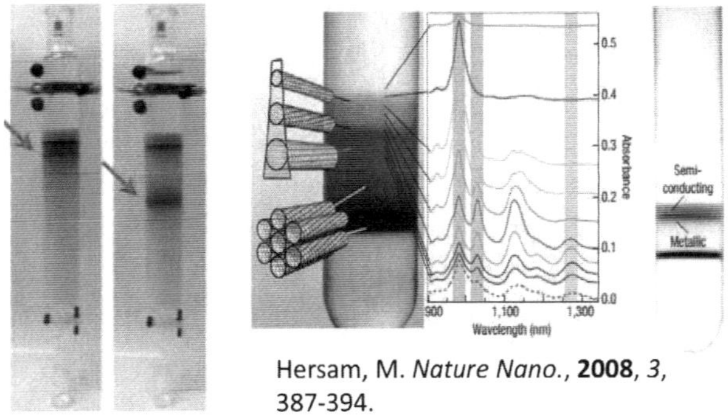

Hersam, M. *Nature Nano.*, **2008**, *3*, 387-394.

Improving electrical conductivity of MWCNTs with chemical doping

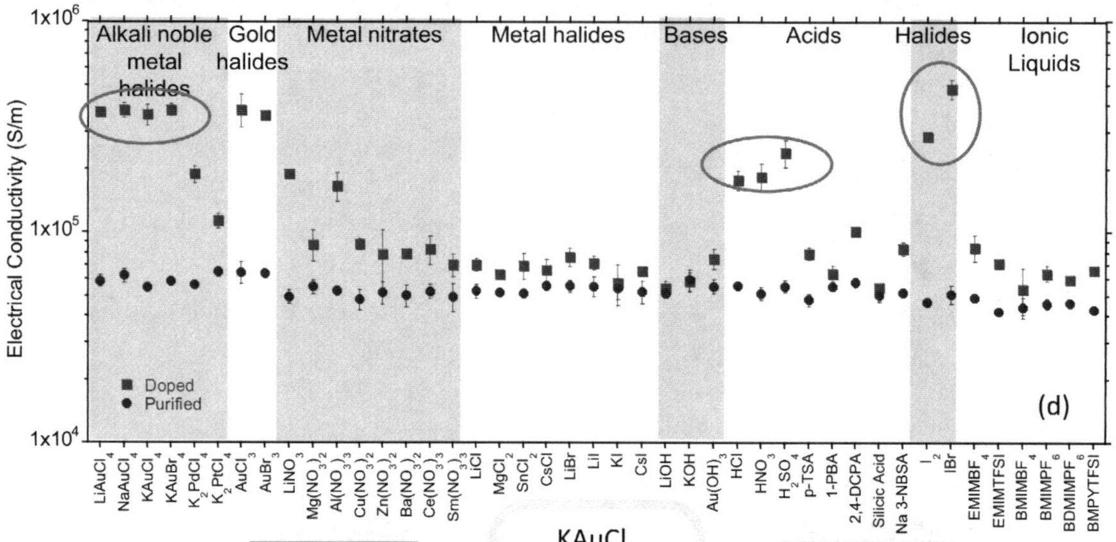

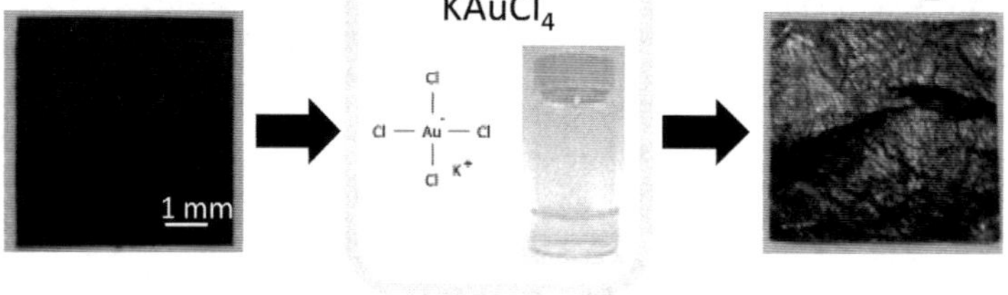

Effects of doping on type separated material

- ❑ Fabricated SWCNT thin films to investigate the doping effects of electronic type separated SWCNT films.
  - ❑ Use superacid (CSA) to disperse commercial NanoIntegris type-separated SWCNTs.
  - ❑ Remove CSA from the as-produced thin-films
  - ❑ Dope thin-films with 4 best performing dopants: I_2, IBr, $KAuBr_4$, CSA.

Thin films fabricated with ~0.1 mg of SWCNTs dispersed in 15ml of CSA.

Type		Thickness
Semi	Iso-Nanotubes-S (99%)	80 nm
Mixed	99% Ultra-Pure SWNT's	100 nm
Metallic	Iso-Nanotubes-M (98%)	70 nm

Films divided in 4 sections to ensure homogeneity

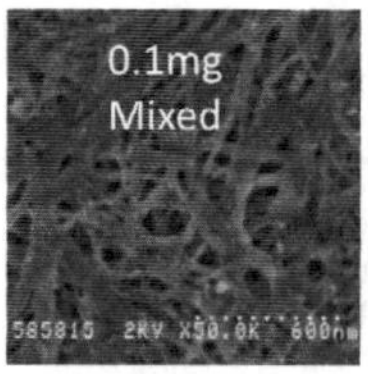

Chemical Doping and Conductivity Improvement

- ❑ CSA removed (30 min, 400°C thermal oxidation*)
- ❑ The purified SWCNT thin-films were doped in:
  - ❑ 10mM $KAuBr_4$(aq)
  - ❑ Ethanol-IBr solution
  - ❑ I_2 vapor
- ❑ The SWCNT thin-film conductivity increase for all dopants:
- ❑ **Semi (2.6-6X)**
- ❑ **Mixed (2.2-4.5X)**
- ❑ **Metallic (1.1x-3.5x)**

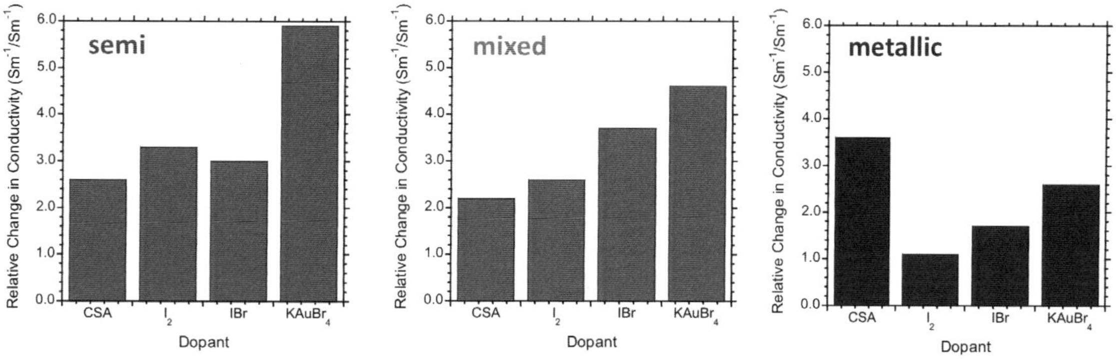

*Decomposition temperature of SWCNTs is > 600°C

With the exception of metallic CSA-doped and mixed IBr-doped films, the same trend is observed within each dopant: semi>mixed>metallic

AICHE 2015 – November 8 *Ivan Puchades, Ph.D.*

Electrical Conductivity Results

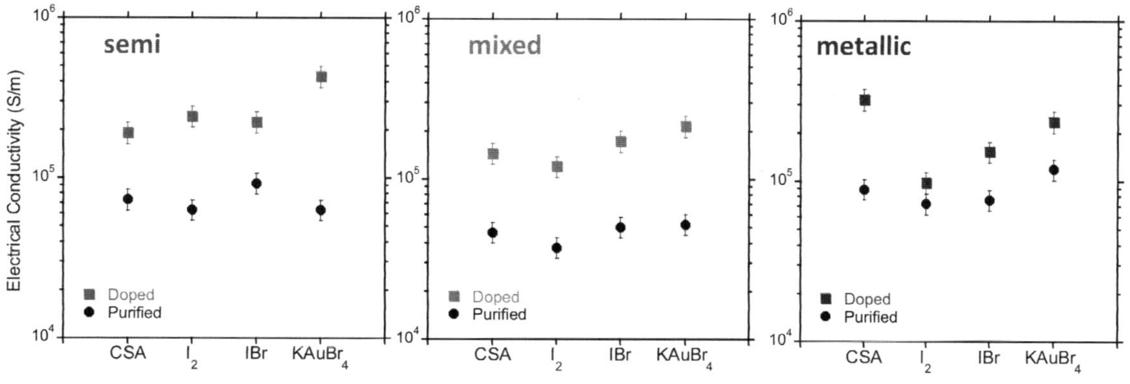

Conductivity of doped semi and mixed SWCNT thin-films can increase beyond the conductivity of doped metallic SWCNT thin-films (with the exception of CSA).

- ❑ Typical values obtained with NanoIntegris type-separated material.
- ❑ Higher electrical conductivity values have been achieved with RIT's LV vaporized SWCNTs (2-5x10^6 S/m)

R·I·T

UV/VIS absorption overlay of doped films by type

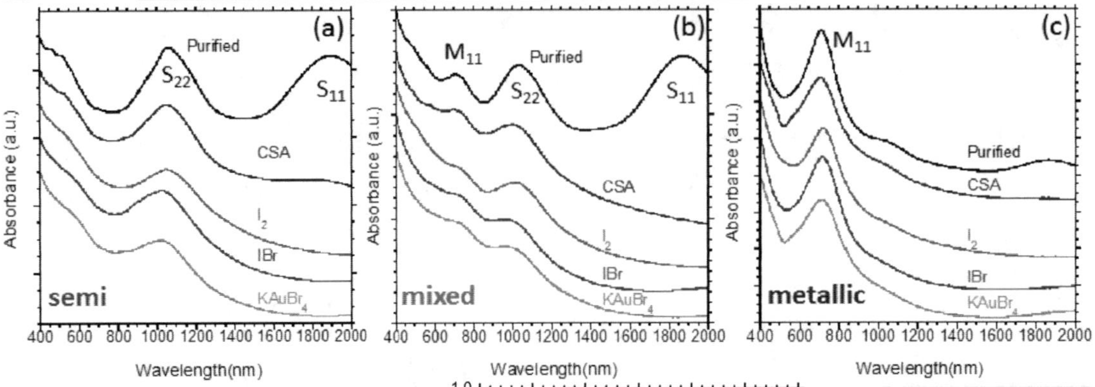

- Light is absorbed by SWCNTs at the available energy transition singularities.
- S_{11} peak is suppressed for all films after doping.
- S_{22} and M_{11} peaks are also quenched.

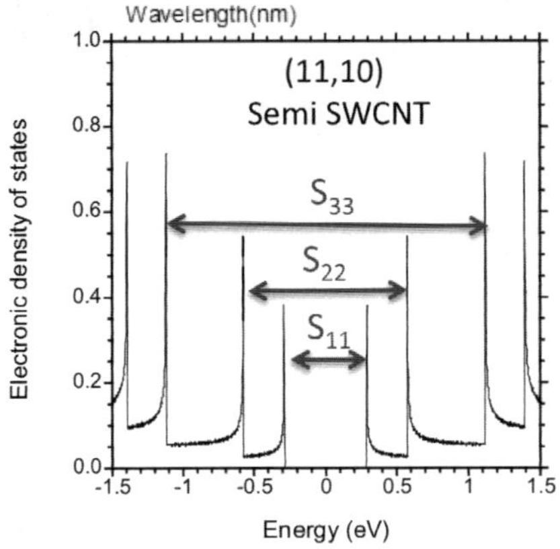

Quenching of transitions peaks determines the level of doping and available density of states being occupied.

R·I·T

Raman G'-band

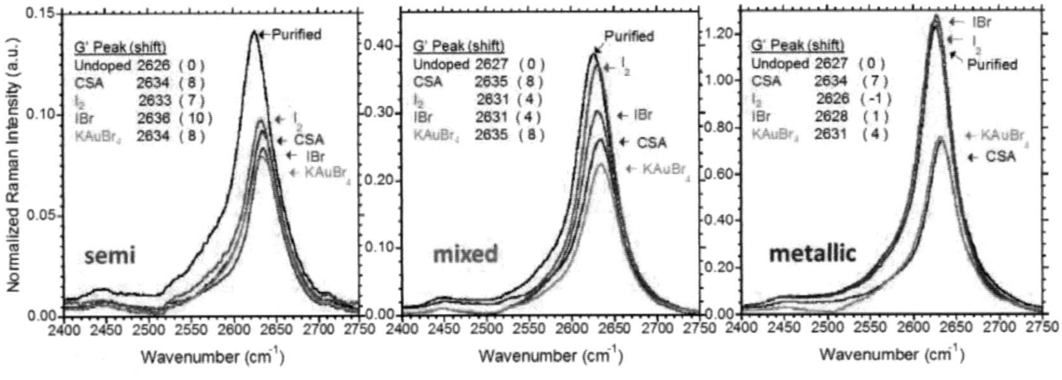

- G' peak magnitude is an indication of available electrons.
- With doping, the G' peak is reduced (electron withdrawal) and its position shifts due to its highly energy-dispersive nature with energy shifts.
- Acceptor type dopants will manifest themselves as a red shift of the G'-band, and donor like dopants will result in blue shifts.

When doped, the G' peak is red-shifted in all cases with the exception of I_2 and IBr doping of metallic films. Generally a trend is seen where I_2>IBr>CSA>$KAuBr_4$

Understanding conductivity increase by type

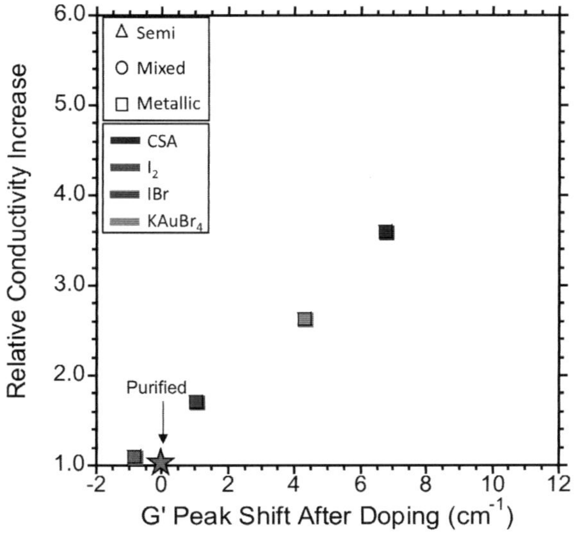

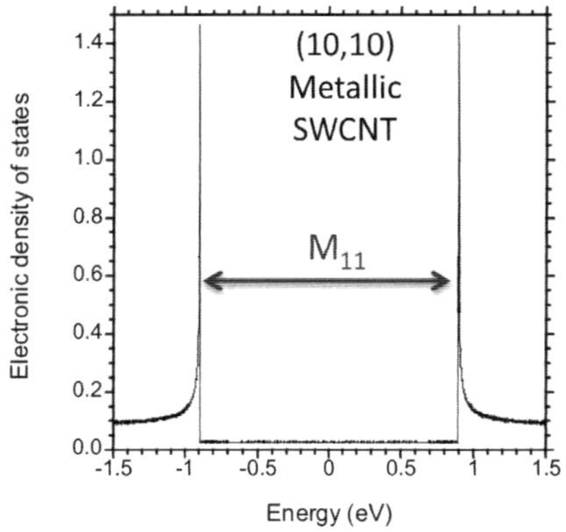

- ❑ Metallic SWCNT films show a linear increase consistent with the low but constantly available DOS at low energies.

AICHE 2015 – November 8 Ivan Puchades, Ph.D.

Understanding conductivity increase by type

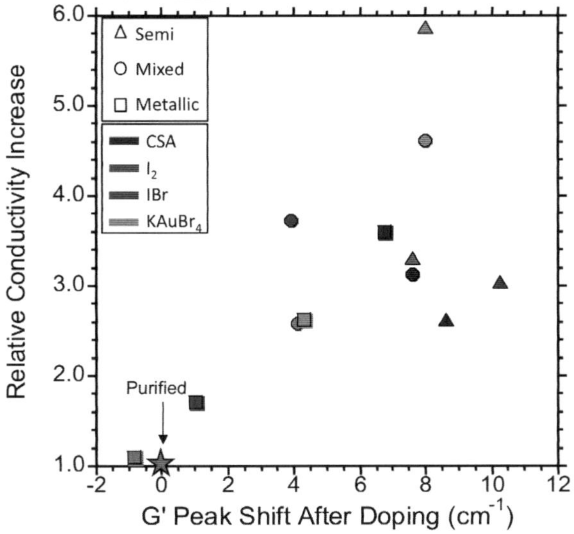

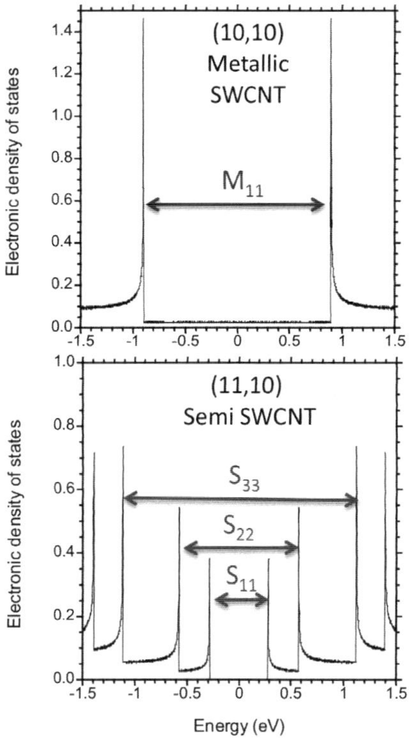

- ❑ Metallic SWCNT films show a linear increase consistent with the low but constantly available DOS at low energies.
- ❑ Semi and mixed SWCNT films are higher but scattered. Step-change in doping level and likely correlate with the **S_{11},S_{22}, and S_{33}** bands.

Doping and redox potentials

The redox potential of dopant solutions with respect to that of SWCNTs could be an indicator of the level of doping achievable.

- Semiconductor SWCNTs need dopants with a lower redox potential difference to be effectively doped down to the first and second valance bands ($S_{11}V$, $S_{22}V$).
- Metal SWCNTs need dopants with a larger redox potential difference ($M_{11}V$).

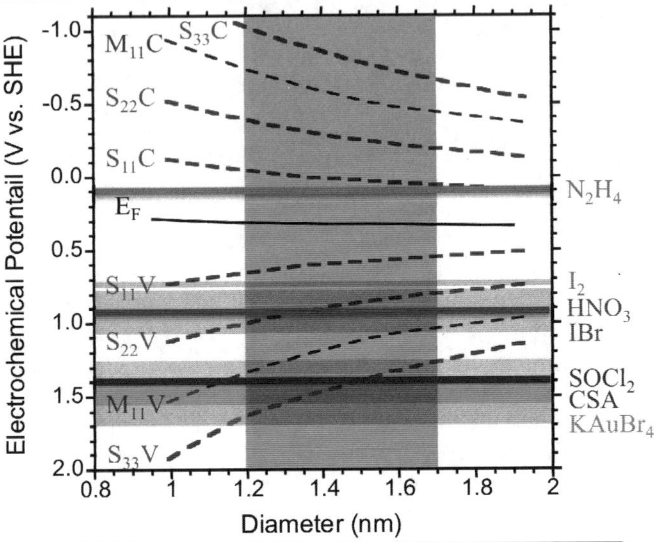

Other Commonly used SWCNT dopants:

N_2H_4 – hydrazine – dedopes SWCNTs
Oxygen – strong dopant
HNO_3 – nitric acid - shifts Fermi level, densify SWCNTs, and removes surfactants
$SOCl_2$ – Thionyl chloride – good thermal stability

Plot of redox potential with respect of standard hydrogen electrode of (1) CNTs, (2) common dopants (N_2H_4, HNO_3, $SOCl_2$), and (3) dopants studied at NPRL (I_2, IBr, CSA, $KAuBr_4$).
A range of redox results for the different chemical constituents of the dopant species.

Redox potential and doping level

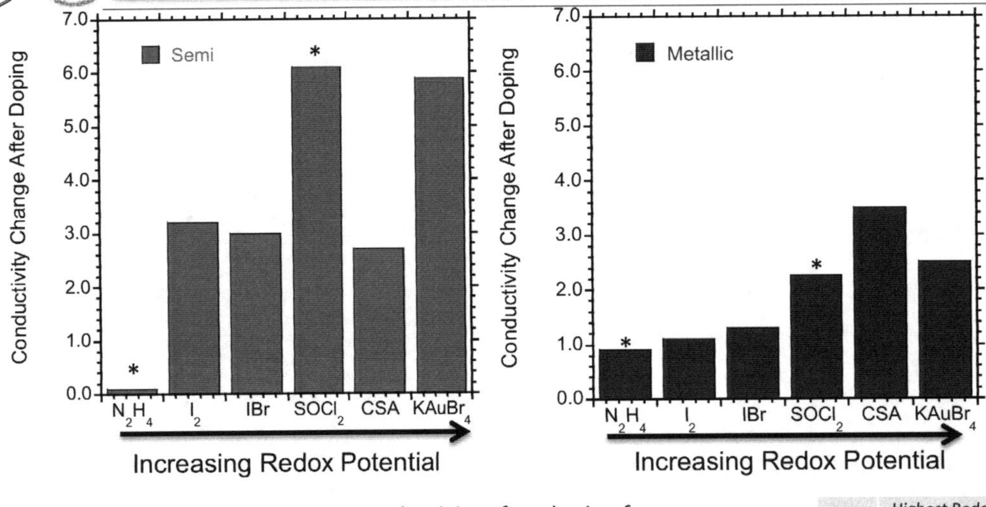

- Plots showing the increase in conductivity after doping for predominately semiconductor and metallic SWCNT films for dopants with increasing redox potentials (from left to right).

* Data for **N_2H_4(hydrazine)** and **$SOCl_2$** (thionyl chloride) from Jackson *et al.*, *ACS Nano* **2010**, on ~50nm Nanointegris SWCNT films. All other data is RIT's and also on ~50nm Nanointegris SWCNT films.

Dopant	Highest Redox Potential (eV)
N_2H_4	0.11
I_2	0.54
HNO_3	0.8
IBr	1.09
$SOCl_2$	1.49
CSA	1.63
$KAuBr_4$	1.83

Stability of type-separated doped SWCNT thin-films

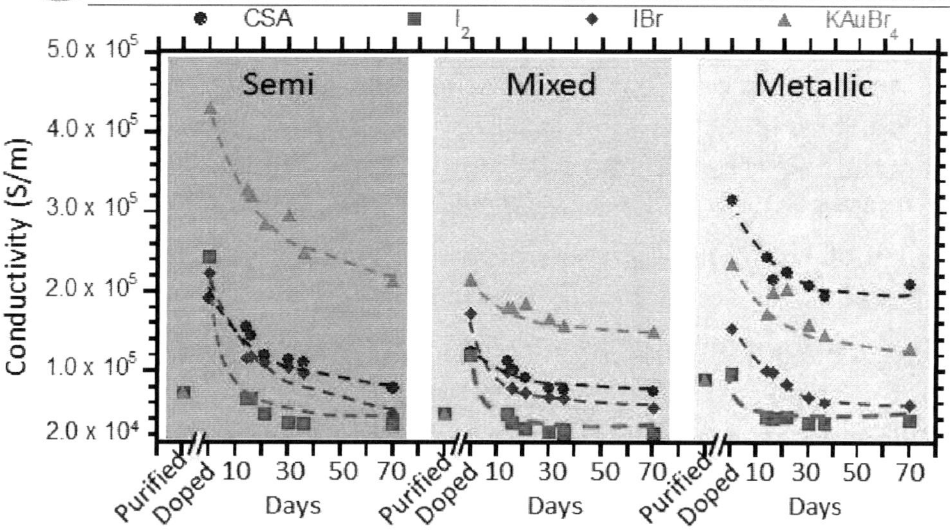

- Semi films show the quickest decrease in conductivity over time, specifically for IBr and I_2 dopants.
- The conductivity of I_2 and IBr decreases below the purified value as a result of residual disorder.
- The time stability of the studied dopant indicates the need for dopant-encapsulation methods.

Optical absorbance and Raman stability

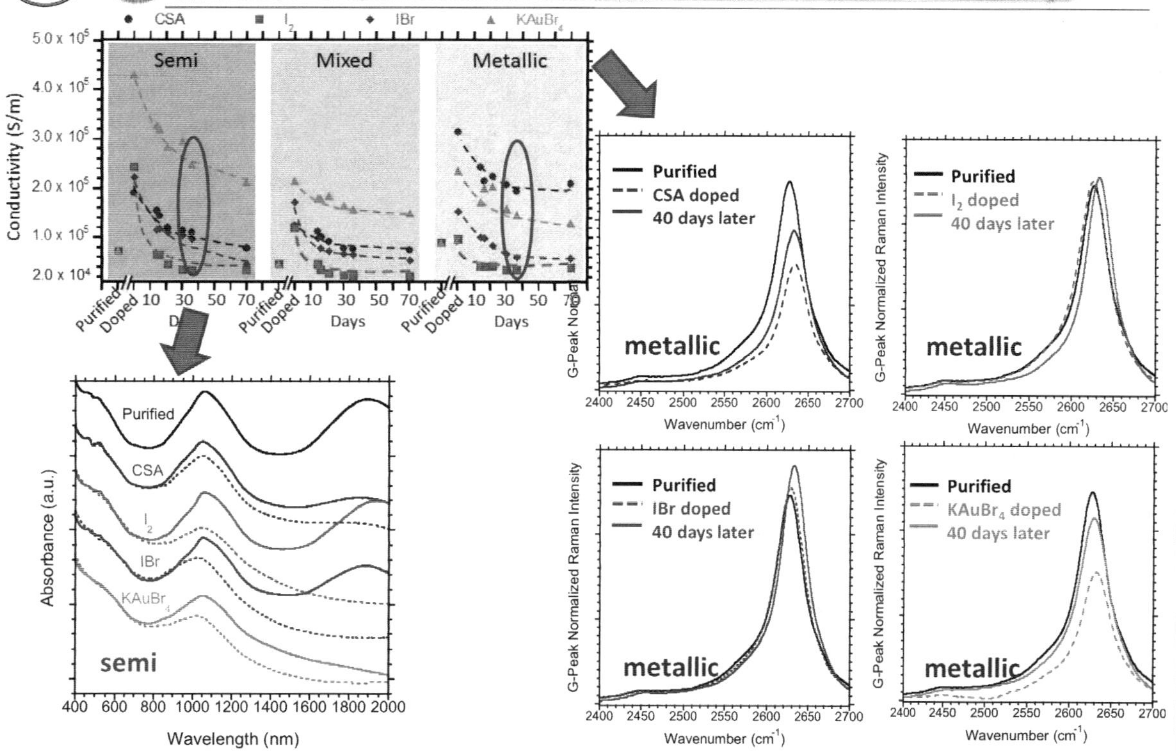

Conclusions for Separated SWCNT Research

- Enhanced electrical conductivity of carbon nanotubes (CNTs) can enable their implementation in a variety of wire and cable applications traditionally employed by metals.
- All dopants doped ***semi and mixed*** films at approximately the same rate while CSA-and $KAuBr_4$-dope ***metallic*** films more efficiently than IBr and I_2
- ***Redox*** potential of doping species explain the efficacy of the different dopants with respect to the type-separated films.
  - The redox potentials of $KAuBr_4$ and CSA are high enough to dope metallic films, while IBr and I_2 are not.
- Time dependent measurements of doped films indicate that metallic films are more stable than semiconducting and mixed films.
- Techniques to encapsulate dopants need to be developed for improved time stability.
- RIT has fabricated high-conductivity ($4x10^6$ S/m) mixed wires doped with $KAuBr_4$.

ADDITIONAL SLIDES

R·I·T

Removing CSA from thin-films - Purification

- Type-separated thin films were prepared from CSA dispersions.
- After deposition, the films were thermally oxidized to purify the films (remove the CSA).
- Raman (a, b) and UV/VIS spectra (c) of the fabricated CSA film and after thermal oxidation at different temperatures compared to commercial NI Ultrapure SWNT paper.

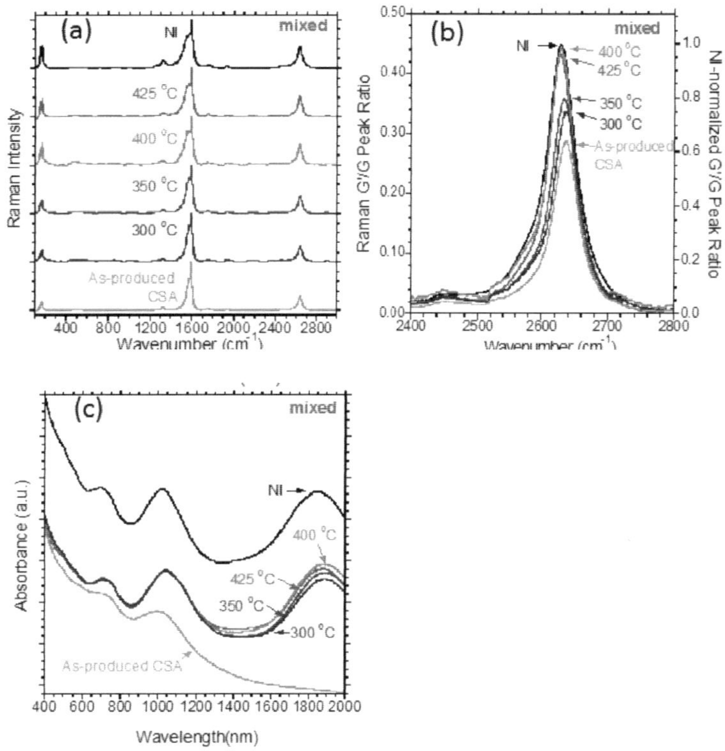

The evolution of the G'/G peak and the S_{11}/SS_{22} ratio (d) indicate the dedoping of the films back to its original estate with a **<u>400 °C thermal oxidation process for 30 min.</u>**

SWCNT decomposition temperature is >600 °C

R·I·T

Thickness and morphology assessment

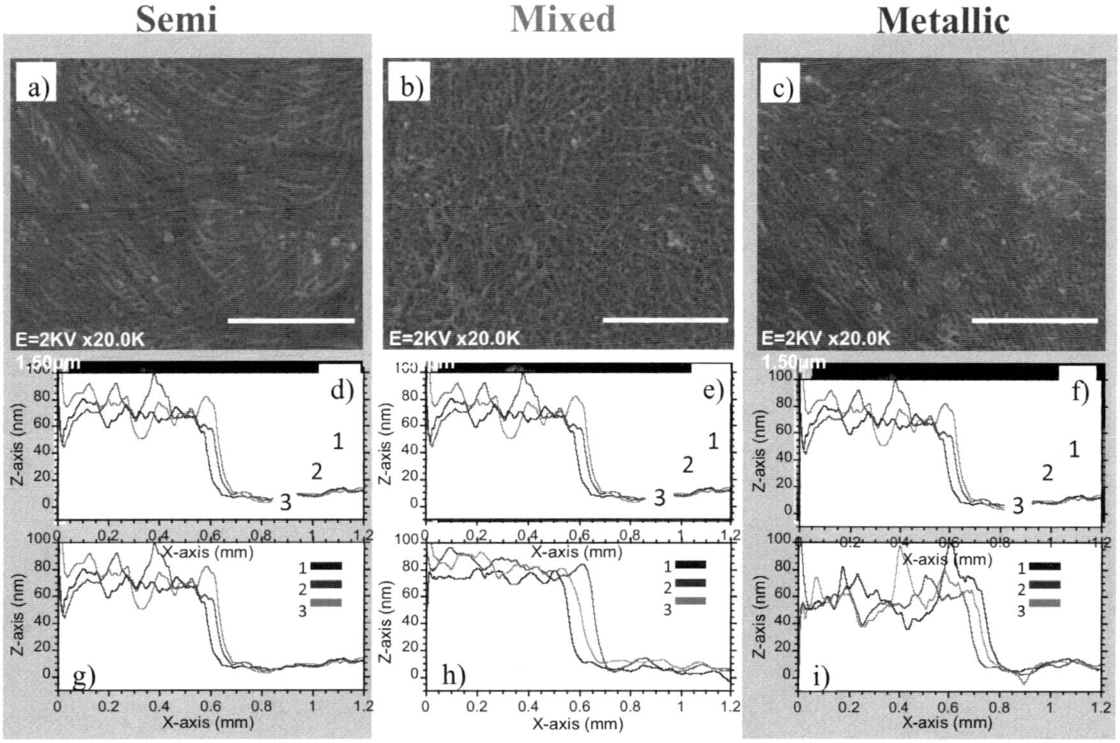

Removing CSA from thin-films - Purification

- Type-separated thin films were prepared from CSA dispersions.
- After deposition, the films were thermally oxidized to purify the films (remove the CSA).
- Raman (a, b) and UV/VIS spectra (c) of the fabricated CSA film and after thermal oxidation at different temperatures compared to commercial NI Ultrapure SWNT paper.

The evolution of the G'/G peak and the S_{11}/SS_{22} ratio (d) indicate the dedoping of the films back to its original estate with a **400 °C thermal oxidation process for 30 min.**

SWCNT decomposition temperature is >600 °C

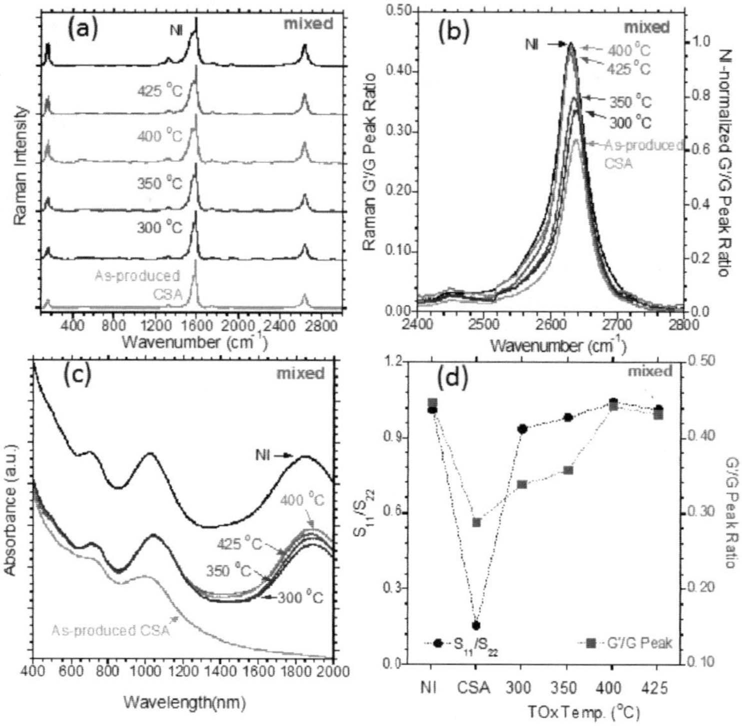

Application needs and goals

- Applications require ***separated*** SWCNTs by electronic type, diameter or chirality.
  - Cables and wires → Metallic and semiconducting SWCNTs separation improves conductivity
  - Nanoelectronic TFTs and Logic circuitry → Semiconducting SWCNTs
  - Photovoltaic applications → Diameter dependent bandgap
  - Batteries → Metallic or doped semiconducting SWCNTs.
- Separated SWCNTs can benefit numerous government programs with impact on ***IC computing, sensing capabilities, and power systems for aerospace missions.***

Nanoelectronics

Wires and Cables

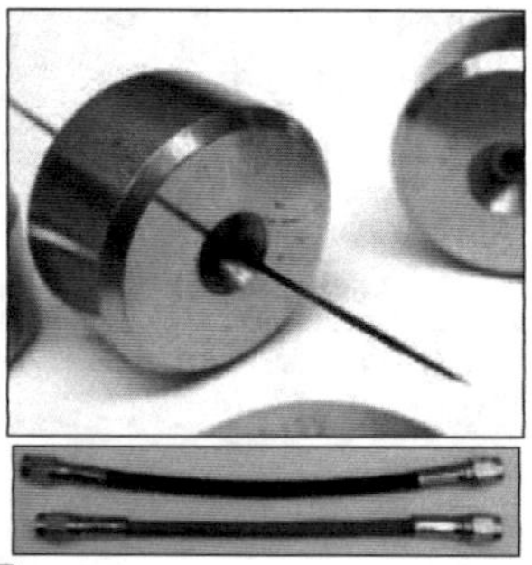

Batteries

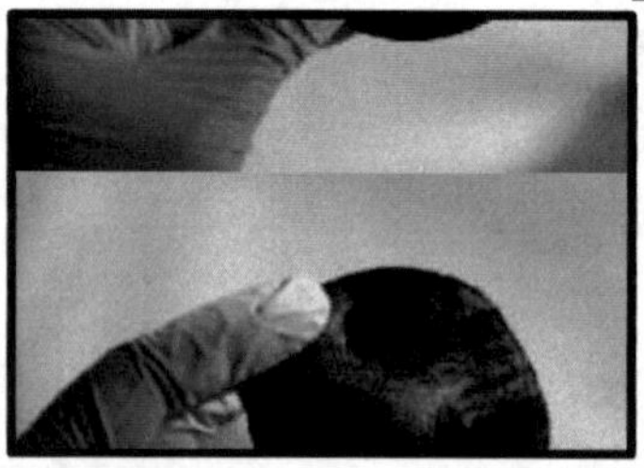

UV/VIS of Type Separated CSA-dispersed Thin Films

- UV-VIS-NIR performed on CSA-dispersed films made from NanoIntegris separated material.
- SEMs show CNT bundling and local alignment but no overall alignment in the films.

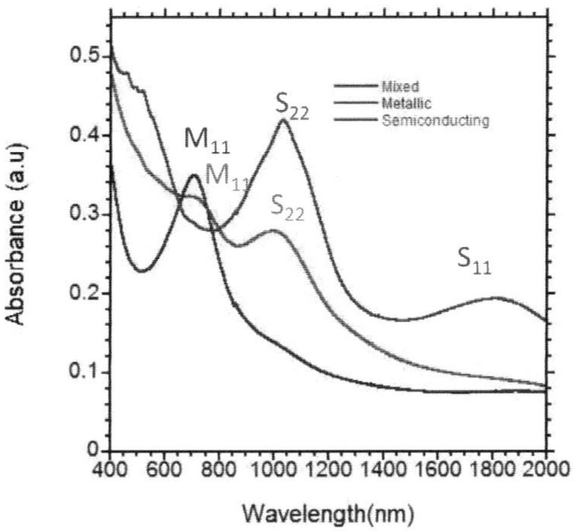

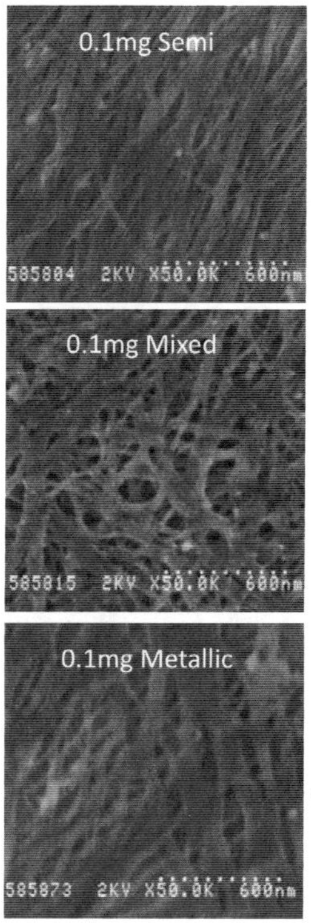

Characterization techniques

Optical Absorbance

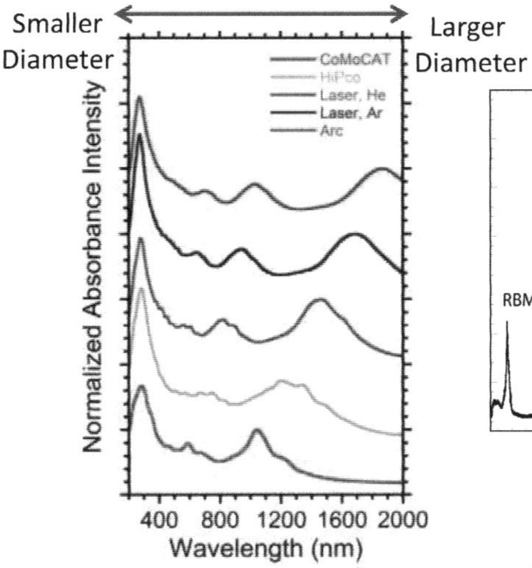

Absorbance at transition energy wavelengths: electronic type and diameter size.

Resonant Raman Spectroscopy

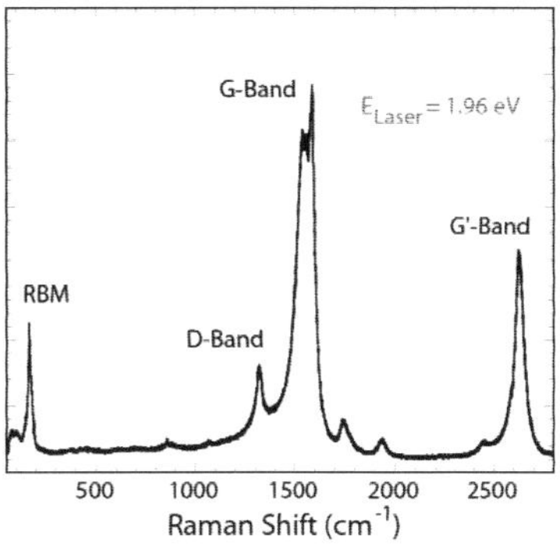

SWCNT resonance determines diameter, defects, electronic configuration, available DOS.

Electrical Conductivity

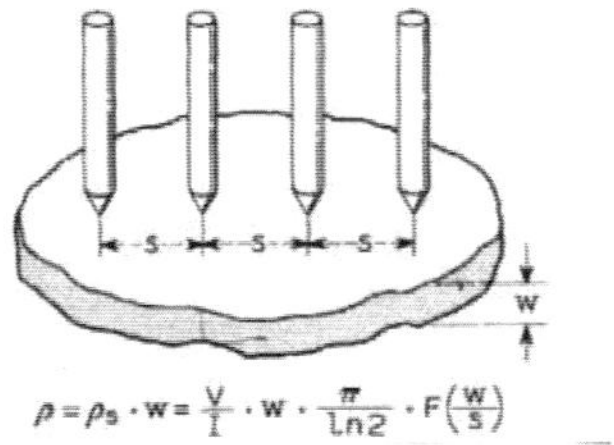

Thin-film Thickness

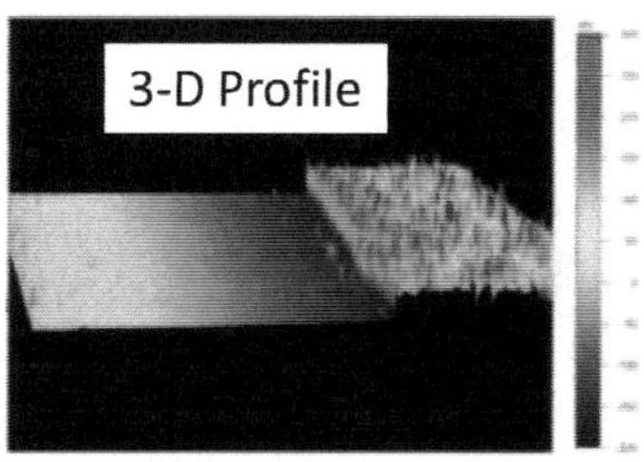

Characterization techniques

Optical Absorbance

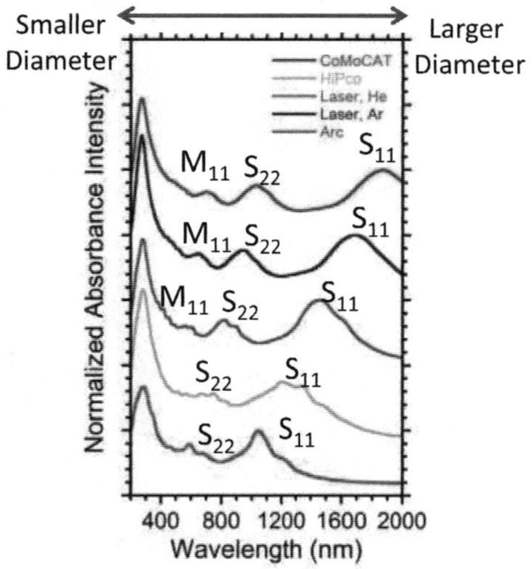

Absorbance at transition energy wavelengths: electronic type and diameter size.

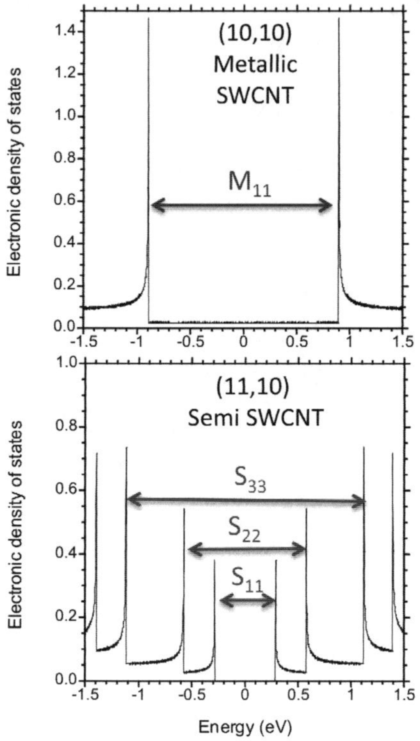

AICHE 2015 – November 8
Ivan Puchades, Ph.D.

Characterization techniques

Resonant Raman Spectroscopy

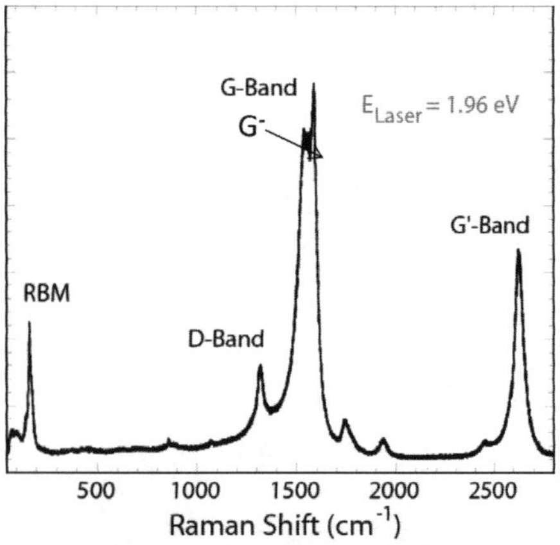

SWCNT resonance determines diameter, defects, electronic configuration, available DOS.

Raman Band	Characteristics
RBM (Radial Breathing Mode)	Radial resonance SWCNT diameter
D (Defects)	Defects
G (Graphite)	Tangential resonance G^- →free electrons
G' (Second order overtone of D-band)	Double resonance in available electronic states. Highly dispersive with energy shifts.

Characterization techniques

Electrical Conductivity

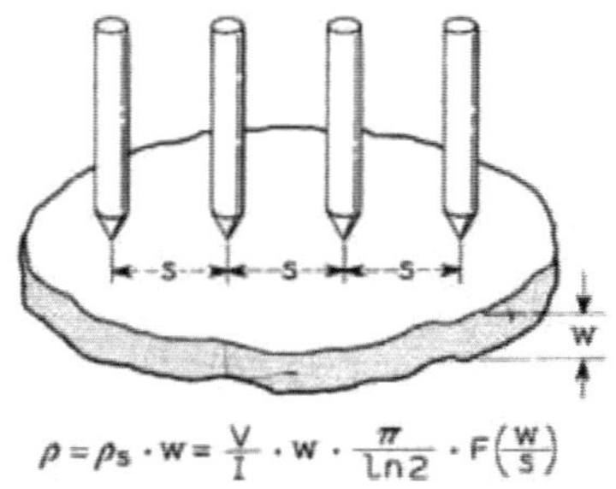

Thin-film Thickness

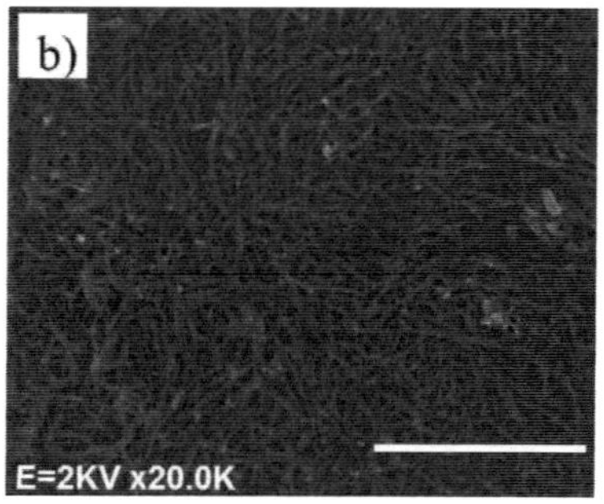

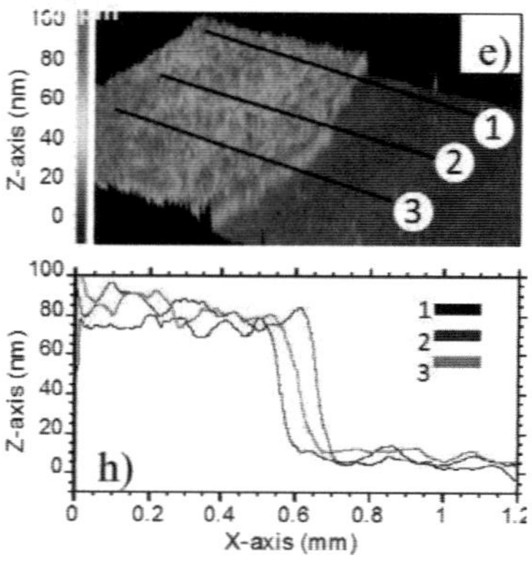

- ❑ The electrical conductivity of the thin-film can be determined from the sample sheet resistance and thickness.
- ❑ A contact-less white-light interferometry method can be used to measure the sample thickness.

Raman Overlay of Doped Films by Type

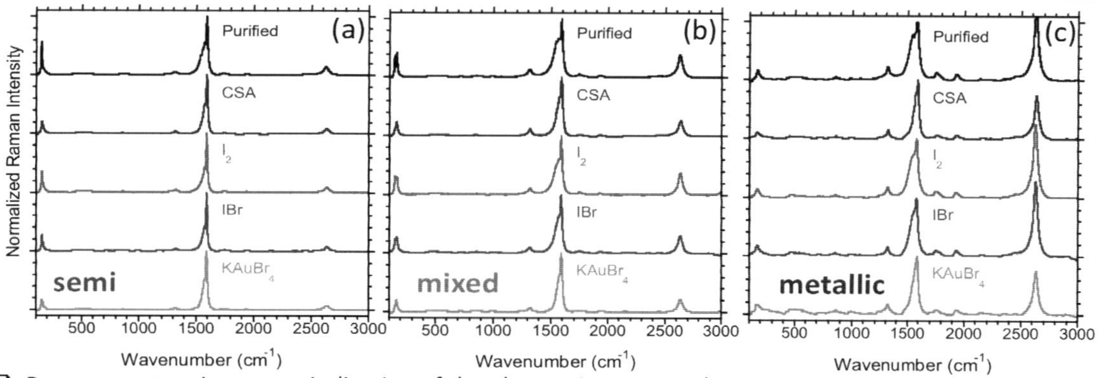

- ❑ Raman spectra gives us an indication of the electronic structure by probing the SWCNT resonance.
- ❑ G' peak magnitude is an indication of available electrons.
- ❑ The G' peak is observed to increase as the metallic content is increased – more free electrons in metallic tubes.
  - ❑ The G'/G peak ratio for purified films increase from 0.14 to 0.39 and 1.23 for semi, mixed and metallic films respectively.

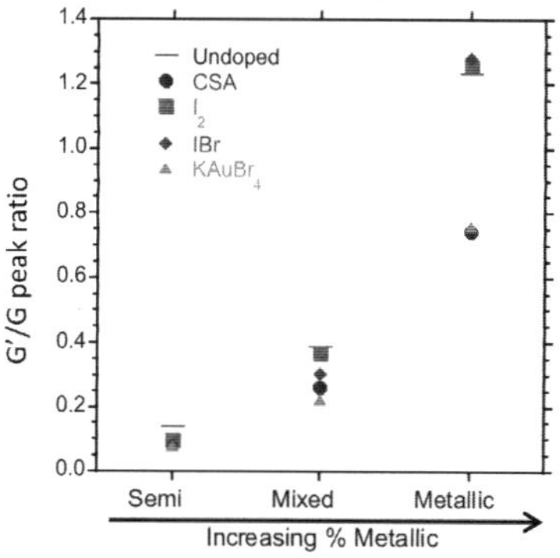

> When doped, the G' peak is reduced (electron withdrawal) in all cases with the exception of I_2 and IBr doping of metallic films.

433434 Local Energetic Analysis of the Interfacial and Surface Energies of Graphene from the Single Layer to Graphite

Abstract:

Recent advances in scanning probe methods that provide direct access to the surface free energy of inorganic layered materials in terms of the Hamaker constant, yield energetic values for monolayer graphene that differ substantially, by a factor of around 0.4, from bulk graphite. The onset of bulk deviating energy values was observed at a multilayer slab thickness of ~3 nm, corresponding to a layer number of 10. The findings, complemented with extractions from water contact angle measurements and calculated interlayer binding energies, find short-range ordinary Van der Waals interactions to be most prominently affected by dimensional constraints and many-body interactions.

411236 Modeling of Graphene-Titania Interfaces

Abstract:

TiO_2-graphene composite photocatalyst materials show promise for a wide variety of environmental applications[1], including pollutant degradation, water treatment, and energy/fuel production. When combined with graphene, TiO_2 shows increased photocatalytic activity, which has been attributed to a variety of reasons[1]. Narrowing of the band gap of TiO_2 into the visible region may occur due to formation of TiO_2-C moieties. The large surface area of graphene may facilitate surface reactions between adsorbates and catalyst. Finally, increased charge separation (which lowers charge recombination rates, the opposite of photoexcitation) may occur as electrons are shuttled away by graphene. Still many questions exist for these materials. For instance, experimental evidence suggests that graphene with low number of defects makes for an ideal TiO_2-graphene photocatalyst[2]. However, many popular production methods involve reduction of graphene oxide, which may produce graphene with many defects. It is crucial to identify the role of such defects in these photocatalysts. Other work also suggests that intimate contact between TiO_2 and graphene will increase catalytic activity[3].

We present density functional theory simulations of TiO_2-graphene systems in order to elucidate the nature of the interface between the two materials. We have specifically focused on TiO_2 clusters up to 1 nm in size and rutile (110) surfaces. We considered graphene as defect-free or with common defects such as hydroxyls, vacancies, or epoxides. Our results indicate that significant binding occurs at defect sites, much more so than over pristine graphene. Most notably defects on the graphene surface lower the band gap of TiO_2, indicating that new composite electronic states may arise and increase photoexcitation yield in such materials. We also analyzed how surface-surface distance (indicative of degree of contact between graphene and TiO_2) affects the binding and TiO_2 band gaps. Our results highlight the potential role of the graphene surface state in TiO_2-graphene photoactive materials, and suggest design strategies for graphene-based composite materials based on the TiO_2-graphene interface structure.

(1) An, X. Q.; Yu, J. C. *RSC Advances* **2011**, *1*, 1426.

(2) Liang, Y. T.; Vijayan, B. K.; Gray, K. A.; Hersam, M. C. *Nano Letters* **2011**, *11*, 2865.

(3) Zhang, Y.; Zhang, N.; Tang, Z.-R.; Xu, Y.-J. *Physical Chemistry Chemical Physics* **2012**, *14*, 9167.

428148 Surface Pressure and Microstructure of Carbon Nanotubes Adsorbed at an Air-Water Interface

Abstract:

Particles of appropriate size and wettability adsorb strongly at fluid-fluid liquid interfaces, lowering the interfacial energy and thereby stabilizing emulsions and foams. In this presentation, we will report the surface pressure and microstructure of two different types of carbon nanotubes (CNTs) at an air-water interface; namely as-produced CNTs (nf-CNTs) and CNTs functionalized with carboxyl groups (f-CNTs)[1]. Both types of CNTs formed 3D aggregates at the interface upon compression. However, f-CNTs showed less degree of aggregation compared with nf-CNTs. This is attributed to the deprotonation of the carboxyl groups within the water subphase, leading to additional electrostatic repulsion between f-CNTs. At high compression, f-CNTs formed aligned CNT domains at the interface. These 2D domains resembled 3D liquid-crystalline structures formed by excluded volume interactions. The denser packing and orientational ordering of f-CNTs also contributed to a compressional modulus higher than that of nf-CNTs. The actual coverage of CNTs during compression was back calculated from the Volmer equation of state and was in agreement with the value obtained independently from optical micrographs. The findings of this work may have a broader impact on understanding the assembly and collective behavior of rod-like particles with a high aspect ratio at an air-water interface. This work is supported by NSF CAREER award #1253613.

(1) Vora, S. R.; Bognet, B.; Patanwala, H. S.; Chinesta, F.; Ma, A. W. K. Surface Pressure and Microstructure of Carbon Nanotubes at an Air–Water Interface. *Langmuir* **2015**.

DEFORMATION OF CLAY-FILLED EPOXY NANOCOMPOSITES

Suresh Ahuja
Xerox Corporation
Webster New York 14580

Email samka_2002@hotmail.com

ABSTRACT

Rheology of cured and uncured epoxy nano-composites were first investigated using a rheometer. Epoxies were mixed with organo-clays, Cloisites. The matrix used in the present study was a two-component epoxy resin system where epoxy resin DGEBA was mixed with Cloisite, commercially available from Southern Clay Products, Inc., USA. It is a natural montmorillonite (MMT) that is organically modified.. Before mixing with epoxy, Cloisite 10A was washed with ethanol and water to remove excess of benzyl dimethyl stearyl ammonium chloride and benzyl dimethyl cetyl ammonium chloride from 7.5% to 0.4% and from 4.0 to 0.2%, respectively. Similarly, Cloisite 30A was washed with ethanol and water to remove bis(2-hydroxyethyl) methyl stearyl ammonium chloride from 7.4 % to 0.4. Vauum was applied to remove bubbles. Epoxies were crosslinked using a crosslinker, hexahydrophthalic anhydride (75%) and Polyether polyol (25%) with 50/50 proportion of crosslinker to epoxy at 90C, 110C and 130C. Coatings of epoxy cross-linked and linear on aluminum surface and Hysitron pyramid indenter was used for different depths.

Results showed nano-clay with higher basal spacing improves the Tg of the composites. SEM was used to distinguish dispersion between different nanoclays in epoxy. Cured epoxy in a nanocomposite can lead to exfoliated morphology and increase in temperature increases the possibility of exfoliation. A three phase model of epoxy, exfoliated clay layers and nano layer clusters developed previously was used to characterize and model the young's modulus of the

INTRODUCTION:

Two different types of nanocomposite structures exist namely, intercalated nano-composites where the polymer chains are sandwiched between silicate layers and exfoliated nano-composites where the layers can be considered individually but remain more or less dispersed in the polymer matrix. One of the clays often used for making nanocomposites is montmorillonite which consists of platelets with an inner octahedral layer sandwiched between two silicate tetrahedral layers. Some of the aluminum atoms in montmorillonite are replaced with magnesium; the difference in valences of Al and Mg creates negative charges distributed within the plane of the platelets that are balanced by positive counter ions, typically sodium ions, located between the platelets or in the galleries. The sodium ions can be exchanged with organic cations, such as those from an ammonium salt, to form an organoclay. The ammonium cation may have hydrocarbon tails and other groups attached and is referred to as a ''surfactant'' owing to its amphiphilic nature. Epoxy-clay nanocomposites have complex micro-structure resulting from the sensitivity of the morphology of clay particles to the dispersion condition. The morphological state is influenced by many factors such as clay content, particle size, cation exchange capacity (CEC), organic surfactant, moisture content and synthesizing condition. Broadly, MMT exist in three typical forms e immiscible, intercalated and exfoliated. For organically treated montmorillonite (OMMT), the co-existence of intercalated and exfoliated ones is frequently observed n samples. The micro-structure of epoxy-clay nanocomposites is regarded as a multi-phase system consisting of bulk matrix, silicate sheets, gallery interfaces and matrix interphase surrounding the silicate sheets as labeled. The focus is on related interfaces including gallery interface and matrix interphase. In polymer-clay nanocomposites the concept of a binding energy is too simplified to be able to account for the presence of multi-phase interfaces. The gallery interface within intercalated clay particles and the interphase region with the polymer matrix are investigated. They are subjected to splitting deformation through molecular dynamics simulations and characterized by their traction separation relationships. Several key parameters including peak strength, fracture energy and final splitting separation distances are qualified from the traction separation curves which can be integrated into continuum models[1].

Tg of the polymer either increase or decrease with the addition of clay particles. The increase in Tg is attributed to the slower segmental motion due the polymer chains being anchored to the surface of the clay [2]. A three phase model of epoxy, exfoliated clay layers and nanolayer clusters was developed by Luo and Daniel [3] to characterize and model the young's modulus of the nanocomposites. The Mori–Tanaka method was used to evaluate the young's modulus of the system as a function various parameters such as exfoliation ratio, clay layer and clay cluster aspect ratios, d spacing and intragallery modulus, matrix modulus and matrix poisson's ratio. it was concluded that (1) the composite modulus of the system increases with increase in the degree of dispersion, (2) exfoliation ratio has a decisive role in the enhancement of modulus, (3) high aspect ratios of the single clay platelet as well as clay clusters are desirable in stiffness enhancement, and (4) larger d spacing is preferred along with high cluster aspect ratio for large improvement in composite modulus.

As the amount of clay is increased, the elastic modulus of epoxy–clay nanocomposites increased up to a maximum of 6% clay .The tensile strength and energy to failure continually increase with clay quantity all the way up to 10%. The amount of increase in a particular property depends on the type of clay used and the surface treatment of that clay. For the rotor speed, the slower speeds were found to be the best, while the higher speeds caused the properties to deteriorate with the critical speed being around 3000 rpm. After 3000 rpm the higher speeds caused the clay layers to actually break rather than separate, making the layers ineffective as nanocomposite fillers. The curing temperature again showed best results at lower temperature, but with higher temperatures causing specimen brittleness and deterioration of its properties. .The optimum temperature range was about 80–100C[4].

EXPERIMENTAL: Rheology of cured and uncured epoxy nano-composites were first investigated using a rheometer. Epoxies were mixed with organo-clays, Cloisites. The matrix used in the present study was a two-component epoxy resin system where epoxy resin DGEBA was mixed with Cloisite, commercially available from Southern Clay Products, Inc., USA. It is a natural montmorillonite (MMT) that is organically modified.. Before mixing with epoxy, Cloisite 10A was washed with ethanol and water to remove excess of benzyl dimethyl stearyl ammonium chloride and benzyl dimethyl cetyl ammonium chloride from 7.5% to 0.4% and from 4.0 to 0.2%, respectively. Similarly, Cloisite 30A was washed with ethanol and water to remove bis(2-hydroxyethyl) methyl stearyl ammonium chloride from 7.4 % to 0.4. Vauum was applied to remove bubbles. Epoxies were crosslinked using a crosslinker, hexahydrophthalic anhydride (75%) and Polyether polyol (25%) with 50/50 proportion of crosslinker to epoxy at 90C, 110C and 130C.

TABLE 1

Materials and their composition are given below

Material	Composition
Epoxy	Diglycidyl ether of bisphenol A
Crosslinked Epoxy	Crosslinked with hexahydrophthalic anhydride (75%) and Polyether polyol (25%)
Cloisite 10A South Clay	Sodium salt of montmorillonite with benzyl dimethyl stearyl ammonium chloride ion
Cloisite 30A South Clay	Sodium salt of montmorillonite Bishydroxy ethyl stearyl ammonium chloride ion

TABLE 2

. Interlayer spacing of various organo-modified MMT and the nanocomposites

Material	Cation	Anion	Interlayer spacing d
Cloisite 30A unwashed	Sodium	Bishydroxy ethyl stearyl ammonium chloride ion	18.334 A
Cloisite 30A washed	Sodium	Bishydroxy ethyl stearyl ammonium chloride ion	17.983A
Nanocomposite of Epoxy and 10% Cloisite 30A	Sodium	Bishydroxy ethyl stearyl ammonium chloride ion	1) 18.034A at 5 (2θ) 2) 11.940 at 7.5(2θ)
Nanocomposite of Epoxy and 10% Cloisite 10A	Sodium	Bishydroxy ethyl stearyl ammonium chloride ion	3) 17. 012A at 5.2 (2θ) 4) 11.385 at 7.7(2θ)

RESULTS AND DISCUSION :

From dynamic viscoelastic measurements at various temperatures, elastic and viscoelastic modulus as function of frequency was used to get relaxation modulus by using transformation functions. The nanocomposite containing organoclay 30A is more temperature dependent than the one containing organoclay Cloisite 10A as shown in Fig.1 possibly due to differences in hydrogen bonding between the surfactant and the montmorillonite clay. Similarly, dynamic viscosity as function of shear rate was fiied into power law functions as shown in Figure2. From preexponent function, dynamic viscosity independent of shear rate was used and graphed against temperature. Temperature dependence of viscosity was used to differentiate the two nanocomposites.

Epoxy was crosslinked at 90C,110C and 130Cusing a crosslinker, hexahydrophthalic anhydride (75%) and Polyether polyol (25%) t 50/50 proportion of Crosslinked to epoxy. Relaxation modulus at different crosslinking level in the nanocomposite shows large difference from nanocomposite with uncrosslinked epoxy. These structures by X-ray diffraction and TEM showed significant exfoliation.

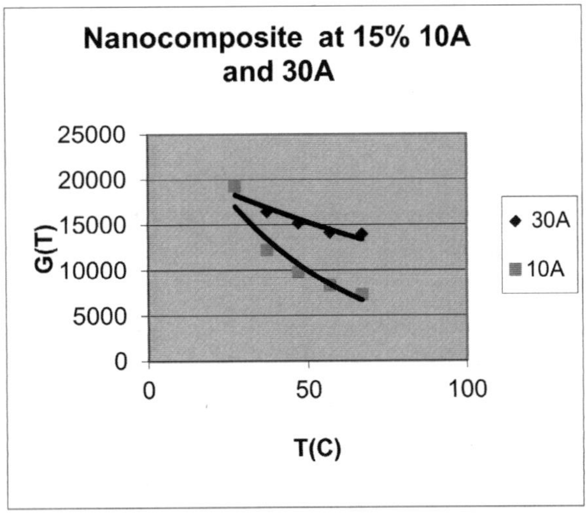

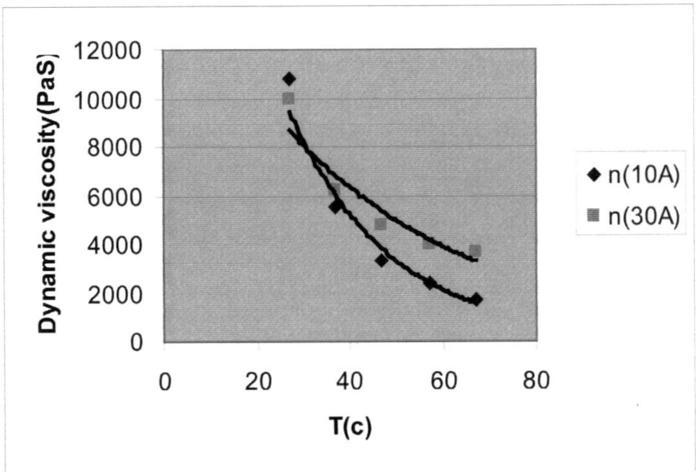

Figure 1 Relaxation modulus at 10% Strain for the Cloisites ***<u>Fgure 2 Dynamic viscosity for the two Cloisites</u>***

In Figure 3 relaxation modulus as function of time is compared with crosslinked and uncrosslinked nanocomposites. From power law dependence, G(0) increases with nanofiller and crosslink temperature. Relaxation time,τ is significantly higher with nano filler while temperature effect is masked by the crosslinked structure. For stronger interactions, characteristic relaxation timeτ increases. This can be explained in terms of an increase of the activation energy, which influences the relaxation time. Table 3 shows G(0) which increases with nanofiller and crosslink temperature.. Also,relaxation time,τ is significantly higher with nano filler while temperature effect is masked by the crosslinked structure.

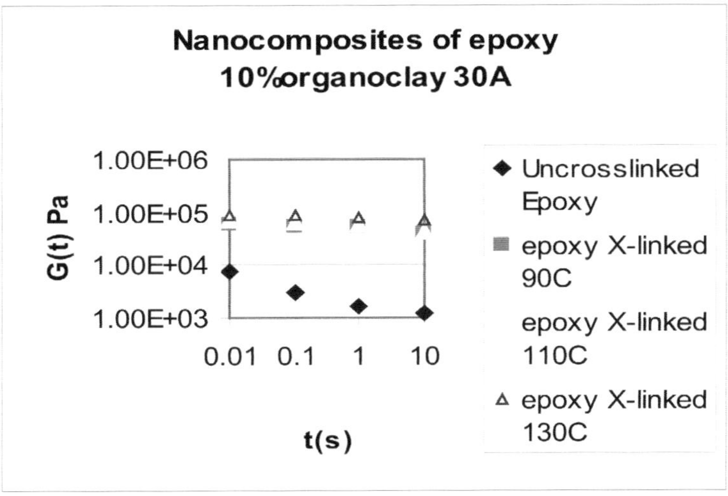

Figure 3 Effect of epoxy crosslinking on relaxation modulus G(t)

Model and simulation analysis of nanocomposites is investigated by using the polymer as a bead spring with interactions between polymer segments and smooth nanoparticles, nanoparticles and nanoparticles are modeled using modified Lennard- Jones functions, which account for the excluded volume of the beads and particles by offsetting the interaction range. The relaxation time is extracted by fitting the conformational relaxation curves with an exponential function at different temperatures, exhibits an Arrhenius relationship with the temperature for the pure and filled systems [5]. Within polymer-particle interactions, hydrogen bond not only improves dispersion but also slows the relaxation process Thus, the kinetics of the adsorption–desorption equilibrium of polymer chains to the clay surfaces (or attachment to or detachment from nanoparticle surfaces) influence the strength of elasticity and dictate the viscoelastic properties. Rheological observations are greatly affected by polymer-particle interactions, especially at low frequencies (i.e., over a long time), which can be proved by the plateau storage modulus at low frequency, stress relaxation over long periods of time, or shear thinning behavior of complex viscosity. Total friction (ζ) onthe monomer is directly related to the number of configurational restrictions. In addition to the monomeric friction coefficient, the longest relaxation time of the Rouse model can indicate segmental mobility or chain immobilization Interestingly, the overall dissipative friction forces along the chain motion can be well understood by the Rouse model [6] The monomeric friction coefficient and longest Rouse relaxation time can be extracted from the zero shear viscosity or dynamic storage modulus

Table 3. Nanocomposites of Epoxy and 10%Cloisite 30A

Sample Description	**G (0),Pa**	**t/τ**
Uncrosslinked epoxy	1934	0.2718
Epoxy Crosslinked at 90C	47118	0.0554
Epoxy Crosslinked at 110C	53835	0.0737
Epoxy Crosslinked at 130C	78567	0.0379

Conclusion

Evaluation of elastic deformation using a three phase model of epoxy, exfoliated clay layers and nanolayer clusters using Mori–Tanaka method is most appropriate. However, epoxy clay composites depending on type and nanofiller concentration behave viscoelastic especially at high clay concentration and when crosslinked. The longest Rouse relaxation time, which was determined from the relaxation modulus time dependence, was employed to get the segmental friction. It is clear that the chain relaxation processwas slowed by the polymer-particle interactions, which created very high monomeric friction. Monomeric friction coefficients as the strength of immobilization may be used to correlate the increasing elasticity, longer relaxation time and molecular confinement quantitatively

References:
[1] Park JH, Jana SC. Mechanism of exfoliation of nanoclay particles in epoxy-clay nanocomposites. Macromolecules 2003;36:2758–68.
[2] Lu HB, Nutt S. Restricted relaxation in polymer nanocomposites near the glass transition. Macromolecules 2003;36:4010.
[3] Luo J-J, Daniel IM. Characterization and modeling of mechanical behavior of polymer/clay nanocomposites. Comp Sci Technol 2003;63:1607–16.
[4] Samandari SS, Khatibi AA, Basic D. An experimental study on clay/epoxy. nanocomposites produced in a centrifuge, Compos Part B-Eng 2007;38:102–107.
[5] 8 Liu, J., Cao, D. , , Zhang, L, and Wang W.,(2009)Macromolecules 9, 42, pp2831-2842
[6] Khatri BS, McLeish TCB (2007) Rouse model with internal friction: a coarse grained framework for single biopolymer dynamics. Macromol 40:6770–6777

418765 Field-Effect Control of Electrochemical Behaviors on Graphene Electrodes

Abstract:

We report here the influence of the field-effect on electrochemical behaviors on graphene electrodes. First, we will show the field-effect enables direct estimation of quantum capacitance and double-layer capacitance of the graphene electrode. The Fermi-level position E_F of the electrode can be tracked by measuring its electrochemical potential with respect to a reference electrode immersed in the electrolyte phase. Thus, the potential δ required to fill the energy band of graphene with charge carriers (*i.e.,* electrons or holes) can be directly measured from the electrochemical potential change while the carrier density in the graphene channel is independently controlled with back-gate bias. In turn, the quantum capacitance C_Q (*i.e.,* the DOS) of graphene can be estimated from this information. Furthermore, when the carriers in the graphene channel are induced through the electrolyte-gate (*i.e.,* the counter electrode immersed in the electrolyte phase), the potential required to charge the electric double-layer at the graphene/electrolyte interface can be conveniently separated from δ by comparing electrochemical potentials of graphene during back- and electrolyte-gating. Furthermore, we will show the rate of electrochemical reactions on graphene surface can be controlled by the field-effect. This *gate-tunable* graphene electrodes are useful testbeds for understanding transport, electronic structure, and electron transfer kinetics on graphene. The approaches we follow here should also be applicable for investigating other 2D materials such as MoS_2 and ultrathin layers of conventional semiconductors.

419322 Enzymatic Cascade Controlled By Organic-Inorganic Microspheres

Abstract:

We report a simple precipitation method for the preparation of multi-enzyme-inorganic hybrid microspheres with a controlled spatial distribution of enzymes. By this localization, the immobilized multi-enzyme system displays obviously enhanced overall activity compared to the corresponding free enzymes and randomly distributed multi-enzyme incorporated microspheres. As a demonstration, the spatially controlled horseradish peroxidase-glucose oxidase incorporated microspheres allow for sensitive detection of glucose in solution.

419334 Imprinted Enzyme Nanogel for Catalysis in Organic Solvents

Abstract:

Nanobiocatalyst with the advantages of high surface area, increased enzyme loading and reduced mass transfer resistance offers great opportunity for the green synthesis of fine chemicals and pharmaceuticals. A lot of enzymatic reactions are carried out in organic solvents because of the increased solubility of substrate and eliminiation of hydrolysis reactions and water-dependent side reactions. However, enzyme has low apparent activity in organic solvents. Here we report a new strategy of 'imprinting' enzyme-encapsulated polyacrylamide nanogels to achieve high enzymatic activities in organic media.

First, lipase from Thermomyces lanuginosus (Lipozyme TL 100L) was encapsulated in polyacrylamide nanogle by a two-step procedure including surface acryloylation and in-situ aqueous polymerization. Then the lipase nanogel was lyophilized in the presence of polyethyleneglycol (PEG) or a substrate, palmitic acid, which was then removed by solvent extraction after lyophilization.

The imprinted lipase nanogel showed a 4.6-fold activity compared to native lipase. Further investigation on the mechanism of the increased enzymatic activity suggested that PEG molecule activates lipase during the imprinting process. Substrate imprinting increases the affinity of polymer network to the substrate and favors the substrate uptake, thus gives a higher apparent activity. The imprinting treatment using PEG and substrate simultaneously gave the maximum increment in enzyme activity and enabled one-step synthesis of chloramphenicol palmitate, an antimicrobial, with an yield of ~99% and purity of ~99% within 6 h, indicating the imprinted lipase nanogel is an appealing catalyst for enzymatic catalysis in organic media.

Magnetic nanoparticle was covalently attached to lipase and then encapsulated in polyacrylamide nanogle thus combines the high stability of lipase nanogel and easy recovery of magnetic nanoparticle. The magnetic enzyme nanogel was then imprinted by PEG and the substrate to increase its catalytic activity in organic solvent. The increased activity, enhanced thermal stability and ease of reuse essentially ensure an expanded application of spectrum of enzymatic catalysis in organic solvents.

References

1. Wang, R.; Zhang, Y.; Huang, J.; Lu, D.; Ge, J.*; Liu, Z.* Substrate imprinted lipase nanogel for one-step synthesis of chloramphenicol palmitate. Green Chemistry 2013, 15, 1155-1158.

2. Wang, R.; Zhang, Y.; Lu, D.; Ge, J.*; Liu, Z.*; Zare, R. N.* Functional protein-organic/inorganic hybrid nanomaterials. WIERs: Nanomedicine and Nanobiotechnology 2013, 5, 320-328.

3. Wang, R.; Zhang, Y.; Ge, J.*; Liu, Z.*. Activation of enzyme nanogel in organic solvents by PEG–substrate joint imprinting. RSC Advances, 2014, 4, 40301-40304.

413204 Elaeis Guineensis (Oil Palm) Kernel Extract Mediated Green Synthesis of Sub-10nm Gold Particles

Abstract:

The present work reports synthesis and characterization of stabilized gold nanoparticles via bioreduction of a gold precursor using *Elaeis guineensis* (oil palm) kernel extracts and without addition of any surfactant or capping reagent. The biosynthesized gold nanoparticles were characterized using UV-Vis, TEM and FTIR measurements. UV-Vis spectra of the gold sol indicate SPR peaks in the range of 530-550 nm. TEM image analysis indicates formation of predominantly spherical nanoparticles with mean diameter of 5.71 ± 1.13 nm. TEM diffraction analysis indicates orthorhombic and polycrystalline structures (2,0,0 and 1,1,0) of the synthesized nanoparticles. FTIR analysis identified that polyphenolic O-H, C-H aromatic and C=O amide moieties in the phytochemicals are responsible for bioreduction of the precursor. Effects of kernel to water mass ratio (1:5-1:20), concentration of gold precursor (0.074-0.294mM), temperature (298-373 K) and pH (3.2-9.3) of the reaction medium on morphology, size distribution, stability and growth kinetics of the gold nanoparticles have been investigated. This study proves the feasibility of using oil palm kernel extract for synthesis of stabilized sub-10 nm gold particles that are potentially cheap and biocompatible.

423499 Enzyme-Polymer Conjugates Synthesized in Reverse Emulsions with Enhanced Activities

Abstract:

A simple approach for synthesizing enzyme-polymer conjugates is developed by using aldehyde-functionalized Pluronic F-127 as the reactive surfactant in the reverse emulsions. The applicability of this method was validated by using horseradish peroxidase (HRP), *Candida rugosa* lipase (CRL) and *Candida antarctic* lipase B (CALB) as model enzymes. Compared to native enzymes, the as-synthesized conjugate shows enhanced activity and stability in organic media and shows a great potential for applications in biocatalysis, biomedical and biosensing.

427133 Ionic Current Rectification in Polyelectrolyte-Modified Conical Nanopores

Abstract:

Ionic current rectification in polyelectrolyte-modified conical nanopores

Hou-Hsieh Wu,* Chih-Yuan Lin, Jyh-Ping Hsu

Department of Chemical Engineering, National Taiwan University, Taipei, Taiwan 10617

ABSTRACT

Due to their specific geometry, conical nanopores are capable of causing ion current rectification (ICR). Extending previous analyses, we consider two types of nanopore are modeled. In type 1 nanopore only the inner surface of a nanopore is functionalized by a PE layer; in type 2 nanopore both the outer and the inner surfaces of a nanopore are functionalized. A thorough numerical simulation is conducted to examine the influences of the double layer thickness and the density of the functional groups of the PE layer of a nanopore on its ICR behavior. We show that type 1 nanopore exhibits a more significant CP than type 2 nanopore does. In addition, as the bulk salt concentration varies, the rectification factors (RF) for types 1 and 2 nanopore are different significantly. If the salt concentration is low, an inversion in RF is observed in type 1 nanopore, which can be explained by the local conductivity near the nanopore tip end.

* Presenting author e-mail: r03524075@ntu.edu.tw

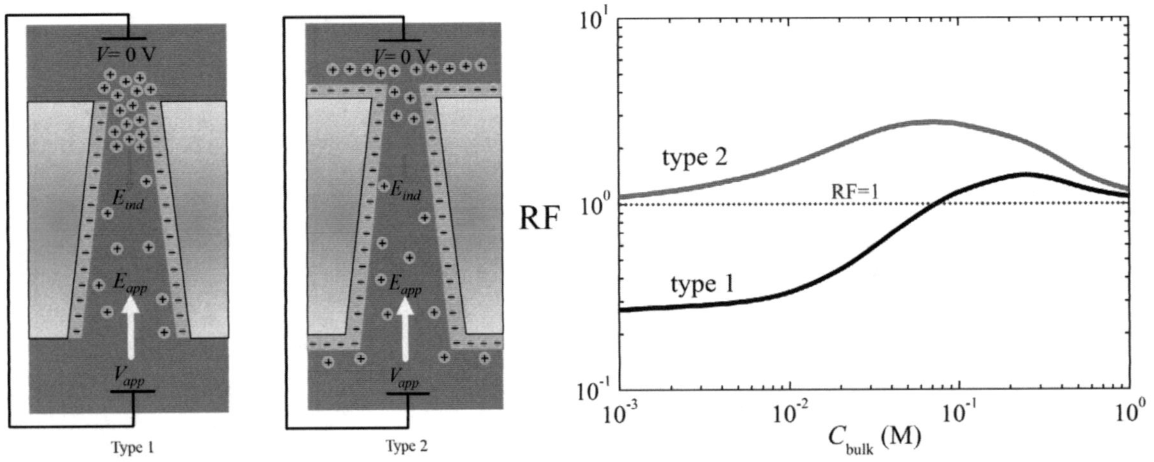

- Abstract Text

420436 Evaluation of the Cancer-Preventive Effect of Resveratrol-Loaded Nanoparticles on the Formation of Tumor Spheroids

Abstract:

Evaluation of the cancer-preventive effect of resveratrol-loaded nanoparticles on the formation of tumor spheroids

Elisa A. Torrico-Guzman[1], Mitchell Gravely[1], Samantha A. Meenach[1,2]

[1]Department of Chemical Engineering, [2]Department of Biomedical & Pharmaceutical Sciences, University of Rhode Island, Kingston, RI

Cancer is the second leading cause of mortality in the U.S. following heart disease. In 2015 it is expected that 1,658,350 new cases of cancer will be diagnosed in the U.S.. Nearly one-third of cancer-related deaths in the U.S. each year are caused by tobacco smoking. Moreover, one-third of the cancer cases in the U.S. are related to obesity, physical inactivity, and/ or poor nutrition, which can be prevented [1]. Chemoprevention is the use of dietary or pharmaceutical agents to reverse or inhibit the carcinogenic process resulting in a decreased cancer risk [2] and has been successfully applied to common malignancies. The FDA has approved some drugs as cancer prevention agents such as: Tamoxifen for breast cancer and the Human Papiloma Virus (HPV) vaccine for cervical intraepithelial neoplasia and cancer. More than 200 other drugs are in clinical trial not yet with reported results [3].

Resveratrol (RSV) is a natural phenol with antioxidant, anti-inflammatory properties, and beneficial effects against cancer. It can be found in many plants including red grapes, peanut butter, dark chocolate and blueberries. A study by Dolfini [4] showed that RSV induces drastic growth inhibition on a human breast cancer cell line *in vivo* and *in vitro*assays. Furthermore, RSV-loaded gel nanoparticles enhanced intracellular uptake and cytotoxicity in a NCI-H460 lung cancer cell line [5]. Moreover, RSV loaded into nanostructures lipid carriers (NLC) exhibit higher penetration into the skin in comparison to solid lipid nanoparticles (SLN) [6], which demonstrates the improvement of RSV in nanoparticle (NP) systems.

In this project we have evaluated the ability of resveratrol-loaded nanoparticles to prevent the growth of tumor spheroids. RSV was loaded into nanoparticles based on a biodegradable polymer acetalated dextran (Ac-Dex). Then, free RSV and RSV NP were tested in two-dimension (2D) and three-dimension (3D) cell culture in order to evaluate the viability of the A549 lung cancer cell line. So far, RSV exerted an inhibitory effect over the growth of cancer cells, showing a promising alternative as a chemopreventive drug, which could palliate the incidence or mortality from cancer.

REFERENCES

1. Siegel, R.L., K.D. Miller, and A. Jemal, *Cancer statistics, 2015.* CA Cancer J Clin, 2015. **65**(1): p. 5-29.

2. Sporn, M.B., et al., *Prevention of chemical carcinogenesis by vitamin A and its synthetic analogs (retinoids).* Fed Proc, 1976. **35**(6): p. 1332-8.

3. Davis, J.S. and X. Wu, *Current state and future challenges of chemoprevention.* Discov Med, 2012. **13**(72): p. 385-90.

4. Dolfini, E., et al., *Resveratrol impairs the formation of MDA-MB-231 multicellular tumor spheroids concomitant with ceramide accumulation.* Cancer Lett, 2007. **249**(2): p. 143-7.

5. Karthikeyan, S., et al., *Anticancer activity of resveratrol-loaded gelatin nanoparticles on NCI-H460 non-small cell lung cancer cells.* Biomedicine & Preventive Nutrition, 2013. **3**(1): p. 64-73.

6. Gokce, E.H., et al., *Resveratrol-loaded solid lipid nanoparticles versus nanostructured lipid carriers: evaluation of antioxidant potential for dermal applications.* Int J Nanomedicine, 2012. **7**: p. 1841-50.

423630 Preparation and Colloidal Stability of Different Binary Mixed Self-Assembled Monolayer Functionalized Au Nanocolloids

Abstract:

Preparation and Colloidal Stability of Different Binary Mixed Self-Assembled Monolayer Functionalized Au Nanocolloids

Aşkın Gizem Çitak*, Özge Çalışkan, Fuad Bayramov, Deniz Yılmaz, Ramazan KIZIL

Istanbul Technical University, College of Chemical and Metallurgical Engineering, Chemical Eng. Dept. Maslak, 34469 Istanbul-Turkey. Contacts: citaka@itu.edu.tr and kizilr@itu.edu.tr

Au nanoparticles with diameters below 20 nm are one of the most versatile nanomomaterials in nanotechnology, since they can easily be incorporated into the systems of photovoltaics, catalysis and nanodevice fabrication. The unique optical and electronic properties of these nanoparticles can be tailored by changing their size, shape or aggregation state and most importantly the surface chemistry via kovalent bonding of organic or polymeric molecules. Chemisorption of alkanethiols on gold surface has long been exploited to form self-assembled monolayers (SAM) on the surface of nanocolloids. Chemical functionalization of Au nanocolloids with alkanethiols significantly changes the rates of flocculation of the dispersion depending on the pH, chain length and terminal functional group of the alkanethiol to be attached covalently to the surface. Thus, colloidal stability of Au nanocolloids depends highly on the chemical nature of capping agents and the pH of dispersion.

Typically, a single type alkanethiol dissolved in alcohol is chemisorbed by colloidal Au in aqueous phase to form stable SAM-capped colloidal dispersion. The most extensively used alkanethiols for SAM formation are 16-mercaptohexadecaonic acid (MHDA) and 11-mercaptoundecanoic acid (MuA), mainly due to their good association characteristics in water. Formation of SAM on Au surfaces from mixed alkanethiol molecules instead of a single one is expected to give an additional chemical functionality to Au nanocolloids, which may offer advantages of precluding steric effects and enabling the particles possess multifunctional reactive terminal groups. However, a potential risk of preparing mixed-SAM on nano-sized colloidal gold is the loss of colloidal stability or fast agglomeration due to the loss of surface charge.

This poster presentation covers investigation of the colloidal stability of Au nanocolloids functionalized both with a single and binary alkanethiols of different chain lengths and function end groups. For this purpose we used three different MHDA, MuA, mercaptohexanol (MHOL) alkanethiols for making single and binary mixed-SAM capped Au nanocolloids prepared by citrate reduction. Alkanethiols were dissolved in ethanol and added dropwise to the aqueous colloidal dispersion of 13 nm Au nanoparticles. Before the chemisorption of alkanethiols, Au nanoparticles were treated with tween solution for physical adsorption of the surfactant molecules onto the metallic surface which are expected to be replaced by alkanethiols by means of chemical interaction. As MuA and MHDA form very stable SAMs on the surface Au nanoparticles when they are used individually, MHOL did not form any SAM on the particles by itself. The colloidal stability of single SAM derivatized Au nanoparticles by MHDA and MuA was satisfactory enough. Mixed-SAM of binary alkanethiols of different functional end groups (MuA+MHOL and MHDA+MHOL) were prepared mixed with an equimolar mixture of two different alkanethiols dissolved in ethanol. UV-vis spectrum suggested that colloidal stability still remains almost the same when binary-mixed SAMs are prepared and interestingly an improved stability was detected for MuA+MHOL mixed-SAM functionalized Au nanocolloids. The amounts of flocculated nanoparticles were quantified by integrating the spectrum curve from 600 to 800 nm using composite Simpson method. We noticed a considerable red shift in the surface plasmon peak of Au nanoparticles when they are functionalized with mixed-SAMs. The pH effect on these binary SAM systems was also investigated to determine the most stable pH range for such colloidal systems. The poster will summarize findings on the collodial stability as a function of pH and the chemical nature of alkanethiols and provide a discussion of the advantages of using binary mixed-SAMs for functionalization of Au nanocolloids.

428363 Determination of Size and Concentration of Unlabeled Nanoparticles Using a Tunable Conical Nanopore

Abstract:

The advance of nanofabrication techniques opens up new opportunities for preparing nanopores with various geometries for potential applications, such as nanofluidic diodes and single-molecule biosensing. In this study, soft nanoparticles, composed of Hydrogenated Lecithin (HL), were prepared by the thin film hydration method and their physicochemical properties, including size, zeta potential and particle concentration, were analyzed using a tunable nanopore made of polyurethane. Compared with conventional analytical tools like TEM and dynamic light scattering, the adopted nanopore sensing technique is suitable for any types of nanoparticles and more sensitive and accurate without any pretreatment procedures because the foundation of measurement relies on the translocation of single nanoparticles through a nanopore. Results show that the HL nanoparticles are slightly negatively charged and their size ranges from 60 and 70 nm, which are consistent with the experimental data measured by the TEM and dynamic light scattering and the experimentally reported data in the literatures.

429082 Tuning Ionic Current Rectification in a Conical Nanopore By Surface Modification of Polyacrylic Acid Brushes

Abstract:

In cell membranes, the ion channels are asymmetrically distributed and possess various biological functions due to the functional groups, such as carboxylic acid, on the channel wall. Moreover, the extracellular and extracellular solution environments, such as pH and salt concentration, in biological cells are typically different, resulting in several unexpected and interesting ion transport phenomena. Inspired by this, the biomimetic, conical nanopores with biomimetic functional groups, polyacrylic acid(PAA) brushes, are prepared through a vapor-based plasma surface modification of a polymer nanopore made of polyurethane, which reveals different charge properties under different solution pHs. Several key factors, including pH, background salt concentration, and pH and salt gradients, on the ionic current rectification (ICR) of the PAA brushes-modified nanopore are investigated. Results show that in addition to pH and background salt concentration, the ICR can be regulated by pH and salt gradients through the nanopore.

428537 Shaping Catalysis through Tailored Nanostructures

Abstract:

Catalyst is a substance that participate chemical reactions by increasing the reaction rate while the catalyst itself is not being consumed. Most catalysts used in industrial processes are based on nanomaterials, because of their high surface-to-volume ratio therefore high material utilization at nanoscale. This presentation covers several recent development of nano-catalysts in energy conversion and storage. The performance of catalysts has been improved through tailoring the sizes, compositions, and morphologies of nanomaterials. This presentation outlines 1) synthesis and preparation of catalytic nanomaterials, including Pt and Pt-alloy catalysts, 2) the connection between shape (morphology) of nanomaterials and the catalytic properties, 3) correlation between electrocatalyst composition profile and electrocatalysis, 4) effective catalyst design strategy, 5) balancing activity and durability by tuning nanostructure, 6) chemical modification of surface to tailor catalytic properties, and 7) superstructures as novel research platform to study catalysis. This presentation emphasizes the novel approaches that improve catalysts performances such as activity and durability through careful tailoring of nanomaterials.

429128 Preparation and Characterization of High-Performance Graphene Oxide and Polyether Amine Composite Films

Abstract:

There are many ways to synthesis graphene material, and graphene has been widespread concern in all sectors because of its excellent performance. Large-scale production of graphene and applied to our natural life is a very difficult problem. We'll focus on the direction of the graphene and polymer composite materials, suction filtration graphene oxide solution for preparing graphene oxide film, and then through the infiltration of polyether amine solution, after high temperature rapid crosslinking, preparing a high-performance composite materials. Through a series of characterization techniques to demonstrate cross-linked, such as X-ray diffraction, Fourier transform infrared spectroscopy and so on. The composite materials can greatly enhance the mechanical properties after crosslinking, mainly in tensile strength and elastic modulus. The preparation of such composite materials provides new ideas for our future graphene materials.

429555 An Optimization Approach for the Selection of CNTs Synthesis Route

Abstract:

Carbon nanotubes (CNTs) are one of the most exciting discoveries in nanoscale sciences because they have superior properties with respect to other materialsdue to their superior properties for use in several range of applications in materials science, electronics, energy management and many other fields. CNTs can be manipulated chemically and physically in very useful ways and their final properties and dimensions strongly depend on the parameters and the nature of the method used for their synthesis, this is why many methods of synthesis have been reported. Controlling the size of a nanomaterial is critical for its application because the characteristic length of a material significantly affects its mechanical, thermal, chemical, and particularly electronic properties. Several key parameters and operating conditions have to be manipulated to produce CNTs with suitable properties for future applications and it is evident that all these methods of synthesis need to be optimized.

This paper presents an optimization approach for determining the best synthesis method and their corresponding operating conditions for synthesizing carbon nanotubes accounting for technical and economic issues as objective functions for a desired application. Proper correlations for the interaction between the considered variables are proposed, and these correlations are based on experimental data taken of the literature reports. The optimization formulation is a mixed-integer nonlinear programming problem. A case study considering three synthesis methods of carbon nanotubes (arc discharge, laser ablation and chemical vapor deposition) is presented. The results show that the proposed model can be useful to determine the best method and operating conditions with the maximum efficiency, quality and safety at the minimum cost. The results provide that at ~100% the cost per gram of carbon nanotubes obtained for arc discharge is 18.645 USD, for laser ablation is 14.438 USD and chemical vapor deposition of 3.419 USD using camphor as carbon source and 0.374 USD using turpentine.

430350 Antibacterial Electrospun Mats Based on Poly(D,L-Lactide) Nanofibers and Zinc Oxide Nanoparticles

Abstract:

PLA nanocomposites with antibacterial properties are of high potential interest as food packaging biomaterials. The investigations carried out in this study include the rheological properties of solutions used for electrospinning the nanofibrous mats as well as, the morphological and antibacterial evaluation of the mats based on poly(D,L-Lactide) nanofibers loaded with different zinc oxide nanoparticles. A detailed rheological analysis of the solutions used for electrospinning was carried out in order to understand the achieved morphology and structure of nanofibers. The effect of ZnO nanoparticles was investigated for various PLA solutions and solvents. The polymer solutions were prepared at various loading level of ZnO: 0.4, 0.8, 1, 3 and 5 wt % to the weight of the polymer. Two solvent systems were used : trifluoroethanol and a mixture of dichloromethane/ trifluoroethanol (50:50 v/v). The PLA solutions with low loading level of ZnO nanoparticles showed higher viscosity (η) than the neat PLA. They also showed a maximum value of η at 3 wt % content. Solutions containing more than 3 wt % ZnO exhibited a notable decrease in viscosity, which could be probably due to the PLA degradation. The morphological-characterization of the electrospun nanofibers was performed using scanning electron microscopy (SEM). The SEM images showed a network of smooth randomly oriented PLA nanofibers. ZnO agglomeration was observed on the surface of the nanofibers. The antimicrobial efficiency of these nanocomposite fibers against *E. coli (DH5 α)* as Gram-negative bacteria and *Staphylococcus aureus* as Gram-positive bacteria was evaluated using the dynamic method. The antibacterial results showed that the addition of zinc oxide nanoparticles reduced slightly bacteria growth on PLA/ZnO nanofibers.

432535 Graphene-Embedded Carbon Nanofibers with High-Loading Surface-Attached Platinum Nanoparticles for Super Efficiency Dye-Sensitized Solar Cells

Abstract:

Dye-sensitized solar cells (DSC cells) have attracted extensive attention since its invention by Grätzel in 1991 as a potential alternative to silicon solar cells due to the ease of fabrication and the relative low cost. . Existing DSC cells generally use Pt as counter electrode (CE) because Pt possesses excellent catalytic activity for I_3^- reduction. Nonetheless Pt can be corroded by electrolyte with time. In the meantime, the high cost and scarcity of Pt makes it a less desirable candidate for use in industry. Carbon based materials including graphite, graphene, carbon black, carbon nanotube, and carbon nanofibers have been investigated as an alternative to Pt because of their strong catalytic effect, high electrical conductivity, and corrosion resistance.

An innovative nanomaterial for CE of DSC cells was developed by integrating graphene into carbon nanofibers through electrospinning followed by growing Pt nanoparticles on nanofiber surface through a redox reaction. Graphene-embedded carbon nanofibers (GCNFs) were produced by electrospinning polyacrylonitrile (PAN) with graphene nanoplatelets followed by stabilization and carbonization. GCNFs with high loading surface-attached platinum nanoparticles (GCNFs-PtNPs) were subsequently prepared by redox reaction and spray-deposited onto fluorine doped tin dioxide (FTO) glass as counter electrode (CE) for dye sensitized solar cells (DSC cells). Graphene in carbon nanofibers and Pt nanoparticles on carbon nanofiber surface demonstrated synergistic effect on performance of DSC cells as this hierarchical nanomaterial GCNFs-PtNPs was employed as CE in DSC cell. Compared to DSC cell with conventional Pt CE, the DSC cell with GCNFs-PtNPs CE showed significant improvement in solar energy conversion efficiency, which reached ~10 %. This is the highest conversion efficiency so far with carbon nanofiber based CE. Our research established an economic approach to acquire decent CE materials for super efficiency DSC

cells.

434014 Patchy Particle Under Electric Fields

Abstract:

We introduce patchy particles with well-defined geometries and investigate their packing and propulsion behavior under AC electric fields. Polymeric colloidal particles equipped with two, three, or four surface patches, which endow the particles with 2-fold, 3- fold, or tetrahedral symmetries, form 1D chains, 2D layers, and 3D packings when polarized by the AC electric field. Two-patch particles pack into the *cmm* plane group and 3D packings with *I4mm* space group symmetry. Three-patch particles form chains having a 2(1) screw axis symmetry, but these chains pair in a manner such that each individual chain has one-fold symmetry but the pair has 2(1) screw axis symmetry. Surprisingly, some chain pairs form unanticipated double-helix regions that result from mutual twisting of the chains about each other, illustrating a kind of polymorphism that may be associated with nucleation from short chain pairs. Larger 2D domains of the three-patch particles crystallize in the *p6m* plane group with alignment (with respect to the field) and packing densities that suggest random disorder in the domains, whereas four-patch particles form 2D domains in which close-packed rows are aligned with the field.

When a layer of gold is site-specifically plated on the patches, each of the aforementioned patchy particles is capable of "swimming" or propelling under electric fields driven by induced-charge electroosmosis. The propulsion behavior, for example the speed of movement and the ability to steer can be regulated by tuning the particles' patch configurations. As a proof of concept, two-patch particles are primarily demonstrated. For particles with bent patches of the same size, the speed of swimming is related to the patch angle and the size of the particle body; for particle with a small and a big patch, the speed of swimming is mainly determined by the big patch, while the steering and curvature is dominated by the size and position of the small patch.

434028 Effects of Temperature and Pore Size on the Nanofluidic Reverse Electrodialysis: Salinity Gradient Power Generation

Abstract:

Salinity gradient power is a promising, challenging, and readily available renewable energy, in human environment. Various methods for harvesting this clean energy have been proposed. Among these, the reverse elecrtodialysis based on nanochannels/nanopores seems promising and has a great potential. For the case of a charged nanopore because the diffusive flux of cations is different from that of anions, electrokinetic energy power can be generated. Since the ionic behaviors depend highly on the temperature, so is the associated power generation processes. Inspired by this, we conduct a thorough numerical simulation on the temperature influence on the nanofluidic reverse electrodialysis (NRED). We show that if the nanopore radius is sufficiently large, the conversion efficiency increases with increasing temperature. However, if the radius of the nanopore is small, the conversion efficiency decreases with increasing temperature. These behaviors are attributed to the effect of double layer overlapping. For large nanopores, the conversion efficiency decreases rapidly with increasing concentration gradient, and might show a local minimum due the contribution of coions. The results gathered in our study provide valuable and necessary information for improving the power generation efficiency in NRED as well as designing relevant units in future renewable energy plants.

434662 Nanostructured Complex Oxides Catalysts for Reduction Soot Ignition Temperature

Abstract:

Nanostructured complex oxides for catalytic oxidation of various fuels are known to be promising and environmentally friendly catalysts that can be potentially used to increase efficiency of coal based power plants. The goal of this study is to identify nanostructured complex oxides catalysts such as $LaCrO_3$, $PrCrO_3$, $La_{0.8}Cr_{1-x}Li_xO_3$, $CeCrO_3$, $Ce(Li)CrO_3$ and $Ce_{1-x}Zr_xO_2$ where x=0.1-0.7 that can be able to reduce coal ignition temperature actually to decrease combustion activation energy. These candidate catalysts were selected due to their high oxygen exchange and storage ability, high temperature stability, the ability to tune their properties by doping, and their low price relative to that of noble metals. An extensive set of complex oxides were synthesized and tested as potential catalyst for soot combustion. We synthesized powders of various complex oxides by two methods (i) Carbon Combustion Synthesis of Oxides (CCSO), which was developed and patented recently [1] and enables a rapid and easy synthesis; and (ii) Solution (metal nitrates-glycine) combustion synthesis (SCS) [2]. The catalytic activity of the as-synthesized powders was tested in a temperature-programmed combustion (TPC) apparatus. It consists of quartz fixed bed inserted in a reactor. The fixed bed was packed with a mixture of 0.1 g of amorphous carbon (45 nm), 1 g of the powdered catalyst and 10 g of silica pellets (0.3–0.7 mm in size). The catalyst/carbon/SiO_2 mixture was confined between two layers of quartz wool. A computer-regulated furnace heated the sample up to 700 °C at a heating rate of 5 °C/min. The effluent gas compositions were determined by a mass spectrometer. The $Ce_{0.5}Zr_{0.5}O_2$catalyst produced by CCSO and solution combustion shows the substantial performance with peak temperature of about 390 °C as opposed to the 670 °C of non-catalytic combustion. However, the preliminary experiments showed that the activity of the oxides synthesized by SCS was somewhat higher than those synthesized by CCSO.

References

1. K.S. Martirosyan and D. Luss, Carbon combustion synthesis of oxides, 2011, US Patent 7,897,135.
2. K.S. Martirosyan, and A.S. Mukasyan, Combustion Synthesis of Nanomaterials, In Dekker Encyclopedia of Nanoscience and Nanotechnology, Third Edition. CRC Press: New York, 2014, pp. 983–1001.

Clear TEM Image of Spherical Micelles of Sodium Conjugated Linoleate

Bei Wei, Ye Fan, Chao Chen, Yongchun Pan, Biao Jiang, Lingyu Bao, Meng Zhu,
Yun Fang*
(Key Laboratory of Food Colloids and Biotechnology, Ministry of Education; School of Chemical & Material Engineering, Jiangnan University, Wuxi, Jiangsu 214122, China)

Small spherical micelles may be the simplest self-assemblies of surfactants, which has become the fundament to further explore complex amphiphilic systems and life phenomenon. Although it is possible to theoretically infer that micellar diameter is about twice stretch length of a hydrophobic chain length of a surfactant (3-5 nm) and laser light scattering experiment provides some indirect evidence of the particle size, there is a lack of a powerful technique to give direct visualization for characterizing surfactant micelles. Due to spherical surfactant micelles having small particle size; unstable cluster structure and weak ability to scatter electrons, even if by a combination technique of the advanced electron microscopy with other special sample preparation techniques (cryogenic, freeze fracture, etc), the images of the small micelles are still unclear. To solve the problems in this paper, a polymerizable anionic surfactant, sodium conjugated linoleate (SCL), was synthesized and self-assembled in pH 13 at the concentration of 3 mmol·L^{-1} to obtain polymerizable SCL micelles, , and then the SCL micelles were tethered by moderated UV irradiation to insure its original fragile micellar structure maintained. The morphology of the self-crosslinked SCL micelles was characterized by transmission electron microscopy (TEM). The clearer TEM image of spherical micelles of surfactants by far was obtained successfully in virtue of polymerized SCL micelles, and provided the direct evidence of the micellar diameter indeed twice stretch length of a surfactant hydrophobic chain (3-5 nm). This investigation gives you a chance of seeing small spherical surfactant micelles clearly for the first time.

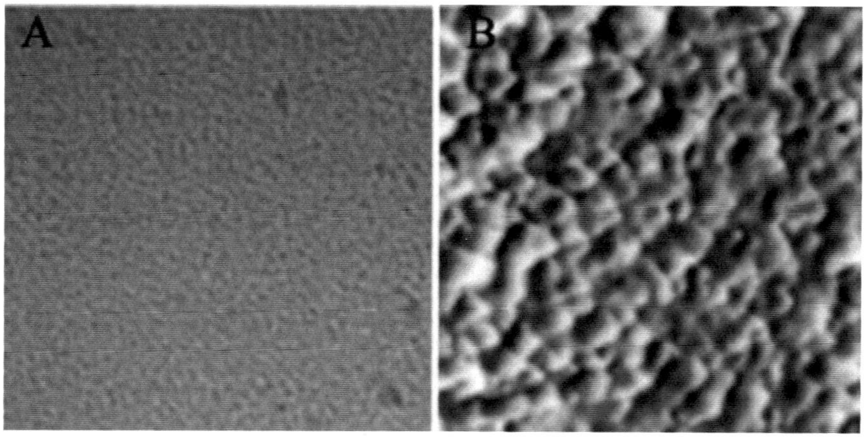

Figure 1. A) Cryo-TEM image of 0.05 mol·L^{-1} N-dodecyl-1, 4-diaminobutane at pH 6.99 (from Ref. 1); B) FF-TEM image of Tween80 and Brij 30 systems (X_{Brij30} = 0.54) in the presence of a common ionic liquid 1-butyl-3-methylimidazolium tetrafluoroborate (0.22 mol·L^{-1}) (from Ref. 2).

Keywords: sodium conjugated linoleate, self-assembly, micelle

References

1. Yang Y, Dong J, and Li X. Micelle to vesicle transitions of N-dodecyl-1, ω-diaminoalkanes: Effects of pH, temperature and salt. Journal of Colloid and Interface Science. 2012, 380(1):83-89.
2. Guo P, and Guo R. Ionic Liquid Induced Transition from Wormlike to Rod or Spherical Micelles in Mixed Nonionic Surfactant Systems. Journal of Chemical and Engineering Data. 2010, 55(9):3590-3597.

420228 Award Submission: Electronic Platform for Real-Time Multi-Parametric Analysis of Cellular Behavior Post Exposure to Single-Walled Carbon Nanotubes

Abstract:

Single-walled carbon nanotubes (SWCNTs) implementation in a variety of biomedical applications from bioimaging, disease targeting to controlled drug and gene delivery, has raised awareness of their potential toxicity. Nanotubes structural aspects which resemble asbestos, as well as their ability to induce cyto and genotoxicity upon interaction with biological systems by generating reactive oxygen species or inducing membrane damage, just to name a few, have led to focused efforts aimed to assess associated exposure risks prior their user implementation. In this study, we employed a non-invasive and real-time electric cell impedance sensing (ECIS) platform to monitor the cellular behavior of human lung epithelial cells upon exposure to a library of SWCNTs with user-defined physico-chemical properties. Using the natural sensitivity of the cells as primary transducers, we evaluated SWCNT-induced cellular changes in relation to cell attachment, cell-cell interactions and cell viability respectively. In particular, our results showed functional differences in the cellular behavior upon exposure to pristine and acid treated SWCNTs according to nanotube's length, surface chemistry, purity and agglomeration state. Our methods have the potential to lead to the development of standardized assays for risk assessment of other nanomaterials as well as risk differentiation based on the nanomaterials physico-chemical properties.

(274b) Award Submission: Silver Nanoparticle-Embedded Polymersome Nanocarriers for the Treatment of Antibiotic-Resistant Infections

Geilich, B. M., Northeastern University
Webster, T. J., Northeastern University

Originally presented on: 11/10/2015 9:00:00 - 9:30:00

Introduction

Antibiotics have been extensively used since their commercialization in the 1930s to treat patients suffering from a wide variety of infectious diseases. Unfortunately, however, these drugs have been used so prevalently over the last 80 years that the bacteria they were designed to kill have begun to evolve and adapt, rendering them ineffective. According to the Center for Disease Control, at least 2 million people acquire serious infections from antibiotic resistant bacteria each year, and over 23,000 die as a direct result. Even when alternative treatments exist, patients with antibiotic-resistant infections have significantly higher mortality rates, and survivors often have increased hospital stays and long-term complications. These infections cost an estimated $20 billion in excess direct healthcare expenses.1 Infections caused by Gram-negative bacteria are particularly difficult to treat because these bacteria have a robust and hydrophobic outer lipopolysaccharide membrane that can impede the influx of drugs into the cell.2 Unfortunately, the number of new antibiotic drugs in the pipeline has also been rapidly decreasing, largely due to the fact that new drugs are extremely expensive to bring to market, and antibiotics are less financially lucrative to develop when compared to treatments for chronic conditions. Thus, the need to develop alternative strategies to treat such antibiotic-resistant bacteria, while still utilizing existing drugs, has never been more urgent than today.

Over the past decade, interest in using nanomedicine-based approaches to combat difficult infections has rapidly grown due to the many advantages offered over conventional treatment with free antibiotics. This study explored encapsulating the drug inside nano-sized structures called polymersomes (that is, artificial vesicles made from biodegradable, high molecular weight, amphiphilic block co-polymers). These vesicles typically display a spherical morphology and are composed of hydrated hydrophilic coronas both at the inside and outside of a hydrophobic polymer membrane. This allows for hydrophilic bioactive materials to be loaded into the particle's aqueous core, and hydrophobic bioactive materials to be loaded into the particle's membrane bilayer. Along these lines, metallic nanoparticles have long been investigated as potential antibacterial agents due to their many unique physiochemical properties which are not present at the macro scale. Among these metals, silver is perhaps the most well-known for its antimicrobial effects. Recent studies have shown that there may even be a synergistic effect when silver nanoparticles and antibiotics are used simultaneously to treat a Gram-negative infection.3 However, there is little information regarding whether a combined treatment is sufficient to overcome bacteria which display genetic antibiotic resistance. Additionally, there has been almost no investigation into the effect of dually encapsulated antibiotics and nanoparticles.

Thus, for all of the above reasons, the objective of the present in vitro study was to design, characterize, and optimize a polymersome nanocarrier to co-localize and deliver both antibiotics and hydrophobic silver nanoparticles, while protecting the drug from hydrolysis by β-lactamase enzymes, into a single "nanoformulation" in order to kill antibiotic- resistant bacteria. Specifically, these particles were then tested for efficacy against a strain of *Escherichia coli* (*E. coli*) which had been genetically modified to be antibiotic- resistant.

Results

Silver nanoparticle-embedded polymersomes (AgPs) were synthesized using a modified stirred-injection technique. Monodispersed hydrophobic silver nanoparticles 5 nm in diameter were suspended in an organic solvent containing dissolved mPEG-PDLLA. This mixture was injected through a syringe atomizer at high speed into actively stirring phosphate buffered saline (PBS, pH 7.4) containing the antibiotic ampicillin. The resulting suspension was allowed to dialyze against PBS to remove the organic solvent and non-encapsulated drug.

Physicochemical characterization was performed to assess AgPs size, surface charge, and loading. Transmission electron microscopy (TEM) revealed polymersomes of highly uniform size and shape with clusters of silver nanoparticles embedded inside (Figure 1A). These silver clusters frequently appeared as a single layer of nanoparticles that was off-center from the nanoparticle core, suggesting that they may be intercalated into the membrane bilayer. Dynamic light scattering (DLS) indicated that the average hydrodynamic diameter was 104.3 nm ± 15.6 nm (Figure 1B). The AgPs surface was found to have a near neutral zeta potential of 0.315 mV ± 1.13 mV at pH 7.4. The number of silver nanoparticles embedded per polymersome was quantified from TEM images. The nanoparticles were shown to load in a normal tailed distribution with an average of 9.29 ± 6.07 silver nanoparticles per polymersome (Figure 1C) . The mass of silver loaded was estimated using the density of silver and the volume of a 5 nm sphere.

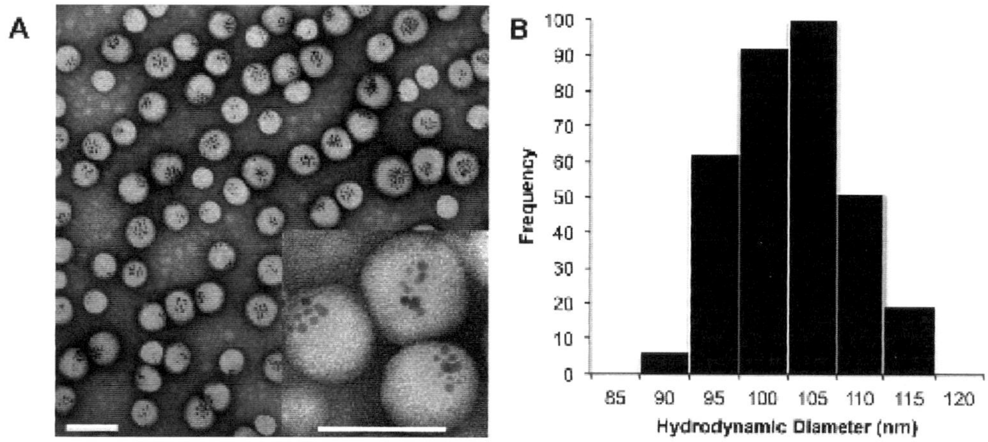

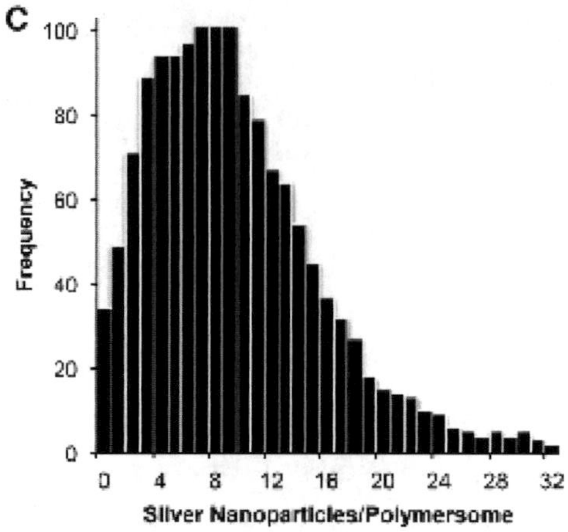

Figure 1: Physiochemical characterization of the AgPs

The growth and proliferation of a 106 colony forming units mL-1 (CFU mL-1) suspension of ampicillin-resistant *E. coli* was examined by measuring the optical density at 600 nm (OD600) for 24 hours following treatment with volumes of AgPs containing a silver : ampicillin (Ag : Amp) ratio of 1:0.28 (Figure 2A), 1:0.44 (Figure 2B), or 1:0.64 (Figure 2C). Ampicillin- loaded AgPs displayed significant bacteriostatic action against the *E. coli*, manifesting as a delay in the time taken to reach exponential growth phase. This response was dose-dependent, with higher concentrations of ampicillin producing a longer delay in bacterial growth. Bacteria treated with ampicillin concentrations above 55 µg mL-1 failed to proliferate within 48 hours. In the absence of silver nanoparticles, no bacteriostatic effect was observed for all ampicillin concentrations tested. This suggests that the presence of silver potentiates the therapeutic efficacy of ampicillin. AgPs without ampicillin like- wise produced no therapeutic benefit. Additionally, no significant differences were observed between bacteria treated with free ampicillin (200 µg mL-1), PBS, AgPs without ampicillin, and ampicillin loaded polymersomes (200 µg mL-1) without silver nanoparticles. When bacteria were treated with sub-optimal concentrations of AgPs, bacterial growth was always observed within 17 hours. The time to exponential phase was found to vary with both silver concentration and ampicillin loading.

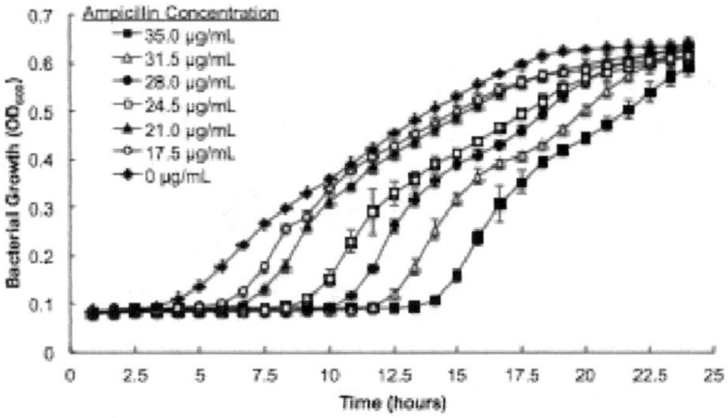

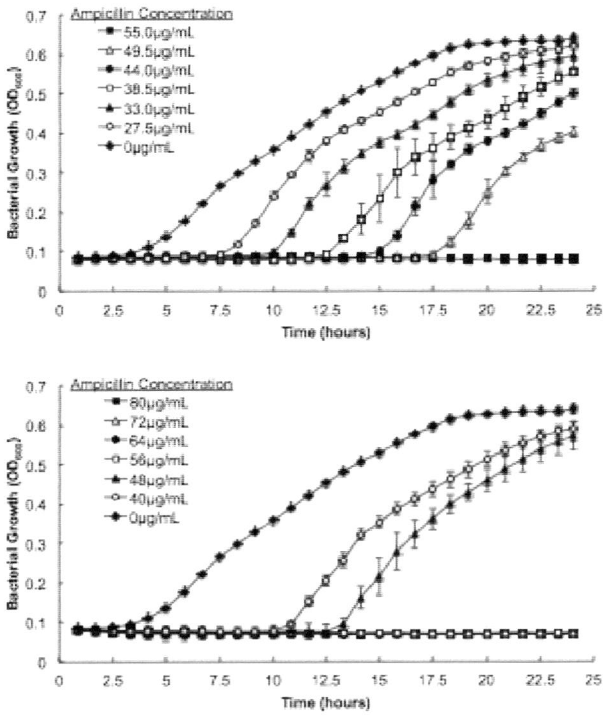

Figure 2: Bacterial growth inhibition from the AgPs

A Bliss Model was utilized to determine the degree of synergy for different silver and ampicillin combinations.4 Drug interactions were found be synergistic ($S > 0$) in all cases where ampicillin was supplied at concentrations of 24 µg mL-1 and above. At lower concentrations, no synergism was observed ($S = 0$). The degree of synergy was dose-dependent and increased with both silver and ampicillin concentrations. The therapeutic benefit of ampicillin reached a plateau at 50 µg mL-1 over a range of silver concentrations due to complete inhibition of bacterial growth. When the silver concentration was held constant, the degree of synergism was directly determined by the amount of ampicillin loaded.

Interactions between *E. coli* and AgPs were visualized using TEM (Figure 3). Indentation of the bacterial cell membrane was observed in regions of AgPs contact. Silver nanoparticles inside AgPs appeared to be polarized in an orientation perpendicular to the bacterial cell membrane, suggestive of hydrophobic interactions with the outer cell membrane. In order to assess physical intracellular changes caused by AgPs, cells treated with an intermediate particle concentration (44 µg mL-1 Amp, 1 Ag:0.44 Amp) for 24 hours were sectioned. Bacteria in contact with AgPs displayed significant protein aggregation and diffuse widening of the cell envelope. This phenomenon has been observed by other researchers following silver ion treatment, and has been shown to correlate with increased membrane permeability and protein misfolding due to disulfide bond disruption.4 Regions of the cell envelope with little to no AgPs contact appeared morphologically normal.

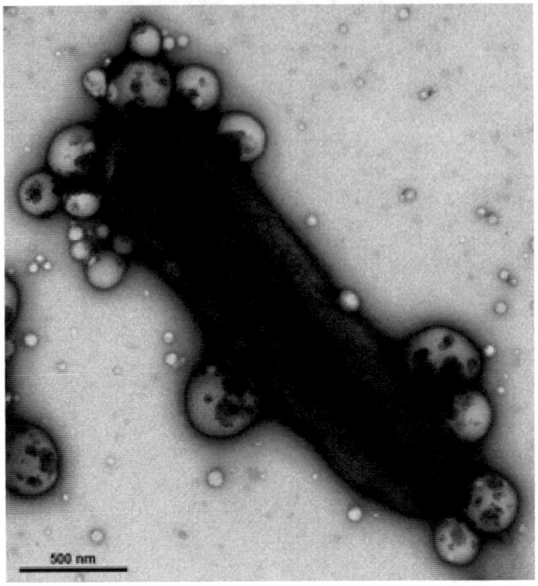

Figure 3: Transmission electron micrograph displaying the AgPs aggregating around an *E. coli* bacterium.

Discussion

This work has many important implications. A long-lasting, single-particle treatment capable of overcoming antibiotic resistance would be extremely beneficial in the clinic. This potential is compounded by the fact that the initial bacteria density found in vivo is rarely as high as the CFU studied here. It has been shown that silver nanoparticles are also effective at treating bacterial persister cells, and therefore AgPs may show promise for biofilm-forming infections.5 One of the most commonly occurring bacterial infections, urinary tract infection, is frequently caused and perpetuated by biofilm-forming (and often antibiotic-resistant) strains of *E. coli*.6 Additionally, the absence of a significant cytotoxic effect from AgPs towards human fibroblasts is a promising sign for toleration by mammalian cells. Reports of toxicity associated with nanoscale silver in vivo have been varied, and toxicity is generally considered to depend on nanoparticle size, concentration, and surface coating. As designed, the AgPs particle is easily loaded with a variety of aqueous drugs, and combinations thereof, opening avenues for the creation of a library of therapeutic particles.

References

[1] Antibiotic Resistance Threats in the United States, Centers for Disease Control and Prevention, 2013.

[2] J. M. Pages et al, The porin and the permeating antibiotic: a selective diffusion barrier in Gram-negative bacteria, *Nat. Rev. Microbiol.*, 2008, 6, 893.

[3] P. Li et al, Synergistic antibacterial effects of β-lactam antibiotic combined with silver

nano- particles, *Nano.,* 2005, 16(9), 1912.

[4] J. R. Morones-Ramirez et al, Silver Enhances Antibiotic Activity Against Gram-Negative Bacteria, *Sci. Transl. Med.*, 2013, 5(190), 190ra81. □

[5] D. Roe et al, Antimicrobial Surface Functionalization of Plastic Catheters by Silver Nanoparticles, *J. Antimicrob. Chemother.*, 2008, 61(4), 869–876.

[6] T. Mah, Biofilm-Specific Antibiotic Resistance, *Future Microbiology*, 2012, 7(9), 1061.

429789 Award Submission: Targeted Delivery of Microrna By Engineered Lipid Nanoparticles for the Treatment of Metastatic Breast Cancer

Abstract:

Introduction: Human epidermal growth factor receptor-2 (HER2) positive breast cancer occurs in 20-30% of breast cancer patients and has been attributed to increased disease aggressiveness, reoccurrence, and overall worse prognosis (Engel & Kaklamani, 2007; Vrbic, Pejcic, Filipovic, Kocic, & Vrbic, 2013). Biological therapies such as monoclonal antibodies/tyrosine kinase inhibitors targeting HER2 have made substantial improvement in patient outcome, but resistance to these treatments over time has represented a real clinical challenge and highlighted a need to develop novel therapeutic approaches (Vrbic et al., 2013). MicroRNAs (miRNAs) are endogenous regulators of gene expression that play a pivotal role in biological processes spanning from global homeostasis to disease onset and progression. The ability to manipulate/ induce cellular re-equilibrium of deregulated miRNA expression profiles by inhibition of oncogenic miRNA or overexpression of tumor suppressor miRNA is a promising cancer strategy, but is currently hindered in application by the lack of nonviral delivery systems. Here, we developed a lipid nanoparticle (LNP) platform surface coated with Hyaluronic Acid (HA) for the targeted delivery of mature tumor suppressor MicroRNA125a-5p as a long term treatment option for breast cancer. We believe the successful implementation of this platform has major implications on future gene therapy regimens for breast cancer.

Methods and materials: LNPs were created via the traditional dry lipid film method, mechanically extruded down to the nanoscale, and covalently surface decorated with HA by crosslinking chemistry. Particle size, charge, and surface characteristics were measured by dynamic light scattering, zeta potential analysis, and transmission electron microscopy. High efficiency entrapment of MicroRNA inside the aqueous core of the HA-LNP was achieved by utilization of the scale up friendly lyophilization-rehydration active loading procedure. Three breast cell lines were employed to show the targeting potential of the nanoparticles: MCF10A (normal breast tissue, CD44 -), SKBR3 (invasive breast cancer, CD44 +), and 21MT-1 (metastatic breast cancer, CD44 ++). Western blot, qRT-PCR, MTT, and wound healing assay were performed to measure protein, mRNA, cellular proliferation, and cellular migration respectively following transfection. Live cell confocal microscopy was applied to validate cytosolic delivery of the nanoparticles, as well as validate the evasion of lyosomal degradation.

Results: The optimized HA-LNP platform was constructed from biocompatible lipids, cholesterol, and a surface coating of HA to achieve a nanocarrier system 160 nm in diameter, a polydispersity index of 0.105, a surface charge of -38 mV, and stability in solution for over 90 days. The HA-LNP delivery platform actively targets patient-derived metastatic breast cancer cells (21MT-1) isolated from the metastatic pleural effusion over normal breast tissue via an intrinsic HA-CD44 mediated endocytosis event, and has the ability to escape from the intracellular endolysosomal pathway for potent gene silencing. Knockdown of the HER2 proto-oncogene at the level of transcription and translation was achieved following HA-LNP mediated transfection with MicroRNA125a-5p. In addition, the PI3K/AKT and MAPK hyperactivated signaling pathways, cellular proliferation, and migration potential were also potently suppressed. Furthermore, the therapeutic efficacy of MicroRNA125a-5p by the HA-LNP platform was demonstrated to be significantly improved by over three fold as compared to the current gold standard commercial transfection reagent Lipofectamine.

Conclusions: Our HA-LNP platform was fabricated from a scale up friendly technique and achieved rapid and targeted intracellular delivery of MicroRNA via a HA-CD44 mediated endocytosis leading to potent oncogene silencing. This study demonstrates the use of a lipid based nonviral carrier to delivery MicroRNA as a promising strategy. In addition, we believe our HA-LNP platform can be used to study the mechanism of action for various MicroRNA and advance the field of MicroRNA therapeutics.

433926 Award Submission: Nanoscale Hydrogel Coatings for Rapid Sorting of Circulating Tumor Cells: Relating Marker Expression to Isolation Yield

Abstract:

A link between the presence of tumor cells in systemic circulation and metastatic cancer progression has long been established. Observations of these circulating tumor cells (CTCs) has prompted efforts to develop a fluid biopsy, with the goal of detecting these rare cells in patient peripheral blood as surrogate markers of metastatic disease for partial replacement or supplement to invasive tissue biopsies. While such an assay could potentially provide valuable prognostic information among numerous other clinical advantages, the implementation of fluid biopsies has been critically hindered by current available technologies for rare cell isolation. Numerous platforms have been designed; most commonly magnetically-activated cell sorting (MACS), fluorescent-activated cell sorting (FACS), and a myriad of antigen-based microfluidic capture technologies. Yet, these have generally failed to support a reliable fluid biopsy assay, due to poor performance parameters such as low throughput, low purity of enriched antigen positive cells, and insufficiently low detection thresholds to detect poor expressed surface markers of target cell populations. In this work, we build on a recently developed rapid cell sorting technique called Antigen Specific Lysis (ASL) based on photo-crosslinked polymer film encapsulation to isolate tumor cells in suspension. Briefly, antibodies direct the specific binding of streptavidin-eosin-5-isothiocyanate (SA-EITC, type II photoinitiator) to surface markers of intact cells, then upon immersion in an aqueous monomer solution, visible light initiates surface-mediated radical crosslinking of nanoscale PEG diacrylate hydrogel films solely around the antigen positive population. Finally, anionic surfactants lyses virtually 100% of uncoated cells while fully encapsulated cells remain protected, and are then easily collected by centrifugation. This sorting process has shown >90% viability of sorted cells, and can be performed in 1-2 hours with inexpensive and disposable labware. Further, because this technique is not dependent on microfluidics or spatial mobility of target cells, we expect it is fundamentally capable of accommodating very large batch sized (10^{11} cells) with only marginal changes in processing time. Here, we focus on determining the response of process yield on the level of photoinitiator loading on target cell. The surface-mediated gelation response has been shown to be dependent on the density of photoinitiator at the initiating surface – therefore, we posit this is a critical parameter to assess for attaining high processing yield, particularly during the lysis-based negative depletion. Further, a lower initiator density threshold is seen in these surface initiated PEG diacrylate films, so that a lower threshold of marker expression for SA-EITC loading will determine successful coating and, ultimately, isolation of target populations. Particularly for CTC isolation, recent evidence suggests that the most highly potent and aggressive CTCs often downregulate epithelial markers that are commonly targeted for cell isolation, further highlighting the need to determine the effects of expression level on performance of this system. To methodically simulate lower expression levels of epithelial markers, we immunolabel Epithelial Cell Adhesion molecule (EpCAM) on a non-small cell lung cancer line H358 with biotin antibodies, then competitively bind incrementally varied ratios of streptavidin and streptavidin-photoinitiators and evaluate the percent yield of isolated cells versus relative photoinitiator loading determined by flow cytometry. We expect these data will inform robust targeting strategies with optimal yield performance to allow the development of ASL as reliable tumor cell isolation platform for clinical use.

422604 Novel Centrifugation-Assisted Preparation (CAP) of Additive-Free Carbon-Decorated Fe3O4 Electrodes with Superior Electrochemical Performances for Lithium-Ion Batteries

Abstract:

2015 AIChE Conference

Novel Centrifugation-Assisted Preparation (CAP) of Additive-Free Carbon-Decorated Fe_3O_4 Electrodes with Superior Electrochemical Performances for Lithium-ion Batteries

Jian Zhu (speaker); K. Y. Simon Ng; Da Deng*

Department of Chemical Engineering and Materials Science, Wayne State University,

5050 Anthony Wayne Dr, Detroit, MI, United States, 48202

Email: da.deng@wayne.edu

Abstract

Lithium ion batteries (LIBs) are the dominant power sources for portable electronic devices, and are penetrating into the market of electric vehicles and sustainable energy grid. There is an increasing demand for higher energy and high powder LIBs. The electrochemical performances of LIBs are mainly determined by the electrodes selected. Taking anode for example, the theoretical capacity of commercial anode of graphite is only 372 mA h/g, and its performance is about to saturate after few decades of development. To further increase the energy density, carbon must be replaced by environmentally friendly and inexpensive materials with high capacity and excellent reversibility. On the other hand, the current conventional procedure in the preparation of electrodes is time consuming involving multiple steps, including materials preparation, slurry formation, coating on to current collectors, and drying in vacuum. Various additives, including binder polyvinylidene fluoride (PVDF), conductive enhancer carbon black (CB), and organic solvent N-Methyl-2-pyrrolidone (NMP), are required in electrode preparation. The complicated procedure and various additives lead to the high cost of electrodes. It would be beneficial but is still challenging to develop facile methods for electrode preparation with simplified procedure, without using any additives, but with superior electrochemical performances.

Herein, we demonstrated a novel and facile centrifugation-assisted preparation (CAP) of additive-free negative electrode of carbon-decorated Fe_3O_4 materials for high energy LIBs (Figure 1a). Notably, the materials were directly synthesized on the current collectors. No additives, such as PVDF binder, carbon black conductivity enhancer and NMP solvent which are typically used in conventional electrodes, were required. Interestingly, the as-prepared of carbon-decorated Fe_3O_4 negative electrodes exhibited superior electrochemical performance with stable specific capacity of 800 mAh/g (more than two times of theoretical capacity of graphite at 372 mAh/g) for at least 240 cycles at 100mA/g; (Figure 1b) no significant capacity decay was observed when the testing currents were increased, up to 1500 mA/g, demonstrating excellent rate performances for high power LIBs (Figure 1c). We believe that our facile CAP method could be used for the preparation of electrodes, including cathodes and anodes with or without additives, based on other active materials.

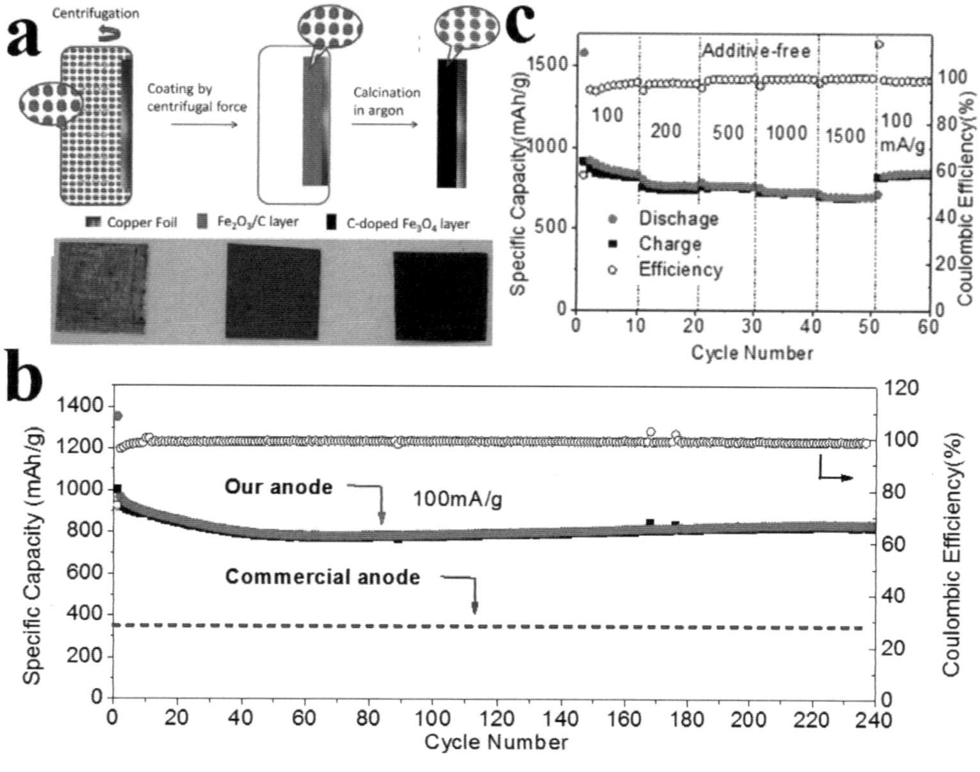

Figure 1. (a) Schematic of three steps of CAP of additive-free electrodes with carbon-decorated Fe_3O_4 on Cu current collector and corresponding optical images on the bottom; (b) excellent cycling performance tested at current of 100 mA/g for 240 cycles; and (c) rate performances and Coulombic efficiency at different currents from 100 to 1500 mA/g of the CAP electrodes of carbon-decorated Fe_3O_4 without any additives.

424677 Stacked SnS2/Graphene Nanocomposites with High Li+ Storage Capacity

Abstract:

There is strong recent interest in using SnS_2/graphene composites for reversible Li^+ storage. While SnS_2 and graphene are 2D nanomaterials; many of the composites reported thus far have a random 3D structure. We prepared a stacked SnS_2/graphene nanocomposite which could retain most of the properties associated with the original 2D nanomaterials and promote the reversible storage of Li^+ via the conversion reaction ($SnS_2+4Li^++4e^-\rightarrow Sn+2Li_2S$). Contrary to the prevailing view that the conversion reaction is irreversible, we will show that reversibility as high as 87.5% is attainable through careful materials design. The increased reversibility also imparted to this composite specific capacity, capacity retention and rate performance which surpassed previously reported SnS_2/graphene composites measured under the same conditions.

431500 Gas-Assisted Electrospray for Ultrauniform, Ultrafast, Instantly-Dry, and Binder-Free Electrode Preparation

Abstract:

Lithium-ion batteries has been a hot research focus in recent years due to their high energy density, flexible and lightweight design, long lifespan, and wide application in portable electronic devices and electric vehicles. Among all the battery-related researches, most focus on electrode materials, a small portion focus on binder, separator, electrolyte, and almost none focus on the preparation of electrode. For the tons of researches, materials preparation could vary a lot, but almost all used either drop cast or slurry coating for final electrode preparation. However, both drop cast and slurry coating methods involve toxic organic solvents and consist of multi steps including mixing, sonication, blading, calendaring, and long-time vacuum drying. Even after all these tedious steps, the as-prepared electrodes still suffer from exfoliation of the active material from current collector, leading to poor cycling stability. Though there are also some free-standing electrodes skipping the traditional electrode preparation procedure, they are not versatile and are often limited to certain kinds of materials.

To simplify the preparation process, and more importantly get better adhesion and high uniformity, we have developed a facile electrospray process from the already industrialized electrospining process. It can be applied to spray slurry (binder, additive carbon, active materials) from both organic and water system. The as-sprayed electrodes not only show no exfoliation upon bending, but also have lower resistance demonstrated by smaller impedance circle in comparison with conventionally made electrodes. The flow rate is 10 or 20 times fast of electrospinning, thus desired mass loading of electrodes can be achieved with only a few minutes of spraying. Excitingly, the electrodes made can be directly used for battery assembly, no extra drying step required. We further advanced the technique to a general binder-free electrode preparation approach where only graphene oxide is added. The graphene oxide sheets not only act as binder and conductive agents, but also act as "pocket" to hold active materials and help resist mechanic stress during charge and discharge, giving outstanding battery stability. In addition to battery electrodes, this approach can be also applied for many other electrode preparations such as supercapacitors, fuel cell, and solar cell etc.

428976 SnO2/Carbon Nanotube–Infiltrated Ni Nanofoams As 3D Anodes for Enhanced Performance of Lithium-Ion Batteries

Abstract:

Metallic nanofoams have been widely investigated as 3D current collectors in lithium-ion batteries (LIBs) due to their high stability and mechanical robustness, as well as the highly conductive 3D framework that can be coated with high-capacity anode materials. However, the extremely high porosities (> 99%) and low specific surface areas (< 10 m^2/g) of metallic nanofoams, limits the amount of active electrode material that can be deposited, thus reducing the maximum achievable areal energy density. Although the weight penalty associated with metallic nanofoams is a disadvantage, their highly conductive interconnected fibers and high void fraction could be exploited for improved LIB performance by infiltrating the nanofoam with a highly conductive, low-density material such as carbon nanotubes (CNTs). Here, we report for the first time, an efficient and scalable gas-phase liquid injection CVD approach using a ferocene/xylene mixture for the infiltration of arrays of high-quality CNTs in pretreated Ni nanofoam. Importantly, the BET specific surface area of the pretreated Ni nanofoam after CNT infiltration increases by a factor of over 20. Further, upon deposition of SnO_2 on the CNT-infiltrated Ni nanofoams by atomic layer deposition, the new nanocomposites showed improved energy capacity and excellent rate capability, as well as high capacity retention during electrochemical charge-discharge measurements.

435122 Synthesis of Layer-By-Layer Thick Mesoporous Titania Films with Vertically Oriented 2D-HCP Nanopores and Their Use in Lithium Ion Batteries As Negative Electrodes

Abstract:

Mesoporous titania films with 2D Hexagonal Close Packed (HCP) cylindrical nanopores have been synthesized by the evaporation-induced self-assembly (EISA) technique with Pluronic surfactant F127 (with average structure $(EO)_{106}(PO)_{70}(EO)_{106}$ where EO is an ethylene oxide unit and PO is a propylene oxide unit) as the structure-directing agent. To provide vertical alignment of the pores, surface modification of substrates with crosslinked surfactant F127 has been used to provide a chemically neutral surface. According to Koganti et. al (***Nano Lett.*** 2006, 6, 2567), thick films do not have perfect vertical alignment of the pores and better vertical HCP orientation is best achieved by sandwiching the films between two modified surfaces. Here, we report the synthesis of relatively thick films with vertically oriented 2D HCP cylindrical nanopores by layer-by-layer synthesis. Such films do not require sandwiching for vertical orientation of the pores. The hypothesis being tested is whether surfactant micelles from an existing layer will interact with micelles from a freshly deposited layer to provide epitaxial self-assembly of vertically oriented cylindrical micelles in registry from layer to layer. Using this technique, titania films with vertically oriented cylindrical nanopores as much as 1 μm thick were obtained. The thickness and vertical orientation were characterized using XRD, SEM and profilometry. These films were then used as negative electrodes in Li-ion batteries because of advantages of titania including good cycling stability, small volume expansion (~3%) during intercalation/extraction and high discharge voltage plateau. The high surface area and small wall thickness of these titania films provide better lithium ion insertion and reduced Li ion diffusion length compared to nonporous titania films or titania nanotube arrays. However, maintaining accessibilty of the full pore space in the thick films is a challenge to be discussed. Furthermore, the mass transport across these films is studied using electrochemical impedence spectroscopy (EIS). Using equivalent circuit modeling, the fraction of the accessible area of the nanoporous film coated electrode was calculated and analyzed as a function of the number of layers deposited.

(306g) Engineered Ionic Diffusion Layers to Increase Rate Capability of NCA Cathode in Lithium-Ion Cells

Dahlberg, K., University of Michigan
Mohanty, D., Oak Ridge National Laboratory
Mahajan, V., XALT Energy, LLC
Lee, M., XALT Energy, LLC
Stevenson, L.,
Stanley, J., XALT Energy, LLC
King, D., PneumatiCoat Technologies
Wood, D. III, Oak Ridge National Laboratory
Albano, F., XALT Energy, LLC
Dhar, S., XALT Energy, LLC
Originally presented on: 11/10/2015 10:30:00 - 10:50:00

High energy and long cycle life are the key technical attributes required for widespread implementation of high-performance electric vehicles (EVs). High rate capability is the key to battery performance for many other applications including stationary grid connected applications and Hybrid Electric Vehicles (HEV's). Despite significant strides made to-date, the current high energy density solutions suffer from degradation and limited cycle life limiting the widespread implementation of NCA cathode materials in Li-Ion batteries. There is growing evidence that several degradation mechanisms influencing these limitations take place at the cathode surface, motivating the use of larger particle sizes to minimize surface area and mitigate degradation. Larger particle sizes, however, decrease the overall rate of diffusion of lithium ions, causing a loss of activation of Li-rich phases and an increased loss of capacity at higher rates. In this paper, we show that Al_2O_3, TiO_2, and other similar ALD coatings applied to NCA powder, continuously covering active material particle surfaces, create an engineered SEI layer which facilitates an improved Li-ion diffusion pathway, increasing the rate capability of NCA cathodes even with larger particle sizes. Electrochemical analyses showed increased rate capability in 95x64 mm pouch cells (~2.5Ah) with coated NCA compared to uncoated, and XRD and TEM were used to characterize the underlying favorable interactions of coatings with the cathode surface. XALT Energy's integrated cell design and manufacturing makes 95x64 mm cell performance representative of large format (216x216 mm) production EV cells, demonstrating the validity and scalability of this approach.

416399 Towards Implantable Miniature Glucose-O2 Biofuel Cells

Abstract:

Miniature glucose/O_2 biofuel cells are of particular interest because they may power implanted medical devices. However, numerous challenges remain to overcome before envisioning such applications

In this talk, I will report our last progress directed to the search of improved biocatalyst for the O_2 reduction as well as the search for adequate electrode materials allowing fast electron transfer and diffusion of substrate. I will also discuss our recent achievement towards implantation in mices and toxicity issues, which are generally raise with the use of Osmium complexes.

420921 Entrapping Cross-Linked Glucose Oxidase Aggregates within a Graphitized Mesoporous Carbon Network for Efficient Electron Transfer

Abstract:

This paper reports a novel method for producing glucose oxidase-nanocomposites by entrapping cross-linked glucose oxidase (GOx) aggregates within a graphitized mesoporous carbon (GMC) network. Entrapment was achieved by utilizing the strong self-aggregation tendency of GMC in aqueous buffer solution to form carbon networks. Using confocal microscopy and TEM, GOx-GMC nanocomposites were visualized. The electrochemical properties of GOx-GMC nanocomposites were studied by means of cyclic voltammograms, chronoamperometric and potentiostatic tests. Results therefrom suggested that the GOx-GMC nanocomposites offer a high electrochemical activity with the maximum electron transfer rate constant estimated at $5.16 \pm 0.61\ s^{-1}$. Furthermore, thermally treating the GOx-GMC nanocomposite and GOx aggregates at 60°C for four hours, both samples maintained 99% of their initial activity, while the free GOx were completely deactivated. These performances suggested that our nanocomposite structure offered both improved electrochemical activity and stability by combining the high electrical conductivity offered by the GMC network with the high enzyme loading and stability offered by the cross-linked GOx aggregates. The GOx-GMC nanocomposite's electrochemical activity towards glucose oxidation was also investigated by using an enzymatic biofuel cell without artificial mediators, producing a power density of up to 22.4 $\mu W/cm^2$ at 0.24V.

412611 Using DNA As a Scaffold for Immobilizing Enzyme Cascades for Biofuel Cells

Abstract:

Enzymatic fuel cells are a class of fuel cells that utilize oxidoreductase enzymes to catalyze the redox reactions at the anode and/or the cathode. Anodes are particularly challenging in enzymatic fuel cells, because multiple enzymes are needed to oxidize most complex biofuels. Recent research has shown that in order to have high flux through enzyme cascade reactions, there is a need for proximity between enzyme active sites. This paper will discuss the use of DNA as a scaffold material for controlling the distance between sequential enzymes in a enzyme cascade-based bioanode for a biofuel cell. The paper will discuss an enzyme cascade for sucrose oxidation involving invertase and glucose oxidase. It will show the importance of decreasing the distance between active sites in sequential enzymes on the performance (current and power density) of bioanodes in biofuel cells. This paper will also discuss methods for utilizing DNA as a scaffold for the redox mediator and DNA origami examples for complex enzyme cascades.

435152 Quantitative Assessment of Channeling Mechanisms in Nanoscale Catalytic Architectures

Abstract:

The throughput and yield of nanoscale catalytic architectures is limited by the ability to retain reaction intermediates and direct them toward downstream reaction steps. Nature has settled on several approaches to such intermediate channelling, including electrostatic attraction, confinement, and surface affinity. This presentation will discuss the limits to such approaches in the context of a steady-state continuum modeling approach. The extent to which intermediate transport limits overall reaction flux is defined by a flux control coefficient, calculated from numerical results. The extent to which each channeling approach can enhance overall flux and/or reaction yield under idealized circumstances is predicted, and the various complications introduced by nonidealities are discussed.

424408 Substrate Channel Evolution and Circular Permutation of an Esterase for the Synthesis of Cilastatin

Abstract:

SUBSTRATE CHANNEL EVOLUTION AND CIRCULAR PERMUTATION OF AN ESTERASE FOR THE SYNTHESIS OF CILASTATIN

The esterase *Rh*Est1 from *Rhodococcus* sp. ECU1013 has been reported for the enantioselective hydrolysis of ethyl (*S*)-(+)-2,2-dimethylcyclopropane carboxylate, producing the building block of cilastatin [1, 2]. Cilastatin can protect the kidneys against potential toxic effects from higher doses of imipenem. Imipenem and cilastatin have been used in combination as a 1:1 mixture, known as Tienam, which showed a broad spectrum activity and fewer side effects for the treatment of severe infections.

In this work [3], error-prone PCR and site-directed saturation mutagenesis were applied to *Rh*Est1 for activity improvement, with the pH-indicator assay as a high-throughput screening method. *Rh*Est1$_{A147I/V148F/G254A}$, with mutations surrounding the substrate access channel, showed a 5-fold increase in its specific activity compared with the native enzyme, as well as a 4-fold increase in protein solubility. Combined with the determination of protein structures and computational analysis, this work shows that the amino acids around the substrate channel play a more important role in the activity evolution of *Rh*Est1 than those in the active site. Mutations surrounding the entrance to the substrate channel allowed for an effective balance between improvement in the activity and a high level of enantioselectivity, which is a big challenge to improve the activity of *Rh*Est1 towards ethyl 2,2-dimethylcyclopropane carboxylate. Circular permutation as a protein engineering approach was also applied to acquire functional variants of *Rh*Est1. 98 CP mutants were constructed from three different sectors that we defined and the active variants mainly located on more flexible turn loops based on structure of *Rh*Est1. The observed results suggest the relocation of protein terminus could be an effective method for protein engineering. The future work will be focused on the further evolution and structure analysis of the permutated *Rh*Est1.

[1] C. H. Liu, J. Pan, Q. Ye and J. H. Xu, *Appl. Microbiol. Biotechnol.*, 2013, **97**, 7659.

[2] Y. Zhang, J. Pan, Z. J. Luan, G. C. Xu, S. H. Park and J. H. Xu, *Appl. Environ. Microbiol.*, 2014, **80**, 7348.

[3] Z. J. Luan, F. L. Li, S. Dou, Q. Chen, X. D. Kong, J. Zhou, H. L. Yu and J. H. Xu, *Catal. Sci. Technol.*, 2015,**5**, 2622-2629.

421086 Recent Advancement in Nanostructured Enzyme Catalysts for Chemical Synthesis

Abstract:

For decades we have been working on novel approaches incorporating enzyme into nanostructures presenting high activity, high stability and ease of recovery and reuse. It is observed, also by our colleagues, that accommodation in a nanometer environment with tailor-made surface properties often lead to greatly improved stability and activity of the confined enzyme, particularly under adverse conditions such as an elevated temperature and the presence of organic solvents. This presentation will summarize our recent progress in the synthesis of nanostructured enzyme catalysts, in particular, the binary enzyme catalysts that performed cascade reactions. We will also show our efforts towards establishing a molecular insight of the interaction among nanostructures and confined enzymes and their impacts on the enzymatic reaction kinetics and process through a complementary input from multi-scale molecular simulation, multidimensional structural characterization and reaction process analysis.

References:
1. Wu XL, Ge J*, Zhu JY, Yong Y and Liu Z*, Chem Commun, 2015, DOI: 10.1039/C5CC01776A
2. Wang R, Zhang YF, Ge J* and Liu Z*, RSC Adv., 2014, 4, 40301
3. Lyu FJ, Zhang YF, Zare RN, Ge J* and Liu Z*, Nano Lett., 2014, 14, 5761.
4. Li ZX, Zhang YF, Su YC, Ouyang PK, Ge J,* and Liu Z*, Chem Commun., 2014, 50,12465.
5. Zhu JY, Zhang YF, Lu DN, Zare RN, Ge J,* and Liu Z*, Chem Commun., 2013, 49, 6090.

433185 Simple and Sensitive Point-of-Care Bioassay System Based on Hierarchically Structured Enzyme-Mimetic Nanoparticles

Abstract:

In this work, we realized enzyme-mimetic nanoparticle-based point of care (POC) bioassay devices. In addition, we quantified signals by using digital images acquired by smart cellular phones, which will further enable the devices to be used for U-healthcare applications. Using this device, quantitative analysis of liquid samples was performed within 15 min with an order of magnitude enhancement of sensitivity compared with conventional Au nanoparticle-based devices. As far as we know, this is the first successful application of enzyme-mimetic nanomaterials for simple and easy-to-use POC devices

413460 Bio-Inspired Nanomaterials for Controlled Release

Abstract:

Bio-inspired nanomaterials for controlled release

Chun-Xia Zhao*, David Wibowo and Anton P.J. Middelberg

Australian Institute for Bioengineering and Nanotechnology, The University of Queensland, St Lucia QLD 4072, Australia.

*Oral presentation and corresponding: z.chunxia@uq.edu.au

Abstract

Engineered nanomaterials have attracted significant research interest during the past decades for various applications. While compared to man-made materials, precision nanomaterials inspired by nature promise to revolutionize many sectors including photonics, coatings, healthcare because of their sophisticated structure. In this talk, I will introduce two different types of bio-inspired nanomaterials for controlled release, 1) oil-core silica nanocapsules for sustained release and 2) core-shell silica nanoparticles for triggered release. Biocompatible silica nanocapsules were developed in our lab using a patented technology - the emulsion and biomimetic dual-templating platform technology. This technology is based on the novel design of bifunctional peptides or proteins (acting as not only biosurfactants but also biomineralizing agents) by modularizing a partial sequence encoding surface activity with a series of amino acids having biosilicification ability. Nanoemulsions were formed in the presence of these bifunctional biomolecules as a result of their surface activity, then a silica precursor such as tetraethyl orthosilicate was added to initiate the biosilicification induced by the catalytic peptide sequence. Oil-core silica nanocapsules were produced under mild conditions, including room temperature, neutral pH and without use of any toxic chemicals. We also demonstrated the facile loading of actives by directly dissolving them in the oil phase, followed by emulsification and biosilicification. Slow release of hydrophobic ingredients encapsulated in the oil phase can be controlled by tuning the shell thickness of the silica nanocapsules. This technology opens a new facile and environmentally friendly strategy for fabricating capsules that are potentially applicable various fields. In the other hand, we developed silica-core and hydroxyapatite shell nanocomposites through a biomimetic mineralization process. The multifunctional nanocomposites composed of a magnetite nanocrystal core and a mesoporous silica shell (Fe_3O_4@$mSiO_2$) end capped with pH stimuli-responsive hydroxyapatite (HAP) nanovalves. The natural nontoxic component hydroxyapatite cap system allows the pH-responsive release of loaded drugs within acidifying intracellular compartments such as endosomes and lysosomes at specific targets. The mineral coated Fe_3O_4@$mSiO_2$ nanocomposites provide efficient pH-responsive nanocarriers for potential applications in simultaneous bioimaging, targeting and pH-responsive release.

(362d) Biocatalytic Coatings of Cytochrome P450-PNIPAM Diblock Copolymers

Obermeyer, A., Massachusetts Institute of Technology
Colant, N.,
Olsen, B. D., Massachusetts Institute of Technology
Originally presented on: 11/10/2015 13:55:00 - 14:20:00

Engineering proteins into functional catalysts enables their application as biosensors and green catalysts. The biocatalyst must maintain mechanical integrity, protein stability and longevity, and access to the protein active site in order to be a viable technology. In addition, a high protein density is needed to achieve high catalytic activity. The self-assembly of proteins in solid-state materials has the ability to achieve each of these requirements. The modification of proteins with synthetic polymers, creating a protein-based block copolymer, can be used to direct both the solution and solid state self-assembly of proteins. As a demonstration of the utility of this technique, cytochrome P450 BM3 was selected as an attractive target for a functional immobilized enzyme. The bacterial P450 contains both oxidase and reductase domains, has been extensively engineered to accept non-native substrates and catalyze novel chemical transformations, and has been evolved to have improved thermostability. Matching these catalytic improvements to improvements in enzyme immobilization technology could enable the use of P450s as biocatalysts. Functionalization of the enzyme with the synthetic polymer poly(*N*-isopropylacrylamide) creates a bioconjugate that can self-assemble to form functional nanostructured films. The self-assembly of the bioconjugate in bulk and concentrated solutions was characterized by small-angle x-ray scattering (SAXS) and depolarized light scattering (DPLS). The synthetic conjugate also enables the enzyme to be flow coated into a thin film that is readily immobilized by a simple chemical crosslinking step. The activity of the self-assembled P450 thin films was assayed for a variety of substrates.

421229 Uniform Protein Hydrogels Via Gelation in Reverse Micelles

Abstract:

Hydrogels are 3D polymer networks with water content up to 90%, making an aqueous environment for the embedded enzyme. Recently, a facile method for forming tetra-PEG hydrogel was established in which a direct mixing of two macromonomer solutions led to the gelation completed within several minutes[1]. Here we extend this method to form enzyme microgels with homogenous nanostructures.

Tetra-amine-terminated PEG (TAPEG) and tetra-NHS-glutarate-terminated PEG (TNPEG) macromonomer solutions containing proteins were mixed and dispersed into cyclohexane to form reverse micelles with diameters from 80nm to 200nm. Gelation inside micelles was triggered by tuning pH from 5.0 to 7.0, resulting in uniform enzyme-incorporated hydrogels with diameter around 1μm.

The versatility of this method was demonstrated using Cyt c, CALB, GOx and HRP, respectively, in which all enzyme hydrogels displayed catalytic activities in aqueous solutions. Moreover Cyt c hydrogel showed a 5-fold increase in peroxidase activity compared to its counterpart in free solution, in addition to an improved stability.

References:

[1] Sakai T, Matsunaga T, Yamamoto Y, et al. [J]. Macromolecules, 2008, 41(14): 5379-5384.

426890 Carbon Dioxide Conversion and Utilization Based on Highly-Stabilized Carbonic Anhydrase

Abstract:

Enzymes have attracted attention rapidly due to their promising applications such as biosensors, biofuel cells, proteomics, enzyme-linked immunosorbent assay (ELISA), antifouling and carbon dioxide conversion. In this work, we developed a enzymatic process for one-pot carbon dioxide (CO_2) conversion and utilization based on conversion of CO_2 to bicarbonate at ambient temperature with no energy input, by using carbonic anhydrase and carboxylated polyaniline nanofibers (cPANFs). Carbonic anhydrase was immobilized on cPANFs via the approach of magnetically-separable enzyme precipitate coatings (Mag-EPC), which is made of covalent enzyme attachment, enzyme precipitation and crosslinking with amine-functionalized magnetic nanoparticles. Mag-EPC showed a half-life of 236 days under shaking (200 rpm), even resistance to 70% ethanol sterilization, and recyclability via facile magnetic separation. For enzymatic one-pot CO_2 conversion and utilization, Mag-EPC was used to accelerate the growth of microalga by supplying bicarbonate from CO_2, representing 1.8-fold increase of cell concentration when compared to the control sample. After two repeated uses via magnetic separation, the cell concentration with Mag-EPC was maintained as high as the first cycle of microalgae cultivation. This one-pot CO_2 conversion and utilization can be an alternative as well as complementary process to adsorption-based CO_2 capture and storage as an environmentally-friendly approach, demanding no energy input based on the stabilized enzyme system.

Multiscale Modeling and Simulation of Nanoimprint Lithography on a Flexible Substrate

Akhilesh Jain, Andrew Spann, Roger T. Bonnecaze
McKetta Department of Chemical Engineering
The University of Texas at Austin

R2R nano-patterning on flexible electronics

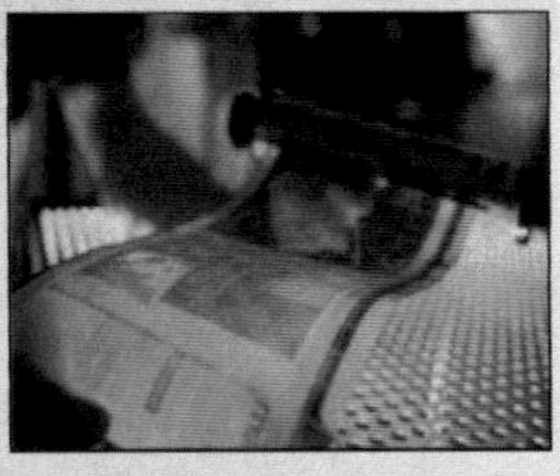

Flexible displays †

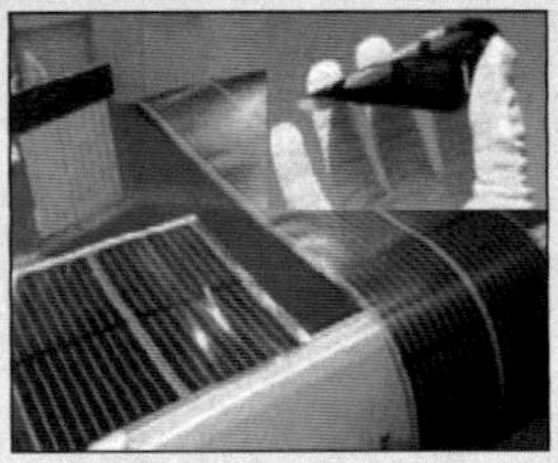

Organic solar cells§

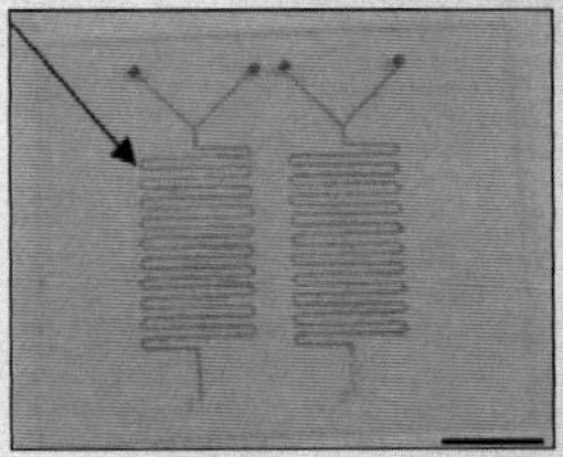

Microfluidic devices‡

Canon SanDisk EVG TOSHIBA Molecular Imprints

SEM image of UV roll imprinted 100 nm dense pillars for bilayer wire grid polarizer **

SEM image of the UV roll imprinted pyramid pattern with pitch 50 nm and height 24 nm for functional optical film application *

† The Flexible Electronics and Display Center, Arizona Display Center
§ Konarka Technologies
‡ L. P. Yeo, Journal of Micromechanics and Microengineering **20** (2010).
* Ahn, S., M. Choi, H. Bae, Jap. Journal of App. Phys. 8B, 5478 (2007)
**Ahn et al, Alternative Lithographic Technologies 8323 (2012).

2

Stepper template based R2R UV-IL

Step 1: Dispense drops

Step 2: Lower template and fill pattern

Step 3: Polymerize S-FIL fluid with UV exposure

Step 4: Separate template from substrate

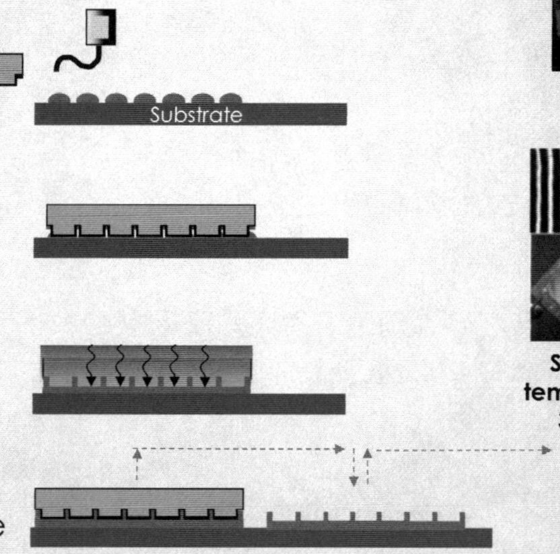

Lithoflex 100, UT Austin

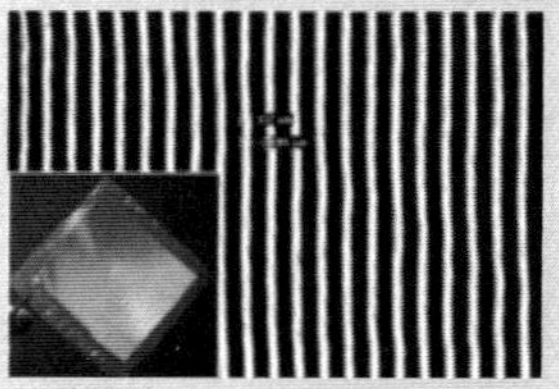

SEM image of imprint template (700 nm period, 350 nm line width) ‡

Roller template based R2R UV-IL

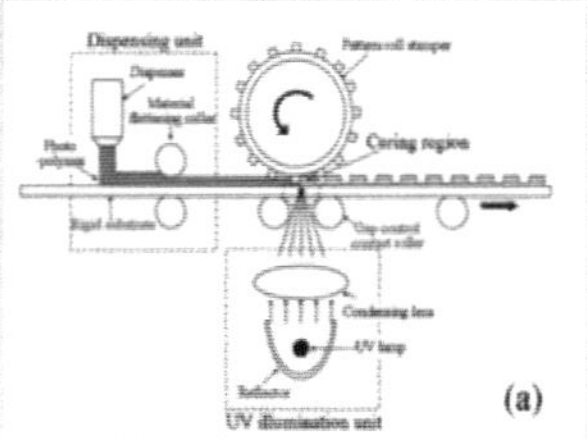

S. Ahn et al, Applied Physics letters **89** (2006).

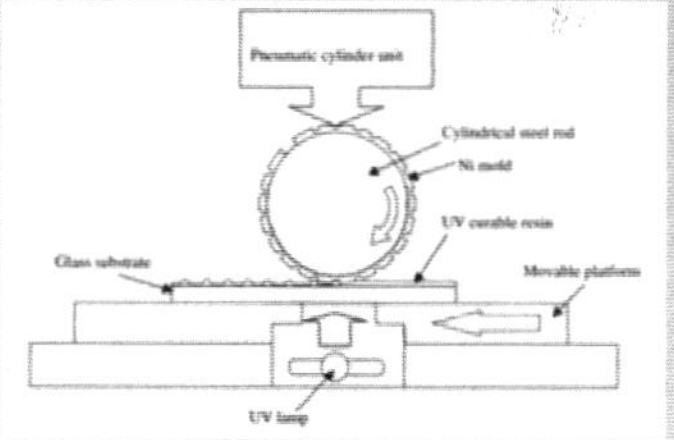

C. Y. Chang et al, Microsystem Technologies, **12**, 754 (2006).

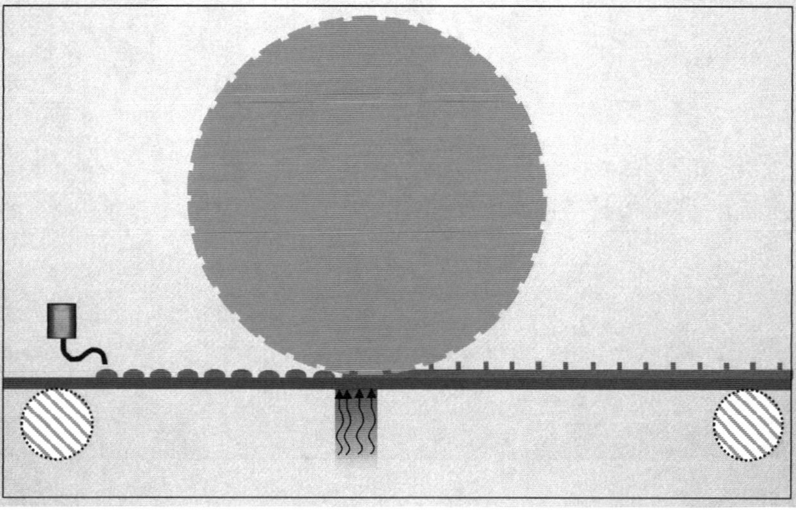

4

Important aspects of R2R UV-IL

- Feature size
- Patterned area
- Web Speed
- Residual layer thickness (RLT)

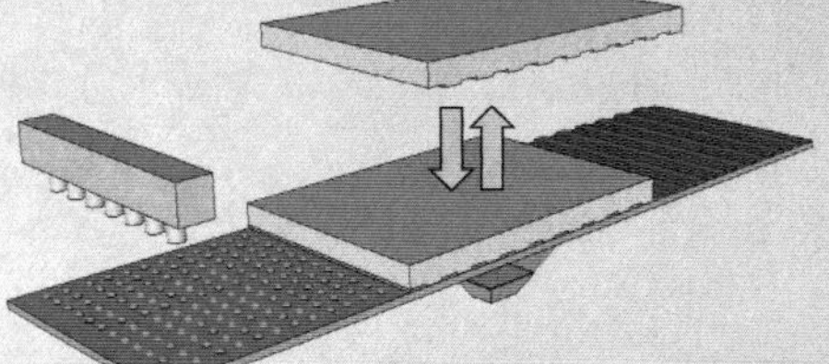

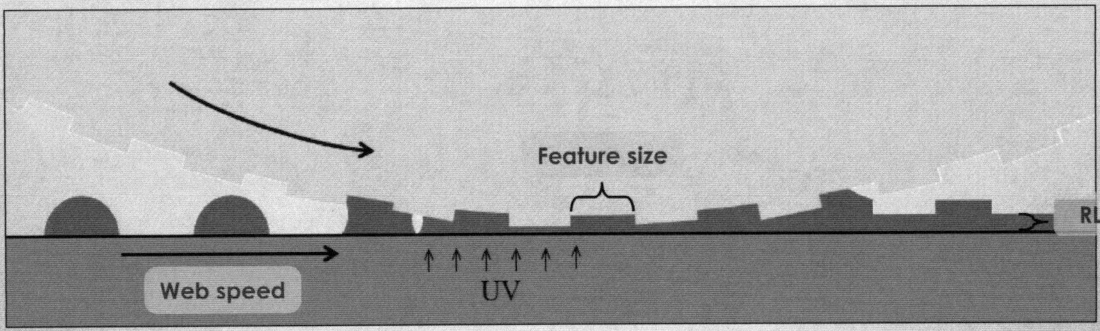

5

State of the art: R2R UV-IL

University of Michigan‡

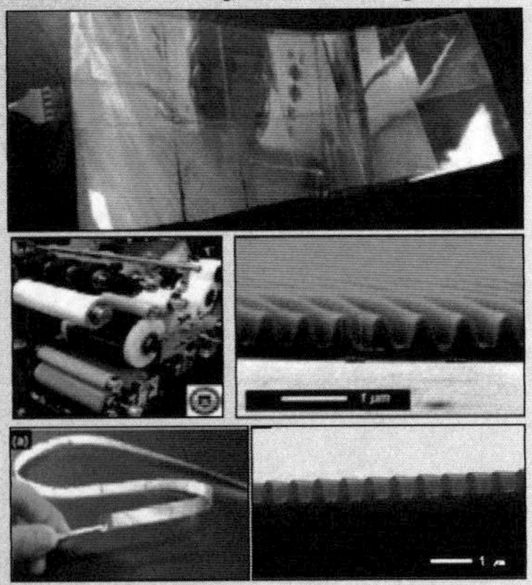

Web speed: 1 m/min
Substrate width: 1 cm
Pattern size: 70 nm line width
RLT: ~ *1μm*
Patterned area: 0.01 m^2/min

Toshiba Machine†

Web speed: 0.2 - 10 m/min
Substrate width: 100 - 230 mm
Pattern size: ?
RLT: ?
Patterned area: 0.75 m^2/min

UT Austin/Molecular Imprints§

Web speed: 3 imprints/min
Substrate width: 25 mm
Pattern size: 50 nm line width
RLT : 12 nm
Patterned area: 0.002 m^2/min

‡L. J. Guo, S. H. Ahn, ACS Nano 3, 2304 (2009)
‡L. J. Guo, S. H. Ahn, JVST B 25, 2388 (2007)
§S. Ahn et al, Alternative Lithographic Technologies 8323 (2012).
†www.toshiba-machine.co.jp

6

Technology roadmap

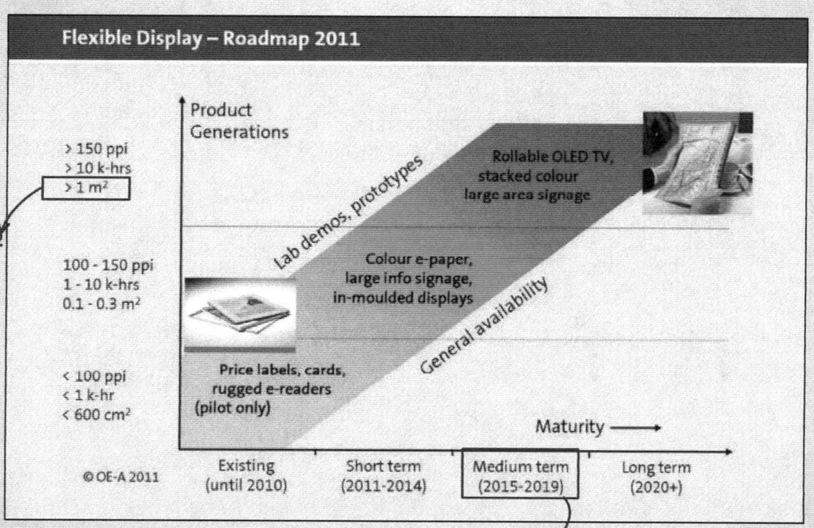

>1 m²

2015-2019

OE-A Roadmap for Flexible Display (Organic and Printed Electronics Association, 2009)

†http://www.evgroup.com

7

Maximum web speed vs RLT for roller based UV-IL

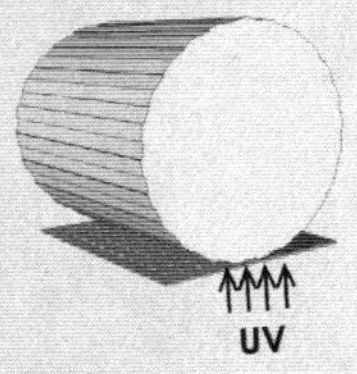

$$U = \frac{I_o L}{J_o h_o}\left(1 - exp\left(-a h_o\right)\right)$$

$L \sim (2Rh_o)^{1/2}$

$R = 2.5$ cm.

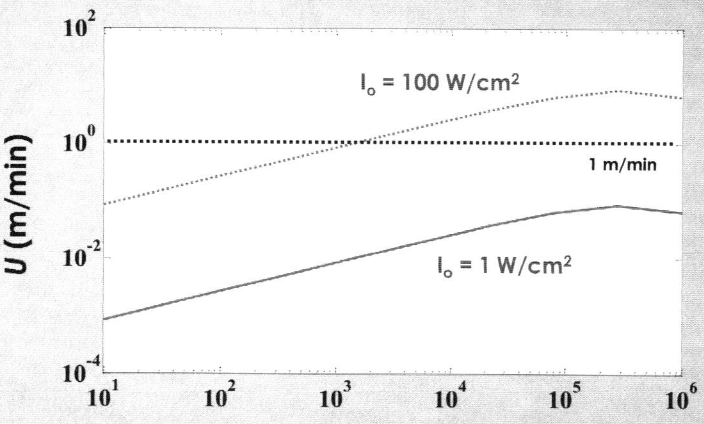

(---): 100 W/cm² (ii) (—): 1 W/cm².

8

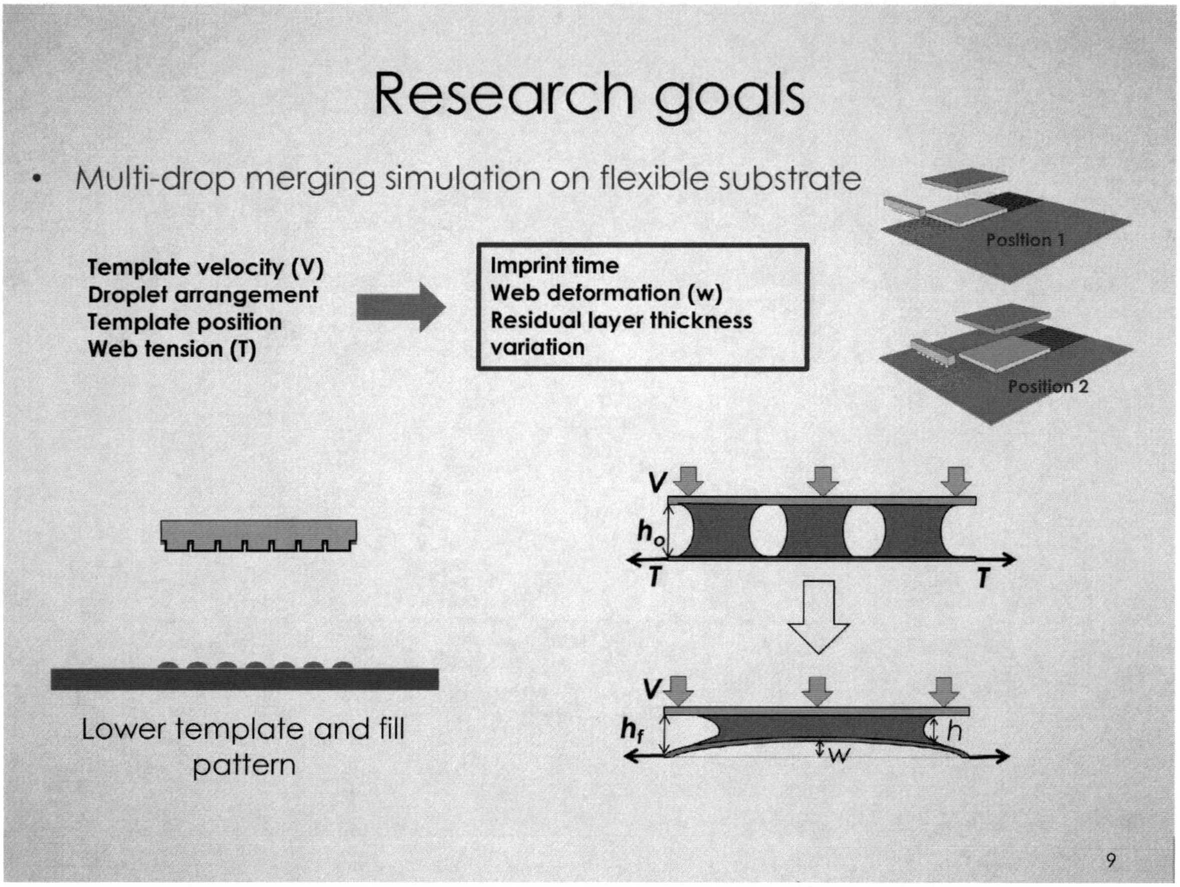
Research goals
• Multi-drop merging simulation on flexible substrate
Template velocity (V)
Droplet arrangement
Template position
Web tension (T)
Imprint time
Web deformation (w)
Residual layer thickness variation
Position 1
Position 2
V
h_o
T
T
Lower template and fill pattern
V
h_f
h
w
9

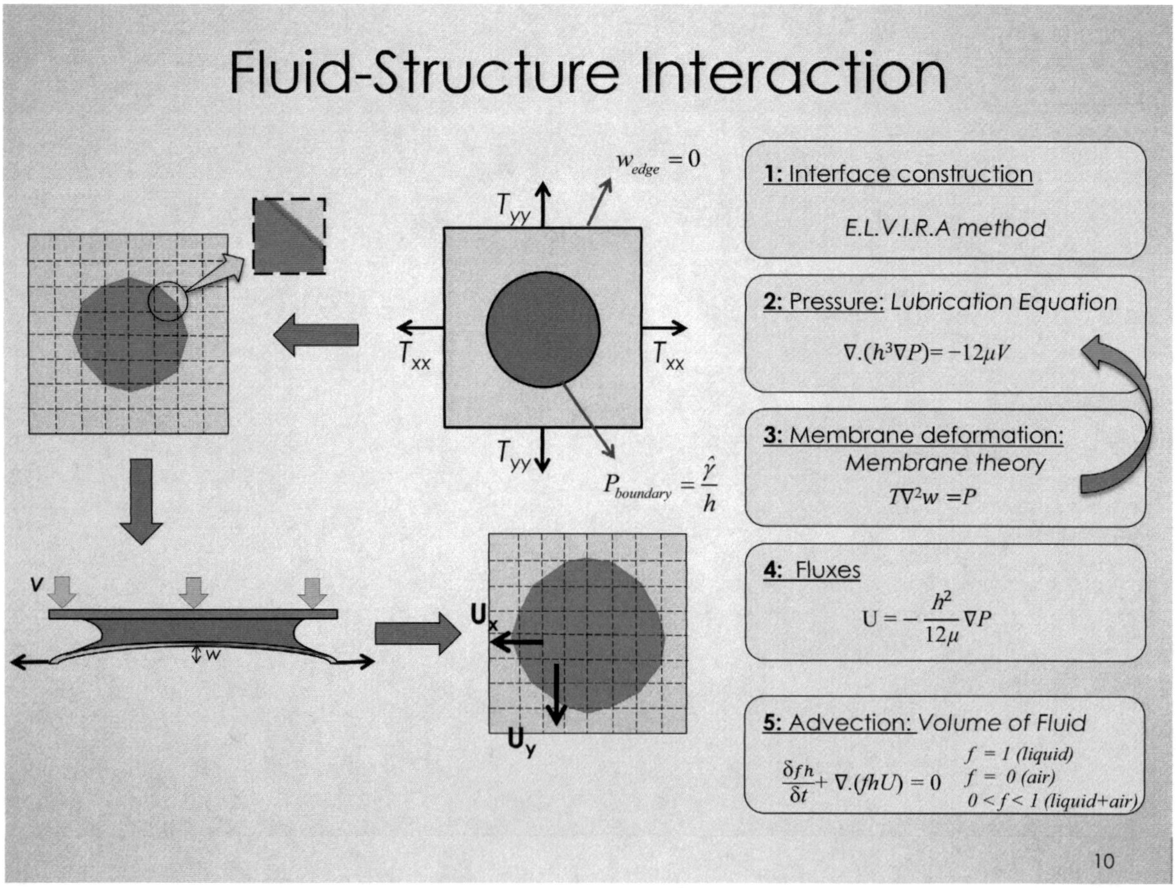
Fluid-Structure Interaction
$w_{edge} = 0$
T_{yy}
T_{xx}
T_{xx}
T_{yy}
$P_{boundary} = \frac{\hat{\gamma}}{h}$
V
w
U_x
U_y
1: Interface construction
E.L.V.I.R.A method
2: Pressure: Lubrication Equation
$\nabla.(h^3\nabla P) = -12\mu V$
3: Membrane deformation: Membrane theory
$T\nabla^2 w = P$
4: Fluxes
$U = -\frac{h^2}{12\mu}\nabla P$
5: Advection: Volume of Fluid
$\frac{\delta f h}{\delta t} + \nabla.(fhU) = 0$
$f = 1$ (liquid)
$f = 0$ (air)
$0 < f < 1$ (liquid+air)
10

	Governing equations	Non-dimensionalized governing equations
Lubrication equation	$\nabla.\left(h^3\nabla P\right)=-12\mu V$	$\nabla.\left(\bar{h}^3\nabla\bar{P}\right)=-1$
Membrane equation	$T\nabla^2 w=P$	$\bar{T}\nabla^2\bar{w}=\bar{P}$
Pressure boundary condition	$P\rvert_{interface}=-\dfrac{2\gamma}{h}$	$\bar{P}\rvert_{interface}=-\dfrac{1}{Ca}\dfrac{1}{\bar{h}}$
Deformation boundary condition	$w\rvert_{edge}=0$	$\bar{w}\rvert_{edge}=0$

$$P_c=\frac{12\mu VL^2}{h_o^3},\ t_c=\frac{6\mu VL^2}{\gamma h_o},\ h_c=h_o,\ w_c=h_o$$

$$\bar{T}=\frac{Th_o^4}{12\mu VL^4},\ Ca=\frac{6\mu VL^2}{\gamma h_o^2}$$

Model verification

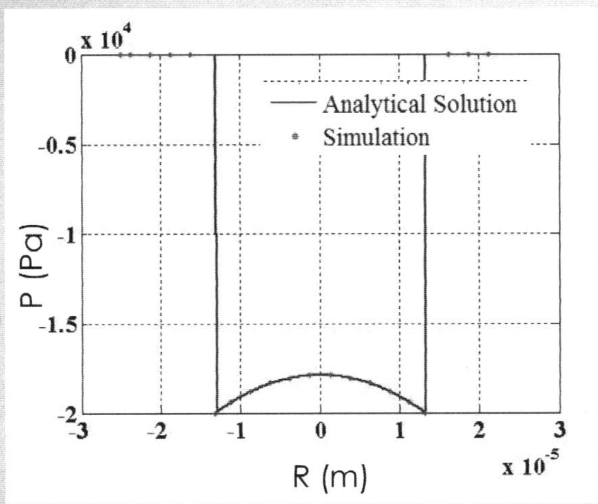

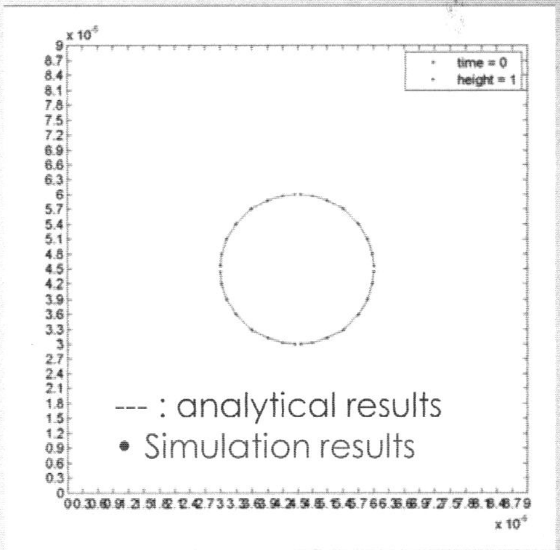

$$P(r)=-\frac{2\gamma}{h}+\frac{3\mu V}{h^3}\left(R^2-r^2\right)$$

$$R(t)=\frac{Q^{1/2}}{\pi}\left[1+16\pi\left(\frac{h_0}{h_f}\right)t\right]^{1/4}$$

Volume Loss < 0.001%

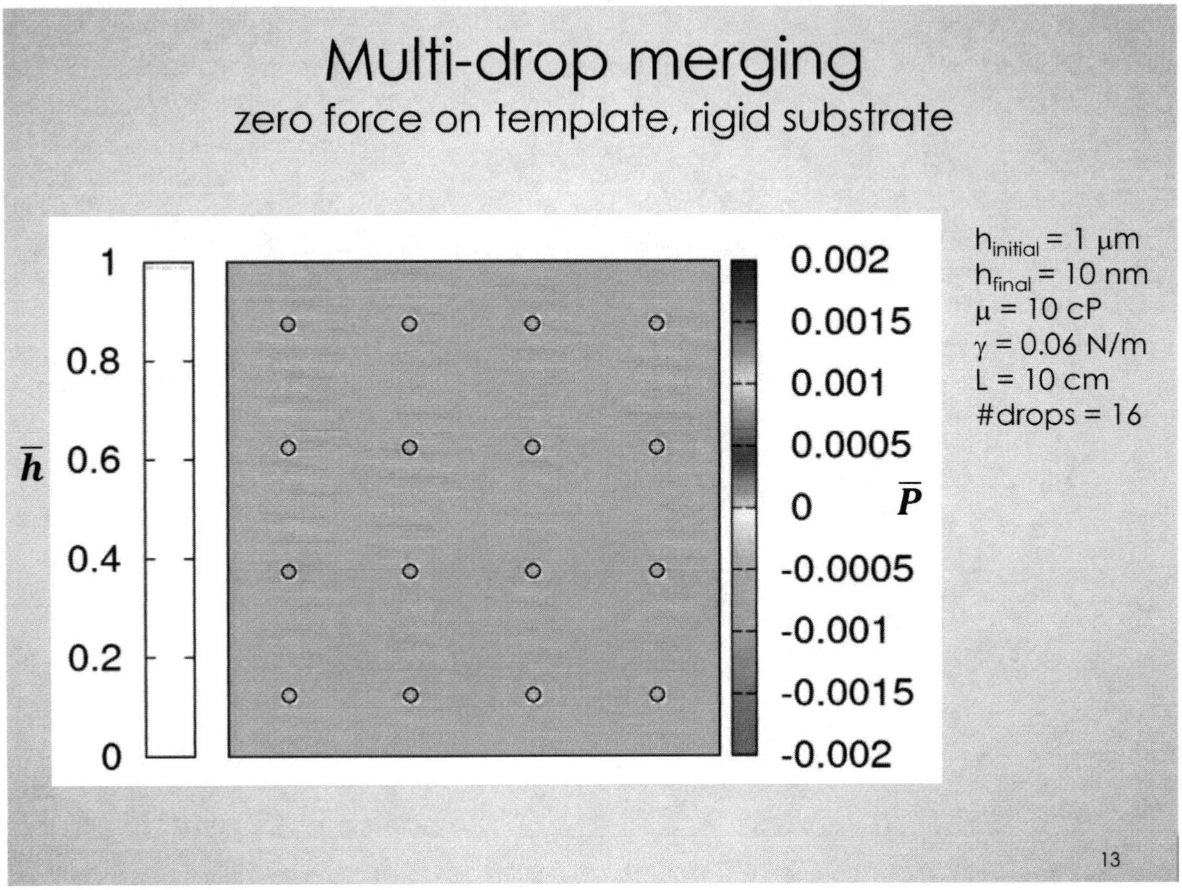

Multi-drop merging
zero force on template, rigid substrate
1
0.8
$\bar{h}$ 0.6
0.4
0.2
0
0.002
0.0015
0.001
0.0005
0 $\bar{P}$
-0.0005
-0.001
-0.0015
-0.002
$h_{initial}$ = 1 μm
h_{final} = 10 nm
μ = 10 cP
γ = 0.06 N/m
L = 10 cm
#drops = 16
13

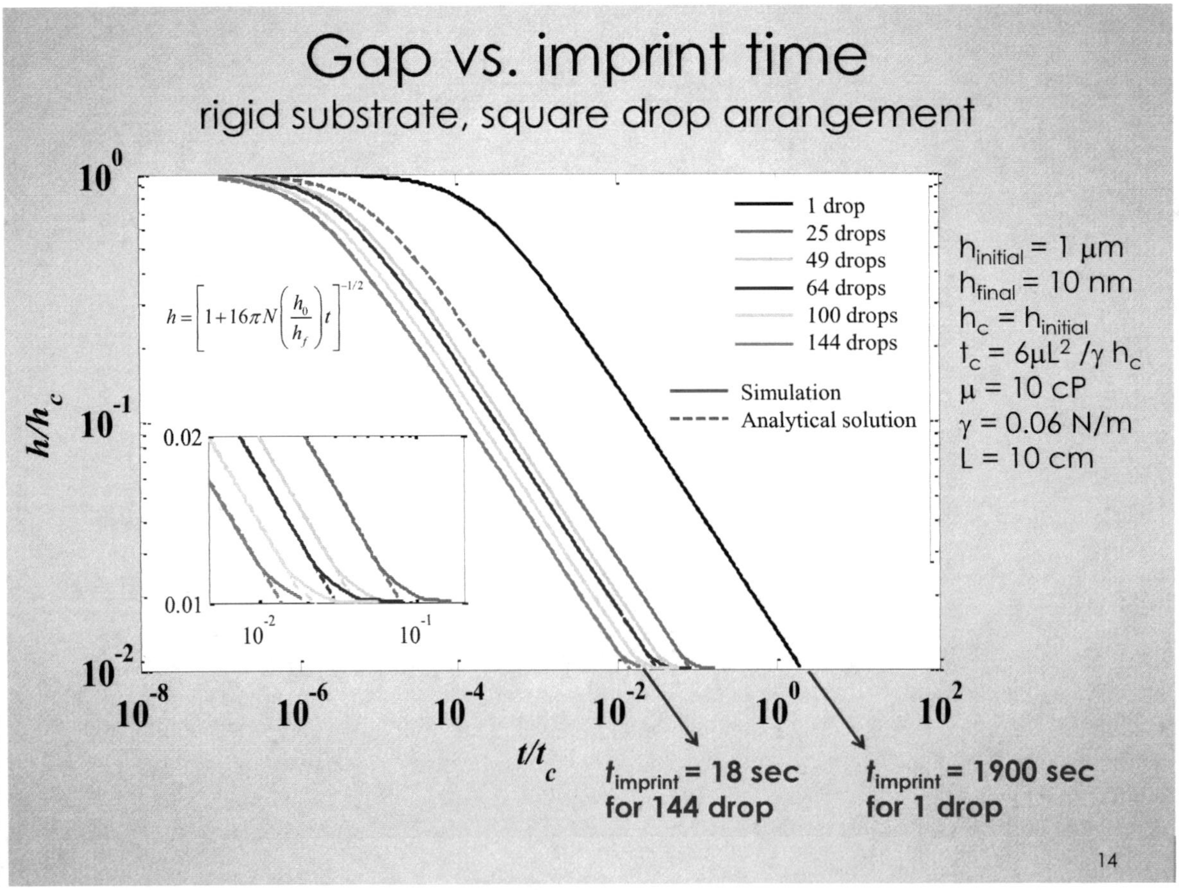

Gap vs. imprint time
rigid substrate, square drop arrangement
1 drop
25 drops
49 drops
64 drops
100 drops
144 drops
Simulation
Analytical solution
$h = \left[1 + 16\pi N \left(\frac{h_0}{h_f}\right) t\right]^{-1/2}$
h/h_c
t/t_c
$t_{imprint}$ = 18 sec for 144 drop
$t_{imprint}$ = 1900 sec for 1 drop
$h_{initial}$ = 1 μm
h_{final} = 10 nm
$h_c = h_{initial}$
$t_c = 6\mu L^2/\gamma h_c$
μ = 10 cP
γ = 0.06 N/m
L = 10 cm
14

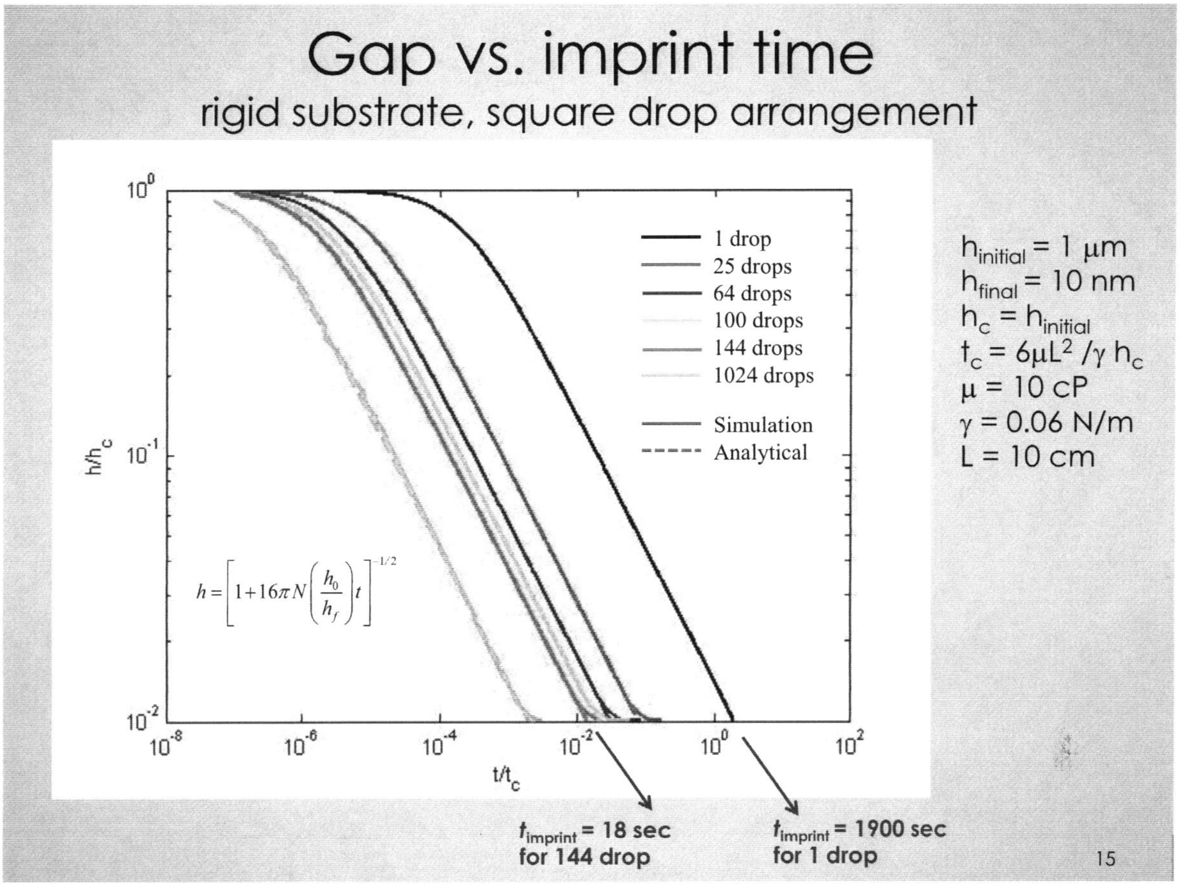

Gap vs. imprint time
rigid substrate, square drop arrangement
1 drop
25 drops
64 drops
100 drops
144 drops
1024 drops
Simulation
Analytical
$h=\left[1+16\pi N\left(\frac{h_0}{h_f}\right)t\right]^{-1/2}$
h/h$_c$
t/t$_c$
$h_{initial}$ = 1 μm
h_{final} = 10 nm
$h_c = h_{initial}$
$t_c = 6\mu L^2/\gamma\, h_c$
μ = 10 cP
γ = 0.06 N/m
L = 10 cm
$t_{imprint}$ = 18 sec for 144 drop
$t_{imprint}$ = 1900 sec for 1 drop
15

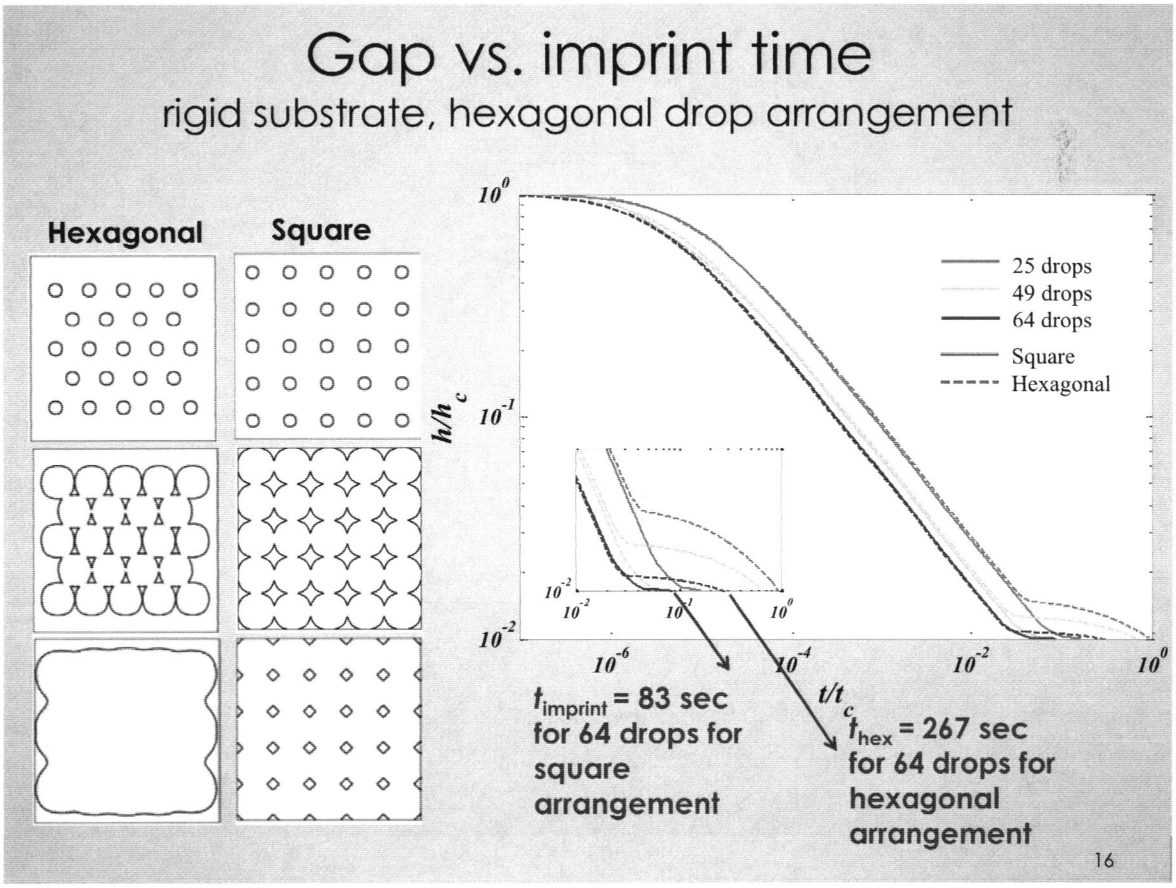

Gap vs. imprint time
rigid substrate, hexagonal drop arrangement
Hexagonal
Square
25 drops
49 drops
64 drops
Square
Hexagonal
h/h$_c$
t/t$_c$
$t_{imprint}$ = 83 sec for 64 drops for square arrangement
t_{hex} = 267 sec for 64 drops for hexagonal arrangement
16

1024 drops on a rigid substrate

zero force, Ca = 1

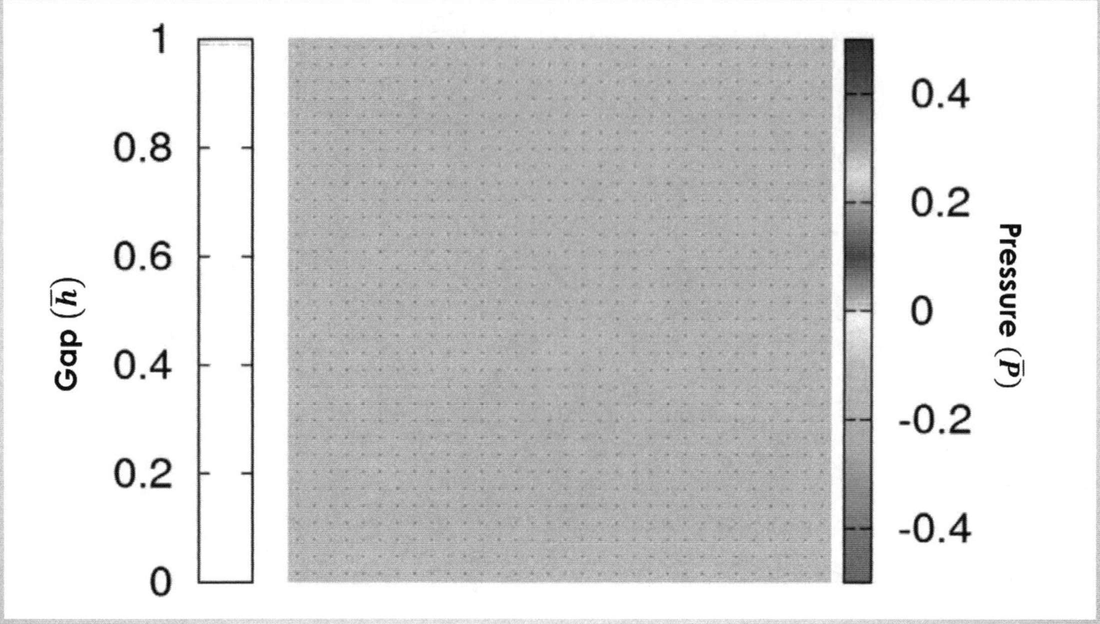

17

100 drops on a flexible substrate

constant velocity, Ca = 0.1, $\overline{T}$ = 10

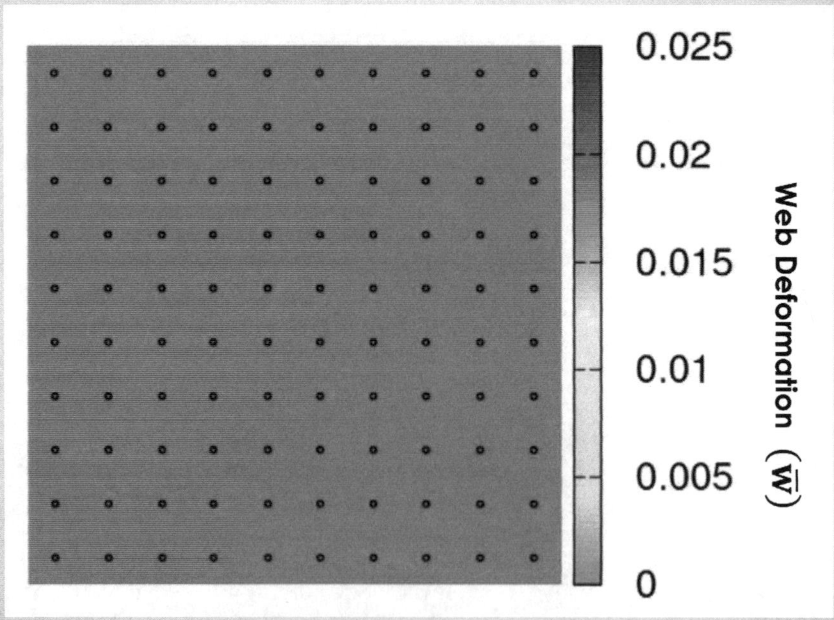

18

Web deformation vs. web tension

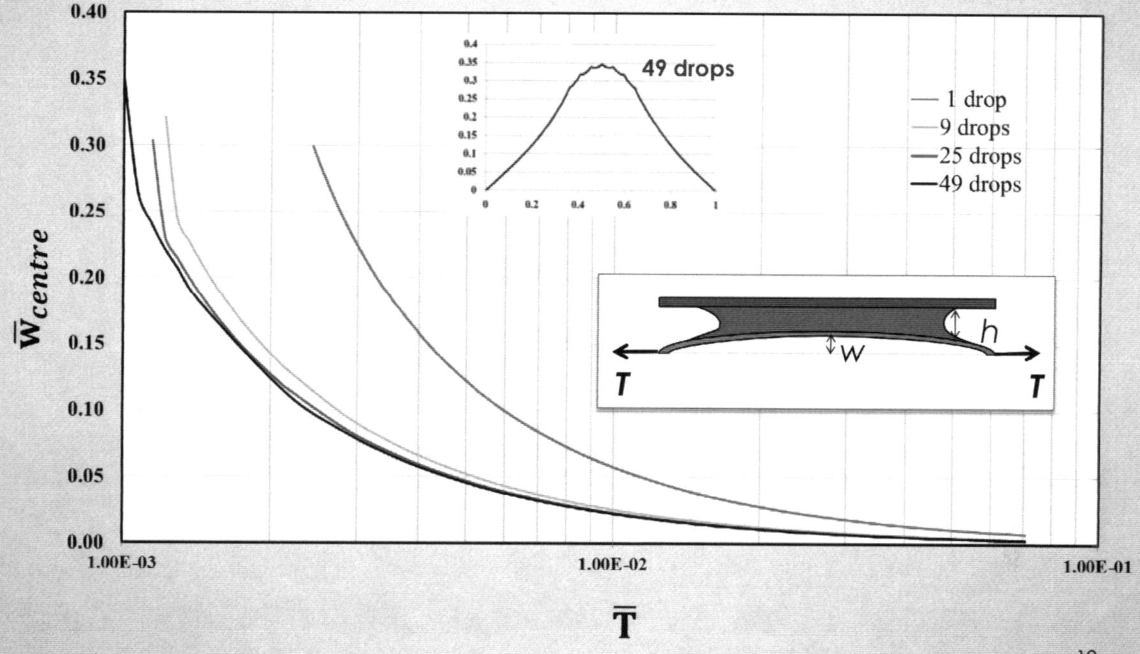

19

Web deformation vs. web tension

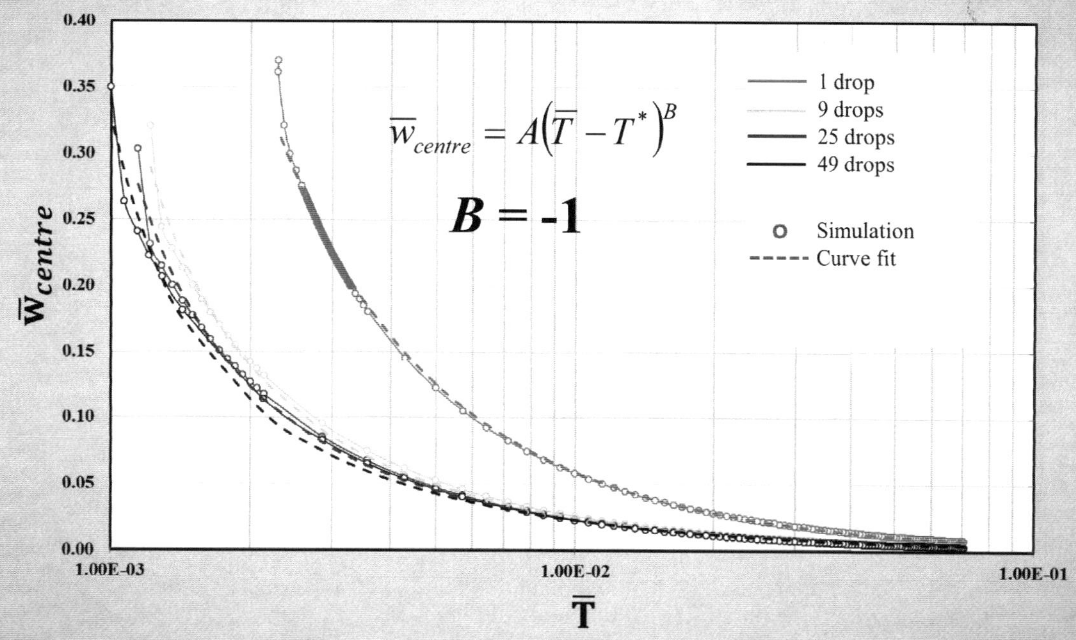

20

Critical web tension

$$\overline{w}_{centre} = A\left(\overline{T} - T^*\right)^{-1}$$

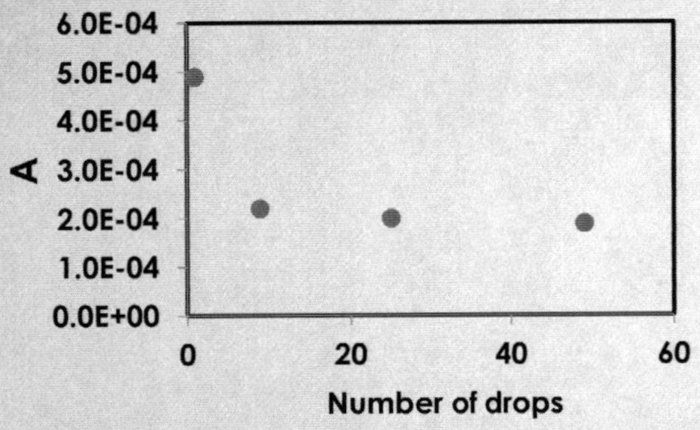

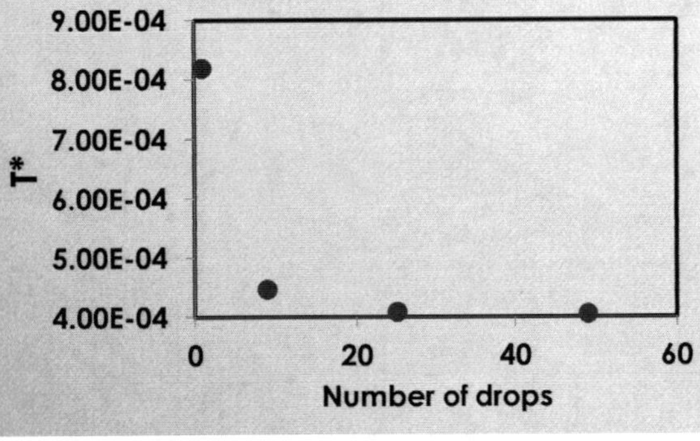

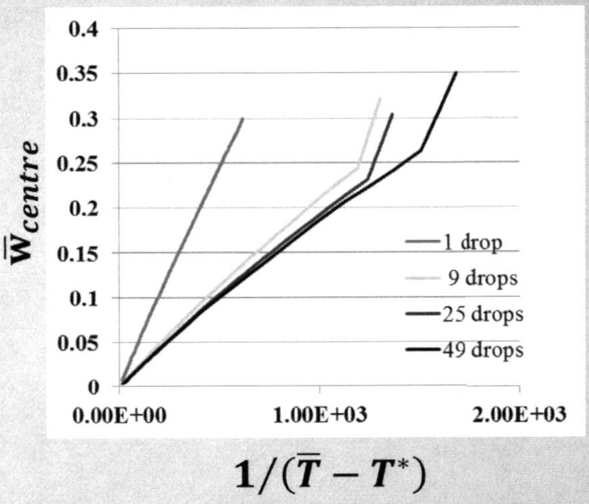

- A critical web tension exists below which the gap between the substrate and template closes.
- Contact between the template and substrate can be prevented by viscous force generated by moving the template.

21

Summary

- **Coupled elasto-hydrodynamic model** for droplet merging in stepper based nanoimprint lithography.
- Simulations possible for **thousands of droplets** with high-performance parallel computing.
- Imprint time and residual layer thickness (RLT) variation **sensitive to**
  - Number of drops
  - Hexagonal vs. square arrangement
  - Web tension, template velocity etc.
- A critical web tension exists below which the gap between the substrate and template closes. **Contact between the template and substrate prevented by viscous force generated by moving the template.**

22

Acknowledgements

- Prof. Randy Schunk, Univ. of New Mexico
- Prof. S. V. Sreenivasan & Dr. Shrawan Singhal, Univ. of Texas at Austin
- NSF Engineering Research Centre : NASCENT
- Texas Advanced Computing Center

NSF

NASCENT
NANOMANUFACTURING SYSTEMS FOR MOBILE COMPUTING AND MOBILE ENERGY TECHNOLOGIES
A NATIONAL SCIENCE FOUNDATION NANOSYSTEMS ENGINEERING RESEARCH CENTER

TACC
TEXAS ADVANCED COMPUTING CENTER

23

Thank you for your attention.

- Questions?

408477 Scaling-up Nano-Material Integration with DNA Brick Crystals

Tuesday, November 10, 2015: 3:35 PM
253B (Salt Palace Convention Center)

Wei Sun and Peng Yin, Wyss Institute for Biologically Inspired Engineering at Harvard University, Boston, MA

De novo design and assembly of functional nano-materials, especially those from chemical synthesis and CVD growth, is a key foundation in nanotechnology and promises diverse applications in photovoltaics, plasmonics, and electronics. As an example, nanoparticle-structured floating gate enables fast data reading plus high information density. However, it is still challenging to integrate the low-dimensional nano-materials into architectures with both sub-10 nm uniformity and large-area programmable device architectures.

Recently, in the field of structural DNA nanotechnology, researchers are using nucleic acids (particularly DNA) to self-assemble sophisticated synthetic structures at 2-nm resolution in a highly scalable manner—billions of identical structures assembling simultaneously. One particular successful strategy is known as single-stranded DNA bricks. By encoding spatial positioning information into the complementary sequence design, our lab has constructed more than 100 2D/3D structures from the modular DNA bricks with the help of computer-aided design software. We further use the modularity of DNA bricks to engineer DNA brick crystals, micron-sized two-dimensional patterns with prescribed thickness up to 80 nm and complex 3D surface features.

Using DNA brick crystals□as templates, we demonstrate the scalable integration of hundred metal nanoparticles into ordered programmable 2D/3D architectures and chiral surface, ranging from several hundreds nanometers to microns. Geometrical information of the prescribed nanoparticle architectures is firstly encoded into the linear sequences of DNA bricks. Under epitaxial growth condition, a DNA brick crystal is folded with the expected periodicity over micron scale. Printing gold nanoparticles, ranging from 5 nm to 30 nm, onto the prescribed positions on the DNA brick crystal templates through surface DNA hybridization produces the prescribed plasmonic architectures with sub-5 nm positioning precision.

Besides metal nanoparticles, carbon nanotubes (CNT) can also be aligned onto DNA brick crystals to form micron scale 2D parallel arrays with prescribed pitch dimensions from around 8 nm to 24 nm at sub-5 nm positioning precision.

Developing DNA brick crystal-based scaling-up approach will bridge current synthetic foundation for functional nano-materials to the emerging nano-electronic applications. By rational control the structural parameters of DNA brick crystals, the architecture of the assembled nano-materials can be de novo designed and reliably fabricated in a scalable manner. High-resolution feature of DNA brick crystal further scales the positioning resolution of nano-materials to around 2 nm.

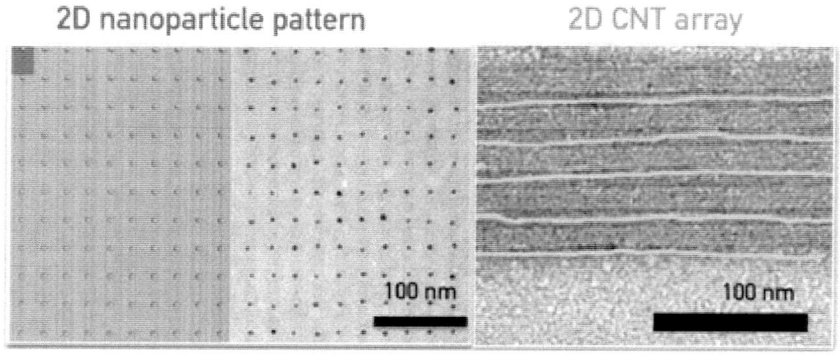

425779 Scalable Processing of 2D Nanosheets into 3D Crumpled Nanoparticles

Abstract:

3D Crumpled graphene was directly obtained from aqueous dispersions of pristine graphene using an industrially scalable spray drying technique. Capillary forces during the water evaporation induced the crumpling of nanosheets to multi-faced dimpled morphology. For the first time, the transition of 2D graphene nanosheets to a 3D crumpled morphology was directly observed inside the spray dryer. Graphene oxide (GO) was spray dried using the same procedure; however, their highly wrinkled final morphology was different than the crumpled pristine graphene nanosheets. The degree of crumpling of the nanosheets was controlled by changing the dimensionless ratio of evaporation rate to diffusion rate. Crumpled particles were redispersed into various solvents to evaluate their morphological changes as a response to rewetting. Crumpled GO nanosheets remained crumpled as a response to hydration, while the pristine graphene nanosheets unfolding behavior was solvent-dependent. This technique is similarly effective for dispersed inorganic nanosheets such as boron nitride and metal dichalcogenides. This study holds significance for both fundamental understanding of the origins of nanosheet crumpling and also for the use of crumpled nanosheets for further material processing.

428259 Stress-Directed Compositional Patterning in Responsive Sige Substrates

Abstract:

The ability to scalably fabricate periodic, large-area assemblies of Ge quantum dots (QDs) on Si or SiGe substrates with high degree of spatial and size uniformity would impact numerous technologies, including nano-/micro-electronics, optoelectronics, nanosensor arrays, and high-density patterned media for data storage. In this talk, we describe a process in which SiGe substrates are compositionally patterned over large areas using spatially-modulated elastic fields applied by a nano-indenter array. The indenter array, which is fabricated by interferometric lithography and dry etching, is pressed against a $Si_{0.8}Ge_{0.2}$ wafer in a custom-made mechanical press. The entire assembly is then annealed at high temperatures, during which the larger Ge atoms are selectively driven away from areas of compressive hydrostatic pressure. Compositional analysis of the substrates demonstrates that this approach leads to a transfer of the indenter array pattern into the near-surface compositional distribution, which in turn induces a patterned surface stress.

We then present a multiscale computer simulation approach of the “stress transfer” process. The model is based on a combination of lattice kinetic Monte Carlo (LKMC) and static energy minimization. The LKMC simulation is propagated using rates for atomic diffusion that depend explicitly on local values of stress, composition, and temperature. The dependence of atomic diffusion on composition is regressed to experimental data while the stress dependence is described using the theory of activation volumes [1]. The stress field is updated quasi-statically using a separate energy minimization routine with forces computed based on a Tersoff interatomic potential for the Si-Ge system [2]. The atomic stresses and identities are then smoothed to generate continuous fields that are used as input into the LKMC simulation. Using our model we show that the residual surface stresses induced by compositional patterning should be large enough to break symmetry during subsequent molecular beam deposition of Ge, and thus may provide an efficient approach for producing highly ordered Ge quantum dot structures via the well-known Stranski-Krastanov heteroepitaxial growth mechanism. We also compare our results to recent experimental measurements.

[1] M. J. Aziz, Applied Physics Letters 70, 2810 (1997).

[2] J. Tersoff, Physical Review B 39, 5566 (1989).

428433 Solvent Evaporation and Drying of Nanofibers in an Electrospinning Process

Abstract:

Electrospinning is a promising technology to produce nano sized fibers for diverse applications including biomedicals, composites, protective coatings, sensors, energy storage, membranes and filters. Electrospun fibers are produced by applying high voltage at the spinning tip where the polymer solution is fed and collecting almost dry polymer fibers on the grounded collector surface. Typically polymer solutions are used in between 2-20% polymer concentration in an electrospinning process depending on their rheological and viscoelastic properties. Solvent evaporation process during the electrospinning of polymer solutions is governed by the convective diffusion of solvent vapor from fiber surface into atmospshere far from the fibre. At least three other phenomena can also affect the evaporation process: i) the surface curvature affects the vapor pressure (Kelvin effect); ii) the charge affects the vapor pressure (Thomson effect); iii) Slip flow due to non-continuos gasous medium (Knudsen correction). The contributions of those factors during the electrospinning of PEO/water solutions will be presented.

415154 Coaxial Electrospun Nanofibers of Polyaniline/Poly(methyl methacrylate) Filled with Graphene

Abstract:

Electrospinning is a simple method to produce nanofibers mats with diameters in the nanometer scale. The technique is based on the application of an electric field on a polymer solution, and it is considered one of the most efficient techniques to fabricate high performance nanofibers mats, with distinct advantages such as very high surface area to volume ratio and porosity. Polyaniline (PANi) is one of the most studied intrinsically conductive polymer, mainly due to its ease of synthesis and interesting doping mechanism. It is however daunting to process PANi compared to most other polymers. As it is common among intrinsically conductive polymers, PANi has a fairly rigid backbone because of its high aromaticity. Thus, electrospinnability of PANi is limited due its insufficient solution viscosity and rigid chemical structure. One approach to make PANi electrospinnable is by its blending with materials that may facilitate its electrospinnability such as easily electrospinnable copolymers (e.g poly(ethylene oxide) and poly(lactic acid)), as previously reported. However, the presence of an insulating copolymer decreases the fibers conductivity due to dilution of the conducting component. The coaxial electrospinning method provides an alternative and effective way of fabricating neat PANi nanofibers. Coaxial electrospinning is an innovative electrospinning method that facilitates the fabrication of unspinable polymer nanofibers with unique core−shell structures. In this technique, two dissimilar solutions are spun simultaneously through a spinneret composed of two coaxial capillaries to produce core−shell structured nanofibers with unspinable polymer at the core section. Recently, graphene, a single-atom-thick two-dimensional sheet of sp^2-hybrized carbon atoms arranged in a honeycomb crystal structure, has attracted a lot of research interest and has been considered as a rising-star carbon material. Graphene has been intensively researched because of its remarkable properties such as high electrical and thermal conductivities, high mechanical strength and large surface area. Given its excellent electrical conductivity and electron mobility at room temperature, a highly conductive graphene-based polymer nanocomposite may be obtained. In this work, we report the preparation of neat polyaniline fibers filled with highly dispersed graphene sheets. The coaxial electrospinning process was used to produce core-shell structured fibers of a poly(methyl methacrylate) shell and a polyaniline core embedded with highly dispersed graphene nanosheets. The insulating shell was selectively removed by solvent etching. SEM, TEM, XPS and electrical conductivity measurements were employed to characterize these nanofibers.

429655 Pharmaceutical Production of Drug-Eluting Nanofibers By Precise Engineering of Needleless Electrospinning with an Oscillating Carriage

Abstract:

Electrospun nanofibers have been broadly investigated as medical fabrics in applications for drug delivery, tissue engineering, and wound healing[1]. These pharmaceutical applications of nanofibers will require a scalable process to produce precise fabric homogeneity and drug loading, which has yet to be demonstrated on a manufacturing scale instrument. Free surface or "needleless" electrospinning is a versatile and scalable method being evaluated for high throughput nanofiber production. A recent development in manufacturing scale needleless electrospinning equipment is the oscillating carriage method for solution entrainment onto a stationary wire electrode, which addresses scalability concerns with the current rotating wire entrainment method[2]. However, a narrow physical understanding of the oscillating carriage method has constrained its applications exclusively to low basis weight nanofiber coatings in the filtration industry[3]. In contrast to filtration coatings, electrospun medical fabrics are more challenging to manufacture due to requirements for fabricating high basis weight, stand alone materials that are needed to realize certain clinical applications.

Here, we present a process to use an oscillating carriage free surface electrospinning platform to fabricate uniform and high basis weight fiber mats loaded with tenofovir (TFV), an antiretroviral drug that inhibits HIV. TFV-loaded nanofibers were electrospun from a blend of polyvinyl alcohol and polyethylene oxide (PVA-PEO) polymers to produce 50-120 grams per square meter basis weight fabrics. We observed a mass accumulation of uniform material of approximately 3 grams per hour, which is a 12-fold increase from our typical process performance. This improvement was realized through isolated optimization of throughput, cross-direction uniformity, and machine-direction uniformity. Statistically designed screening experiments for the oscillating carriage entrainment method showed drug loading, polymer molecular weight, electrode distance, electric field intensity, and collecting electrode geometry are all significant in explaining process throughput and content uniformity. For the rotating wire geometry, significant variables only emerged to describe throughput whereas a strategy for optimizing content uniformity was not identified. Using these response outcomes and empirical observations of the electrospinning process, we developed a model describing the nanofiber mass deposition profile as a function of jet spacing, entrainment volume, jet initiation time, jetting time, polymer concentration, carriage speed, and wire length. This model allows us to derive a general optimization approach for any material composition. We expect our findings and analysis of the oscillating carriage entrainment method for needleless electrospinning will advance the pharmaceutical production of other novel medical fabrics.

---

[1] Meinel, A. J.; Germershaus, O.; Luhmann, T.; Merkle, H. P.; Meinel, L. *European Journal of Pharmaceutics and Biopharmaceutics* **2012**, *81*, 1.

[2] Forward, K. M.; Rutledge, G. C. *Chemical Engineering Journal* **2012**, *183*, 492.

[3] Petrik, S.; Maly, M. In *Materials Research Society*; Tanioka, A., Ed. Boston, MA, USA, 2009; Vol. 1240

432750 Microstructure of Two-Dimensional Grains in Copper Thin Films By Laser Induced Rapid Lateral Solidification

Abstract:

Solidification is a common process used in the manufacture of bulk metals, and is fundamentally a phase transformation from liquid to solid that has been studied for many years. In the case of rapid solidification, which is characterized by rapid heat loss and a high interface velocity, non-equilibrium conditions prevail and are not yet understood very well [1-3]. In this research, we present experimental results of RLS (rapid lateral solidification) under the newly geometric heat flow and growth direction.

A pulsed excimer laser is applied as an energy source to melt copper thin films. This leads to rapid resolidification upon cooling, and unique solidification microstructures [4-8]. Using this technique, we are able to successfully prepare RLS microstructures that are reproduceable and controllable through the laser process parameters. Much of the work presented here is an investigation of the rapid solidification process indirectly by examination of the post-solidified microstructure using electron microscopy. We have quantitatively studied several aspects of the microstructure, and address critical questions associated with the current theories. This includes identification of four-zone microstructure and quantification of mechanisms governing defect generation and texture formation.

Copper films with thickness of 100 nm, 200 nm, 500 nm and 1000 nm have been melted by single pulse of 248 nm laser, and the induced RLS areas are consisted with partial two-dimensional ribbon grains with low defect density and strong texture.

The length and width of the RLS grains will increase with the thickness of the thin films. The 200 nm thick copper thin films are characterized to reach the maximized length of 22 um with about 1.2 um width, while the 1000 nm thick ones reach 40 um in length with about 1.6 um width. Furthermore, the resolidified grains will form very strong textures during RLS, and the textures increase with the thickness too. The 200 nm thick copper thin films are characterized of strong <100> texture in the growth direction, while the 1000 nm thick ones have <100> textures in the three geometric directions. The defect density is not found to have relationship with the thickness. However, in all of the copper thin films with different thickness, the defect density is characterized to be high only at the beginning RLS and at the end of RLS. The mainly parts of PLS areas are found to be perfect with very low defect density.

This unique microstructure of RLS thin films is expected to be better in electronic properties than deposited thin films [9-10], and is potentially applied in electronic nano-device in near future.

1. Jones BL, Weston GM. Orientation Relations Between Dendrites Grown in a Highly Undercooled Melt. *Journal of Crystal Growth.* 1970;7(2):143-146.
2. Turnbull D, Cech RE. Microscopic Observation of the Solidification of Small Metal Droplets. Journal of Applied Physics. 1950;21(8):804-811.
3. Tenwick MJ, Davies HA. The Melt Spinning and Multilayer Deposition of Pure Copper. *Materials Science and Engineering.* 1984;63(1):L1-L4.
4. Kline JE, Leonard JP. Suppression of dewetting phenomena during excimer laser melting of thin metal films on SiO_2. *Thin Solid Films.* 2005;488(1-2):306-33.
5. Zhong R, Kulovits A, Wiezorek JMK, Leonard JP. Four-Zone Solidification Microstructure Formed by Laser Melting of Thin Copper Films. *Applied Surface Science.* 2009;256(1):105-111.
6. Kulovits A, Zhong R, Wiezorek JMK, Leonard JP. Electron Microscopy of Geometrically Confined Copper Thin Films after Rapid Lateral Solidification. *Thin Solid Films.* 2009;517(13):3629-3634.
7. Zhong R, Wiezorek JMK, Leonard JP. Texture of Rapid Solidified Cu Thin Films Studied by SEM EBSD and TEM. *15th International Conference on Textures of Materials (ICOTOM 15).* 2008;201:35-46.
8. Zhong R, Kulovits A, Wiezorek JMK, Leonard JP. Laser-Induced Microstructural Modification of Polycrystalline Cu and Ag Films Encapsulated in SiO_2. *Materials Research Society Symposium B Proceedings.* 2007;990:0990-B09-07.
9. Karim S, Ensinger W, Cornelius TW, Neumann R. Investigation of Size Effects in the Electrical Resistivity of Single Electrochemically Fabricated Gold Nanowires. *Pysica E.* 2008;40(10):3173-3178.
10. Wang M, Zhang B, Zhang GP, Yu QY, Liu CS. Effects of Interface and Grain Boundary on the Electrical Resistivity of Cu/Ta Multilayers. *Journal of Materials Science & Technology.* 2009;25(5): 699-702.

436692 Optically Directed Mesoscale Assembly of Organic-Inorganic Hybrid Nanomaterials

Abstract:

Different desired characteristics of materials are often in conflict in materials design. For example, materials that have low density also have low mechanical strength, and vice versa. The fabrication of materials with high density/low Young's modulus or low density/high Young's modulus are important for applications such as flexible electronics. Assembling hybrid materials to combine the conflicting physical, chemical, mechanical, optical and electrical properties of pure components remains a major challenge in materials design. We intoroduce an optically directed method to synthesize new materials that are assembled from nanometer scale constituents into components, systems, or materials that are microns to millimeters in size. Optically-Directed Mesoscale Assembly and Patterning (ODMAP) is a recent Argonne-patented process that simultaneously assembles and patterns the nanoscale building blocks into novel mesoscale/macroscopic structures. The optically-directed methodology is used for the synthesis and patterning of hybrid materials such as those comprising organic (carbon) and inorganic (gold) nanoparticles (NPs). The new materials will provide functionality in the form of *flexible*, high conductivity, robust microcircuitry pertinent to a broad range of energy, sensory, diagnostic, and communications functions.

Self-assembly in aqueous solution and nano-morphology of a dual-responsive hydrophilic random copolymer poly(NIPAM-co-AA)

Lu Yang, Hanting Chen, Yun Fang*

1.School of Chemical and Material Engineering, Jiangnan University, Wuxi, China,

2.MOE Key Laboratory of Food Colloids and Biotechnology, Jiangnan University, Wuxi, China

Key words: hydrophilic random copolymer, stimuli-responsive, self-assembly, nano-morphology

ABSTRACT: A dual-responsive hydrophilic random copolymer (HRC) poly(NIPAM-co-AA) is synthesized in this paper using AIBN as a free radical initiator in *t*-butanol at 70 °C. The solvent affinity of NIPAM chain units and AA chain units in poly(NIPAM-co-AA) molecules could be reversed in aqueous solution depending on pH and/or temperature change, to produce poly(NIPAM-co-AA) nano-objects with various morphology including spheres[1,2], rectangles, and rods[3-6]. In the present study, particular attention is paid to nano-morphology and nano-structure of these assemblies. Such nano-object exhibits thermo-responsive behavior: heating to 40 °C or 60 °C (above the lower critical solution temperature, LCST, of the copolymer) and keeping pH 4.0 causes self-assembly due to the onset of a hydrophilic-to-hydrophobic transition of NIPAM chains. TEM image of the self-assembly reveals a sphere that consists of NIPAM-rich segments mainly inside and AA-rich segments almost outside, and that is because the NIPAM chain units are transformed from hydrophilic to hydrophobic and decreasing the hydrogen bonding interactions between them and water. And such nano-object also exhibits pH-responsive behavior: decreasing to pH 0.9 or increasing to pH 6.0 and keeping the temperature at 20 °C (below LCST) causes self-assembly due to the onset of a hydrophilic-to-hydrophobic (and the latter hydrophobic-to-hydrophilic) transition of AA chains. TEM image of the self-assembly corresponding to pH 0.9 reveals a rectangle that consists of AA-rich segments mainly inside and NIPAM-rich segments almost outside. However, TEM image of the self-assembly corresponding to pH 6.0 reveals a rod that consists of NIPAM-rich segments mainly inside and AA-rich segmengs almost outside. Particular attention is such nano-object exhibits pH and thermo dual-responsive behavior: increasing to pH 6.0 and heating to 60 °C (above LCST) also causes self-assembly due to the onset of both the hydrophilic-to-hydrophobic transition of NIPAM chain units and the hydrophobic-to-hydrophilic transition of AA chain units. TEM is used to investigate the nano-morphology changes that occur during the sphere-to-rectangle-to-rod duo to the pH and/or temperature responsive behavior. In addition, AuD-TEM (in situ reduction and gold doping-TEM coupling technology)[7] is used to investigate the nano-structure. The rectangle (pH 0.9, 20 °C) and rod (pH 6.0, 20 °C) exhibits a sponge nano-structure due to the hydrophilicity of a small quantity of NIPAM chain units or AA chain units inside.

REFERENCES

(1) Kim, S. K.; Jung, E.; Goodman, M. D.; Schweizer, K. S.; Tatsuda, N.; Yano, K.; Braun, P. V. *Acs Appl Mater Inter* **2015**, *7*, 9128.

(2) Wang, J. L.; Wang, S. H.; Zhou, Y. Q.; Wang, X. G.; He, Y. N. *Acs Appl Mater Inter* **2015**, *7*, 16889.

(3) Doane, T. L.; Alam, R.; Maye, M. M. *Nanoscale* **2015**, *7*, 2883.

(4) Chen, C.; Qian, Y. C.; Sun, C. B.; Huang, X. J. *Soft Matter* **2015**, *11*, 6266.

(5) An, T. H.; La, Y.; Cho, A.; Jeong, M. G.; Shin, T. J.; Park, C.; Kim, K. T. *Acs Nano* **2015**, *9*, 3084.

(6) Zhang, Y.; Liu, Q. K.; Mundoor, H.; Yuan, Y.; Smalyukh, I. I. *Acs Nano* **2015**, *9*, 3097.

(7) Fang, Y.; Pang, P. P.; Lai, Z. Y. *Chemistry Letters* **2011**, *40*, 1074.

429928 Ultrastable Core-Shell Structured Nanoparticles Directly Made from Zwitterionic Polymers

Abstract:

Colloid stability is crucial for nanoparticles (NPs) to remain nano-effects in various functioning media. Various methods had been utilized to modify hydrophobic core to render nanoparticle with colloid stability, especially in bio-related environment. However, these methods were either involving complex chemical synthesis or low in efficiency due to unpredicted NP surface environment. Here, rather than modify the nanoparticle, for the first time we directly make self-assembled core-shell nanoparticles from zwitterionic polymer using green chemistry. Zwitterionic polymers are novel generation biomaterials showing excellent non-fouling properties. NPs protected by zwitterionic polymers have been found with long blood circulation characteristics and particularly suitable for drug delivery applications. Characterization confirmed the core-shell structure and resulting zwitterionic nanoparticles had superior colloid stability in highly concentrated protein/PBS buffer solution and even freeze-drying condition. This work initiates a new concept of directly making stable nanoparticles through green chemistry and will inspire new strategy and green chemistry to be applied in nanotechnology.

431222 The Electrorheological Effect for Polyhedral Silsesquioxane Cage Structures with Cyanopropyl Functional Groups

Abstract:

Previous research has shown that sulfonated polyhedral silsesquioxane (s-POSS) material when mixed with silicone oil (PDMS) exhibits a significant ER effect. Here we show that cyanopropyl POSS (c-POSS) particles in silicone oil also show significant ER activity. The effects of the functional group attached to the nanocages (POSS) on the rheological properties of the suspension is demonstrated through steady flow and oscillatory tests. Furthermore the electrorheological properties of these mixtures is also shown in these tests. The yield stress, t_y, scales with the electric field, E, as $\tau_y \sim E^{1.15}$. The elastic storage modulus of the suspension at small strains (γ=0.10) scales with the electric field E as $G' \sim E^{0.99-1.76}$. The research also contributed answers to the effects of changing the concentration of c-POSS particles. The preyield, yield and postyield characteristics of the ER fluid were determined using the flow curves, small amplitude oscillatory tests, and thixotropic stress loop tests.

432348 Observation of Large Nematic Domain in Suspension of Nano-Plate

Abstract:

Observation of large nematic domain in suspension of nano-plates

Zhengdong Cheng, Abhijeet Shinde

Concentration dependent liquid crystalline phase transitions in aqueous suspensions of plate-like nanoparticles are of interest to scientists[1, 2] and engineers[3]. From metastable suspension, nematic tactoids nucleate, grow and settle due to gravity. As the time passes, gravity keeps compressing the nematic tactoids thus increasing their concentration. tactoids with different orientations merge and develop a uniform domain spanned over the entire suspension except the top isotropic phase. Role of gravity in the formation of the large domain is investigated. It is also found that growth of the domain is faster in suspensions containing low aspect ratio (thickness/size) nano-plates. Alignment of nanoplates in the large domain is studied using polarized optical microscope, circular dichroism and synchrotron X-ray.

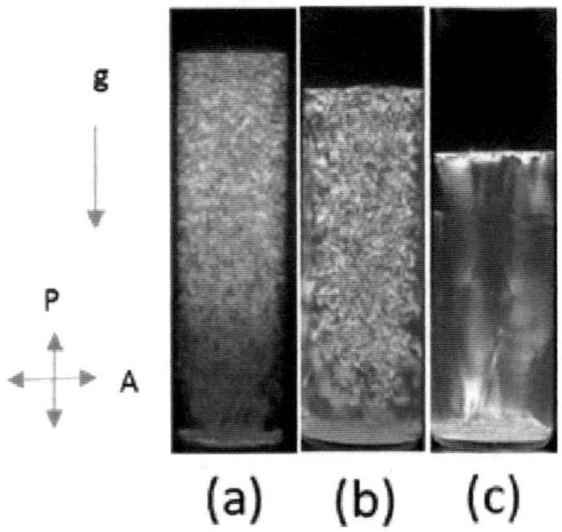

Figure. Cross polarizer observations of vertical tube containing suspension of nano-plates of same concentration but at different times. (a) Isotropic (dark) and Nematic (colorful) phases separation observed in suspension 1 day after its preparation. Different colors in nematic phase appear due to different orientations of nematic tactoids. (b) After 70 days, large domain developing near side wall and bottom of the tube. (c) Large domain with some defects developed in the suspension after 160 days. Directions of polarization in polarizer and analyzer are indicated with lines 'P' and 'A' respectively. Gravity is vertically downwards and is indicated by arrow 'g'.

1. van der Kooij, F. M.; Kassapidou, K.; Lekkerkerker, H. N. W. Liquid crystal phase transitions in suspensions of polydisperse plate-like particles. *Nature* 2000, 406, 868-871.

2. Mejia, A. F.; Chang, Y. W.; Ng, R.; Shuai, M.; Mannan, M. S.; Cheng, Z. D. Aspect ratio and polydispersity dependence of isotropic-nematic transition in discotic suspensions. *Phys. Rev. E* 2012, 85, 12.

3. Wong, M.; Ishige, R.; White, K. L.; Li, P.; Kim, D.; Krishnamoorti, R.; Gunther, R.; Higuchi, T.; Jinnai, H.; Takahara, A.; Nishimura, R.; Sue, H. J. Large-scale self-assembled zirconium phosphate smectic layers via a simple spray-coating process. *Nat. Commun.* 2014, 5, 12.

432978 Liquid-Phase Exfoliation of Phosphorene: Design Rules from Molecular Dynamics Simulations

Abstract:

Liquid-Phase Exfoliation of Phosphorene: Design Rules from Molecular Dynamics Simulations

Vishnu Sresht[1], Agilio A. H. Pádua[1,2], Daniel Blankschtein[1]

[1]Department of Chemical Engineering, Massachusetts Institute of Technology, Cambridge, Massachusetts 02139, United States

[2]Institut de Chimie de Clermont-Ferrand, Université Blaise Pascal and CNRS, 63171 Aubière, France

Session: NANOSCALE SCIENCE AND ENGINEERING FORUM, 22002 Self and Directed Assembly at the Nanoscale

The liquid-phase exfoliation of phosphorene, the two-dimensional derivative of black phosphorus, in the solvents dimethylsulfoxide, dimethylformamide, isopropyl alcohol, *N*-methyl-2-pyrrolidone, and *N*-cyclohexyl-2-pyrrolidone is investigated using three molecular-scale "computer experiments". We modeled solvent–phosphorene interactions using a new atomistic force field, based on *ab-initio* calculations and lattice dynamics that accurately reproduces experimental mechanical properties. We probed solvent molecule ordering at phosphorene/solvent interfaces and discovered that planar molecules such as *N*-methyl-2-pyrrolidone preferentially orient parallel to the interface. We subsequently employed a novel simulation technique to peel a single phosphorene monolayer from a stack of black phosphorus, and analyzed the role of "wedges" of solvent molecules intercalating between phosphorene sheets in initiating exfoliation. The exfoliation efficacy of a solvent is enhanced when either molecular planarity "sharpens" this molecular wedge, or strong phosphorene–solvent adhesion stabilizes the newly exposed phosphorene surfaces. Finally, we examined the colloidal stability of exfoliated flakes by simulating their aggregation, and showed that dispersion is favored when the cohesive energy between the molecules in the solvent monolayer confined between the phosphorene sheets is high (as with DMSO), and is hindered when the adhesion between these molecules and phosphorene is strong; the molecular planarity in solvents like DMF enhances the cohesive energy. Our results are consistent with, and provide a molecular context for, experimental exfoliation studies of phosphorene and other layered solids, and our molecular insights into the significant role of solvent molecular geometry and ordering should complement prevalent solubility-parameter-based approaches in establishing design rules for effective nanomaterial exfoliation media.

433148 Characterization and Control of the Structure and Properties of Low-Density, Linker-Mediated Nanocrystal Gels

Abstract:

Owing to their open, porous structure and high surface area, aerogels have found application in a variety of areas, including insulation, water purification, and catalysis. The use of nanocrystals to form aerogels or other low density gels is a particularly promising strategy to form novel materials: the high surface area and porosity allow the material to retain any unique properties (enhanced catalytic activity, for example) that are often inherent to nanocrystals, while creating a material with macroscopic dimensions. One strategy to form such gels is to mediate attractions between the nanocrystals via a linker species, e.g., a cation that is coordinated by negatively charged ligands. A variety of low-density gels have been formed experimentally from nanocrystals using this strategy, but how to control the physical properties of these gels is not well understood in general. Using a combination of theoretical calculations and Monte Carlo simulations, we investigate gelation in a system where the attractions are provided by an intermediate linker species, focusing on elucidating the effect of varying (1) linker concentration, and (2) the effective interactions between nanocrystals. We connect the underlying phase behavior of the linker-mediated nanocrystalline system to the resulting properties of the gel, and we compare to other systems known to form low-density gels.

428598 Application of Sub-Microfluidic Devices in Water Filtration Processes to Investigate the Removal of Pathogens

Abstract:

Access to clean water has become a critical issue as the water resources became scarce. There are several filtration techniques used in water treatment facilities such as reverse osmosis, and sand and gravel filters. Slow sand filtration is an economical filtration step used to obtain drinking water in cities. It filters particles and bacteria as water flows through the sand. Studies revealed that bacteria could alter their shape and fit into structures that are narrower than their sizes. The effect of flow on bacterial morphology in pores is not well documented. Here, we are applying nanobiotechnology to investigate the bacterial behavior in narrow structures as they are exposed to fluid flow. Miniaturized systems are great tools to mimic pores, and observe bacterial cells at a single cell level. This study will provide a pressure profile for pathogenic bacteria to improve the filtration processes. We designed sub-microfluidic devices that have 950 nm tall constrictions with widths ranging from 1.5-3 μm. The constrictions connect the inlet channel, where the pressure is applied, to an outlet channel, which is open to atmosphere.

The microfluidic devices were fabricated in the clean room. Constrictions were patterned on a silicon wafer with electron beam lithography, and chromium metal was sputtered to determine the height of the constrictions. Inlet and outlet microchannels were aligned onto the constrictions with photolithography. The completed master wafer was used as a mold to obtain poly-dimethylsiloxane (PDMS) devices, which were then permanently bonded to glass cover slips. PDMS is a polymer that is widely used in microfluidics field to study cells. Here, we studied two microorganisms: *Escherichia coli* (*E. coli*) and *Pseudomonas aeruginosa* (*P. aeruginosa*). *E. coli* is an indicator of contamination when detected in water. The bacteria were initially grown overnight in lysogeny broth at 37 °C. The cultures were diluted to avoid accumulation of bacterial cells at the entrance of the constrictions. The cells were flowed through the constrictions as the pressure was applied at increasing increments by using Fluigent microfluidics flow control system, which was connected to a house airline. The images were taken after each pressure application.

Our results show that filtration of species varies with the applied pressure. When the cells are exposed to the same applied pressure, the physiology of bacteria becomes a critical factor to determine their behavior in narrow structures. *E. coli*, which is larger in diameter and less motile than *P. aeruginosa*, entered the constrictions at higher pressure values. As the constriction size became narrower than the diameter of both bacteria, the pressure applied significantly increased. Further analysis is critical to determine optimum parameters for separating pathogens to improve the filtration processes.

430313 Electrocatalytic Inactivation of E. coli in Point of Use Drinking Water Applications Using TiO2 Nanotubes

Abstract:

A flow reactor device initiating electroporation was developed for bacterial disinfection in lab-synthesized and natural waters. Complete inactivation of bacterial contaminated waters with flow rates of 50 mL/min was observed. The device consisted of immobilized titanium dioxide nanotube arrays, which applied a bias of up to 5 V without noticeable water splitting. Water chemistry, applied voltage, and bacterial concentration were all found to impact disinfection rates. Complete inactivation of *E. coli* W3110 (~8 x 10^3 CFU/mL) occurred in 15 seconds in the device with an applied voltage of 3 V in a 100 ppm NaCl solution. Inactivation rates varied widely as a function of ionic concentrations. Disinfection in natural water was inhibited by the presence of inorganic ions and other constituents which are commonly found in natural water. A higher voltage of 6 V was required to reach 100% efficiency because of this. An applied bias to the nanotube flow through disinfection chamber shows promise as an effective bacterial disinfection device for *E. coli* inactivation without the need for solar irradiation.

STABILITY OF CARBON DOTS IN NATURAL WATERS: IMPLICATIONS FOR SENSING AND THE ENVIRONMENT

María M Fidalgo, Jingjing Dai, Austin Zambrana and Chloe Rees, Department of Civil and Environmental Engineering, University of Missouri, Columbia, MO, US.

Introduction

Interest for both fundamental research and applications has surged dramatically carbon nanodots (C-dots)[1-3]. Thus, it is inevitable such materials will find their way to natural water sources where, unlike the more well-studied fullerenes and carbon nanotubes, their environmental impact and fate is mostly unknown.

C-dots constitute a fascinating class of recently discovered nanocarbons that comprise discrete, quasi-spherical nanoparticles with sizes below 10 nm[1]. Because C-dots display size and excitation wavelength-dependent photoluminescence behavior, can be derived from inexpensive and abundant materials, and have presumably lower environmental impact, they are attractive alternatives to metal-based quantum dots (such as CdS and CdTe) especially for bio applications and aqueous-phase sensing. Early studies have shown C-dots are biocompatible[4] and less cytotoxic than Cd-based quantum dots[5] and may even inhibit cancer cells[6]. The increase in activity for these emerging carbon nanomaterilas make their appearance in natural waters inescapable.

Due to their large surface area and activity, adsorption is expected to play an important role in the fate and transport of C-dots [7]. It has been shown that adsorption may alter the surface characteristics of nanoparticles, so that they become dispersed and mobile in aqueous media, or aggregate and settle rapidly [8]. Organic macromolecules, such as natural organic matter, polysaccharides and proteins, are ubiquitous in natural waters and therefore readily available to interact with nanomaterials [9].

The objective of this work is to investigate the stability of C-dots, an emerging carbon nanomaterial, in natural waters (i.e. variable salinity) and the effect of the transformations they undergo under such conditions on the fluorescence properties.

Methods and Materials

Synthesis of C-dots:

Carbon Dots were synthesized following a modification of a previously published method [10]. A 70% w/w glycerol solution (Sigma, >99.5%) was added to a 20% w/w solution of NaH_2PO_4 (Fisher Scientific). The solution was heated for 20 minutes at 900W power in a domestic microwave. The product was allowed to cool to room temperature and the C-dots resuspended by addition of water and continuous stirring until a homogenous colored suspension was obtained.

Amine capped C-dots (AC-dots) were also fabricated: 2.10 g of a 10 mmol solution of citric acid (Sigma-Aldrich ACS reagent, >=99.5%) was added into 10 mL of Polyethylenimine, ethylenediamine branched (Sigma-Adrich). The reaction was heated from room temperature to 170 °C in silicone oil bath (Fisher Scientific) for 2h. Then the reaction flask was naturally cooled to room temperature and deionized water was added. The resulting solution was dialyzed against water

(Spectrumlabs, MWCO 1.0 kDa) for 2 days for purification, replacing the water was every 6 hours. Finally, AC-dots were freeze-dried to yield a yellow solid.

Size and surface charge measurements

Size measurements were performed by Dynamic Light Scattering (DLS) using a Zetasizer Nano ZS, Malvern, UK. Laser Doppler velocimetry was applied to characterize the electrophoretic mobility (EPM) of the nanomaterials in the various electrolyte solutions using a Malvern Zetasizer Nano ZS. Measured EPMs were converted to *zeta*-potential using the Smoluchowski equation [11]

$$U = \varepsilon\zeta / \mu$$

where U is the electrophoretic mobility, ε is the dielectric constant of the solution, μ is its viscosity, and ζ is the zeta potential. Disposable folded capillary cells were employed. The samples were taken from the freshly prepared suspensions, but measured after at least one hour, when aggregation equilibrium was assumed.

Fluorescence measurements

Fluorescence readings were conducted in a Hitachi F-4500 Fluorescence Spectrophotometer. For C-dots emission scans, an excitation wavelength of 350nm was used, corresponding to the peak from the excitation scan. For excitation scans, the emission wavelength used was 450nm for C-dots, and 460nm for AC-dots. These values correspond to the emission peaks. A slit width of 5nm was used for all measurements. Scans were conducted in a wavelength range of 300nm to 700nm, at a speed of 1200 nm/minute.

Results and Discussions

Two synthesis routes for C-dots were used and investigated. In the first approach, a solution of glycerol is thermally treated in the presence of a catalyst to yield the nanoparticles (C-dots). Surface functionalization is commonly performed after C-dots synthesis, either to give special functionality to the particle itself or to provide sites for reaction with other chemicals. Amino groups are frequently used for this last strategy. The second synthesis method used in this work yielded amino-capped carbon dots (AC-dots).

The transport behavior of the nanomaterials in the environment is expected to be controlled partially by their chemical composition but mostly by their surface chemistry. Although chemically similar with respect to their core composition, these two nanoparticles have the potential to react very differently to changes in natural water chemistries, and the effect on their stability and inherent properties (fluorescence) may also be modified.

Fresh waters (rivers, lake) are characterized by low ionic strength (1mM -5 mM) while seawater ionic strength can be as high as 0.5 M. Groundwater has intermediate, variable levels. The effect of increasing ionic strength on the size and surface potential, characterized by the zeta potential, of the carbon dots was investigated and shown in Figure 1 and figure 2 respectively; NaCl and $CaCl_2$ were used as electrolytes. The dots were resuspended in the electrolyte solutions from their "as-prepared" dry forms, resulting in a pH of circumneutral for the C-dots and a pH of 9 for AC-dots.

Increasing ionic strength is expected to have a destabilization effect in charged particles due to the compression of ionic double layer that turns electrostatic repulsion forces shorter range and allows for the net interaction energy to be attractive.

That trend was weakly observed for both materials in NaCl solutions; AC-dots became very unstable at ionic strengths higher than 50 mM and a reliable value for their size could not be obtained by DLS due to the presence of large aggregates (> 5 microns) and fast settling. The effect was even less clear in the divalent cation solutions. One possible explanation for the apparent indifference with respect to ionic strength may be their very small size, which may result in other mechanisms than surface charge (e.g. diffusion) being the ones controlling the aggregation phenomenon in the range of ionic strengths investigated.

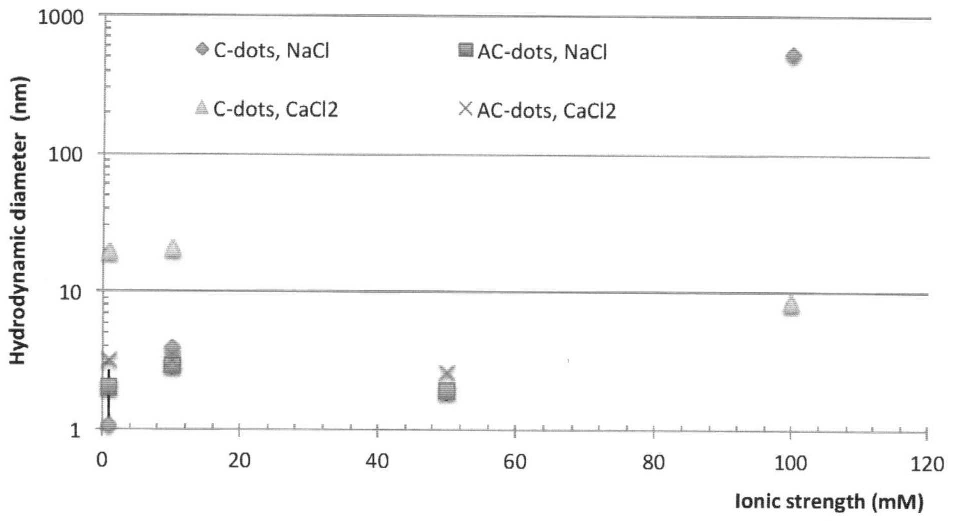

Figure 1. Hydrodynamic diameter of C-dots at various ionic strength concentrations; C-dots: glycerol derived carbon dots; AC-dots: amino capped carbon dots.

However, the effect of ionic strength in the zeta potential of the particles was important. As expected, C-dots showed a negatively charged surface wile amine carbon dots were positively charged; the absolute value of the zeta potential in deionized water was significant in both cases and resulted in a strong stabilization effect. The zeta potential of C-dots followed the expected trend of decreasing absolute values with increasing salt concentration. The effect was even more pronounced in the presence of Ca^{2+}. Monovalent cations (Na^{+}) are considered to indifferent with respect to the surface, while some degree of interaction due to specific adsorption may occur in the presence of divalent cations. In particular, calcium is well known for promoting aggregation of dissolved natural organic matter (NOM). Due to chemical similarities of NOM with the C-dots, we hypothesized that calcium may also play a role in the destabilization of carbon nanomaterials. Calcium ions seemed to adsorb onto the negatively charged C-dots, giving a lower surface charge, in absolute values. While the decrease in zeta potential for the Na^{+} solutions could be entirely attributed to charge shielding due to double layer compression, an additional effect of specific adsorption was clearly present in the Ca^{+2} suspensions. Interestingly, the opposite was observed for the positively charged AC-dots. Increasing concentrations of NaCl electrolyte produced a significant decrease in surface charge, the presence of divalent cations resulted in a reduced decrease in charge. This observation hints to some degree of adsorption of the Ca^{2+} to the positively charge surface. In this way, there is an additional contribution to charge in the particle surface that compensates, at least partially, the expected zeta potential decrease due to higher electrolyte concentrations.

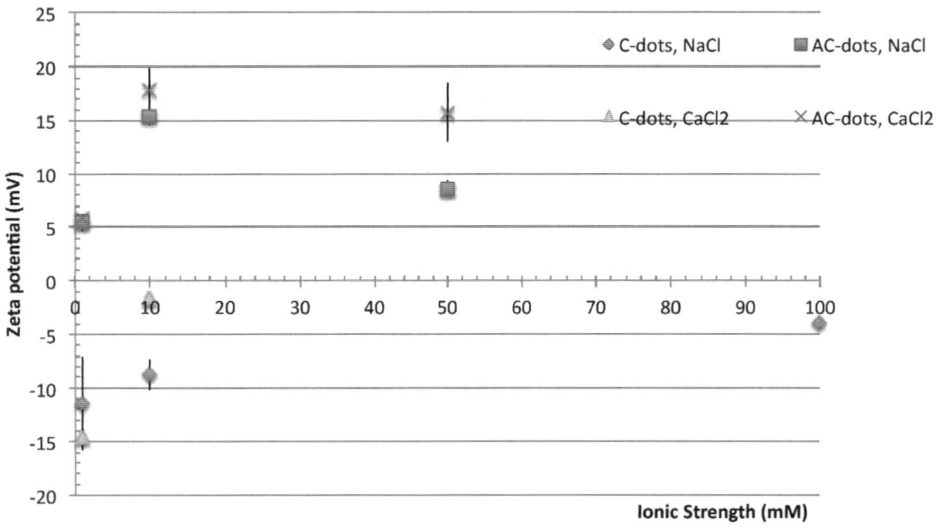

Figure 2. Effect of ionic strength on the surface charge of C-dots; C-dots: glycerol derived carbon dots; AC-dots: amino capped carbon dots.

Fluorescent nanoparticles are especially interesting in sensing applications in environmental science as well as medical imaging, due to the high emission intensity and stability. Quantum dots have been extensively investigated but their toxicity in aqueous systems makes them fairly unattractive. Carbon dots are a benign alternative fluorescent particle with a huge potential in both fields. In order to be useful, their properties need to remain stable over the range of conditions normally encountered in natural waters as well as physiological mediums.

We investigated the fluorescence emission intensity of both nanomaterials in three levels of ionic strength, given by two electrolytes: NaCl and $CaCl_2$. The results are presented in Figure 3. Since aggregations negatively affects the emission, we focused our work on those ionic strength conditions in which we did not observed significant particle size growth in our DLS determinations. Fluorescence emission intensity increased with ionic strength for both C-dots and AC-dots in the presence of an indifferent electrolyte. However, different effects were observed for the suspensions prepared in $CaCl_2$. Lower emission intensities were observed for C-dots with increasing concentration of Ca^{+2} in comparison with the same ionic strength level given by the monovalent ion. However, the presence of calcium resulted in a negligible effect on the AC-dots. The specific adsorption of Ca^{+2} was much more noticeable on the negatively charged C-dots that in the AC-dots; there was, nevertheless some adsorption as indicated by the differences in zeta potential detected for the latter, but it was not strong enough to have an influence on the fluorescence properties of the carbon dots.

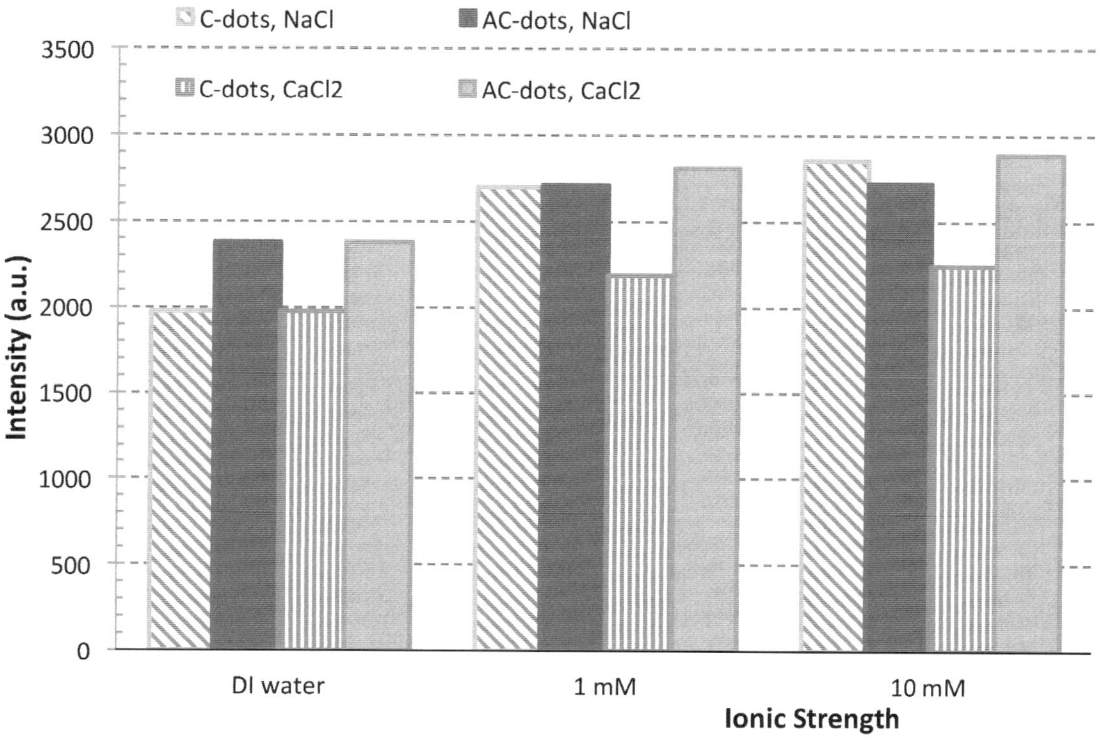

Figure 3. Fluorescence intensity peak for carbon dots at various levels of ionic strength. C-dots: glycerol derived carbon dots; AC-dots: amino capped carbon dots.

Conclusions

The carbon nanomaterials investigated, C-dots and AC-dots, resulted in stable suspensions up to an ionic strength level of 50mM, both in the presence of monovalent and divalent cations. Calcium ions can adsorb specifically to both materials, regardless of the fact that the AC-dots are positively charged: however, the adsorption and its effect on stability and fluorescence was much more noticeable for the negatively charged C-dots.

The increase in electrolyte concentration produced a decrease in absolute value of the zeta potential of the particles; in the presence of calcium ions, this charge screening effect is accentuated for C-dots and counteracted for AC-dots. Given the stabilization effect of surface charge, we can expect C-dots to become unstable and aggregate in calcium rich waters, while AC-dots would exhibit increased stability and therefore their higher tendency to stay in the water column and be transported with the flow.

The emission intensity exhibited a small increase over the ionic strength levels investigated in the presence of a non-adsorbing electrolyte. However, in the calcium rich suspensions, a decrease of fluorescence intensity was evident for the C-dots only. It is unclear if the reason for the observed variations is related to some surface interactions with the electrolyte or if the somewhat larger particle size observed for C-dots in $CaCl_2$ suspensions (i.e. moderately aggregated state) may have produced the lower emission observed.

References

1. Baker, S.N. and G.A. Baker, *Luminescent carbon nanodots: Emergent nanolights.* Angewandte Chemie - International Edition, 2010. **49**(38): p. 6726-6744.
2. Li, D. and R.B. Kaner, *Materials science: Graphene-based materials.* Science, 2008. **320**(5880): p. 1170-1171.
3. Rao, C.N.R., A.K. Sood, K.S. Subrahmanyam, and A. Govindaraj, *Graphene: The new two-dimensional nanomaterial.* Angewandte Chemie - International Edition, 2009. **48**(42): p. 7752-7777.
4. Qu, S., X. Wang, Q. Lu, X. Liu, and L. Wang, *A biocompatible fluorescent ink based on water-soluble luminescent carbon nanodots.* Angewandte Chemie - International Edition, 2012. **51**(49): p. 12215-12218.
5. Song, Y., D. Feng, W. Shi, X. Li, and H. Ma, *Parallel comparative studies on the toxic effects of unmodified CdTe quantum dots, gold nanoparticles, and carbon nanodots on live cells as well as green gram sprouts.* Talanta, 2013. **116**: p. 237-244.
6. Hsu, P.C., P.C. Chen, C.M. Ou, H.Y. Chang, and H.T. Chang, *Extremely high inhibition activity of photoluminescent carbon nanodots toward cancer cells.* Journal of Materials Chemistry B, 2013. **1**(13): p. 1774-1781.
7. Chowdhury, I., M.C. Duch, N.D. Mansukhani, M.C. Hersam, and D. Bouchard, *Colloidal properties and stability of graphene oxide nanomaterials in the aquatic environment.* Environmental Science and Technology, 2013. **47**(12): p. 6288-6296.
8. Lanphere, J.D., C.J. Luth, and S.L. Walker, *Effects of saolution chemistry on the transport of graphene oxide in saturated porous media.* Environmental Science and Technology, 2013. **47**(9): p. 4255-4261.
9. Saleh, N.B., L.D. Pfefferle, and M. Elimelech, *Influence of biomacromolecules and humic acid on the aggregation kinetics of single-walled carbon nanotubes.* Environmental Science and Technology, 2010. **44**(7): p. 2412-2418.
10. Wang, X., K. Qu, B. Xu, J. Ren, and X. Qu, *Microwave assisted one-step green synthesis of cell-permeable multicolor photoluminescent carbon dots without surface passivation reagents.* Journal of Materials Chemistry, 2011. **21**(8): p. 2445-2450.
11. Elimelech, M., J. Gregory, X. Jia, and R.A. Williams, *Particle Deposition and Aggregation.* Colloid and Surface Engineering: Applications in the process industries. Controlled Particle, Droplet and Bubble Formation, ed. R. Williams. 1995, Linacre House, Jordan Hill, Oxford, UK: Butterworth-Heinemann, Ltd.

EXPERIMENTAL STUDIES USING ULTRAVIOLET-VISIBLE SPECTROSCOPY ON SYNTHESIZED NANOPARTICLES

Engr. Dr. M. N. Idris* and **A. Abdullahi**
Energy & Process Refining Technology Group
Department of Chemical Engineering
University of Maiduguri
Borno State, Nigeria
Tel: +234 (0) 705 877 1096
Email: idrismn@unimaid.edu.ng, muhinu@hotmail.co.uk

ABSTRACT

In this study, silver nanoparticles were synthesized by two main chemical reduction methods. Silver nitrate was adopted as the main precursor, and reduced by sodium borohydride and trisodium citrate to produce particles of different size regimes. UV-Visible spectrophotometer was used to monitor the formation of the nanoparticles. The synthesis of nanoparticles via sodium borohydride was carried out at both ambient temperature and ice-bath temperature to compare the effect of temperature on the quality of final synthesis. The stability of the silver nanoparticles dispersed in aqueous medium was investigated with respect to time.
A yellow hue was given off by the silver nanoparticles sols that, a typical surface Plasmon resonance absorption provide maxima of silver nanoparticles at 370-450 nm spectra. The sizes of the nanoparticles were found to be dependent on the reduction routes. The sodium borohydride and trisodium reduction routes resulted in silver nanoparticles with a diameter of around 12 nm and 35 nm, respectively. However, sodium borohydride sols were found to be more concentrated and had lower polydispersity. In the borohydride synthesis, lower synthesis was found to be helpful in formation, purity and stability of the nanoparticles colloids. Also, an increase in the volume ratios of aqueous precursor solutions was found to result in nanoparticles colloids that were more concentrated, less distributed and had higher colloidal stability.

Keywords: *Colloidal, Characterization UV-visible, Synthesis & Polydispersity*

Correspondence about this article*: Please for all correspondence on this article, kindly contact the principal author as stated above. Thank you.*

1.0 INTRODUCTION

A particle is defined as a small object that behaves as a whole unit in terms of its transport and properties [1]. Particles are further classified according to size in terms of diameter [1]. Course particles cover a range between 10,000 and 2,500 nm. Fine particles are sized between 2,500 and 100 nanometers. Ultrafine particles or nanoparticles are sized between 1 and 100 nm. The basis of the 100 nm limit for nanoparticles is the fact that novel properties that differentiate them from the bulk material typically develop at a critical length scale of under 100 nm [2, 3].

The quest to fabricate, characterize and understand small noble metal molecules is among the outstanding challenges of contemporary material science, as such nanoparticles bridge the atomic and the bulk worlds [4]. A bulk material should have constant physical properties regardless of its size, but at the nano-scale size, dependent properties are often observed. Thus, the properties of materials change as their size approaches the nanoscale and as the percentage of atoms at the surface of the material becomes significant. The interesting properties of nanoparticles are therefore, largely due to the larger surface area, which dominates the contributions made by the small bulk of the materials. [5].

Nanoparticles are small enough to confine their electrons and produce quantum effects and can absorb much light than material in thin films of continuous sheets. Those fascinating properties make them Physico-chemically superior to both atomic and bulk materials of the same element [6]. For instance, copper which is opaque at macro scale becomes completely transparent to visible light at the nanoscale [7]; stable materials like aluminum turn combustible [8]; Gold, which is rarely soluble becomes more soluble [9].

In this research, samples of silver nanoparticles are synthesized and their formation is studied using UV-Visible spectrophotometer.

1.1 Problem Statement

Metal nanoparticles, due to their special properties and also small dimensions, find important applications in optical, magnetic, thermal, sensoric devices, catalysis, etc. Many metal nanoparticles are under active research because they possess interesting physical properties differing considerably from that of the bulk phase [10]. It comes from the small sizes and high surface/volume ratio. The most common method employed for the synthesis of metal nanoparticles is the reduction of metal ions in solution [11, 12]. However, the control of size and shape at the nanometer level is a real problem, as the mechanism of size/shape control is still left largely unresolved.

In this work, an attempt has been made to analyze formation of silver nanoparticles from silver salt as well as after storing for variable time in atmospheric condition. The effect of time after synthesis on characteristics of these particles is investigated Using UV/Visible spectrophotometer. Further, the effect of synthesis temperature and using an alternate reducing agent is analysed.

1.2 Ultraviolet-Visible Spectroscopy

Spectroscopy is a technique used to quantify the light that is absorbed and scattered by a sample (a quantity known as the extinction, which is defined as the sum of absorbed and scattered light). In its simplest form, a sample is placed between a light source and a photo detector, and the intensity of a beam of light is measured before and after passing through the sample. These measurements are compared at each wavelength within UV-Visible range (300-700 nm) to quantify the sample's wavelength dependent extinction spectrum. The data is typically plotted as extinction as a function of wavelength. Each spectrum is background corrected using a "blank" – a cuvette filled with only the dispersing medium – to guarantee that spectral features from the solvent are not included in the sample extinction spectrum [13].

Nanoparticles have optical properties that are sensitive to size, shape, concentration, agglomeration state, and refractive index near the nanoparticle surface, which makes UV/Vis spectroscopy a valuable tool for identifying, characterizing, and studying these materials. Nanoparticles made from certain metals, such as gold and silver, strongly interact with specific wavelengths of light and the unique optical properties of these materials is the foundation for the field of plasmonics [13].

1.3 Aims and Objectives

The aim of this project is to study the formation of silver nanoparticles using Ultraviolet-visible spectrophotometer.

The objectives are:

- To synthesize samples of silver nanoparticles.

- To study the effect of varying precursor volumes on the resulting nanoparticles.
- To study the effect of synthesis temperature on the resulting nanoparticles.
- To study the effect of an alternate reducing agent on the resulting nanoparticles.
- To study the effect of storage for variable times in the atmosphere.

1.4 Scope of the Research

The scope of this research is synthesis of colloidal silver nanoparticles by reduction method (using two reducing agents) at two different temperatures; and characterization of all samples synthesized using UV-Visible spectrophotometer.

2.0 BRIEF BACKGROUND ON NANOPARTICLES

Until the Middle Ages, the reputation of soluble gold was based mostly on its fabulous curative powers of various diseases, for example, heart and venereal diseases, dysentery, epilepsy, and tumors; it was also used in the diagnosis of syphilis. The history of the nanoparticles from ancient times to the Middle Ages has been summarized by Daniel and Astruc [15]. The first book on colloidal gold was published in 1618 by the philosopher and medical doctor Francisci Antonii. This book includes considerable information on the formation of colloidal gold sols and their medical uses, including successful practical cases. The book noted that soluble gold appeared around the fifth or fourth century B.C. in Egypt and China. On the other hand, industrial manufacturing of stained glass with colloidal particles was established by Kunckel in the seventeenth century (1676). He also published a book whose Chapter 7 was concerned with 'drinkable gold that contains metallic gold in a neutral, slightly pink solution that exerts curative properties for several diseases' [16]. He concluded that gold must be present in aqueous gold solutions to a degree of contamination such that it is not visible to the human eye. A colorant in glasses, that is, the 'Purple of Cassius', was a colloid resulting from the presence of gold particles and tin dioxide and was highly popular in the seventeenth century. A complete treatise on colloidal gold was published in 1718 by Helcher [17].

In 1857, in a well-known publication, Michael Faraday [18] reported the formation of deep red solutions of colloidal gold by reduction of an aqueous solution of chloroaurate ($AuCl_4^-$) by phosphorus in CS_2 (a two-phase system). He also investigated the optical properties of thin films prepared from dried colloidal solutions and observed reversible color changes of the films upon mechanical compression (from bluish-purple to green). Since that pioneering work, thousands of scientific papers have been published on the synthesis, modification, properties, and assembly of metal nanoparticles, using a wide variety of solvents and other substrates [19].

Nanotechnology is easily evident in various old churches. A well-known application of early nanotechnology is the ruby red color that was used for stained glass windows during the Middle Ages. Beautiful examples of these applications can be found in glass windows of many Gothic European cathedrals, among which the León Cathedral (Spain) represents one of these unique masterpieces, located on the medieval French pilgrimage path to Santiago de Compostela (Spain); its impressive 2000 m^2 colored windows offer a unique view that certainly warrants a visit. Of course, the medieval artisans were unaware that they were using nanotechnology. They just knew that a particular process produced a beautiful effect. For example, the stained glass of a wonderful rose can be seen at the world heritage Cathédrale Notre-Dame de Chartres in France. Later, chemistry clarified the reasons behind the generation of the color. These vivid colors were controlled by the size and the form (or shape) of the nanoparticles of gold and silver. The relation between particles and their associated colors has been discussed recently by Jin and coworkers [20]. In an article of 22 February 2005, the New York Times summarized the relationship between the color of stained glass and the size/shape of the nanoparticles [21].

After several decades, the ingredients present in the stained glass (colored glass) of various churches were clarified subsequent to the development of analytical instruments. People without professional expertise in nanotechnology are also increasingly contributing to the technology. This continues as we speak, with various nanostructures being proposed and discovered, and their applications described [19].

2.1 Properties of Nanoparticles

2.1.1 Uniformity

The chemical processing and synthesis of high-performance technological components for the private, industrial, and military sectors requires the use of high-purity ceramics, polymers, glass-ceramics, and composite materials. In condensed bodies formed from fine powders, the irregular particle sizes and shapes in a typical powder often lead to non-uniform packing morphologies that result in packing density variations in the powder compact [22].

Uncontrolled agglomeration of powders due to attractive van der Waals forces can also give rise to in microstructural inhomogeneities [23]. Some pores and other structural defects associated with density variations have been shown to play a detrimental role by growing and thus limiting end-point densities. Differential stresses arising from inhomogeneous densification have also been shown to result in the propagation of internal cracks, thus becoming the strength-controlling flaws [24]. It would, therefore, appear desirable to process a material in such a way that it is physically uniform with regard to the distribution of components and porosity, rather than using particle size distributions that will maximize the green density. The containment of a uniformly dispersed assembly of strongly interacting particles in suspension requires total control over inter-particle forces. Monodisperse nanoparticles and colloids provide this potential. Inert gas evaporation and inert gas deposition are free many of these defects due to the distillation (cf. purification) nature of the process and having enough time to form single crystal particles, however even their non-aggregated deposits have lognormal size distribution, which is typical with nanoparticles [25].

2.1.2 Large Specific Surface Area

Nanoparticles possess strong surface area effects (e.g. reactivity, high energy of surface area, adsorption, higher solubility, lower melting point, etc). For bulk materials larger than one micrometer (or micron), the percentage of atoms at the surface is insignificant in relation to the number of atoms in the bulk of the material, while the properties of materials change as their size approaches the nanoscale and as the percentage of atoms at the surface of a material becomes significant. Example, the high surface area to volume ratio of nanoparticles provides a tremendous driving force for diffusion, especially at elevated temperatures. Sintering can take place at lower temperatures, over shorter time scales than for larger particles [25].

2.1.3 Quantum Effects

Nanoparticles often possess unexpected optical properties as they are small enough to confine their electrons and produce quantum effects [26]. For example gold nanoparticles appear deep-red to black in solution. Nanoparticles of yellow gold and grey silicon are red in color; Gold nanoparticles melt at much lower temperatures (~300 °C for 2.5 nm size) than the gold slabs (1064 °C); Absorption of solar radiation is much higher in materials composed of nanoparticles than it is in thin films of continuous sheets of material. In both solar PV and solar thermal applications, controlling the size, shape, and material of the particles, it is possible to control solar absorption [27, 28].

Other size-dependent property changes include quantum confinement in semiconductor particles, surface plasmon resonance in some metal particles and superparamagnetism in magnetic materials. What would appear ironic is that the changes in physical properties are not always desirable. Ferromagnetic materials smaller than 10 nm can switch their magnetisation direction using room temperature thermal energy, thus making them unsuitable for memory storage [29].

2.2 Types and Origin of Nanoparticles

There are a number of ways, based on different aspects, that NPs can be classified into groups. Considering their origin, they can be divided into natural and anthropogenic (man-made) which, in turn, can be subdivided into accidental and engineered or manufactured NPs or they can be further separated, based on their chemical composition, into organic (carbon containing) and inorganic [30]. As the name clearly implies, natural NPs are a group of nanosized materials which are generated by the action of natural processes and have been present in the natural environment for a very long time since the origin of the planet Earth. In addition to natural NPs, earlier human activities like mining, agriculture, forest burning and constructions have tremendously increased the concentration of accidental nanoparticles into the environment especially into the atmosphere. Despite the fact that NPs have been present in the environment for millions of years as natural and later as accidental NPs, it was not until quite recently that their interest in both pure and applied sciences have been extensively exploited due to the discovery of their promising, novel and applicable properties which make them attractive for a vast number of products in a wide range of scientific researches and for many sectors of the modern technology.

2.2.1 Natural Nanoparticles

Natural NPs are omnipresent and they are distributed throughout the atmosphere, oceans, soil systems, terrestrial water systems (groundwater and surface water), and in and/or on most living organisms both at the micro and macro levels and have been present in the environment forever [31]. They have been formed mainly by the effect of naturally occurring physical, chemical and biological processes such as hydrolysis, erosion, weathering, volcanic eruption, sea spray, photochemical reactions, growth nuclei in super-saturated fluids, plants roots on rocks, minerals and microorganisms. Forest fires, as volcanic eruptions, for instance, can spread ash and smoke over thousands of miles and lead to an increase in particulate matters, including NPs, exceeding ambient air quality standards [32].

2.2.2 Accidental Nanoparticles

Human activities have been releasing different forms of nanoparticulate matter for millennia as by-products of some activities which were essential for their survival such as agriculture, construction, food cooking, mining and mineral processing. However, accidental NPs input to the environment has risen sharply and dramatically since the beginning of the Industrial Revolution in the mid-18th century and after due to manufacturing emissions, nuclear waste generation and the combustion of fossil fuels [33]. Diesel and automobile exhausts are the primary source of atmospheric nano-and microparticles in urban areas (2002). Since their productions are spontaneous, uncontrolled accidental nanoparticles are more likely to be polydisperse /heterogeneous and have irregular shapes. They contain sulfide, sulphate, nitrate, ammonium, organic carbon, elemental carbon and trace metals [33].

2.2.3 Engineered Nanoparticles

Nanostructured materials behave differently and are often superior in properties to bulk materials due to two primary factors: surface effects (enhanced surface phenomena caused by the very high surface area to volume ratio, most atoms are on or near the surface thus they can be weakly bonded and more reactive) and quantum effects (discontinuity in behavior caused by the confined delocalised electrons on the surface of the NPs due to their small dimensions) [34]. Those properties affect the overall physicochemical properties of the nanoscale materials in a way that they demonstrate novel and unprecedented properties.

Nanotechnology takes the advantage and exploits these aforementioned unique properties of nanoscale matter to design, translate and apply science to plenty of useful products [34]. The modern nanotechnology exploits the novel nanoscale properties of nanoparticles which has led to the production of a vast amount of engineered NPs. Engineered NPs are deliberately manufactured by human activities to serve for special purposes and they are different from both incidental NPs that are produced as a side product of human activity, for example, from industrial processes or transport and from natural NPs, for example, humic substances, produced from weathering, microbial action or chemical hydrolysis [35].
Engineered NPs can be divided into a number of classes and not as a single homogeneous group. Based on their core materials, these manufactured nanomaterials can be classified into organic and inorganic. Organic NPs can be further defined as fullerenes (C60 and C70 and derivatives) and carbon nanotubes (multi-walled or single walled CNTs), while inorganic NPs can be sub-divided into metal oxides (of iron, zinc, titanium, cerium etc), metals (mainly Silver, gold and copper) and quantum dots (such as cadmiumselenides) [36].

2.2.4 Production of Engineered Nanoparticles

Many different techniques have been developed and employed to generate metal nanoparticles, including silver nanoparticles. Those techniques for preparing nanoparticles have advanced rapidly over the recent decades and continue to evolve leading to more and improved control over the size and shape of the particles generated. Two fundamentally different approaches towards the controlled generation of nanostructures have evolved irrespective of the field or discipline [37]. The bottom up method (the chemical approach), where the atoms (produced from reduction of ions) are assembled to generate nanostructures, and the opposite approach, the top down method, also known as the physical method, where material is removed from the bulk material through grinding, milling, chemical methods or volatilisation of solid material followed by condensation of the vapour components, leaving only the desired nanostructures. Both approaches can be implemented in either gas, liquid, supercritical fluids, solid states, or in vacuum. Most of the manufacturers are interested in the ability to control one or more of the following aspects of the nanoparticles [38]: Particle size, Particle shape, Size distribution, Particle composition and Degree of particle agglomeration.
An important aspect of both approaches is the stabilisation of the particles to avoid aggregation and coalescence [38]. Stabilization can occur in many different ways but primarily aggregation is prevented by electrostatic repulsion or steric hindrance. Two of the most important issues confronting nanocrystal synthesis are obtaining a purposeful control over the nanocrystal mean size and routinely producing narrow size distributions [39].

2.3 Silver Nanoparticles

Among the different types of metal, nanoparticles silver is by far the most studied in the literature due to its wide and rapidly growing application in a number of scientific areas and in consumer products [43] and its known toxic effects on the environment and human health [44]. Its areas of application vary from textile engineering, catalysts [45], water treatment [46] and disinfecting to electronics and biotechnology [47].

2.3.1 Synthesis Silver of Nanoparticles

Colloidal silver is made by adding an excess of the reducing agent to silver salt. for example, by adding an excess of the reducing agent sodium borohydride, $NaBH_4$ to silver nitrate, $AgNO_3$, nanoparticles that are about 10-14 nm in diameter are produced [48].

$$AgNO_3 + NaBH_4 \rightarrow Ag + 1/2H_2 + 1/2B_2H_6 + NaNO_3 \qquad 2.2$$

2.3.2 Procedure for Synthesis of Silver Nanoparticles

A large excess of $NaBH_4$ was needed to reduce the ionic silver and to stabilize the silver nanoparticles that were formed. Different volumes of 0.001M silver nitrate were added drop wise (about 1 drop per second) to 30 ml of 0.002 M sodium borohydride solution that had been chilled in an ice bath. The reaction mixture was stirred vigorously on a magnetic stirrer. The solution turned to light yellow after the addition of 2 ml of silver nitrate and to brighter yellow when all of the silver nitrate had been added [48].

3.0 MATERIALS AND METHODS

3.1 The Chemical Synthesis of Silver Nanoparticles

Many chemical reduction methods have been used to synthesize silver nanoparticles from silver salts. Two main synthesis methods were adopted in the work of this thesis to synthesize silver nanoparticles, using sodium borohydride and trisodium citrate as reducing agents. These were primarily chosen on their possibilities to produce particles of different size regimes, and also because of their synthesis simplicity. In addition, variations in relative quantities of reagents, temperature and durations are performed. The concentration, size, shape and stability of the nanoparticles depend upon these conditions.

3.2 Chemicals

The chemicals used for the experiments include silver nitrate ($AgNO_3$, M = 169.87 g/mol); trisodium citrate ($Na_3C_6H_5O_7.2H_20$, M = 294.10 g/mol); Sodium borohydride ($NaBH_4$, M = 37.83).

3.3 Reduction of Silver Nitrate by Sodium Borohydride

One of most common methods to synthesize silver nanoparticles is by the use of sodium borohydride to reduce silver nitrate.

$$AgNO_3 + NaBH_4 \longrightarrow Ag + \frac{1}{2}H_2 + \frac{1}{2}B_2H_6 + NaNO_3 \qquad (1)$$

3.4 Precaution

Great care has been taken during the synthesis to obtain colloidal silver, especially when dealing with a strong reducing agent such as sodium borohydride. Proper cleaning of laboratory equipment in advance is crucial, and the silver nitrate must by no means be exposed to the aqueous sodium borohydride except during the controlled addition.

3.5 Colloidal silver in water- Synthesis details

All equipment was cleaned thoroughly with distilled water. Reactions were done in a 250 ml conical flask. A given amount of aqueous silver nitrate contained in measuring cylinder was added drop wise to an aqueous silver nitrate, using a handheld dropper. The solution was stirred vigorously on a magnetic stir plate throughout the whole reaction (stirring speed = 600).

3.6.1 Experiment 1: Synthesis at Temperature of 27^{0}C

27.5 mg of $AgNO_3$ (molar weight = 169.87g/mol) was dissolved in 200 ml of deionised water to prepare a 0.8 mM aqueous silver nitrate solution. Analogously, a 1.6 mM aqueous sodium borohydride solution was made by dissolving 36.32 mg of $NaBH_4$ (molar weight = 37.83 g/mol) to 600 ml of deionised water. The compounds dissolved quite easily within 1 minute. The concentration of the sodium borohydride was twice that of the aqueous silver nitrate to ensure colloid stability fixed by the reaction stoichiometry.

The sodium borohydride is both a reducing agent and a stabilizing agent. Hence, different volumes of the precursors were reacted in order to get the ratio at which the sodium borohydride will effectively produce the nanoparticles and stabilize them. An overview of the compositions and durations of the experiment is given in **Table 1.**

Table 1: Details for the reaction of $AgNO_3$ with borohydride at 27^0C.

Sample	$AgNO_3 : NaBH_4$ ratio (ml)	Adding time (Minutes)
A1	14 : 50	2: 48
A2	10 : 50	2:02
A3	20 : 50	3:04
A4	5 : 50	1:13

All attempts to carryout synthesis above 20:50 $AgNO_3$:$NaBH_4$ ratio yielded instantly aggregated samples. However, only the sols listed in the above table were used in further characterization.

3.6.2 Experiment 2: Synthesis at Temperature of 4^0C

The synthesis was performed with solutions of equal concentration as those of the first experiments. For these samples, the solutions were cooled down to 4^0C. However, the initial volumes of the two solutions were half of the first experiment. This was done for better temperature stability at 4^0C.

13.75 mg of $AgNO_3$ was dissolved in 100 ml of deionised water to prepare a 0.8 mM aqueous silver nitrate solution. Analogously, a 1.6 mM aqueous sodium borohydride solution was made by dissolving 18.16 mg to 300 ml of deionised water. Both solutions were kept in ice. The experiment was started when the temperature of the solutions stabilized at 4^0C. An overview of the compositions and durations of the experiment is given in **Table 2.**

Table 2: Details for the reaction of silver nitrate with borohydride at 4^0C

Sample	$AgNO_3 : NaBH_4$ ratio (ml)	Adding time (Minutes)
B1	14 : 50	2:39
B2	10 : 50	2: 00
B3	20 : 50	3:14
B4	5 : 50	1:16

3.7 Reduction of $AgNO_3$ by Trisodium Citrate

Sodium borohydride is a very strong reducing agent. As a result of its reducing strength, a side reaction could jeopardize the synthesis of nanoparticles due to their sensitivity. The samples produced could also be harmful to the environment if not properly disposed. However, the synthesis of silver nanoparticles is achievable via an alternative route through an organic reducing agent sodium citrate which is less expensive and offers less potential hazards. However, vigorous conditions are needed for this synthesis.

3.7.1 Experiment 3: Using Trisodium Citrate

90 mg of $AgNO_3$ was dissolved in 500 ml deionized water to give 1.06 mM silver nitrate solution. Additionally, 1% (38 mM) trisodium citrate was easily prepared by dissolving 1g of $Na_3C_6O_7.2H_2O$ in 100 ml of deionized water. In a typical experiment, a beaker containing 50 ml of the aqueous silver nitrate was placed directly on top of a heater and heated up to boiling temperatures. Solution of trisodium citrate was added directly on top of the heater using a handheld dropper.

Different solutions ratios were used to investigate the effect on stability and the resulting nanoparticles characteristics. The details for the synthesis are given in **Table 3.**

Table 3: Details for the reaction of silver nitrate with trisodium citrate

Sample	$AgNO_3$:Citrate (ml)	Adding time (Minutes)	Post-addition stirring (Minutes)
C1	50: 25	4:19	5:20
C2	50: 35	5:15	5:20
C3	50: 45	7:13	6:05

3.8 Nanoparticles Characterisation

The characterization was performed by nano-drop with an inherent UV/Visible Spectrophotometer. The cover of the nano-drop was opened and the cuvette was filled with distilled water and blanked. About 1.5mil of a sample was poured into the cuvette, the cover was closed.

The optical properties (absorbance) of solution (sample) were evaluated with UV-VIS spectrophotometer at a wavelength range of: 200 nm –700 nm. It measures the intensity of light passing through a sample (I), and compares it to the intensity of light before it passes through the sample (I_o). The ratio I/I_o is called the transmittance, and is usually expressed as a percentage (%T). The absorbance, A, is based on the transmittance: $A = -\log (\%T)$. The scanning process was conducted using a computer system connected to the nano-drop. After the scanning is over a chromatograph of the absorbance versus the wavelength is displayed.

While scanning the sample a platform for naming the sample will pop out, this allows the user to name the sample. The chromatograph is stored inside the computer and subsequently, chromatographs of the same sample at variable times are added to the workbook.

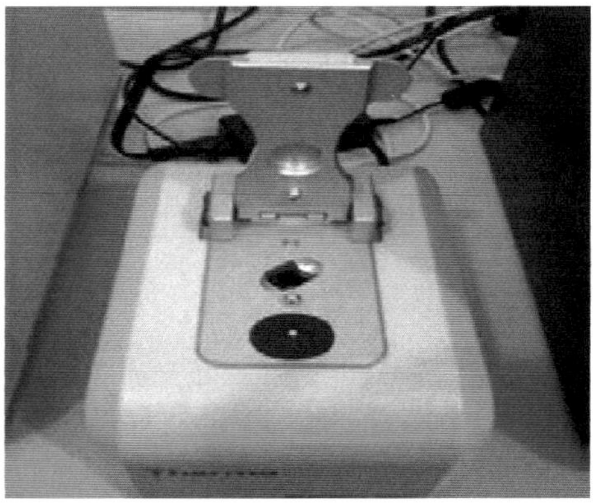

Figure 1: Nano-drop used for UV/Visible spectroscopy

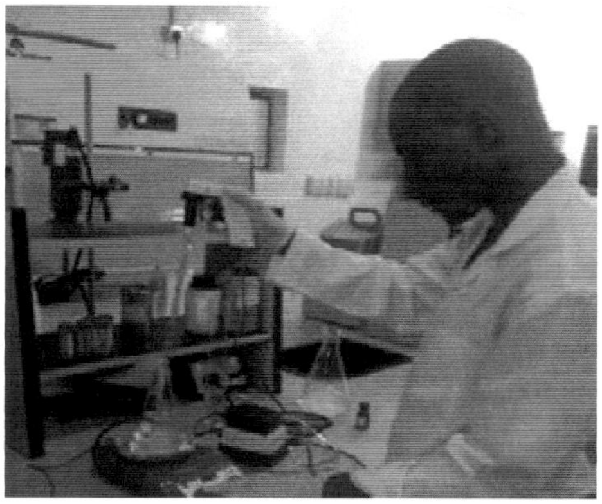

Figure 2: Material Laboratory, North-East zonal Biotechnology Centre, University of Maiduguri, Nigeria

4.0 EXPERIMENTAL RESULTS

This chapter presents the result of all the experiments performed. Eleven samples were successfully synthesized at varying conditions and the spectrum of each sample was recorded three times: the day of synthesis, the third day and the seventh day. The quest was to find the best synthesis condition in terms of size, shape, concentration, size distribution, and stability.

4.1 Sample Synthesized from Sodium Borohydrate at 27^{0}C

Four samples (A1, A2, A3, A4) were synthesized. The synthesis details are given in Table 1.

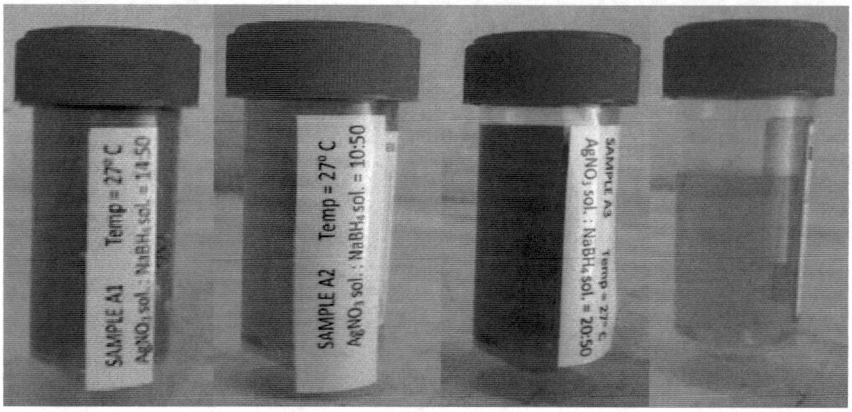

a b c d

Figure 3: Colloidal solutions of synthesis at 27^{0}C: (a) A1 (b) A2 (c) A3 (d) A4 [$NaBH_4$:$AgNO_3$ for A1, A2, A3, A4 are respectively 14:50, 10:50, 20:50 and 5:50]

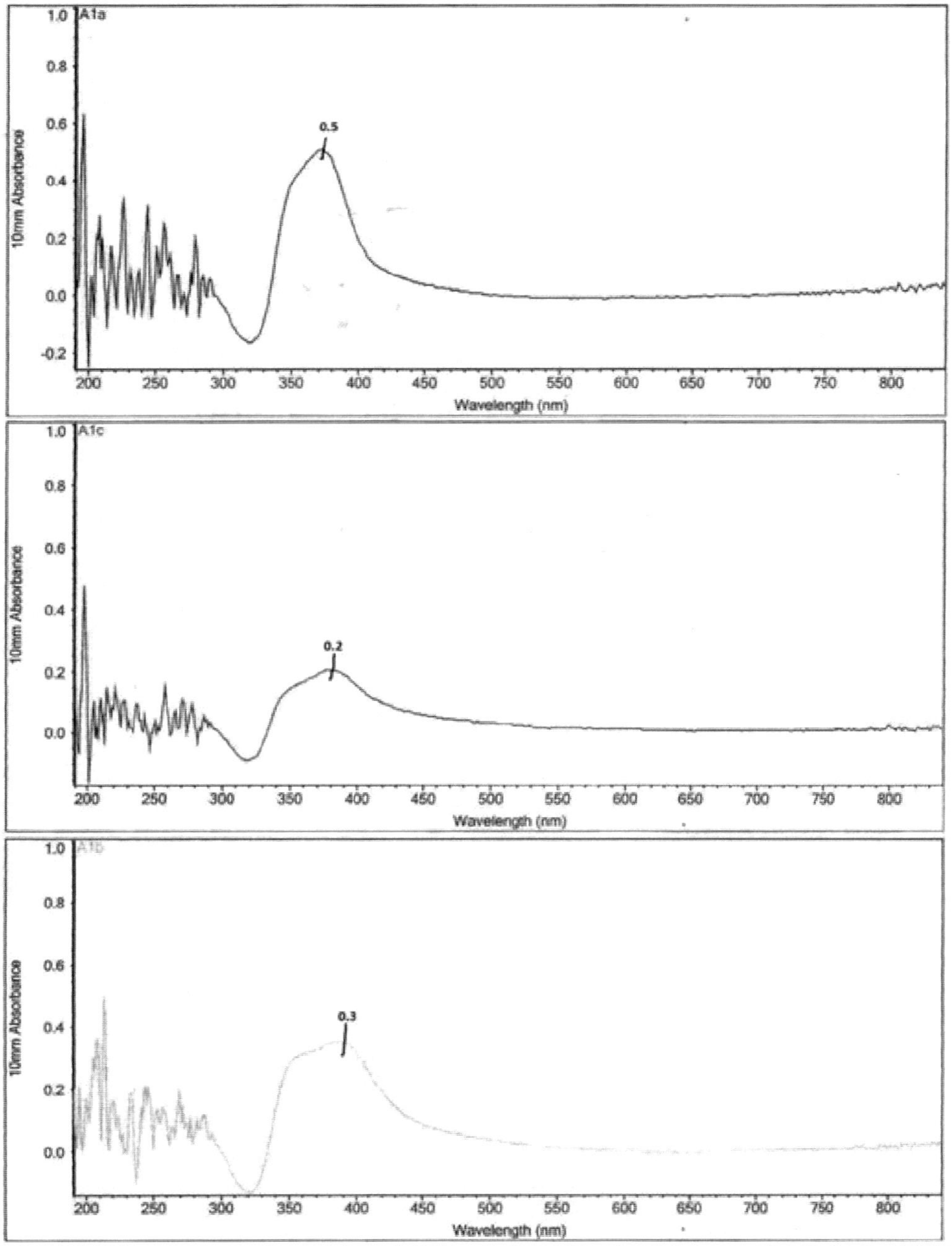

Fig. 4.1a: UV-Vis spectra of sample A1 [Time of synthesis: (a) 1 day (b) 3 days, (c) 7 days]

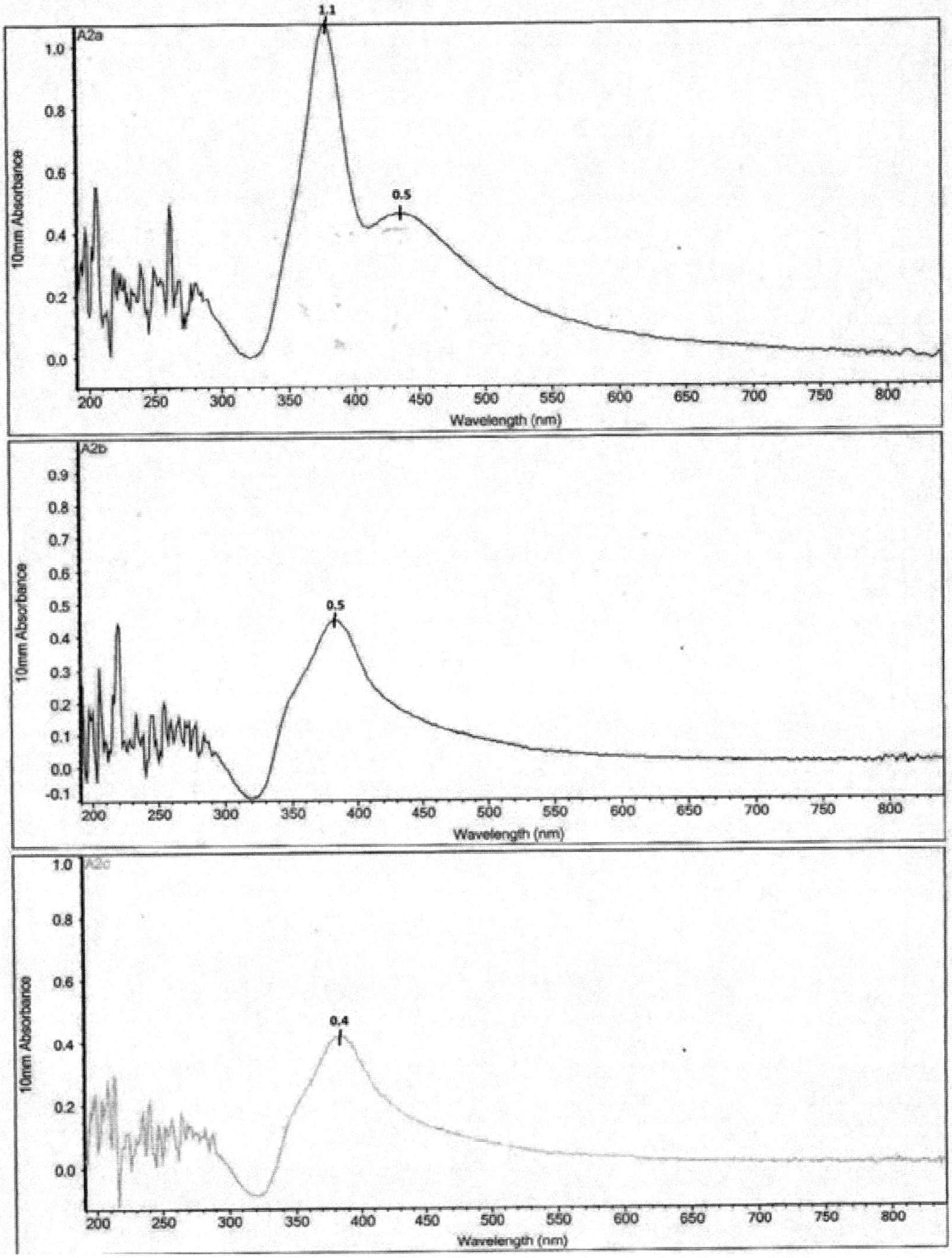

Fig. 4.1b: UV-Vis spectra of sample A2 [Time of synthesis: (a) 1 day (b) 3 days, (c) 7 days]

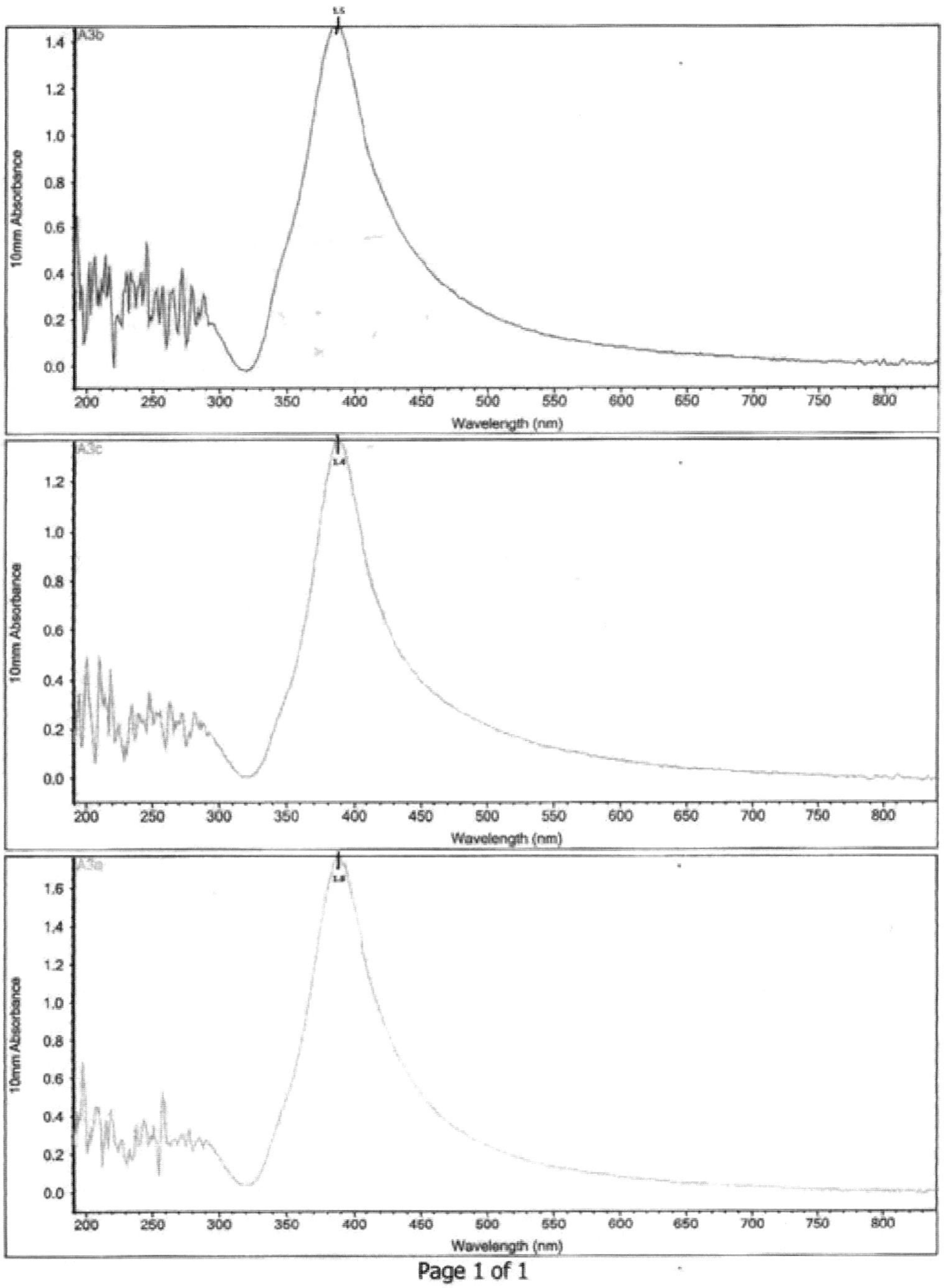

Fig. 4.1c: UV-Vis spectra of sample A3 [Time of synthesis: (a) 1 day (b) 3 days, (c) 7 days]

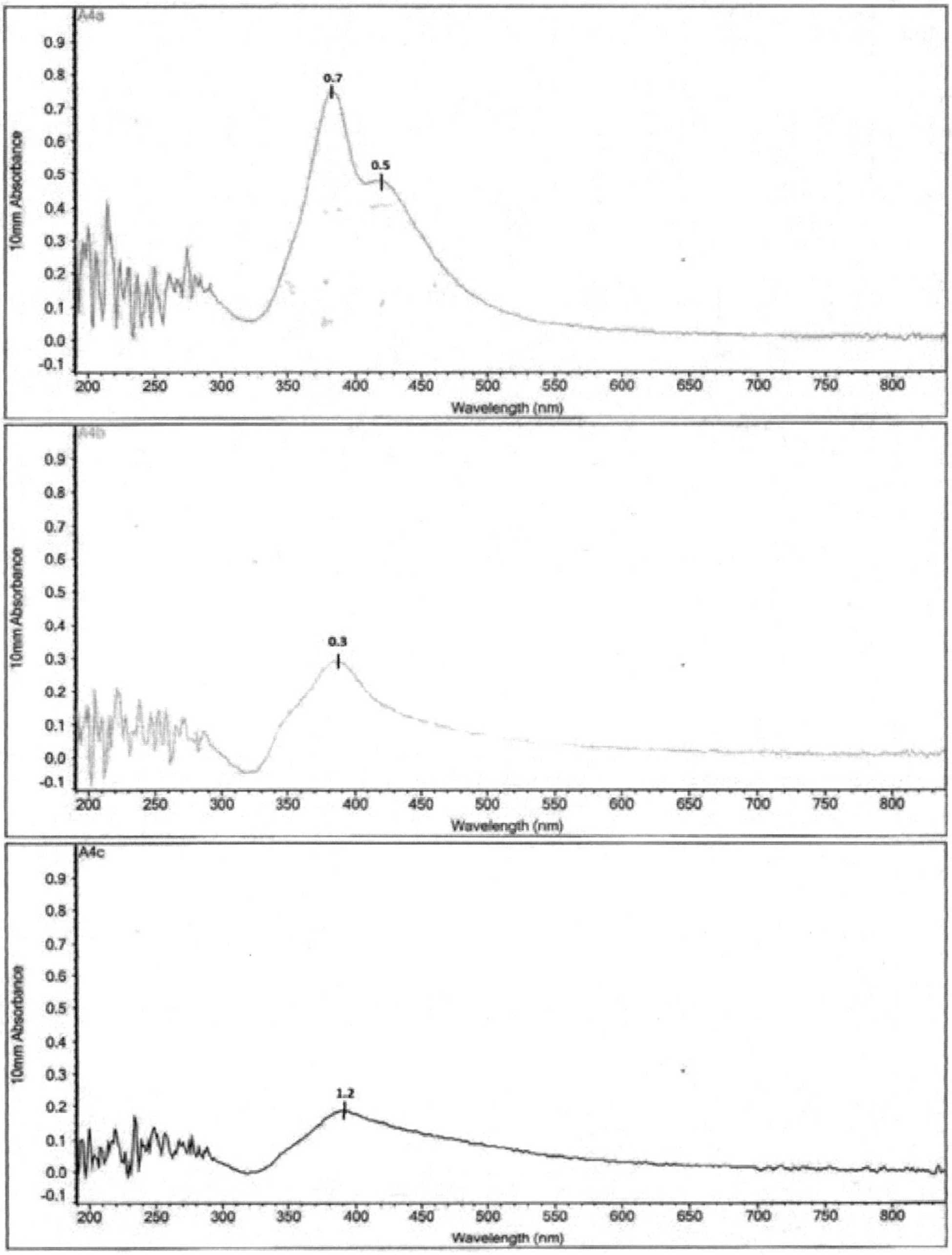

Fig. 4.1d: UV-Vis spectra of sample A4 [Time of synthesis: (a) 1 day (b) 3 days, (c) 7 days]

4.2 Sample Synthesized at 4^{0}C

Four samples (B1, B2, B3, and B4) were synthesized at 4^{0}C and their spectra were obtained at three different times (the first day, the third day and the seventh day). The details of the synthesis are given in Table 3.2. Larger ratios was attempted, but all attempts yielded an aggregated solution. Only the successfully synthesized solutions were characterized.
Figure 4.2 shows the appearance of the samples.

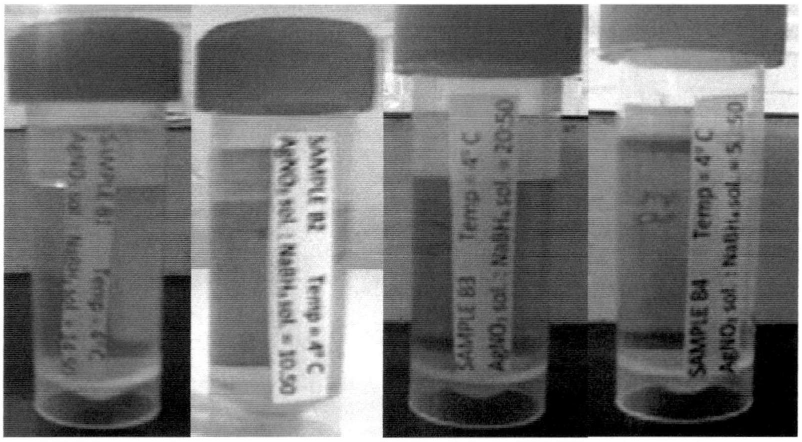

a b c d

Fig. 4.2: Colloidal solutions of synthesis at 4^0C (a) B1 (b) B2 (c) B3 (d) B4
[$NaBH_4$:$AgNO_3$ for B1, B2, B3, B4 are respetively 14:50, 10:50, 20:50 and 5:50]

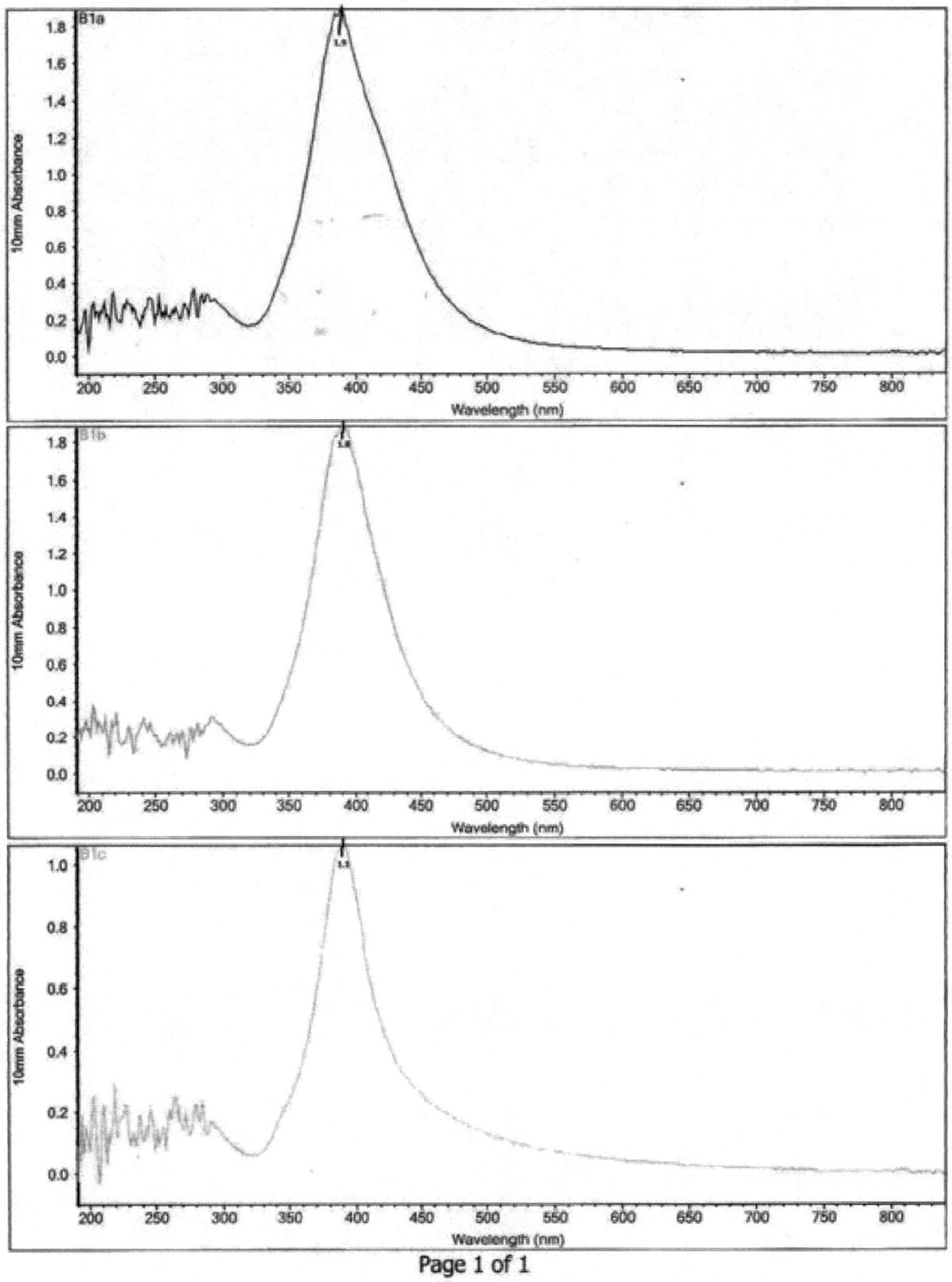

Fig. 4.2a: UV-Vis spectra of sample B1 [Time of synthesis: (a) 1 day (b) 3 days, (c) 7 days]

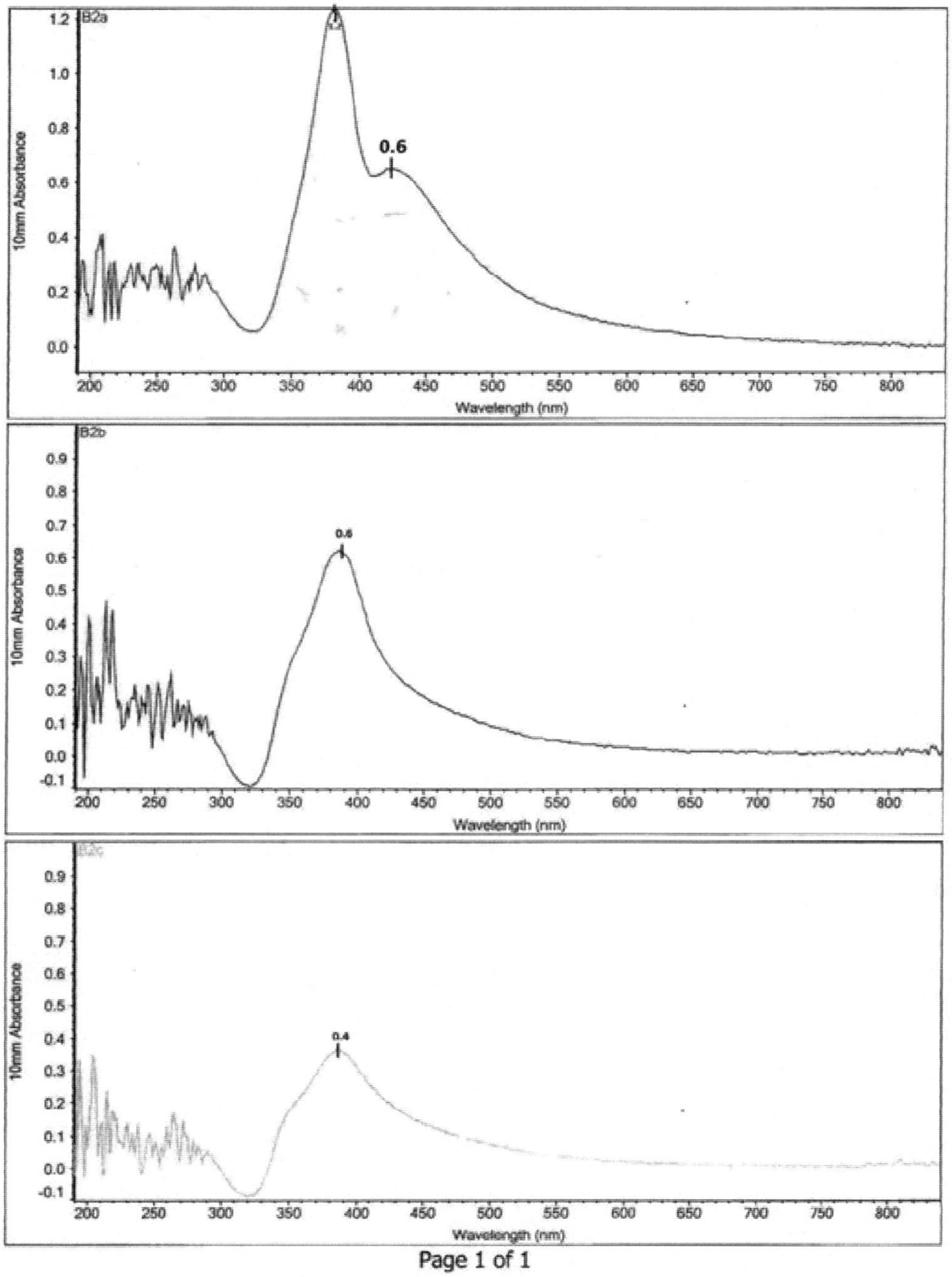

Fig. 4.2b: UV-Vis spectra of sample B2 [Time of synthesis: (a) 1 day (b) 3 days, (c) 7 days]

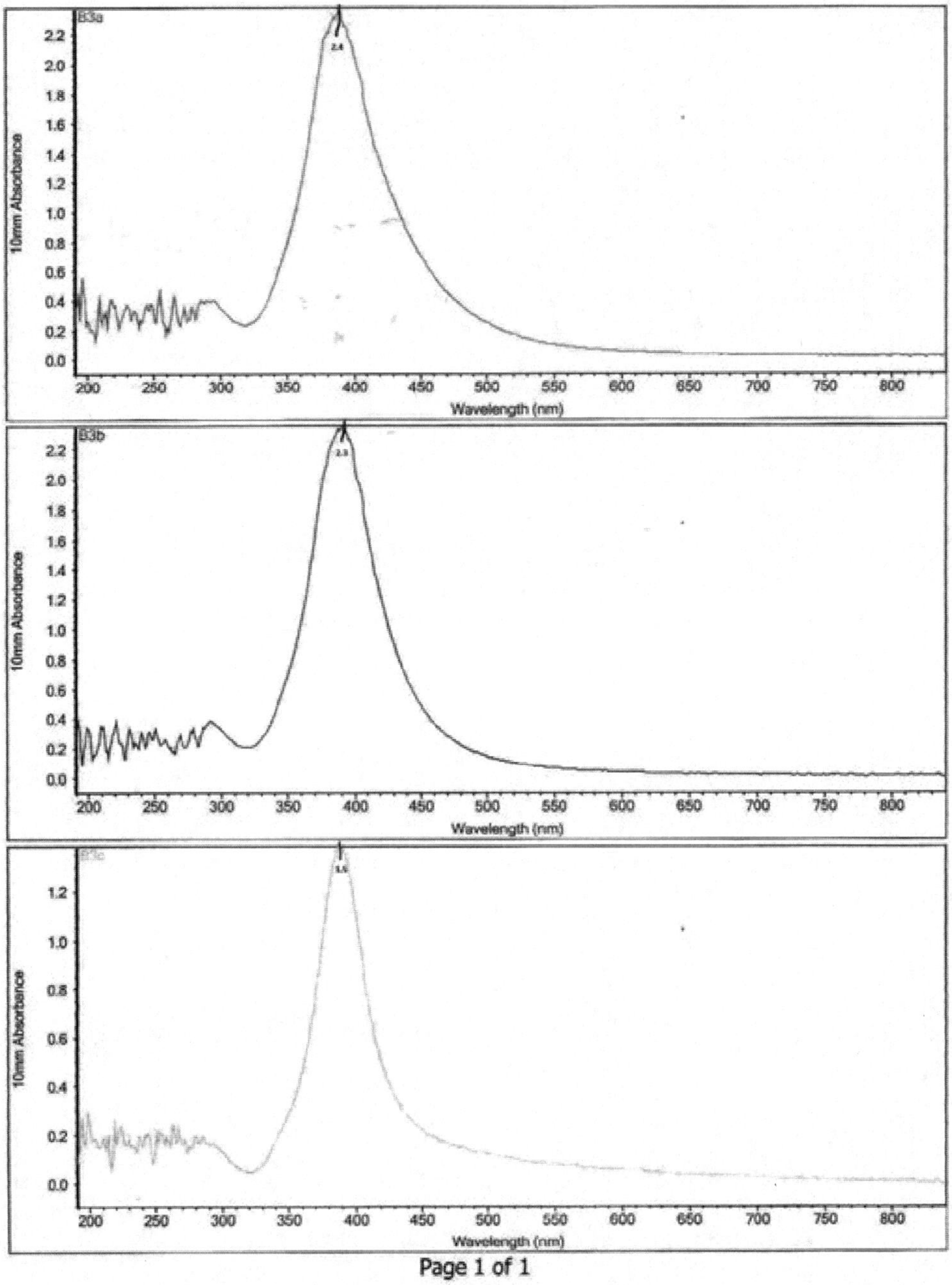

Fig. 4.2c: UV-Vis spectra of sample B3 [Time of synthesis: (a) 1 day (b) 3 days, (c) 7 days]

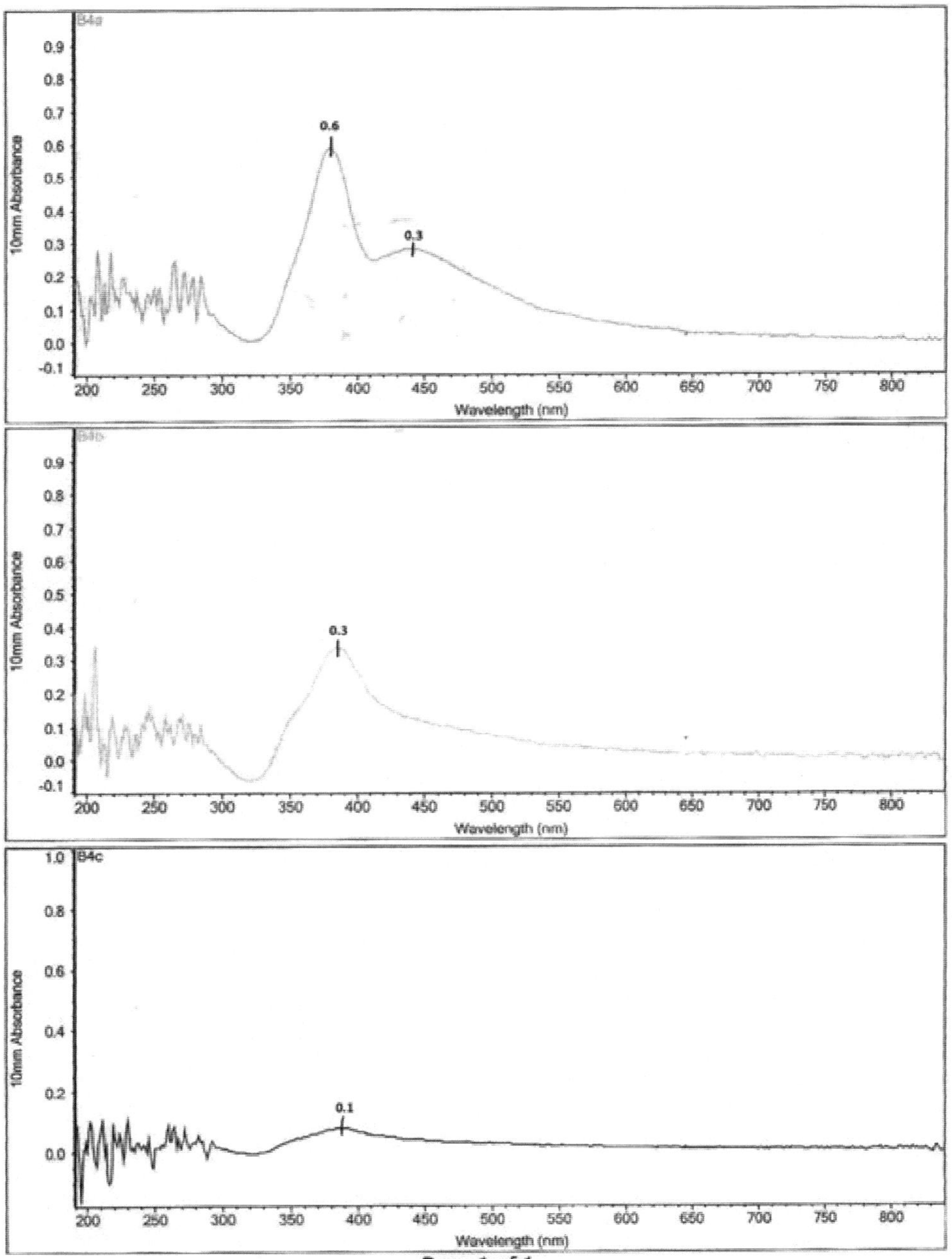

Fig. 4.2d: UV-Vis spectra of sample B4 [Time of synthesis: (a) 1 day (b) 3 days, (c) 7 days]

4.3 SAMPLES SYNTHESIZD BY SODIUM CITRATE

Three samples (C1, C2 and C3) were synthesized by reduction of silver salt ($AgNO_3$) using trisodium citrate. The reaction was carried out at the tempreture of boiling water. The solution was vigorously stirred during and after addition of the silver salt to citrate salt. The details of the experiment is given in Table 3.3

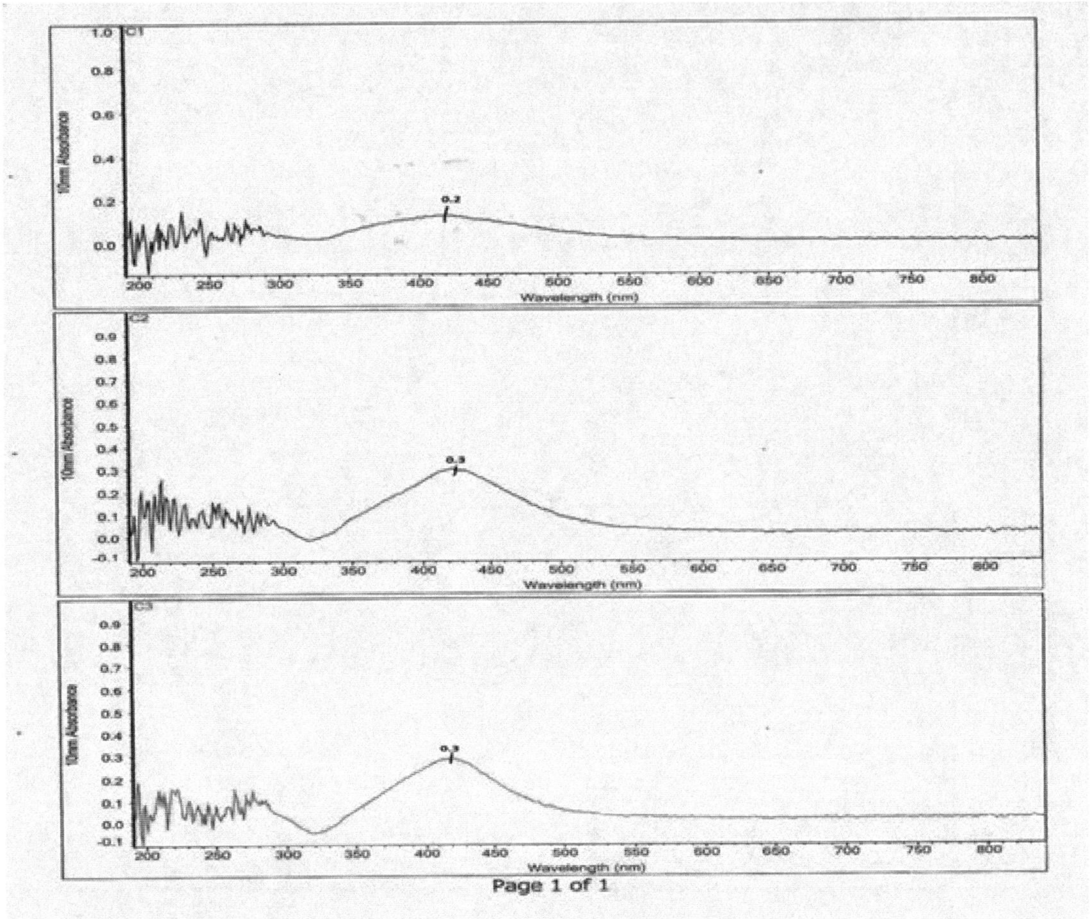

Fig. 4.3: UV-Vis spectra of samples synthesized by trisodium citrate: (a) C1 (b) C2, (c) C3

5.0 DISCUSSION OF RESULTS

In order to find out the effect of various synthesis conditions on the resulting nanoparticles, the nanoparticles were characterized. For the characterization of synthesized NPs, special attention was devoted on the quality of the product of the synthesis methods by comparing the properties of the nanoparticles in terms of size, size distribution, monodispersity, sphericity, concentration and stability of the NPs.

5.1 Nanoparticles and UV-Visible Spectroscopy

The dispersions of silver nanoparticles display intense colors due to the Plasmon resonance absorption. Metals with free electrons possess Plasmon resonances in the visible spectrum, which give rise to such intense colours [61]. These properties are mainly observed in Au, Ag, and Cu because of the presence of free conduction electrons. The electric field of the incoming radiation induces the formation of a dipole in the nanoparticle. A restoring force in the nanoparticle tries to compensate for this, resulting in a unique resonance wavelength. Therefore metallic nanoparticles have characteristic optical absorption spectrum in the UV-Visible region [62]. This makes UV-Visible spectroscopy an indispensable tool for characterization of nanoparticles.

5.2 Synthesis and Growth of $AgNO_3$ Nanoparticles

The synthesis of silver nanoparticles was carried out using three different approaches. The first samples with codes A1, A2, A3, A4 (see Table 1 for synthesis details) were synthesized with sodium borohydride at room temperature. When the reducing agent ($NaBH_4$) was introduced to the solution of $AgNO_3$, the initial colourless solution turned instantaneously to yellow, which indicates that the silver ions were spontaneously reduced to silver atoms, which in turn, simultaneously clumped into silver nanoparticles, as reported [63]. The yellow colour formed within 2 minutes remained unchanged, this indicates the formation of stable silver nanoparticles [63]. In similar manner, the colour change and the growth of samples B1, B2, B3, B4 (see Table 2 for synthesis details) synthesized with sodium borohydride at room temperature manifested the same pattern as the above, although their formation was less drastic which indicates that the reaction rate was faster at higher temperatures. However, the colour changes of samples C1, C2, C3 using sodium citrate were observed after heating the solution to about 100^{0}C with vigorous stirring and waiting to about 15 minutes. Total reaction time of these last samples was about 30 minutes. This slow change can be attributed to the weak reducing effect of the citrate salt [64].

5.3 PRESENCE OF SILVER NANOPARTICLES

In all the experiments summarized above and described in Chapter three and subsections, the colours of the solutions have changed from colourless to yellow, due to silver ions from precursor to silver nanoparticles, which is characteristic formation of silver in nanoscale [62, 63]. The colour change manifested is caused by surface plasmon resonance (SPR) of the forming silver NPs [62, 63]. Therefore, the presence of silver NPs in the final suspensions was further confirmed by Uv-Vis absorbance measurement which shows the NPs localised surface Plasmon resonance (LSPR) behaviours. The spectra are given in chapter five and clearly indicate the characteristic sharp absorbance peak around 370-440 nm for all samples, which is a characteristic of silver nanoparticles [65].

Electromagnetic radiation looks white when all wavelengths of radiation from 350 nm to 750 nm are present. White light is composed of various hues of red, orange, yellow, green, blue, indigo, and violet light. When a specie absorbs selected portions of the visible light spectrum, color results. For example, the solution of silver nanoparticles mentioned above is a yellow solution that strongly absorbs light at around 400 nm.

Table 4: Absorbance and Complementary Colour

Wavelength	Absorbed Colour	Complementary Colour
650-780	**red**	**blue-green**
595-650	**orange**	**greenish bue**
560-595	**yellow-green**	**purple**
500-560	**green**	**red-purple**
490-500	**bluish green**	**red**
480-490	**greenish blue**	**orange**
435-480	**blue**	**yellow**
380-435	**violet**	**yellow-green**

Source: Owen, (1999)

From Table 4, the yellow color of the solution is the result obtained when the violet component of white light is removed. The color of the solution, therefore, represents a composite of all colors transmitted (i.e., not absorbed) by the particles. It is often possible to determine the color (or wavelength) of light absorbed by a specie based on the color of the solution and vice versa. In simple cases, where a solution absorbs light in only one region of the visible spectrum, the color of the solution is complementary to the color of light absorbed by the solution.

5.4 Particle Size Characterisation

Table 5: Spectroscopy and Size of Nanoparticle

Particle size nm	Λ_{max} nm	PWHM nm
10-14	385-400	50-70
35-50	420	100-110
60-80	438	140-150

(International Journal of Engineering Research & Technology, 2013)

The position of the maximum absorbance measured with Uv-Vis depends on a number of factors including the shape, size and the surrounding medium. As a general rule, as the particles become larger, the absorption maximum shifts to longer wavelength [66].

All samples synthesized from sodium borohydride irrespective of synthesis temperature [Fig 4.1 (a, b, c and d) and Fig 4.2 (a, b, c and d)] showed absorption maximum at wavelength of around 387 nm, as reported by other researchers [67]. However, samples synthesized from sodium citrate (Fig 4.3) showed absorption maximum at around 420 nm. Accordingly, from Table 5.2, the sizes all samples synthesized from sodium borohydride were around 12 nm as reported by other researches [67]. Also, nanoparticles synthesized from trisodium citrate were around 35 nm, as also reported [68].

Similarly, symmetrical particles absorb at longer wavelength [66] which indicates that particles from sodium borohydride were perfectly spherical and particles from sodium borohydride were roughly spherical.

However, sodium borohydride samples synthesized with lower $NaBH_4$:$AgNO_3$ solution ratios [A2 and B2 with $NaBH_4$:$AgNO_3$ (10:50); A4 and B4 with $NaBH_4$:$AgNO_3$ (5:50)] showed double peaks at 387 nm and 440 which indicates distribution of two nanoparticles of sizes 12 nm and 37 nm.

5.5 Stability

The optical properties of silver nanoparticles change when particles aggregate and the conduction electrons near each particle surface become delocalized and are shared amongst neighbouring particles. When this occurs, the surface Plasmon resonance shifts to lower energies, causing the absorption and scattering peaks to red-shift to longer wavelengths [66].

To monitor the stability of the silver colloids, we have measured the absorption of the colloids at different periods of time (Day 1, Day 3, and Day 7).

The UV-VIS spectrums recorded after different time intervals shown Figure 4.1 (a, b, c and d) and Figure 4.2 (a, b, c and d) are summarized in Table 5.3a and Table 5.3b respectively.

Table 5.3a: Summary of absorbance and corresponding wavelength recorded at three different times for samples prepared from sodium borohydride at 27^{0}C

	Time of Synthesis					
	Day1		Day2		Day3	
Sample	**Maximum Absorbance**	**λ_{max}**	**Maximum Absorbance**	**λ_{max}**	**Maximum Absorbance**	**λ_{max}**
A1	**0.5**	**385**	**0.4**	**387**	**0.2**	**388**
A2	**1.1**	**385**	**0.5**	**388**	**0.5**	**388**
A3	**1.8**	**386**	**1.5**	**387**	**1.4**	**387**
A4	**0.7**	**386**	**0.3**	**388**	**0.2**	**389**

Table 5.3b: Summary of absorbance and corresponding wavelength recorded at three different times for samples prepared from sodium borohydride at 4^{0}C

	Time of Synthesis					
	Day1		Day2		Day3	
Sample	Maximum Absorbance	λ_{max}	Maximum Absorbance	λ_{max}	Maximum Absorbance	λ_{max}
B1	1.9	386	1.8	387	1.1	387
B2	1.2	387	0.6	388	0.5	388
B3	2.4	387	2.2	387	1.5	387
B4	0.6	386	0.3	388	0.1	390

The extinction peaks for the samples have shifted to longer wavelength, decreased in intensity and have broadened. However, this change was more drastic for samples synthesized at 27^{0}C. This was also observed from the colour changes of the samples as the yellow darkens, turns violet, then grayish as the particles settle out within days of synthesis. However, samples synthesized with higher $NaBH_4$:$AgNO_3$ (B3, A3) showed an appreciable stability as their spectrum stabilized and their colour persisted for weeks after synthesis.

5.6 Effects of Difference Synthesis Condition

For the synthesis of silver nanoparticles, multi-approach was applied so as the see the effects of different synthesis conditions on the resulting nanoparticles. These conditions are examined in the coming sub sections.

5.6.1 Effect of Variable $NaBH_4$:$AgNO_3$ Solution Ratios

Before reducing agent is added to the silver solution, there are only silver ions. The reducing agent, sodium borohydride, will force silver atoms to form in solution. The concentration of silver atoms will rise rapidly until supersaturation is achieved, at which aggregation will occur [69]. The concentration of particles will depend on the number of nucleation sites: the more the particles [69]. The amount of nucleation sites is dependent on the amount of reducing agent. As the reducing agent is increased, there will be more nuclei and thus more nanoparticles. With too much reducing agent, the nucleation sites will form at different times and destabilize much quicker [69].

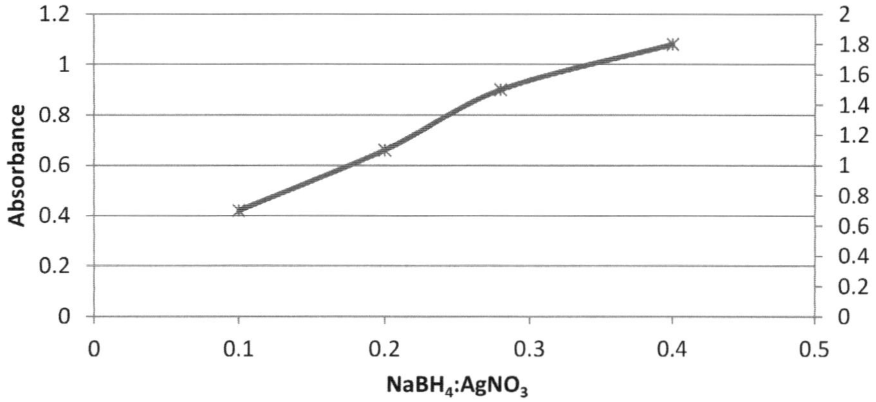

Fig 5.1a: Absorbance Versus $NaBH_4$:$AgNO_3$ ratios for synthesis at 27^{0}C

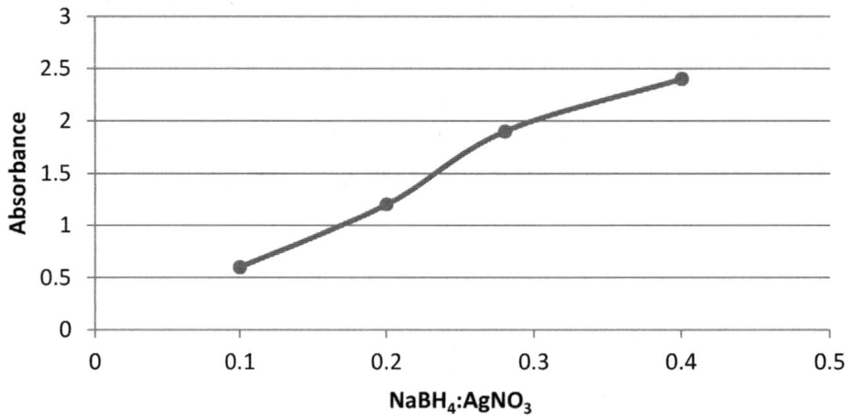

Fig 5.1b: absorbance versus $NaBH_4$:$AgNO_3$ ratios for synthesis at 4^0C

The absorbance of samples increase steadily with increase in $NaBH_4$:$AgNO_3$ for both synthesis at 4^0C and 27^0C as shown in fig 5.4a and fig 5.4b. This shows that conversion increase with an increase in amount of the reducing agent. However, an excess of the nuclei will be generated when $NaBH_4$ added is too high, which results in agglomeration. This was confirmed in the experiment as all attempts to perform the synthesis at $NaBH_4$:$AgNO_3$ > 20:50 yielded an instantly aggregated solution.

5.6.2 Effect of Synthesis Temperature

Chemical reactions are faster at higher temperature. However, due to a very high reducing effect of $NaBH_4$, impurities can easily be easily brought into the system as product of side reactions.

Samples synthesized at 4^0C were clear yellow (Fig 4.1) while samples synthesized at 27° were yellowish-brown (Fig 4.2) which indicates the possibility of presence of impurities.

Fig 5.5 shows that the concentration of samples synthesized at 4^0C is higher than that of samples synthesized at 27^0C. However, the effect of synthesis temperature is more drastic at higher $NaBH_4$:$AgNO_3$.

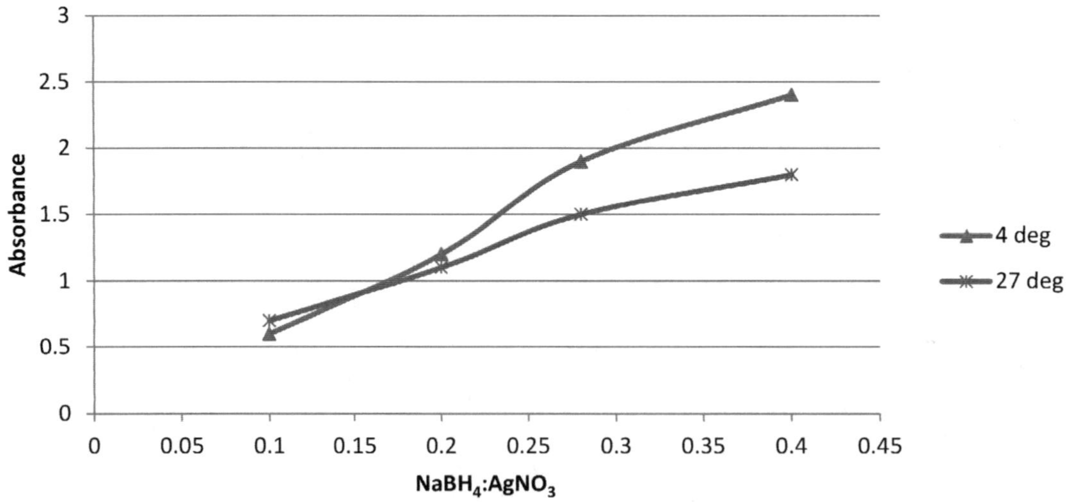

Figure 5.2: Effect of synthesis temperature

To study the effect of temperature on concentration and stability of the nanoparticles, Fig 5.6a and Fig 5.6b give an overview of synthesis at 4^0C and 27^0C.

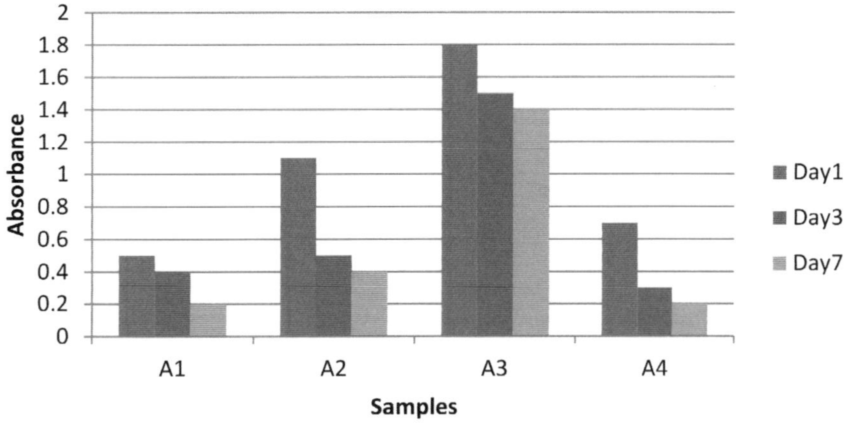

Figure 5.3a: An Overview of Synthesis at 27°C

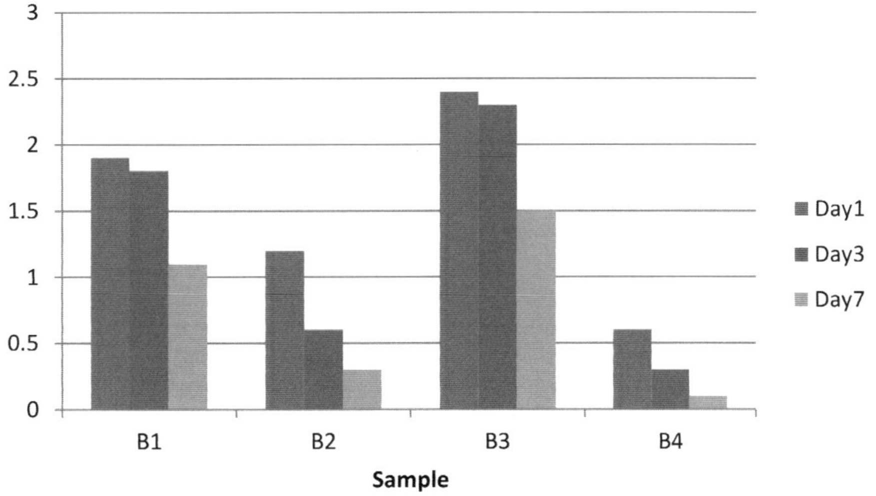

Figure 5.3b: An Overview of Synthesis at 4°C

It can be seen that samples prepared at 4^0C were more concentrated and stable than samples prepared at 27^0C.

5.5.3 Effects of Using Trisodium Citrate As Reducing Agent

Three samples were successfully prepared from sodium citrate (the details of the synthesis are given in Table 3.3). The reaction lasted for about 15 minutes with vigorous stirring and heating. The resulting solutions were cloudy yellow which showed that the conversion of silver ions to silver nanoparticles were moderate. This was also confirmed by their spectra (Fig 4.3): the peaks were less intense, broader than that of samples from sodium borohydride.

Similarly, the sizes of particles were bigger (about 35 nm) and structurally less regular than particles obtained from sodium from borohydride. However, the samples was found to be stable under atmospheric conditions, this could be attributed to the repelling actions of citrate ions [66].

6.0 CONCLUSION

Silver nanoparticles were synthesized by reducing Ag^+. Two reducing agents were used. In the first and second experiment, sodium borohydride was used at synthesis temperature of 27^0C and 4^0C respectively. In the last experiment, sodium borohydride was replaced with trisodium citrate. In all the three experiments, variable reactants' volumes were used. The absorption spectrum of each sample synthesized was determined using UV-Visible spectrophotometer at three different times. Below were findings of the research:

During the synthesis, the first sign for the production of the nanoparticles was manifested in colour change from colourless solution to yellow. The formation of silver nanoparticles was confirmed by the characteristic surface Plasmon resonance for silver which is between 370-450 nm. Synthesis of silver nanoparticles via sodium borohydride take 1-2 minutes due to high reducing strength of sodium borohydride. Irrespective of synthesis temperature, colloids produced via sodium borohydride have absorption peak at 387 nm, which suggests that the particles were 12 nm in diameter. Low reaction temperature is helpful in optimum nanoparticles formation. Low-size, uniform silver nanoparticles with low polydispersity are obtainable at higher $NaBH_4$:$AgNO_3$ volume ratios. Nanoparticles synthesized from sodium citrate showed maximum absorption at 440 nm, which indicated the presence of nanoparticles of sizes around 35 nm. Nanoparticles synthesized by sodium borohydride were smaller, more concentrated and less distributed than nanoparticles synthesized by trisodium citrate. The best synthesis conditions for synthesis are: Using sodium borohydride as reducing agent at low temperature and high $NaBH_4$:$AgNO_3$ volume ratios.

REFERENCE

1. Kiss, L. B.; Söderlund, J.; Niklasson, G. A.; Granqvist, C. G. (1999). "New approach to the origin of lognormal size distributions of nanoparticles." *Nanotechnology* **10**: 25.

2. Alemán, J.; Chadwick, A. V.; He, J.; Hess, M.; Horie, K.; Jones, R. G.; Kratochvíl, P.; Meisel, I.; Mita, I.; Moad, G.; Penczek, S.; Stepto, R. F. T. (2007). "Definitions of terms relating to the structure and processing of sols, gels, networks, and inorganic-organic hybrid materials (IUPAC Recommendations 2007)."*Pure and Applied Chemistry* **79** (10).

3. Vert, Michel; Doi, Yoshiharu; Hellwich, Karl-Heinz; Hess, Michael; Hodge, Philip; Kubisa, Przemyslaw; Rinaudo, Marguerite; Schué, François (2012). "Terminology for biorelated polymers and applications (IUPAC Recommendations 2012)." *Pure and Applied Chemistry* **84** (2): 377–410.

4. Whetten, R. L.; Khoury, J. T.; Alvarez, M. M.; Murthy, S.; Vezmar, I.; Wang, Z. L.; Stephens, P. W.; Cleveland, C. L.; Luedtke, W. D.; Landman, U. Adv. Mater.1996, 8, 428−433.

5. Zheng, J.; Nicovich, P. R.; Dickson, R. M. Annu. Rev. Phys. Chem. 2007, 58, 409−431.

6. Christof, M. N. (2001). "Nanoparticles, Proteins, and Nucleic Acids: Biotechnology Meets Materials Science." *Angewandte Chemie International Edition*, 40, 4128-4158.

7. Shanmin, G. & et al. (2003). "Preparation of $CuAlO_2$ nanocrystalline transparent thin films with high conductivity." *Nanotechnology*, 14, 538.

8. Shafirovich, E.; Diakov, V. A. (2006). "Combustion of novel chemical mixtures for hydrogen generation." *Combustion and Flame*, 144, 415-418.

9. Paolo Pengo, S.P.; Marino, Battagliarin; Lucia, Pasquato; Paolo Scrimin (2003). "Synthesis, characterization and properties of water-soluble gold nanoparticles with tunable core size." *Journal of Materials Chemistry*, 2471 -2478.

10. Chattopadhyay, D.P.; Patel B.H (2009). "Improvement in physical and dyeing properties of natural fibres through pre-treatment with silver nanoparticles." *Indian Journal of Fibre & Textile Research*, 34, 368-373.

11. Chattopadhyay, D.P.; Patel B.H. (2010). "Effect of nanosized colloidal copper on cotton fabric." *Journal of Engineered Fiber Fabrics*, 1-6.

12. Jana, N.R.; Wang Z.L., Sau T.K. (2000). "Seed-mediated growth method to prepare cubic copper nano particles." *Current Science*, 79, 1367-1370.

13. Owen, Tony (1996). "Fundamentals of Modern UV-Visible Spectroscopy." *Hewlett-Packard.*

15 Daniel, M.C.; Astruc , D. (2004) *Chem. Rev.* , 104 , 293 – 346 .
16- Helcher, H.H. (1718). "Aurum Potabile Oder Gold Tinstur." J. *Herbord Klossen,* Germany
17. Helcher, H.H. (1718). "Aurum Potabile Oder Gold Tinstur." *J. Herbord Klossen*, Germany.
18. Faraday, M. (1857) *Philos. Trans. R. Soc.* London , 147 , 145 – 181 .
19. Satoshi, Horikoshi; and Nick, Serpone (2013). "Microwaves in Nanoparticle Synthesis." First Edition, *Wiley-VCH Verlag GmbH & Co*, p3.
20. Jin, R.; Cao, Y.; Mirkin, C.A.; Kelly, K.L.; Schatz, G.C. ; Zheng, J.G. (2001) *Science* , 294 , 1901 – 1903 .
21. For the details about The New York Times article of February 22, 2005 see: http://www.nytimes.com/imagepages/2005/02/21/science/20050222_NANO1_GRAPHIC.html.
22. Onoda, G.Y.; Hench, L.L. (1979). "Ceramic Processing Before Firing.*" New York: Wiley & Sons.* ISBN 0-471-65410-8.
23. Aksay, I.A., Lange, F.F., Davis, B.I. (1983). "Uniformity of Al_2O_3-ZrO_2 Composites by Colloidal Filtration." *J. Am. Ceram. Soc.* **66** (10): C–190.
24. Fahlman, B. D. (2007). *Materials Chemistry*. Springer. pp. 282–283. ISBN 1-4020-6119-6.
25. Taylor, Robert A; Otanicar, Todd; Rosengarten, Gary (2012). "Nanofluid-based optical filter optimization for PV/T systems." *Light: Science & Applications* **1** (10): e34.
26. Taylor, Robert; Coulombe, Sylvain; Otanicar, Todd; Phelan, Patrick; Gunawan, Andrey; Lv, Wei; Rosengarten, Gary; Prasher, Ravi; Tyagi, Himanshu (2013). "Small particles, big impacts: A review of the diverse applications of nanofluids." *Journal of Applied Physics* **113**: 011301. doi:10.1063/1.4754271.
27. Taylor, Robert A; Otanicar, Todd; Rosengarten, Gary (2012). "Nanofluid-based optical filter optimization for PV/T systems." *Light: Science & Applications* **1** (10): e34.
28. Taylor, Robert A; Phelan, Patrick E; Otanicar, Todd P; Adrian, Ronald; Prasher, Ravi (2011). "Nanofluid optical property characterization: Towards efficient direct absorption solar collectors." *Nanoscale Research Letters* **6** (1): 225
29. Alemán, J.; Chadwick, A. V.; He, J.; Hess, M.; Horie, K.; Jones, R. G.; Kratochvíl, P.; Meisel, I.; Mita, I.; Moad, G.; Penczek, S.; Stepto, R. F. T. (2007). "Definitions of terms relating to the structure and processing of sols, gels, networks, and inorganic-organic hybrid materials (IUPAC Recommendations 2007)." *Pure and Applied Chemistry* **79** (10): 1801
30- Nowack, B.; Bucheli, T. D. (2007). "Occurrence, behavior and effects of nanoparticles in the environment." *Environmental Pollution*, 150, 5-22.
31- Theng, B.K.G.; Yuan, G. (2008). "Nanoparticles in the Soil Environment." *Elements*, 4, 395-399.
32. Sapkota, A.; Williams, D. A.; Buckley, T. J. (2005). "Tollbooth Workers and Mobile Source-Related Hazardous Air Pollutants: How Protective Is the Indoor Environment?" *Environmental Science & Technology,* 39, 2936-2943.
33. Wiesner, M. R.; Lowry, G. V.; Jones, K. L., Hochella, J. (2009). "Decreasing Uncertainties in Assessing Environmental Exposure, Risk, and Ecological Implications of Nanomaterial." *Environmental Science & Technology*, 43, 6458-6462.
34. Roduner, E. (2006). "Size matters: why nanomaterials are different." *Chemical Society Reviews*, 35, 583-592.
35. Lead, J. R.; Wilkinson, K. J. (2006). "Aquatic Colloids and Nanoparticles: Current Knowledge and Future Trends." *Environmental Chemistry*, 3, 159-171
36. Ju-nam, Y.; Lead, J.R. (2008). "Manufactured nanoparticles: An overview of their chemistry, interactions and potential environmental implications." *Science of the Total Environment*, 400, 396-414.
37. Shenhar, R.; Rotello, V.M. (2003). "Nanoparticles: Scaffolds and Building Blocks." *Accounts of Chemical Research*, 36, 549-561.
38. Ju-nam, Y.; Lead, J. R. (2008). "Manufactured nanoparticles: An overview of their chemistry, interactions and potential environmental implications." *Science of The Total Environment*, 400, 396-414.
39. Shields, S. P.; Richards, V. N.; Buhro, W. E. (2010). "Nucleation Control of Size and Dispersity in Aggregative Nanoparticle Growth. A Study of the Coarsening Kinetics of Thiolate-Capped Gold Nanocrystals. " *Chemistry of Materials*.
40. Faraday, M. (1857). "The Bakerian Lecture: Experimental Relations of Gold (and Other Metals) to Light." *Philosophical Transactions of the Royal Society of London*, 147, 145-181
41. Turkevich, P.C.; J., Hiller (1951). "A study of the Nucleation and Growth Processes in the Synthesis of Colloidal Gold." *Discussions of the Faraday Society*, 11, 55-75.

42. Murry, H.H. (2000). "Gold: Progress in Chemistry, Biochemistry and Technology Edited by Hubert Schmidbaur (Technical University of Munich, Germany)." *Journal of the American Chemical Society*, 122, 4534-4534.
43. Tolaymat, T. M.; El Badawy, A. M.; Genaidy, A.; Schenekel, K.G.; Luxton, T.P.; Suidan, M. 2010. "An evidence-based environmental perspective of manufactured silver nanoparticle in syntheses and applications: A systematic review and critical appraisal of peer-reviewed scientific papers." *Science of The Total Environment*, 408, 999-1006.
44. Khaydarov, R.R.; Khaydarov, R.A.; Evrhafova, S.; Wagner, S.; CHO, S. Y. (2011). "Environmental and Human Health Issues of Silver Nanoparticles Applications." *Environmental Security and Ecoterrorism*. In: ALPAS, H., BERKOWICZ, S. M. M. & ERMAKOVA, I. (eds.). Springer Netherland.
45. Lewis, L.N. (1993). "Chemical catalysis by colloids and clusters." *Chemical Reviews*, 93, 2693-2730.
46. Solo 'E , A.; Pote Hina, T.; Cherno A, I.; Basin, B. (2007). "Track membrane with immobilized colloid silver particles." *Russian Journal of Applied Chemistry*, 80, 438-442.
47. Niemeyer, C. M. (2001). "Nanoparticles, Proteins, and Nucleic Acids: Biotechnology Meets Materials Science." *Angewandte Chemie International Edition*, 40, 4128-4158.
48. David, D.; Evanoff, Jr.; George, C. (2005). "Synthesis and optical properties of silver nanoparticles and arrays." *Chem. Phys. Chem.* 6, 1221-1231.
49 Kazakevich, P.V.; Voronov, V.V.; Simakin, A.V.; Shafeev, G.A. (2004). "Production of copper and brass nano particles upon laser ablation in liquids." *Quantum Electron*, 34 (2004), 951-956.
50. Hunsberger, Lynn (1999). "UV-Vis Absorption Spectroscopy General Chemistry Laboratory Manual"
51. Shelby, Hatch; George, Schatz (2008). "Synthesis and analysis of Silver/Gold nanoparticles." Northwestern University, Evanston, IL 60208.
52. http://www.nanocomposix.com/services/aerosol-facility
53 Schasfoort, R.B.M.; Tudos, A.J. (2008). "Handbook of Surface Plasmon Resonance." *The Royal Society of Chemistry Publishing*, Cambridge, UK.
54- Pockrand, I.; Swalen, J. D.; Gordon II, J. G.; Philpott, M. R. (1978). "Surface Plasmon spectroscopy of organic monolayer assemblies." *Surface Science*, 74, 237-244.
55- Gordon II, J. G.; Ernst, S. (1980). "Surface plasmons as a probe of the electrochemical interface." *Surface Science*, 101, 499-506.
56. Li, W. R.; Xie, X. B.; Shi, Q. S.;, Zeng, H. Y.;, OuYang, Y. S.; Chen, Y. B. Appl Microbiol Biotechnol. 2010, 85(4), 1115-22
57- Ying, Jackie (2001). *Nanostructured Materials*. New York: Academic Press. ISBN 0-12-744451-3.
58- Hanley, C; Thurber, A; Hanna, C; Punnoose, A; Zhang, J; Wingett, DG; Thurber; Hanna; Punnoose; Zhang; Wingett (2009). "The Influences of Cell Type and ZnO Nanoparticle Size on Immune Cell Cytotoxicity and Cytokine Induction". *Nanoscale Res Lett* **4** (12): 1409–20.
59. Wigginton, N. S., Haus, K. L. & Hochella JR, M. F. (2007). "Aquatic environmental nanoparticles." *Journal of Environmental Monitoring*, 9, 1306-1316.
60. Domingos, R. F.; Baalousha, M. A.; Ju-nam, Y., Reid, M. M.; Tufenji, N.; Lead, J. R.; Leppard, G. G.; Wilkinson, K. J. (2009). "Characterizing Manufactured Nanoparticles in the Environment: Multimethod Determination of Particle Sizes." *Environmental Science & Technology*, 43, 7277-7284.
61. David, D.; Evanoff, Jr.; George, C. 2005. "Synthesis and optical properties of silver nanoparticles and arrays." *Chem. Phys. Chem.* 6, 1221-1231.
62. Henglein, A. (1993). "Physicochemical properties of small metal particles in solution: microelectrode reactions, chemisorptions, composite metal particles and the atom-to-metal transition" *J. Phys. Chem.*, 97 (1993), 5457–5471.
63. P. Mulvaney, Surface plasmon spectroscopy of nanosized metal particles, Langmuir, 12 (1996), 788-800.
64. Mamun Ur Rashid (2009). "Synthesis of Silver Nano Particles (Ag-NPs) and their uses for Quantitative Analysis of Vitamin C Tablets."
65. Solomon, Sally D., Mozghan Bahadory, Aravindan V. Jeyarajasingam, Susan A. Rutkowsky, and Charles Boritz (2012). "Synthesis and Study of Silver Nanoparticles." *Journal of Chemical Education* 84.2 : 322-25. ACS Publications. Web. 9 Mar. 2012
66. Yusuf, Nur (2011). "Gold Nanoparticles: Synthesis, Characterisation and their Effect on Pseudomonas Flourescense." Thesis, University of Birmingham of Birmingham, (p-93)
67. Bonsak, Jack (2010). "Chemical Synthesis of Silver Nanoparticles for Light Trapping Applications in Silicon Solar Cells." Thesis. University of Oslo, 2010. Web. 9 Mar. 2012.

68. Mathew Wynn (2010). "The Synthesis and Characterization of Gold and Silver Nanoparticles in Formal and Informal Settings." Materials Engineering Department California Polytechnic State University, San Luis Obispo.
69. Das, R., S. S. Nath, D. Chakdar, G. Gope, and R. Bhattacharjee (2012). "Preparation of Silver Nanoparticles and Their Characterization." . <http://www.azonano.com/article.aspx?ArticleID=2318>.

436278 Immobilization and Catalytic Activity of Lipase on ModifiEd MWCNT for Oily Wastewater Treatment

Abstract:

We report here on the results of a study of immobilization of Lipase enzyme from *Candida Rugosa* on Multiwall Carbon Nanotubes (MWCNT) locally produced in a Chemical Vapor Deposition (CVD) reactor. Covalent bonds was formed between the enzyme and MWCNTs using organic cross linkers (NHS and EDC). FTIR results confirmed the covalent binding of the amide functional group of the lipase molecules onto the surface of MWCNT. The produced carbon nanotubes proved to be a good matrix for the lipase loading and immobilization. TGA results and calculations confirmed that 30% wt% of the enzyme was successfully attached to the functionalized MWCNTs. The nanocomposite material revealed high thermal stability and good hydrophobic characteristics which is considered an important property for commercial applications. Findings revealed higher than generally reported biological activity of the protein was successfully retained. Results confirmed that Immobilized lipase on MWCNTs present an excellent opportunity for treatment of oily wastewater. Teste with synthetic water/oil emulsion resulted with oil degradation reached up to 98%.

QUALITATIVE SYNTHENSISATION OF NANOPARTICLES USING FTIR SPECTROSCOPIC ANALYSIS

Engr. Dr. M. N. Idris* and **K. M. Gwary**
Energy & Process Refining Technology Group
Department of Chemical Engineering
University of Maiduguri
Borno State, Nigeria
Tel: +234 (0) 705 877 1096
Email: idrismn@unimaid.edu.ng, muhinu@hotmail.co.uk

ABSTRACT

In this work, FTIR spectrometry was used to analyse the spectra configurations on the synthesised silver nanoparticles were synthesized using two different reducing agents (1) sodium borohydride ($NaBH_4$), at two temperature conditions of 27^{0}C and 4^{0}C (room temperature and ice cold temperature), and (2) trisodium citrate (3NaCi) at 100^{0}C (boiling temperature) by chemical reduction method.

The resulting nano-sized colloids have been characterized by FTIR spectroscopy. A clear and yellow colouration was obtained from all the synthesised samples, which indicates the presence of formed nanoparticles with strong absorption peaks obtained at wave length range of 1370 - 1380 cm^{-1} and 840 - 841 cm^{-1} proving the presence of silver nanoparticles based on the functional group. The elements found to be present at such peak were the nitro and N-O functional groups.

Keywords: *Synthesis, FT-IR, Nanoparticle, reducing agent and functional group*

Correspondence about this article*: Please for all correspondence on this article, kindly contact the principal author as stated above. Thank you.*

HYDROGEN PEROXIDE STABILITY IN SILICA HYDROGELS

Fulya Sudur, Nancy Goodyear, Nese Orbey, University of Massachusetts Lowell, Lowell, MA, USA

Abstract

Hydrogen peroxide (H_2O_2) is an environmentally friendly oxidant whose current applications are limited due to its instability. Silica hydrogels obtained by sol-gel process have the potential of increasing its stability due to strong hydrogen bonding between the H_2O_2 molecules and the silanol groups (Si-OH) at the silica surface. The pH of the sol determines the structure of the hydrogel; in previous studies it is controlled by adding an acidic or alkaline solution. In the present study, the pH of the sol is adjusted by varying the sodium content of sodium silicate. The effect of sodium on the kinetics of sol-gel formation and the structure of the resulting hydrogels are studied. The surface area of the hydrogel/xerogel determines the availability of sites for H_2O_2 bonding in the presence of sodium ions. Therefore, the decomposition rate of H_2O_2 is studied at different sodium concentrations to understand its stability at room temperature. Samples are characterized by measuring the surface area, pore volume and average pore size using Brauner-Emmet-Teller (BET) analysis and scanning electron microscopy (SEM). The antimicrobial activity of the hydrogels containing 10 and 20 wt% H_2O_2 was evaluated against *Escherichia coli* and *Staphylococcus aureus* and the results demonstrated that hydrogels exert a potent microbicidal efficacy.

Introduction

Hydrogen peroxide (H_2O_2) is a green oxidant and an antimicrobial agent. It has broad range of industrial applications, such as pulp and textile bleaching, teeth whitening, skin disinfection, and water treatment.[1-4] The use of H_2O_2 in current applications is limited due to its instability according to the following reaction:

$$H_2O_{2(aq)} \rightarrow H_2O_{(l)} + \tfrac{1}{2}O_{2(g)} \qquad (1)$$

The stability of H_2O_2 can be increased by entrapping it in silica xerogels because hydrogen peroxide forms strong hydrogen bonds with the silica gel surface.[5,6] The xerogels are formed via sol-gel chemistry by using an aqueous silicate precursor, sodium silicate.[7] The first step of synthesizing silica gels is to form a colloidal suspension, also defined as sol, by removing sodium ions from the sodium silicate with a cation exchange resin. The sodium content of the sol can be adjusted using an ion exchange process,[8-10] as shown below,

$$Na^+ + RzSO_3^-H^+ \rightarrow H^+ + (RzSO_3)Na \qquad (2)$$

where Rz is the styrene-divinylbenzene copolymer and SO_3H is the sulfonic acid group. Ion exchange occurs easily because the ionic bond strength of a sodium ion is stronger than that of a hydrogen ion.[10]

Further bonding between the silanol groups leads to the formation of a three-dimensional interconnected network resulting in a "wet gel" or "hydrogel". Xerogels are formed simply by drying the hydrogels. The pH of the sol affects the inorganic polymerization, and, hence influences the properties of the silica network.[11]

Silica xerogels containing H_2O_2 have been reported in the literature. Żegliński el al.[5] synthesized xerogels containing 3.8 to 68 wt% H_2O_2. The stability of 3.8 wt% H_2O_2 in xerogel was 71% after 63 days of storage at room temperature and as high as 94% after storage at 3 °C. Bednardz et al.[6] used H_2O_2 containing silica xerogels as oxidants in organic synthesis. The xerogels were stored at

5 °C and half of the initial H_2O_2 loading of 8 wt% was retained. Room temperature stability of H_2O_2 was not reported.

In the present work, the effect of sodium on the kinetics of sol-gel formation in the presence of H_2O_2, its consequent effect on its stability and the structure of the resulting hydrogels are studied. H_2O_2 stability in silica hydrogels at room temperature was monitored at periodic intervals by potassium permanganate ($KMnO_4$) titration. Hydrogels were characterized by scanning electron microscopy (SEM) and Brunauer-Emmett-Teller (BET) analysis. The antimicrobial activity of the hydrogels at various H_2O_2 concentrations was examined.

Experimental Procedure

Materials

Sodium silicate solution (reagent grade, 10.6% Na_2O, 26.5% SiO_2); Amberlite IR120 (hydrogen form); hydrogen peroxide (contains inhibitor, 30% wt in H_2O, ACS reagent); orthophosphoric acid (85% wt in H_2O, 99.99% trace metals basis); potassium permanganate (ACS reagent >99%) and sulfuric acid (99.999%). All chemicals were purchased from Sigma Aldrich Chemical Co. and used without further purification.

Hydrogel Synthesis

Details of the hydrogel synthesis procedure are described elsewhere.[12] Briefly, as received sodium silicate solution was diluted with deionized water and combined with 30 wt% H_2O_2 solution and 7 μL of phosphoric acid (H_3PO_4). The sodium content of the sodium silicate solution was controlled by adding different amounts of ion exchange resin (Amberlite IR 120, hydrogen form) to the sol. The mixture was stirred with an overhead mixer, the resin was allowed to settle, and the sol was transferred to a glass Petri dish (D = 100 mm, h = 15 mm) and placed in a BlueM gravity oven at 40 °C until gelation was observed.

The effect of sodium concentration on the stability of H_2O_2 was studied keeping the initial H_2O_2 concentration constant at 20 wt% and changing the sodium concentration between 0.027 and 0.557 wt%. The effect of initial H_2O_2 concentration on the stability of H_2O_2 was also studied at concentrations between 10 and 20 wt% while keeping the sodium concentration constant at 0.027 wt%.

Characterization

The sodium content of the sol before gelation was measured using a flame atomic absorption spectrometer (240 AA, Agilent Technologies). The H_2O_2 content of the hydrogels was determined by $KMnO_4$ titration immediately after gelation and at periodic intervals. Three repeated titrations were performed on each sample to check reproducibility.

The morphology of the hydrogels was studied using a JEOL JSM 7401F scanning electron microscope. The surface area, pore size, and pore size distribution were obtained through BET analyses by first converting into aerogels using a critical-point dryer (Tousimis SAMDRI-795) to preserve the solid network and retain the pore structure.

Antimicrobial activity was determined by agar well diffusion. Briefly, *Escherichia coli* (ATCC 29214, American Type Culture Collection, Manassas, VA) and *Staphylococcus aureus* (ATCC 6538) were grown overnight at 37°C in Tryptic Soy Broth (Becton Dickinson Co, Sparks, MD). The overnight growth was standardized spectrophotometrically to ~1.5 x 10^8 colony forming units/mL by dilution in sterile phosphate buffered saline (PBS). The standardized bacterial solution was inoculated to duplicate Tryptic Soy Agar plates (Becton Dickinson Co, Sparks, MD) to achieve an even lawn of growth over the entire plate. A 6 mm well cutter was used to punch a single well in the middle of each inoculated plate. The wells were filled to the top with hydrogels containing 10 wt% and 20 wt% H_2O_2 at same sodium content (0.027 wt%). The control sample contained equivalent amount of silica hydrogel without H_2O_2. After overnight incubation at 37°C, the diameters of the zones of inhibition were measured in mm and the average and standard deviation of the duplicates calculated.

Results

Sodium Content

In this study, the amount of residual sodium in the silicic acid determined the pH of the sol whereas in the previous studies the pH was controlled by adding acidic or alkaline solution.[13]

Figure 1 shows a decrease in the pH of the sol at lower sodium contents because sodium ions were replaced with hydrogen ions as a result of ion exchange. A sharp decline in pH was observed at sodium contents below 0.1 wt%. We believe this is due to the high buffer capacity of the sol above 0.1 wt%.[14] In other words, the low pH indicates the low sodium content of the sol.

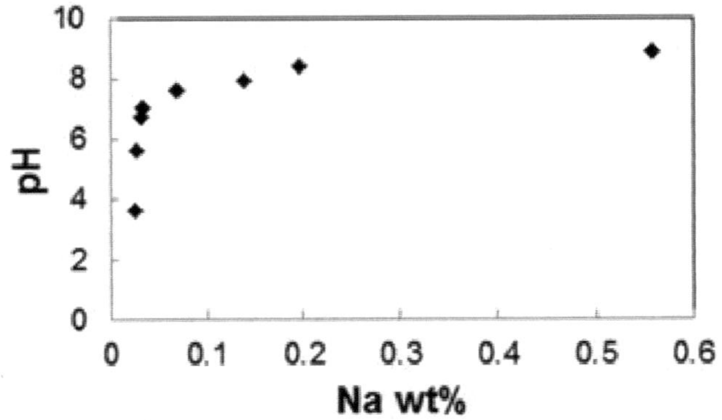

Figure 1. Relationship between the pH and Na content of the sol (initial H_2O_2 concentration 20 wt%)

Relationship between gel time and pH

In this study, the gel time was determined based on the procedure reported in the literature. The moment when the gel broke away from the wall of the Petri dish, instead of flowing as a liquid when the dish was tilted, was considered as the gelation point.[15, 16]

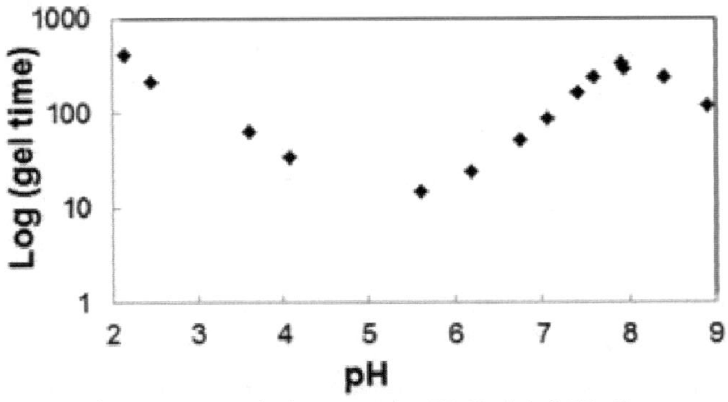

Figure 2. Relationship between gel time and pH (initial H_2O_2 concentration 20 wt%)

The effect of pH on the gel time of silicic acid-water-H_2O_2 was studied and the results are plotted in Figure 2. Low concentration of hydroxyl ions at pH range 2-4 resulted in a slow gel formation. As pH was increased from 4 to 6, the net charge of the sol was zero resulting in a rapid gelation. After the sol passed through the isoelectric point (pH 6–8), the concentration of hydroxyl ions increased and longer gel times were obtained. The presence of sodium ions affected the gelation above a pH of 8 resulting in a decrease in gel time.[17]

When compared to the ones reported in the literature, our results shifted slightly towards the acidic side due to the presence of H_2O_2 in the sol.[18-20]

The effect of H_2O_2 concentration on the gel time was also studied and found that the gel time decreased as the H_2O_2 concentration increased.

Stability of H_2O_2 at room temperature

Decomposition of H_2O_2 followed a first-order rate law during short term storage (10 days):

$$-\frac{d[H_2O_2]}{dt} = k_{H_2O_2}[H_2O_2] \quad (3)$$

$$\ln\left(\frac{[H_2O_2]}{[H_2O_2]_0}\right) = -k_{H_2O_2}t \quad (4)$$

where $k_{H_2O_2}$ is the observed first-order rate constant, and $[H_2O_2]$ and $[H_2O_2]_0$ are the H_2O_2 concentrations in the hydrogels at any time t and time zero, respectively.

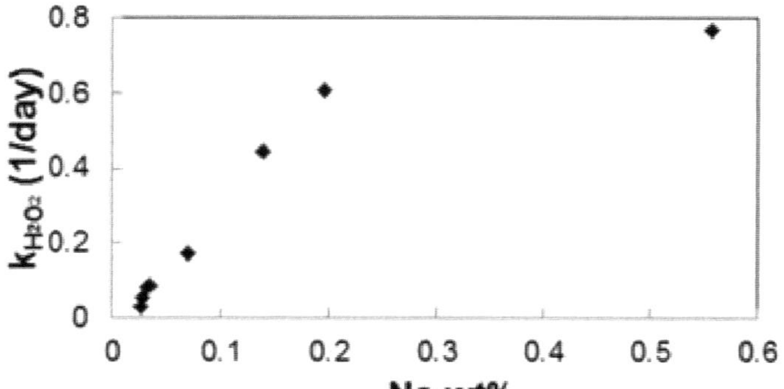

Figure 3. Dependence of the decomposition rate constant on the sodium content of the hydrogels (at 20 wt% H_2O_2)

Figure 3 shows that as the sodium content decreased, the H_2O_2 decomposition rate also decreased. The decomposition rate constant was the same for the hydrogels having different H_2O_2 concentrations at a given sodium content for short term storage.

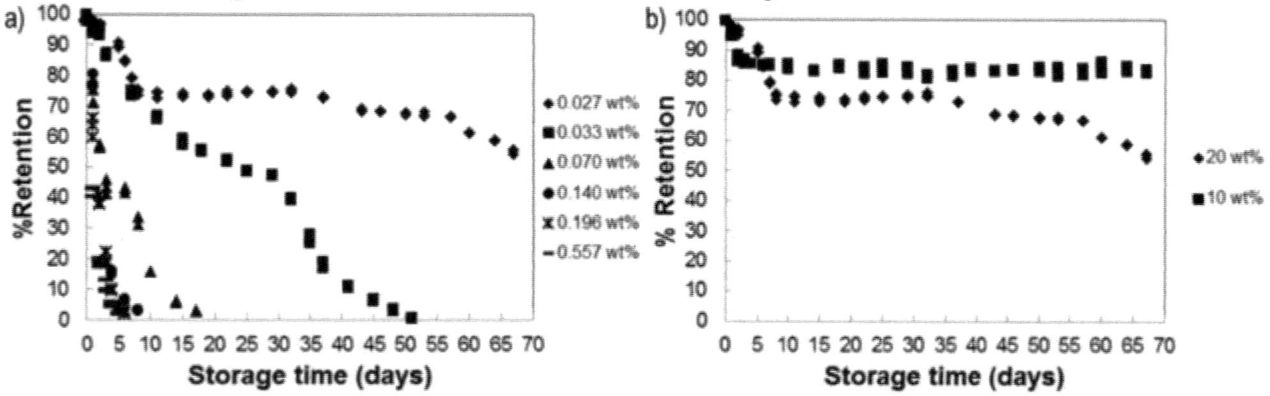

Figure 4. Room temperature stability of H_2O_2 as a function of storage time at changing Na content (initial H_2O_2 concentration 20 wt%) and at changing H_2O_2 concentration (Na content 0.027 wt%)

The stability of entrapped H_2O_2 in hydrogels containing different sodium and H_2O_2 was monitored for 70 days at room temperature. Figure 4a shows that the highest sodium content hydrogel resulted in the fastest H_2O_2 decomposition (5 days) whereas the lowest sodium content hydrogel retained more than 50% of H_2O_2 after 70 days of storage. On the other hand, at low sodium concentration when H_2O_2 concentration decreased to 10 wt%, the retention values were significantly enhanced (up to 89%) as shown in Figure 4b. The optimized sodium content and gelation conditions (time and temperature) resulted in higher stability of H_2O_2.

Morphology and structure of hydrogels

BET and SEM analyses revealed that sodium affected the structure of hydrogels. The samples were prepared by forming aerogels using supercritical CO_2. The effect of sodium content on the surface area, pore volume, and average pore diameter obtained from BET analysis are reported in Figure 5.

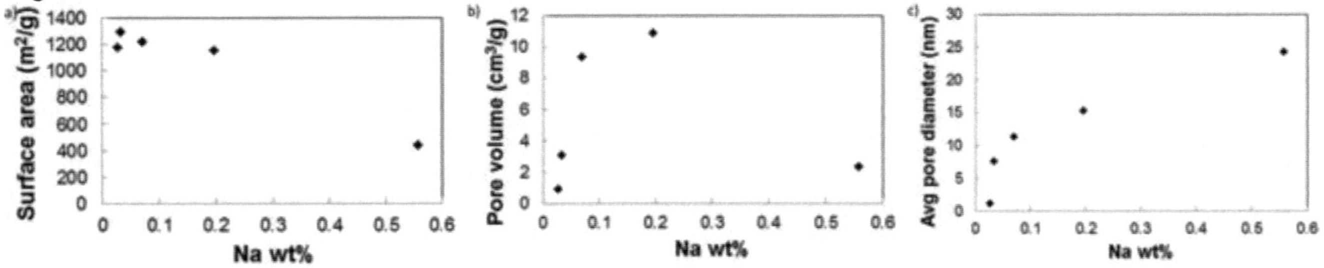

Figure 5. Structural properties of silica hydrogels with changing sodium content (initial H_2O_2 concentration 20 wt%)

Figure 5a shows the surface area measurements where the low sodium concentrations did not have an appreciable effect. However, at higher sodium contents a sharp decrease in the surface area was observed. We believe that silica dissolves and reprecipitates at high sodium content (and at high pH) caused a reduction in the surface area and formation of pores with larger diameter.[21, 22]

We observed an initial increase and then a decrease in the pore volume as a function of sodium content (Figure 5b). Pore volume increased with the sodium content resulting in a decrease in H_2O_2 stability. H_2O_2 stability was also low when the pore volume was low at high sodium contents. These results indicate that sodium has a more significant effect on stability than the pore volume.

Figure 5c shows that sodium content and the pore diameter have a linear relationship. The dissolution and reprecipitation of silica along with the interparticle condensation as a result of high sodium content resulted in larger pores. Formation of smaller sized pores was observed due to the low dissolution rate of silica at low sodium concentrations.

The effect of the H_2O_2/water ratio on the surface area, pore volume and average pore diameter was much less pronounced than the effect of sodium content.

Figure 6. SEM images of hydrogels containing different sodium contents: a) 0.557 wt%, b) 0.196 wt%, c) 0.027 wt% (initial H_2O_2 concentration 20 wt%)

The morphology of the hydrogels with different sodium content is shown in Figure 6. The SEM images indicate that silica network formation was affected by the sodium presence due to changes in hydrogen bonding capacity and strength. At low sodium contents, more compact and denser gel structure was obtained.

Antimicrobial Activity

The test results obtained for the efficacy of our hydrogels containing 0, 10 and 20 wt% H_2O_2 with *E. coli* and *S. aureus* are presented in Table 1 where the average diameters of zones of inhibition (ZOI) are listed. Figure 7 shows the results obtained for *S. aureus* only.

Table 1. Average diameters of zones of inhibition (ZOI) for each hydrogel tested

	Average ZOI Diameter in mm (SD)	
Silica hydrogels	***E. coli***	***S. aureus***
with water	16 (±1.4)	19 (±1.4)
with 10 wt% H_2O_2	43 (±1.4)	51.5 (±2.1)
with 20 wt% H_2O_2	49 (±1.4)	59.5 (±2.1)

The results indicate that bacterial growth was inhibited for both *E. coli* and *S. aureus* and that the ZOI diameter increases with increasing H_2O_2 content in the hydrogel. Since the structure of the hydrogels containing 10 and 20 wt% H_2O_2 was very similar to each other (based on the discussion in the previous section), we believe that the difference obtained from the results is due to the variation in H_2O_2 concentrations. The hydrogels containing water only also exhibit antibacterial properties. We think that the presence of the gel by itself inhibits the bacterial growth.

The mean ZOI values for *S. aureus* were slightly higher than those for *E. coli*. The suppression of the growth of *S.aureus* was attributed to the structural differences between the two bacteria.[23]

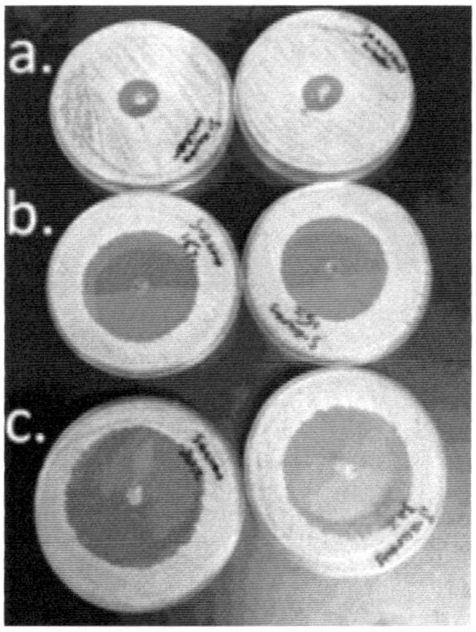

Figure 7. Growth of *S.aureus* when incubated in silica hydrogel a) with water, b) with 10 wt% H_2O_2, c) with 20 wt% H_2O_2

Conclusion

H_2O_2 was successfully stabilized in a silica hydrogel matrix. Our results indicate that the sodium content of the sol has a significant effect on the structure of hydrogels. Mesoporous (3-24 nm) hydrogels with a large surface area (1000-1400 m^2/g) were obtained. Sodium content also affected the retention and stability of H_2O_2 and its decomposition followed a first-order reaction. Optimizing the sodium content and gelation protocol (time and temperature) resulted in higher retention and better stability of H_2O_2 at room temperature. The hydrogels showed antimicrobial activity.

References

1. Colodette, J. L.; Rothenberg, S.; Dence, C. W. Factors Effecting Hydrogen Peroxide Stability in the Brightening of Mechanical and Chemimechanical Pulps. *J. Pulp. Pap. Sci.* 1989, 15, 3-10.
2. Sulieman, M.; Addy, M.; MacDonald, E.; Rees, J. S. The Effect of Hydrogen Peroxide Concentration on the Outcome of Tooth Whitening: An In Vitro Study. *J. Dent.* 2004, 32, 295-299.
3. Kim, J.C.; Lee, K. U.; Shin, W. C.; Lee, H. Y.; Kim, J. D.; Kim, Y. C.; Tae, G.; Lee, K. Y.; Lee, S. J.; Kim, J. D. Monoolein Cubic Phases Containing Hydrogen Peroxide. *Colloids Surf., B* 2004, 36, 161-166.
4. Patil, A. R.; Kausley, B. S.; Balkunde, L. P.; Malhotra, P. C. Comparative Study of Disinfectants for Use in Low Cost Gravity Driven Household Water Purifiers. *J. Water Health* 2013, 11.3, 443-456.
5. Żegliński, J.; Cabaj, A.; Strankowski, M.; Czerniak, J.; Haponiuk, J.T. Silica Xerogel-Hydrogen Peroxide Composites: Their Morphology, Stability, and Antimicrobial Activity. *Colloids Surf., B* 2007, 54, 165-172.
6. Bednardz, S.; Rys, B.; Bodgal, D. Application of Hydrogen Peroxide Encapsulated in Silica Xerogels to Oxidation Reactions. *Molecules* 2012, 17, 8068-8078.
7. Baccile, N.; Babonneau, F.; Thomas, B.; Coradin, T. Introducing Ecodesign in Silica Sol-Gel Materials. *J. Mater. Chem.* 2009, 19, 8537-8559.
8. Bergna, H. E.; Robertw, W. O. *Colloidal Silica: Fundamentals and Applications*; CRC Press: Florida, 2005.
9. Van Dokkum, H.P.; Hulskotte, H. J.; Kramer, K. J. M; Wilmot, J. Emission, Fate and Effects of Soluble Silicates (Waterglass) in The Aquatic Environment. *Environ. Sci. Technol.* 2004, 38, 515-521.
10. Kim, T.; Hwang, S.; Hyun, S. Development of a Continuous Manufacturing Process for Silica Sols via the Ion Exchange of a Waterglass. *Ind. Eng. Chem. Res.* 2008, 47, 6941-6948.
11. Quarch, K.; Kind, M. Inorganic Precipitated Silica Gel. Part 1: Gelation Kinetics and Gel Properties. *Chem. Eng. Technol.* 2010, 33, 1034–1039.
12. Sudur, F; Pleskowicz, B.; Orbey, N. Hydrogen Peroxide Stability in Silica Hydrogels. *Ind. Eng. Chem. Res.* 2015, 54(6), 1930-1940.
13. Iler, R. K. *The Colloidal Chemistry of Silica and Silicates*; Cornell University Press: New York, 1955.
14. Hamouda, A. A.; Amiri, H. A. A. Factors Affecting Alkaline Sodium Silicate Gelation for In-Depth Reservoir Profile Modification. *Energies* 2014, 7, 568-590.

15. Merrill, R. C.; Spencer, R. W. The Gelation of Sodium Silicate. Effect of Sulfuric Acid, Ammonium Sulfate, and Sodium Aluminate. *J. Phys. Chem.* 1950, 54, 806-812.
16. Munoz-Aguado, M. J.; Gregorkiewitz, M. Sol-Gel Synthesis of Microporous Amorphous Silica from Purely Inorganic Precursors. *J. Colloid Interface Sci.* 1997, 185, 459-465.
17. Berrier, E.; Courtheoux, L.; Bouazaoui, M.; Capoen, B.; Turrell, S. Correlation between Gelation Time, Structure and Texture of Low Doped Silica Gels. *Phys. Chem. Chem. Phys.* 2010, 12, 14477-14484.
18. Iler, R. K. *The Colloidal Chemistry of Silica and Silicates*; Cornell University Press: New York, 1955.
19. An-Pang, T. A Theory of Polymerization of Silicic Acid. *Sci. Sin.* 1963, 12, 1311-1320.
20. Brinker, J. C.; Scherer, G. W. *Sol-Gel Science: The Physics and Chemistry of Sol-Gel Processing;* Academic: San Diego, 1990.
21. Baxter, S.; Bryant, K.C. Silica Sols. Part III. Accelerated Gelation and Particle Size. *J. Chem. Soc.* 1952, 74, 3024-3027.
22. J.; Deshpande, R.; Smith, D. M; Brinker, C. J.; Assink, R. A. Pore Structure Evolution in Silica Gel During Aging/Drying. IV. Varying Pore Fluid pH. *J Non-Cryst. Solids* 1994, 167, 295-306.
23. Shilpakala, S.R.; Prathiba, J.; Malathi, R. Susceptibilities of *Escherichia coli* and *Staphylococcus aureus* to *Aloe barbadensis*. *Eur. Rev. Med. Pharmacol. Sci* 2009, 13, 461-464.

427199 Micro Structural and Electrical Properties of Ceria-Yttria Stabilized Zirconia Nanocomposites

Abstract:

Yttria stabilized zirconia (YSZ) is widely accepted nanomaterial electrolyte for solid oxide fuel cells (SOFCs). But, high operational temperature (~1000°C) is required in 8 mol. % YSZ (8YSZ) electrolytes for achieving acceptable levels of ionic conductivity (~0.1S/cm). Alternately, doped cerium-oxide (ceria) show promise as electrolytes capable of operating at reduced temperatures (~600°C). Enhanced ionic conductive interface is observed in doped ceria and YSZ composites in comparison to that of YSZ alone. Thus, it would be advantageous to synthesize ceria-8YSZ composites nanomaterials with 0, 5 and 10 wt. % ceria addition in 8YSZ, for attaining enhanced ionic conductivity. The novelty of current proposal is to synthesize ceria-8YSZ nanocomposite via spark plasma sintering (SPS, 1200 and 1300°C, 5min. 30 MPa), and compare the resulting properties (mechanical properties, ionic conduction, etc) of ceria-8YSZ composites. This will allow correlation of processing temperature, microstructure, and ionic conduction of ceria-8YSZ, so that a mechanism for enhanced ionic conductivity can be explained. XRD analysis revealed that all samples exhibit both cubic and tetragonal structuressynthesized via SPS. The impedance spectroscopy analysis shows that ionic conductivity isincreased by the increasing CeO_2content in 8YSZ (which gave the clear evidence for ionic conduction not electronic as ceria content increases); this is also attributed to the nano size effect, high ionic conducting pathways in dense CeO_2-YSZ nanocomposite to promoting fast oxygen ion transportation. Thus optimal selection (physical, micro structural, mechanical and ionic conduction) of ceria-YSZ electrolytes will strongly assist in designing solid oxide fuel cells with enhanced performance.

427592 Nanostructured Robust Cobalt Alloy Based Anode Electro-Catalysts with Superior Electrochemical Activity for Proton Exchange Membrane Fuel Cells

Wednesday, November 11, 2015: 8:48 AM
251F (Salt Palace Convention Center)

Prasad P. Patel[1], Moni Kanchan Datta[2,3], Oleg Velikokhatnyi[2,3], Prashanth Jampani[2] and Prashant N. Kumta[1,2,3,4,5], (1)Department of Chemical & Petroleum Engineering, University of Pittsburgh, Pittsburgh, PA, (2)Department of Bioengineering, University of Pittsburgh, Pittsburgh, PA, (3)Center for Complex Engineered Multifunctional Materials, University of Pittsburgh, Pittsburgh, PA, (4)Department of Mechanical Engineering and Materials Science, University of Pittsburgh, Pittsburgh, PA, (5)McGowan Institute of Regenerative Medicine, University of Pittsburgh, Pittsburgh, PA

The rapid depletion of fossil fuels and increased environmental pollution due to colossal consumption of ubiquitous fossil-fuel is a major impetus for efficient use of energy and exploration of renewable and clean energy sources.[1, 2] Generation of electricity from renewable energy sources, such as solar, wind without producing carbon dioxide-an undesirable green house pollutant, offers enormous potential for meeting the future energy demands.[3] Development of novel and efficient technology to store electrical energy is thus extremely important for meeting the global energy demand and environmental concerns. Fuel cell technology has garnered increasing attention over the years as it provides promising and sustainable approach for the production of continuous power with potential for reduced greenhouse gas emissions and higher efficiencies compared to hitherto combustion based technologies. In particular, proton exchange membrane fuel cells (PEMFCs) are considered to be suitable power sources for automobiles, consumer electronic devices and auxiliary power units due to the advantages of using hydrogen as fuel which is light-weight, clean and has a low operating temperature. Hydrogen also offers quick start-up, extended durability of system components, high power density with low weight and volume due to elimination of additional steps needed for fuel reformation. The simple system design would be reflected as an ease in operation, reduced cost and high reliability. However, capital cost of the system is a major constraint limiting commercialization of PEMFCs due to use of expensive noble metal based Pt/C catalyst. Hence, development of non-noble metal based catalysts with high electrochemical activity and durability is of specific interest to replace Pt/C and thus, lower the cost of PEMFC system.

The present study explores $Co_{1-x}(Ir_x)$ (x=0.2, 0.3, 0.4) solid solution alloys as anode electro-catalyst for hydrogen oxidation reaction (HOR) for PEMFC applications. **Fig. 1** shows the SEM micrograph of $Co_{0.7}(Ir_{0.3})$ with x-ray mapping of Co and Ir. The x-ray mapping shows homogeneous distribution of Co and Ir without segregation on any specific site. Electrochemical characterization has been carried out in H_2 saturated 0.5 M sulfuric acid (H_2SO_4) as an electrolyte, employing Pt wire as the counter electrode and Hg/Hg_2SO_4 as the reference electrode (+0.65 V with respect to normal hydrogen electrode, NHE), using a scan rate of 10 mV/sec at a temperature of 40°C. The solid solution electro-catalyst $Co_{1-x}(Ir_x)$ (x=0.3, 0.4) exhibit superior electrochemical activity than Pt/C with electrochemical stability matching to that of Pt/C. In addition, to obtain a better understanding of the fundamental electrochemical activity of $Co_{1-x}(Ir_x)$ electro-catalyst, first-principles calculations of the total energies and electronic structures of the model systems with chemical compositions similar to those of the experimentally synthesized materials have been carried out to complement the present experimental study.

The electrochemical study conducted in half-cell configuration and single PEMFC full cell results shows the potential of $Co_{1-x}(Ir_x)$ (x=0.3, 0.4) as replacement of Pt/C, owing to the excellent electrochemical performance and stability. These results portend significant reduction in the overall capital cost of PEMFCs. The results of structural characterization and electrochemical activity of these electro-catalysts will be presented and discussed.

References

1. C. Liu, F. Li, L.-P. Ma and H.-M. Cheng, *Advanced Materials*, 2010, 22, E28-E62.

2. J. R. Miller, *Science*, 2012, 335, 1312-1313.

3. M. S. Whittingham, *Mrs Bull*, 2008, 33, 411-419.

Acknowledgement

Research supported by the CBET, National Science Foundation, Grant-NSF-0933141. PNK also acknowledges the Edward R. Weidlein Chair Professorship funds and the Center for Complex Engineered Multifunctional Materials (CCEMM) for partial support of this research.

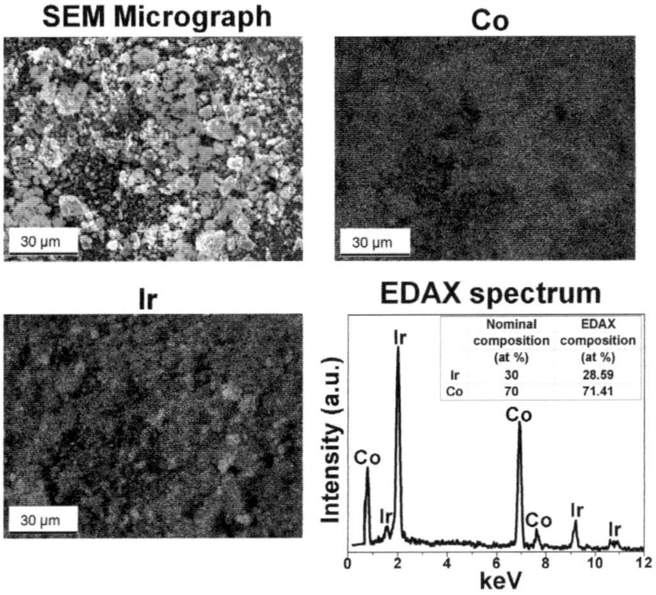

Fig 1: SEM micrograph with EDAX spectrum and x-ray mapping of $Co_{0.7}(Ir_{0.3})$ electro-catalyst

429924 Pathways Towards Defect-Tolerate, Electrochemically Stable Solar-Hydrogen Membranes

Abstract:

Defect tolerance in electrochemical corrosion protection is an important concept in chemical engineering devices. In particular, this concept has been applied to artificial photosynthetic devices that mimics nature's photosynthesis and use semiconductors to take sunlight and splits water into H_2 and O_2. Previously, more than half a century of photoelectrochemical (PEC) water splitting research mainly stayed at model systems of semiconductor – non-aqueous electrolyte interfaces, because of oxidative instability of essentially all technologically important small band gap semiconductors.

In this work, a general corrosion protection strategy for stabilizing semiconductor–electrolyte interfaces for water oxidation and hydrogen production in pH 14 base will be presented. This has been achieved by introducing complex metal-oxide-semiconductor interfaces from opto-electronics research to photoelectrochemistry. This not only brings back an entire class of semiconductors that were previously rejected, but also extends PEC theory by introducing new electrochemical models and band diagrams of stabilized semiconductor-electrolyte interfaces. Furthermore, nanowire-shape semiconductors with conformal protective coatings showed unprecedented > 2200 hours of stability for continuous water oxidation. This discrete nanowire geometry not only enables manufacture of flexible, nanowire-embedded solar-hydrogen membranes, but also confine defects within the affected region by virtue of self-passivating mechanism under the discrete nanowire geometry. Finally, I will give an outlook of the materials challenges and opportunities towards scalably manufacturable artificial photosynthetic devices that directly store solar energy into chemical fuels.

433837 PtCo/Coox Nanocomposites As Bifunctional Electrocatalysts for Oxygen Reduction and Evolution Reactions Synthesized Via Tandem Laser Ablation Synthesis in Solution-Galvanic Replacement Reactions

Abstract:

PtCo/CoO_x Nanocomposites as Bifunctional Electrocatalysts for Oxygen Reduction and Evolution Reactions synthesized via Tandem Laser Ablation Synthesis in Solution-Galvanic Replacement Reactions

Sheng Hu[2, 3], Gabriel Goenaga[2], Thomas A. Zawodzinski[2, 3], Dibyendu Mukherjee[1, 2, 3] *

[1]Department of Mechanical, Aerospace, and Biomedical Engineering; [2]Department of Chemical and Biomolecular Engineering; [3]Sustainable Energy Education and Research Center (SEERC); University of Tennessee, Knoxville, Tennessee, 37996

Intermetallic nanostructures extend the range of bulk metallic properties by virtue of their manifold compositional, structural, phase, and morphological variations. The ability to tune structure-property characteristics of metal/intermetallic nanoparticles (NPs), via controlled synthesis techniques, can enable unique physico-chemical properties that find wide-spread scientific applications in new classes of energetic and catalytic materials. In the field of proton exchange membrane fuel cell (PEMFC), efficient yet, low-cost electrocatalysts are indispensable for electrochemical oxygen reduction reactions (ORR) and oxygen evolution reactions (OER). We present a one-step, one-pot solution-phase experimental route that, for the first time, incorporates laser ablation synthesis in solution in tandem with galvanic replacement reaction (LASiS-GRR) to synthesize nanocomposites (NCs) of PtCo nanaoalloy (NA) embedded in CoO_x matrices, which act as bifunctional electrocatalysts. High resolution TEM reveals PtCo NAs of mean sizes ~8.5-17.7 nm embedded in sponge-shaped CoO_x matrices. Electron diffraction and X-ray diffraction data along with energy dispersive X-ray spectrometry (EDX) confirm the NC compositions and crystalline structures. Detailed electrochemistry data indicates outstanding ORR and OER activities for the PtCo/CoO_x NCs while exhibiting better stability than the respective standard nanocatalysts. Such activities are ascribed to the shrunken lattice constants of alloyed PtCo that promote oxygen adsorption and the synergic "spillover" effect from high surface area CoO_x matrices that accelerate both ORR and OER through symbiotic adsorption/desorption of intermediate species. Furthermore, the CoO_x matrix prevents aggregation and/or dissolution of the embedded NAs in alkaline medium. We report a combined overpotential of 756 mV vs. RHE for the PtCo/CoO_x NC with 33.3 % (molar) Pt content which is the best value ever reported for catalysts supported by carbon black. The enhanced bifunctional catalytic activities of the NCs are attributed to their unique heteronanostructures tailored by our one-pot, "green" synthesis route of LASiS-GRR.

435861 Activity of Electroless Deposited Transition Metal-Catalyst for the Alcoholysis of Ammonia Borane

Abstract:

Catalysts based on electroless nickel and bi-metallic nickel-cobalt nanoparticles were synthesized for the alcoholysis of ammonia borane for hydrogen generation. The catalysts were synthesized by polymer-stabilized Pd nanoparticle-catalyzation and activation of TiO2 substrate and electroless Ni or Co-Ni plating of the substrate for selected time lengths. Catalytic activity of the synthesized catalysts was tested for the alcoholysis of ammonia borane (NH_3BH_3) for hydrogen generation. The effects of electroless plating time lengths, temperature and NH_3BH_3 concentration on hydrogen generation rates were analyzed and discussed. Compositional analysis and surface morphology were carried out for nano-metallized TiO2 using Scanning Electron Micrographs (SEM) and Energy Dispersive X-Ray Microanalysis (EDAX). The activity of the electroless-deposited transition metal-catalysts - both under stirred and unstirred conditions - will be analyzed and discussed. The kinetics and the activation energy of the hydrogen generation alcoholysis using the catalyst will be presented.

409668 Sub-Cell Size Spiky Nanoparticles for Drug Delivery

Abstract:

Array of high aspect-ratio nanowires (NWs) have attracted great interest for biomedical research owing to their one dimensional architecture and unique properties, yet bio-applications with solution-based delivery, such as neccessary in vivo, are limited due to the inherent attachment of NWs to a planar substrate. Celluar uptake of free NWs through endocytosis generally suffers from ineffective endosomal escape and low efficiency of cytosolic access. Here we explore synthesis of nanoparticles covered with stiff NWs, namely "spiky particles", which combine the advantages of NWs' cell penetration capability, and aqueous delivery of suspended nanoparticles. In this work, ZnO NWs covering suspended SiO_2 nanoparticles were developed using a solution-based synthesis approach. These spiky particles can be fabricated with tunable dimensions or sizes, and were successfully dispersed in solution, a challenge for other solid-phase synthesis techniques. Possible applications for materials delivery to cells was explored including the cell-particle interface, spiky particle cytotoxicity, and spiky particle-mediated DNA transfection.

424040 Agglomeration of Nanoparticles Evaluated Via a Constant Number Monte Carlo Simulation

Abstract:

Nanoparticles (NPs) tend to agglomerate in aqueous media which then affects their fate and transport in the environment and behavior in biological systems. Given the rapid growth of the nanotechnology industry, experimental mapping of the agglomeration behavior of the rapidly rising large number of NPs, over the range of NPs properties and water chemistry that may be of interest is a daunting task. In order to provide a platform for assessing the agglomeration behavior of NPs, a computational model was developed based on a constant-number dynamic simulation Monte Carlo (DSMC) simulation of particles in a box to essentially solve the Smoluchowski equation of particle agglomeration. In this modeling approach, NPs which are in Brownian motion collide and the attachment efficiency of colliding particles is dictated by the interaction energy between particles. In the present work, the total particle-particle interaction energy was quantified by an extended-DLVO theory that accounts for electrostatic and hydration repulsions and van der Waals attraction. Model predictions were assessed relative to experimental dynamic light scattering (DLS) size measurements for TiO_2, CeO_2, SiO_2, and a-Fe_2O_3 (hematite) demonstrated that the hydration repulsion energy can be significantly greater than the electrostatic repulsion energy at high ionic strength (IS) and low absolute zeta potential. Under such conditions the hydration repulsion increases the repulsion energy thus leading to lowering of the resulting agglomerate diameter. The present analysis suggests that quantification of the hydration repulsion over a broad range of suspension and NPs would be beneficial to improving predictions of the agglomeration of NPs in aqueous suspensions.

433635 Nanoscale Hydrogel Coatings for Rapid Sorting of Circulating Tumor Cells: Relating Marker Expression to Isolation Yield

Abstract:

A link between the presence of tumor cells in systemic circulation and metastatic cancer progression has long been established. Observations of these circulating tumor cells (CTCs) has prompted efforts to develop a fluid biopsy, with the goal of detecting these rare cells in patient peripheral blood as surrogate markers of metastatic disease for partial replacement or supplement to invasive tissue biopsies. While such an assay could potentially provide valuable prognostic information among numerous other clinical advantages, the implementation of fluid biopsies has been critically hindered by current available technologies for rare cell isolation. Numerous platforms have been designed; most commonly magnetically-activated cell sorting (MACS), fluorescent-activated cell sorting (FACS), and a myriad of antigen-based microfluidic capture technologies. Yet, these have generally failed to support a reliable fluid biopsy assay, due to poor performance parameters such as low throughput, low purity of enriched antigen positive cells, and insufficiently low detection thresholds to detect poor expressed surface markers of target cell populations. In this work, we build on a recently developed rapid cell sorting technique called Antigen Specific Lysis (ASL) based on photo-crosslinked polymer film encapsulation to isolate tumor cells in suspension. Briefly, antibodies direct the specific binding of streptavidin-eosin-5-isothiocyanate (SA-EITC, type II photoinitiator) to surface markers of intact cells, then upon immersion in an aqueous monomer solution, visible light initiates surface-mediated radical crosslinking of nanoscale PEG diacrylate hydrogel films solely around the antigen positive population. Finally, anionic surfactants lyses virtually 100% of uncoated cells while fully encapsulated cells remain protected, and are then easily collected by centrifugation. This sorting process has shown >90% viability of sorted cells, and can be performed in 1-2 hours with inexpensive and disposable labware. Further, because this technique is not dependent on microfluidics or spatial mobility of target cells, we expect it is fundamentally capable of accommodating very large batch sized (10^{11} cells) with only marginal changes in processing time. Here, we focus on determining the response of process yield on the level of photoinitiator loading on target cell. The surface-mediated gelation response has been shown to be dependent on the density of photoinitiator at the initiating surface – therefore, we posit this is a critical parameter to assess for attaining high processing yield, particularly during the lysis-based negative depletion. Further, a lower initiator density threshold is seen in these surface initiated PEG diacrylate films, so that a lower threshold of marker expression for SA-EITC loading will determine successful coating and, ultimately, isolation of target populations. Particularly for CTC isolation, recent evidence suggests that the most highly potent and aggressive CTCs often downregulate epithelial markers that are commonly targeted for cell isolation, further highlighting the need to determine the effects of expression level on performance of this system. To methodically simulate lower expression levels of epithelial markers, we immunolabel Epithelial Cell Adhesion molecule (EpCAM) on a non-small cell lung cancer line H358 with biotin antibodies, then competitively bind incrementally varied ratios of streptavidin and streptavidin-photointiators and evaluate the percent yield of isolated cells versus relative photoinitiator loading determined by flow cytometry. We expect these data will inform robust targeting strategies with optimal yield performance to allow the development of ASL as reliable tumor cell isolation platform for clinical use.

419297 Artificial Micromotors in the Mouse's Stomach: A Step Towards in Vivo Use of Synthetic Microbots

Abstract:

Artificial micromotors, operating on locally supplied fuels and performing complex tasks, offer great potential for diverse biomedical applications, including autonomous delivery and release of therapeutic payloads and cell manipulation. Various types of synthetic motors, utilizing different propulsion mechanisms, have been fabricated to operate in biological matrices. However, the performance of these man-made motors has been tested exclusively under in vitro conditions (outside the body); their behavior and functionalities in an in vivo environment (inside the body) remain unknown. Herein, we report the first in vivo study of artificial micromotors in a living organism using a mouse model. Such in vivo evaluation examines the distribution, retention, cargo delivery, and acute toxicity profile of synthetic motors in mouse stomach via oral administration. Using zinc-based micromotors as a model, we demonstrate that the acid-driven propulsion in the stomach effectively enhances the binding and retention of the motors as well as of cargo payloads on the stomach wall. The body of the motors gradually dissolves in the gastric acid, autonomously releasing their carried payloads, leaving nothing toxic behind. This work is anticipated to significantly advance the emerging field of nano/micromotors and to open the door to in vivo evaluation and clinical applications of these synthetic motors.

431218 REAL-Time Evaluations of Binding Events Between Streptavidin and Its Ligands

Abstract:

In the cells, protein-ligand specific recognition involves association and dissociation processes controlled by the affinity of the two partners under chemical harvesting of adenosine triphosphate energy. Fundamental knowledge of selected specific recognition events is currently translated in synthetic environment and widely used in biosensors, immunoassays and diagnosis applications, or in pharmaceutical development. However, in order to advance such fields, one needs to determine the bond lifetime by characterizing the binding efficiency function of the rupture force of the two partners, as well as the energy landscape parameters or the bond and barrier width energy.

In this study, contact mode analysis of rupture forces between streptavidin and its ligands, namely biotin and anti-streptavidin antibody, were evaluated in real-time using atomic force microscopy (AFM) to provide insights of their relative strength. Our approach used 3-Aminopropyltriethoxysilane (APTS) and glutaraldehyde (GA) to covalently functionalize monolayers of streptavidin with controlled protein packing at glass interfaces, while the AFM tip was used to immobilize biotin/anti-streptavidin antibodies. Analysis revealed that the rupture events between the protein and its individual ligands are characteristics of multi- and single bonds of different strengths and energy landscapes as determined by the structural packing, the symmetry and conformational changes of the individual complexes. Our results also revealed that understanding the importance of the rupture forces between individual protein and its ligand could serve as the first step to protect weak bindings in on-off switches employed in biomedical research applications where specificity and selectivity is foremost sought.

431224 A Nanostructured Hybrid Bioanode Hydrogel for Deep Oxidation of Glycerol

Abstract:

There has been a significant interest lately in the complete oxidation of biofuels to CO_2. The use of multiple biocatalysts allows for deeper biofuel oxidation and thus maximizes the energy density that can be extracted per molecule of biofuel. However, there are still many biofuels for which a suitable biocatalyst is not known. We have recently taken an alternative approach to combine the high catalytic activity of enzymatic biocatalysts with the promiscuous nature of small molecule organic catalysts. We demonstrate the complete electrochemical oxidation of the biofuel glycerol to CO_2 using a hybrid enzymatic and small-molecule catalytic system. Combining an enzyme, formate dehydrogenase (FDH), and an organic oxidation catalyst, TEMPO, we are able to electrochemically oxidize glycerol at a carbon electrode, while collecting as many as 16 e^-s per molecule of fuel. Additionally, we have developed a method to immobilize both FDH and the TEMPO catalyst in a cross-linked hydrogel film at the surface of a carbon electrode which allows for a dramatic increase in the rate of electrocatalytic oxidation of several sugars and short-chain alcohols. A TEMPO derivative was covalently attached to a linear poly(ethylenimine) backbone, and the modified polymer was infused with carbon nanotubes and cross-linked onto the surface of a glassy carbon electrode in the presence of FDH. The use of a nanostructured hydrogel film allows for a porous network through which substrate can freely diffuse and the use of a polyamine provides a high localized concentration of base necessary for the catalytic cycle of TEMPO.

434833 Detergent-Mediated Photosystem I (PS I) Based Proteoliposome Formation: A First Step Towards Bio-Mimetic Device Fabrication

Abstract:

We investigate the role of detergent-mediated protein reconstitution in incorporating photosynthetic membrane proteins, Photosystem I (PS I), into biomimetic lipid vesicles. The underlying motivation behind our research effort is to fabricate Photosystem I (PS I) based bio-hybrid opto-electronic devices. Our recent investigations on the interaction of two nonionic detergents *n-dodecyl*-β-D-maltoside (DDM) and Triton X-100 (TX-100) with two types of phospholipids, namely DPhPC (1,2-diphytanoyl-*sn*-glycero-3-phosphocholine) and DPPG (1,2-dipalmitoyl-*sn*-glycero-3-phospho-(1'-*rac*-glycerol)), revealed the ideal detergent/lipid candidates for proteoliposome formation. By utilizing these detergent/lipid combinations, the current work examines the nature of PS I-lipid interactions in colloidal solutions with the aid of scanning probe microscopy, dynamic light scattering and cryo-transmission electron microscopy imaging. In turn, the aforementioned study provides valuable insight into mimicking the naturally occurring protein-phospholipid interactions in photosynthetic thylakoid membranes. In future, these bio-mimetic systems shall facilitate easy incorporation of directionally oriented proteoliposomes into novel bio-hybrid devices.

418963 Parametric Study, Sensitivity Analysis, and Optimization of Polyol Synthesis of Silver Nanowires

Abstract:

The effects of process parameters in polyol synthesis of silver nanowires (AgNWs) on the nanostructure morphology and silver nanowire yield were investigated in order to optimize yield. The AgNWs were synthesized in a batch polyol salt-mediated surfactant-assisted process. The AgNW morphology and yield were characterized by scanning electron microscopy (SEM). The effects of different factors including reagent concentration and molecular weight (MW), temperature, reagent solution preparation, argon purging and stirring rate were investigated. Based on this parametric study, the most effective and important factors on AgNWs yield were determined. A design of experiment (DOE) technique was used to evaluate the coupled effects of process parameters on yield.

416749 Mechanisms of Metal Mineralization on Biotemplates for Nanowire Synthesis

Abstract:

The underlying fundamental mechanisms governing biomineralization of palladium on virus biotemplates were studied using different virus-palladium systems. The Barley Stripes Mosaic Virus (BSMV) has been applied as a new biotemplate for the synthesis of palladium nanowires. BSMV-metal templating has subsequently been compared to those of the Tobacco Mosaic Viruses (wild type and mutants) to attain insight into the mechanisms of adsorption and reduction of precursor onto biotemplates. Mineralization was performed in a 3mL chemical reactor, using the recently discovered facile synthetic method known as the hydrothermal synthesis. Comparative kinetic and adsorption studies were performed using UV-Vis spectroscopy with the different virus-metal systems under different conditions of pH, temperature and reactant ratios. BSMV is shown to be an expedient and alternate template suitable for production of very high quality nanowires which are uniform, controllable and monodisperse. Such properties make this template a suitable/superior alternative to many current biotemplates. Furthermore, the elucidation of pertinent fundamental parameters provide a more robust understanding of mechanisms crucial for scaleup and directed synthesis in the hydrothermal synthesis on TMVs and BSMV.

434137 Design and Synthesis of Nanowire Heterostructures for Thermoelectric Applications

Abstract:

To design superior thermoelectric materials the minority carrier blocking effect in which the unwanted bipolar transport is prevented by the interfacial energy barriers in the heterogeneous nanostructures has been theoretically proposed recently. The theory predicts an enhanced power factor and a reduced bipolar thermal conductivity for materials with a relatively low doping level, which could lead to an improvement in the thermoelectric figure of merit (ZT). Here we show the first experimental demonstration of the minority carrier blocking in lead telluride–silver telluride (PbTe–Ag2Te) nanowire heterostructure-based nanocomposites. The nanocomposites are made by sintering PbTe–Ag2Te nanowire heterostructures produced in a highly scalable solution-phase synthesis. Compared with Ag2Te nanowire-based nanocomposite produced in similar method, the PbTe–Ag2Te nanocomposite containing 5 atomic % PbTe exhibits enhanced Seebeck coefficient, reduced thermal conductivity, and 40% improved ZT, which can be well explained by the theoretical modeling based on the Boltzmann transport equations when energy barriers for both electrons and holes at the heterostructure interfaces are considered in the calculations. For this p-type PbTe–Ag2Te nanocomposite, the barriers for electrons, that is, minority carriers, are primarily responsible for the ZT enhancement. By extending this approach to other nanostructured systems, it represents a key step toward low-cost solution-processable nanomaterials without heavy doping level for high-performance thermoelectric energy harvesting.

405227 Unusual Nanoscale Electronic Effects and Carrier Dynamics in Heterojunction Nanowires

Abstract:

The unique electronic properties of quantum-confined semiconducting nanowires hold great promise for their incorporation in next-generation transistors, circuits, and electronic devices. This reduction in dimensionality results in a dramatic change in their carrier dynamics and electronic structure, [1-2] leading to novel properties such as ballistic transport and conductance quantization. One area of particular interest is the formation and understanding of electron gases in heterostructured nanowires which provides new avenues for exploring enhanced carrier dynamics in these materials.

In order to tailor these nanostructures with the desired physical properties, we must first understand their electronic properties as a function of size and material composition. To this end, we have developed a series of self-consistent predictive computational methods [1-3] to calculate the properties of heterojunction electron gases in quantum-confined core-shell nanowires. Under certain conditions (depending on doping density and spatial geometry), we find that quasi-one-dimensional electron gases can localize at the corners of the nanowire, leading to carrier dynamics that are dramatically different than analogous bulk heterojunctions. In addition, we highlight several areas where many-body quantum effects play a significant role in these low-dimensional structures. In particular, we surprisingly find that simple theoretical approaches can (1) considerably overestimate the number of occupied electron levels, (2) overdelocalize electrons, and (3) significantly underestimate the relative energy separation between electronic subbands. Our results allow a guided understanding of electron carrier dynamics in heterostructure nanowires and further indicate that electron gases in free-standing nanoscale systems are qualitatively different from their bulk counterparts.

[1] B.M. Wong, F. Leonard, Q. Li, and G.T. Wang, Nano Letters, 11, 3074 (2011).

[2] M. Fickenscher, T. Shi, H.E. Jackson, L.M. Smith, J.M. Yarrison-Rice, C. Zheng, P. Miller, J. Etheridge, B.M. Wong, Q. Gao, S. Deshpande, H.H. Tan and C. Jagadish, Nano Letters, 13, 1016 (2013).

[3] A.W. Long and B.M. Wong, AIP Advances, 2, 032173 (2012).

(488e) Determining the Time Window for Dynamic Nanowire Cell Penetration Processes

Xie, X., Massachusetts Institute of Technology
Melosh, N., Stanford University
Originally presented on: 11/11/2015 10:10:00 - 10:35:00

Abstracts:

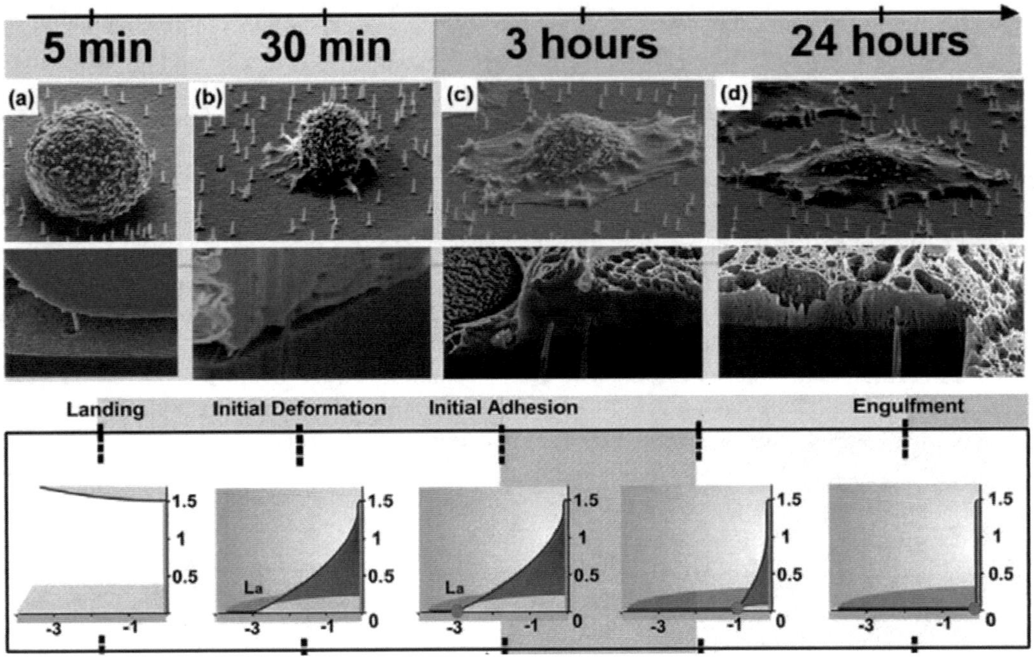

418931 Rheological Behavior of Silver Nanowire Screen Printable Conductive Ink I: Effect of Silver Nanowire Content on the Rheological Behavior of the Ink II: Modeling of the Build-up of the Ink Structure

Abstract:

Screen printing of metal nanostructure inks is a promising technology for the fabrication of current collectors on solar voltaic panels. Among metal nanostructures, silver nanowires (AgNWs) have high potential as conductive fillers in screen-printable ink. Inks suitable for screen printing should exhibit thixotropic rheological behavior with both time dependence and shear thinning. This rheological behavior permits charging of the ink at low shear onto the screen, followed by printing through the screen at high shear. Rapid recovery of the low-shear viscosity after printing is necessary to obtain sharp line definition. The purposes of this work are investigation of the rheological behavior of silver nanowire inks, evaluation of the effect of silver nanowire content on the rheological behavior of the ink, and modeling of the build-up of the ink structure after screen printing. The critical parameter here is the characteristic time for viscosity recovery and its dependence on temperature. Aqueous inks consisting of silver nanowire with dispersant and binder were formulated. The rheological behavior of the inks was characterized by Stress Sweep Step (SSS), Steady State Flow Step (SSFS), Frequency Sweep Step (FSS) and Peak Hold Step (PHS) experiments with a parallel plate rheometer. Rheological measurements under conditions that mimic the screen printing process were done on the conductive inks through the PHS test to assess viscoelastic properties induced by flow alignment of the wires. The stretched exponential model (SEM) was used to model the recovery process after screen printing in order to obtain the characteristic time of recovery for various ink formulations. The characteristic time was determined at several temperatures to assess the temperature dependence of the characteristic time. The recovery mechanism could be due to either Brownian motion or non-Brownian motion. The domination of Brownian motion or non-Brownian motion behavior can be characterized by a Peclet number, which is the ratio of shear rate to the rotational diffusion coefficient. The Peclet number was calculated for different inks at recovery process to illustrate the mechanisms of the build-up after screen printing process.

411761 Modulation of the Protein Corona, Deposition Efficiency, and the Nano-Bio Interface Under Physiologically Relevant Conditions

Abstract:

Colloidal gold nanoparticles (AuNPs) are being increasingly utilized in biomedical applications, such as drug/gene delivery and bio-imaging techniques. However, the effectiveness of these procedures are highly dependent upon sustained, strong interactions between AuNPs and the surrounding environment; referred to as the nano-bio interface. Recently, it has been established that this interface is reliant on the formation of a protein-NP complex. The protein corona surrounding NPs is dynamic by nature and is dependent upon numerous factors including the environmental composition and specific physicochemical properties of the particles. Previous studies have demonstrated that the corona structure mediates NP deposition efficiency, membrane exchanges, and mechanism of endocytosis; making it a critical factor in the use of AuNPs for drug and gene delivery applications. However, little is currently known about how physiological variables, such as fluid flow and biological fluids outside of blood, modulate the protein-NP complex and subsequent particokinetics.

Firstly, we demonstrated that the addition of shear stress, introduced through dynamic fluid movement within the cellular system, severely disrupted the protein corona surrounding 13 nm AuNPs. Furthermore, within a dynamic environment, both the NP behavior and the nano-bio interface were modulated; culminating in a dramatic drop in AuNP deposition efficiency versus a static system. As drug/gene delivery require high AuNP delivery to a target, this loss of NP transport potential highlights a current limitation within the field. Moreover, circulating NPs have a strong likelihood of encountering multiple physiological environments, such as interstitial and lysosomal fluids; of which the influence on the nano-bio interface has yet to be explored. We determined that when dispersed in these biological fluids, both the protein corona and the deposited AuNP dose was altered; as a function of original surface chemistry. These alterations to the nano-bio interface can be directly correlated to NP behavior in accurate physiological fluids. Furthermore, we identified a unique synergistic response to the cellular system during exposure to tannic acid coated AuNPs in conjunction with interstitial fluid: the induction of cytotoxicity.

Given these results, it was clear that under the influence of physiologically relevant variables both the protein corona and the nano-bio interface were disrupted; resulting in a loss of AuNP deposition efficiency. As effective use of AuNPs for drug/gene delivery requires specific cellular targeting, to a high degree, we next sought to utilize the protein corona as a means of improving AuNP-cellular interactions. This was accomplished by generating a corona comprised solely of epidermal growth factor (EGF) around AuNPs prior to cellular introduction. These EGF-AuNPs demonstrated an increase in cellular deposition over stock AuNPs, presumably through augmented binding to predominant EGF receptors on the cellular surface. This study demonstrated that AuNP pretreatment with target proteins could improve cellular deposition, providing a potential means for improving nano-based delivery mechanisms.

(512b) M2e Ligated to Elastin like Polypeptide Induces Protective Immunity Against Influenza A Virus

Ingrole, R.,
Tao, W., Texas Tech University
Gill, H. S., Texas Tech University
Originally presented on: 11/11/2015 12:45:00 - 13:00:00

M2e ligated to elastin like polypeptide induces protective immunity against influenza A virus

With seasonal influenza epidemics leading to remarkable mortality and morbidity throughout the globe, and posing a continuous threat of another pandemic, the influenza virus is considered one of the most deadly human pathogen. In contrast to current vaccines, which are based on the two dominant influenza virus surface glycoproteins - hemagglutinin (HA) and neuraminidase (NA), recent effort has been devoted towards development of a universal influenza vaccine based on the highly conserved sequence of the extracellular region of an influenza membrane protein called M2e. An M2e-based vaccine could offer broader protection against different influenza strains, in contrast to the narrow protection provided by HA and NA-based vaccines. Research done in the past has shown that immunogenicity of M2e can be greatly enhanced by linking M2e, a comparatively smaller antigen, to an appropriate carrier. With this in mind, we wanted to investigate the use of elastin-like polypeptides (ELPs) as an approach to enhance the immunogenicity of M2e. ELPs contain a repeat sequence of ?GXGVP', where 'X' can be any amino acid except proline, and have been shown to be biocompatible and biodegradable in nature. ELPs also exhibit an inverse transition behaviour wherein they precipitate above a certain critical temperature. Thus we hypothesized that M2e-ELPs when designed to precipitate below the body temperature could provide a particulate stimulus to the immune system, and might offer a better alternative to certain carriers such as human papillomavirus, hepatitis B virus core subunit, and keyhole limpet hemocyanin.

In this study we synthesized two different M2e-ELP constructs, the first a monoblock ELP with a hydrophobic guest residue, and the second a diblock ELP with a hydrophilic and hydrophobic guest residue. The resulting molecules were characterized via western blot and MALDI. The inverse transition temperatures were also determined. It was found that attachment of M2e increased the transition temperatures of the ELP molecules, but it was still below 37 C. Upon vaccinating mice with M2e ligated to ELPs, along with CpG as an adjuvant, it was found that a significantly high level of anti-M2e antibodies could be stimulated in the serum. Subsequent challenge of vaccinated mice with lethal influenza A virus (PR8) led to full protection and lower body weight loss. Overall this study provides a proof-of-concept that ELPs could serve as a carrier for M2e and have potential for development of a universal influenza vaccine.

410315 Tumor-Penetrating Aerosol Nanocomposite Microparticles for the Treatment of Lung Cancer

Abstract:

One of the major public health problems in the Unites States is cancer, where over 1 in 4 deaths is due to this disease each year. As of 2010, lung and bronchus cancers are the leading types of cancer-related deaths in all ages for males and females in the US. The five-year relative survival is 54% for localized cancer, 26% for regional, 4% for distant, and 17% in all stages of lung cancer diagnosed [1]. Therefore, there is an unmet need for the development of more effective forms of treatment for lung cancer.

A chemotherapeutic treatment often applied for lung cancer is intravenous (I.V.) paclitaxel (PTX) in the form of Taxol. In general, I.V. therapy exerts negative systemic side effects that can impact the entire body as it travels to its destination via the bloodstream, thereby leading to a need for improved delivery methods. Methods used to treat respiratory diseases such as asthma and chronic obstructive pulmonary disease (COPD) successfully involve the use of aerosols devices and the first use of aerosolized chemotherapy was reported in 1968 [2]. The advantages of using aerosols treatment include targeting of the therapeutic directly to the lungs and the reduction of systemic side effects provoked by high doses of drug in the body.

Cancer cells tend to form solid tumors that are characterized for having hypoxic cells and a necrotic core, which gives them high resistance to the treatments. One reason for the failure of cancer treatments is the low concentration of drug in tumors due to distance and limited access [3]. Recently, it was identified that for tumors expressing αv integrins on their surface that the use of the tumor-homing and penetrating peptide iRGD (CRGDKGPDC) as a targeting moeity to integrins helps to overcome the penetration limitations into these tumors via targeting and tumor penetration [4].

In this project we have developed a dry powder nanocomposite microparticle (nCmP) aerosol containing PTX-loaded nanoparticles comprised of acetalated dextran (Ac-Dex). Ac-Dex is easily biodegradable and results in fast release of therapeutic agents under acidic conditions as seen in tumor tissues [5]. In addition to the drug, a tumor-penetrating peptide (iRDG) was conjugated to Ac-Dex to help the targeting and penetration into the inner layers of solid tumors. Finally, we formulated nCmPs in mannitol via spray drying. The physicochemical properties of the nano- and microparticles (size, charge, drug loading) were evaluated. The effectiveness of the complex drug produced was tested in lung cancer cells (A459 lung adenocarcinoma) in two-dimensional (2-D) cell culture, which was followed by three-dimensional (3-D) cell culture studies that reflect many of the properties of solid tumors and mimic better all the barriers to drug diffusion, transport and distribution in tumors [3, 6]. Overall, the systems shows promise in the improved delivery of chemotherapeutic agents to the lung to allows for the more effective treatment of lung cancer.

REFERENCES

1. Siegel, R., et al., *Cancer statistics, 2014.* CA Cancer J Clin, 2014. **64**(1): p. 9-29.

2. Shevchen.It and G.E. Resnik, *Inhalation of Chemical Substances and Oxygen in Radiotherapy of Bronchial Cancer.* Neoplasma, 1968. **15**(4): p. 419-&.

3. Minchinton, A.I. and I.F. Tannock, *Drug penetration in solid tumours.* Nat Rev Cancer, 2006. **6**(8): p. 583-92.

4. Teesalu, T., K.N. Sugahara, and E. Ruoslahti, *Tumor-penetrating peptides.* Front Oncol, 2013. **3**: p. 216.

5. Kauffman, K.J., et al., *Synthesis and characterization of acetalated dextran polymer and microparticles with ethanol as a degradation product.* ACS Appl Mater Interfaces, 2012. **4**(8): p. 4149-55.

6. Sutherland, R.M., *Cell and environment interactions in tumor microregions: the multicell spheroid model.* Science, 1988. **240**(4849): p. 177-84.

412414 In Situ Formation of pH-Sensitive Polymer-Gold Nanohybrid System for Cancer Treatments

Wednesday, November 11, 2015: 1:30 PM
253B (Salt Palace Convention Center)

Wenjing Lin, Na Yao and Lijuan Zhang, School of Chemistry and Chemical Engineering, South China University of Technology, Guangzhou, China

In situ formation of pH-sensitive polymer-gold nanohybrid system for cancer treatments

Wenjing Lin, Na Yao, Lijuan Zhang*

School of Chemistry and Chemical Engineering, South China University of Technology, Guangzhou 510640, P. R. China

E-mail: lin.wenjing@mail.scut.edu.cn

Tremendous polymer materials have been intensively exploited as innovative nanovectors for cancer therapy in recent years. The current trend focus on the elaboration of multi-functional nanovectors which can reach to the diseased site, release a drug in a controlled manner and act as an imaging agent for both targeted therapy and diagnosis[1,2]. Most of these nanocarriers can take advantage of each function and integrate synergistically into one single system, which would otherwise be difficult to accomplish in a single-component system.

In this study, we have designed an in situ formed pH-sensitive polymer-gold nanohybrid system as a promising dual-functional nanoplatform for antitumor drug delivery and CT imaging. The general approach is to synthesize β-CD-(PLA-PDMAEMA-PEtOxMA)$_{21}$ by the combination of ROP and ARGET ATRP techniques), employ the polymer as template for in situ formation of Au nanoparticles without any reducing agent, and encapsulate doxorubicin (DOX). The formed β-CD-(PLA-PDMAEMA-PEtOxMA)$_{21}$/Au nanoparticles were characterized by UV-vis, DLS, TEM, microplate reader and CT imaging system. Varying block ratios, polymer/Au salt molar ratios and pH, they showed controllable maximum absorption wavelengths, uniform sizes, excellent colloidal stability, negligible cytotoxicity and enhanced X-ray attenuation property. After the loading of DOX, DOX@β-CD-(PLA-PDMAEMA-PEtOxMA)$_{21}$/Au revealed pH-controlled release because of the protonation of the PDMAEMA, high accumulation and good antitumor activities confirmed by *in vitro* release test and biodistribution studies. Therefore, we believe this pH-sensitive polymer-gold nanohybrid system can be used as a latent nanocarrier for cancer theranostic application.

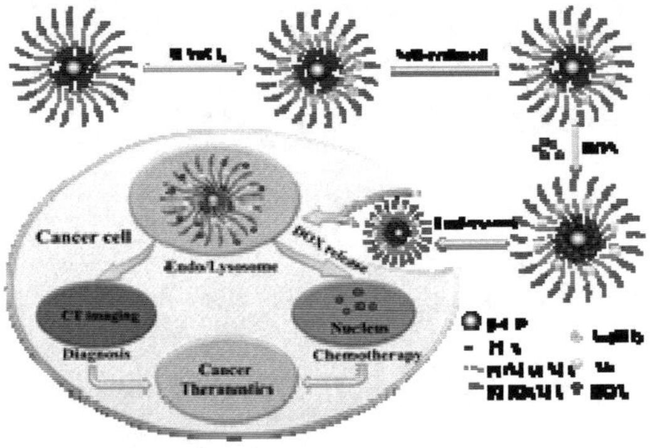

Fig. 1 The scheme of in situ formed pH-sensitive polymer-gold nanohybrid system.

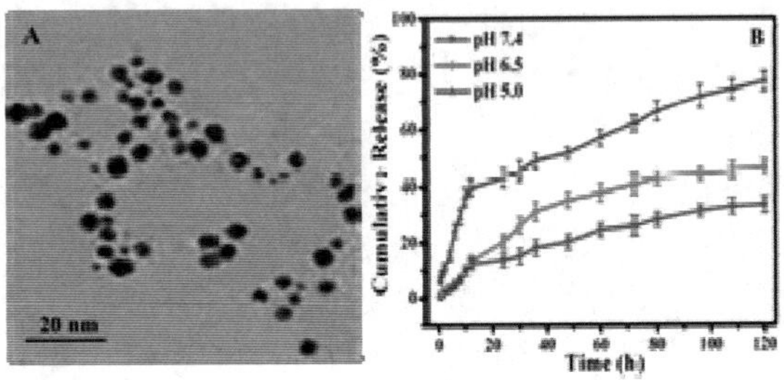

Fig. 2 (A) TEM image of polymer-stabilized gold nanoparticles and (B) pH-controlled DOX release from the polymer.

References

1. Y. Cheng et al. *Advanced Drug Delivery Reviews* 2014, 66: 42-57.

2. G. Chen et al. *Biomaterials* 2015, 47: 41-50.

429656 Intelligent Nanogels for the Concurrent Delivery of Hydrophilic and Hydrophobic Chemotherapeutic Agents

Abstract:

Intelligent Nanogels for the Concurrent Delivery of Hydrophilic and Hydrophobic Chemotherapeutic Agents

Angela M. Wagner[1] and Nicholas A. Peppas[1,2,3,4]

[1]Department of Chemical Engineering, [2]Department of Biomedical Engineering, [3]College of Pharmacy, and [4]Institute for Biomaterials, Drug Delivery, and Regenerative Medicine, The University of Texas at Austin, Austin, TX 78712

Author Correspondence: angelamwagner@utexas.edu, peppas@che.utexas.edu

The lack of specificity in traditional chemotherapeutic administration typically leads to significant systemic toxicity and requires patients to wait for long periods between treatments. During this time, cancerous cells have an opportunity to recover from the treatment. One method to improve treatment efficiency is to use nanoparticle systems as carriers for chemotherapeutic agents. Hydrogel nanocarriers have demonstrated their suitability as a customizable multicomponent system for the intracellular delivery of therapeutic agents, and the physical properties of these carriers can be designed to take advantage of passive targeting (via the enhanced permeability and retention effect). Our current work aims to develop a responsive hydrogel platform to enhance treatment efficiency through the synchronous delivery of hydrophilic and hydrophobic anticancer agents for combination therapy.

Our lab has previously shown success using cationic hydrogels to deliver hydrophilic therapeutics such as insulin and siRNA. Here, poly(2-(diethylamino)ethyl methacrylate)-g-poly(ethylene glycol methyl methacrylate) nanogels have been developed for the intravenous delivery of multiple chemotherapeutic agents. The nanogels entrap anticancer agents at physiological conditions and release the therapeutic agents in response to the intracellular environment. To improve hydrophobic drug-polymer interactions, we modified the P(DEAEMA)-g-PEGMA nanogel via copolymerization with a series of n-alkyl methacrylate hydrophobic monomers and investigated the impact of nanogels physical characteristics.

Nanogels were synthesized using an oil-in-water emulsion UV-initiated polymerization using tetra ethylene glycol dimethacrylate (TEGDMA) as a crosslinking agent with a crosslinking density of 2.5 mole %. The impact of n-alkyl methacrylate monomer inclusion was investigated through systematic variation of monomer functionality and chain length. Monomers included methyl methacrylate, ethyl methacrylate, isopropyl methacrylate, tert-butyl methacrylate, phenyl methacrylate, n-hexyl methacrylate, and 2-ethylhexyl methacrylate. The physical properties of the resulting nanogels were then compared, utilizing dynamic light scattering, zeta potential, titration, pyrene fluorescence, and hemolysis assays to understand the influence of polymer composition on swelling ratio, surface charge, pKa, relative hydrophobicity and hydrophile-hydrophobe phase transition, and membrane disruption capabilities. Varying the type and ratio of n-alkyl methacrylate monomer allows for precise control over the nanogel pKa. Loading and release experiments were performed to determine the improvement of loading efficiency and simultaneous release of hydrophilic (Cisplatin) and hydrophobic (Paclitaxel) chemotherapeutic agents. The rational design and characterization of P(DEAEMA)-g-PEGMA networks was necessary for tailoring the cationic nanogel system as a platform for the synchronous delivery of multiple chemotherapeutic agents.

Acknowledgements: This work was supported by a grant from the National Institutes of Health (R01-EB-000246-22). We also acknowledge the assistance of Rishabh Shah and Balark Chetan.

425256 TPP-Modified Dendrimer-siRNA Complexes and Their Aerosol Formulation for siRNA Delivery to the Lungs

Abstract:

The regulation of genes utilizing the RNA interference (RNAi) mechanism via the delivery of synthetic siRNA as a therapeutic molecule has great potential in the treatment for a variety of prevalent diseases that affect the lungs including lung cancer, asthma, and cystic fibrosis. However, there are many challenges in delivering siRNA to the lungs including the poor bioavailability of free siRNA and difficulty in formulating biomacromoleucles in portable inhalers. The use of nonviral vectors such as cationic dendrimers as siRNA nanocarriers has shown promise to improve siRNA bioavailability, but has had limited due to low effective transfection efficiency. However, the modification of dendrimers with various targeting ligands such as triphenylphosphonium (TPP), a mitochondrial targeting agent, has improved targeting, biocompatibility, and transfection efficiency of dendrimers. The delocalized positive charge and hydrophobic nature of TPP enhances cellular uptake and efficient membrane fusion as well as efficient escape from endolysosomes to target mitochondria. The presence of TPP on dendrimer surfaces as helps modulate the interaction between nucleic acid (siRNA) and dendrimer, which may allow for the bioavailability of free siRNA once delivered into cells and thus increase its transfection ability. Furthermore, TPP can passively target cancerous cells due to the increased mitochondrial membrane potential within these cells, which may benefit cancer treatment strategies.

In this work we investigate the ability of a triphenylphosphonium (TPP) modified generation 4 poly(amdoamine) (PAMAM) dendrimers (G4NH2-TPP) as a nanocarrier of siRNA, and the aerosol formulation of the complexes in pressurized-metered dose inhalers (pMDIs) and dry powder inhalers (DPIs). We will discuss the synthesis and characterization of the dendrimers, and the preparation and characterization of the complexes of siRNA and G4NH2-TPP (dendriplexes). The complexation efficiency via gel electrophoresis and *in vitro* gene knockdown ability in stably eGFP expressing lung alveolar epithelial (A549) cells were assessed as a function of the number density of TPP conjugated to G4NH2, as well as the N/P ratio of the dendriplexes. G4NH2-TPP dendriplexes were prepared with TPP densities of 0, 4, 8, and 12 per dendrimer and increasing N/P ratios of 5, 10, 20, and 30. An increase in TPP density and N/P ratio led to an increase *in vitro* gene knockdown with G4NH2-12TPP dendriplex at N/P ratio 30 showing the highest *in vitro* gene knockdown of all groups. To assess the potential of G4NH2-TPP dendriplexes for pulmonary use, we also develop micron particle technologies for both pMDI and DPIs, and determine their aerosol characteristics utilizing an Andersen Cascade Impactor (ACI). The proposed particle engineering strategy is shown to be effective in producing aerosol properties of suitable for deep lung deposition and respirable fractions up to 66%, as well as showing no impact on the *in vitro* gene knockdown efficiency of the siRNA present in those formulations. This work demonstrates the potential benefits of utilizing TPP-conjugated dendrimers in the formation of dendriplexes and in the aerosol formulation of siRNA for the pulmonary delivery to the lungs using portable inhalers.

429492 Targeted Delivery of Microrna By Engineered Lipid Nanoparticles for the Treatment of Metastatic Breast Cancer

Abstract:

Introduction: Human epidermal growth factor receptor-2 (HER2) positive breast cancer occurs in 20-30% of breast cancer patients and has been attributed to increased disease aggressiveness, reoccurrence, and overall worse prognosis (Engel & Kaklamani, 2007; Vrbic, Pejcic, Filipovic, Kocic, & Vrbic, 2013). Biological therapies such as monoclonal antibodies/tyrosine kinase inhibitors targeting HER2 have made substantial improvement in patient outcome, but resistance to these treatments over time has represented a real clinical challenge and highlighted a need to develop novel therapeutic approaches (Vrbic et al., 2013). MicroRNAs (miRNAs) are endogenous regulators of gene expression that play a pivotal role in biological processes spanning from global homeostasis to disease onset and progression. The ability to manipulate/ induce cellular re-equilibrium of deregulated miRNA expression profiles by inhibition of oncogenic miRNA or overexpression of tumor suppressor miRNA is a promising cancer strategy, but is currently hindered in application by the lack of nonviral delivery systems. Here, we developed a lipid nanoparticle (LNP) platform surface coated with Hyaluronic Acid (HA) for the targeted delivery of mature tumor suppressor MicroRNA125a-5p as a long term treatment option for breast cancer. We believe the successful implementation of this platform has major implications on future gene therapy regimens for breast cancer.

Methods and materials: LNPs were created via the traditional dry lipid film method, mechanically extruded down to the nanoscale, and covalently surface decorated with HA by crosslinking chemistry. Particle size, charge, and surface characteristics were measured by dynamic light scattering, zeta potential analysis, and transmission electron microscopy. High efficiency entrapment of MicroRNA inside the aqueous core of the HA-LNP was achieved by utilization of the scale up friendly lyophilization-rehydration active loading procedure. Three breast cell lines were employed to show the targeting potential of the nanoparticles: MCF10A (normal breast tissue, CD44 -), SKBR3 (invasive breast cancer, CD44 +), and 21MT-1 (metastatic breast cancer, CD44 ++). Western blot, qRT-PCR, MTT, and wound healing assay were performed to measure protein, mRNA, cellular proliferation, and cellular migration respectively following transfection. Live cell confocal microscopy was applied to validate cytosolic delivery of the nanoparticles, as well as validate the evasion of lyosomal degradation.

Results: The optimized HA-LNP platform was constructed from biocompatible lipids, cholesterol, and a surface coating of HA to achieve a nanocarrier system 160 nm in diameter, a polydispersity index of 0.105, a surface charge of -38 mV, and stability in solution for over 90 days. The HA-LNP delivery platform actively targets patient-derived metastatic breast cancer cells (21MT-1) isolated from the metastatic pleural effusion over normal breast tissue via an intrinsic HA-CD44 mediated endocytosis event, and has the ability to escape from the intracellular endolysosomal pathway for potent gene silencing. Knockdown of the HER2 proto-oncogene at the level of transcription and translation was achieved following HA-LNP mediated transfection with MicroRNA125a-5p. In addition, the PI3K/AKT and MAPK hyperactivated signaling pathways, cellular proliferation, and migration potential were also potently suppressed. Furthermore, the therapeutic efficacy of MicroRNA125a-5p by the HA-LNP platform was demonstrated to be significantly improved by over three fold as compared to the current gold standard commercial transfection reagent Lipofectamine.

Conclusions: Our HA-LNP platform was fabricated from a scale up friendly technique and achieved rapid and targeted intracellular delivery of MicroRNA via a HA-CD44 mediated endocytosis leading to potent oncogene silencing. This study demonstrates the use of a lipid based nonviral carrier to delivery MicroRNA as a promising strategy. In addition, we believe our HA-LNP platform can be used to study the mechanism of action for various MicroRNA and advance the field of MicroRNA therapeutics.

424129 Co-Delivery of miRNA Inhibitor and Doxorubicin for Enhanced Anticancer Therapy

Abstract:

Co-delivery of miRNA inhibitor and doxorubicin for enhanced anticancer therapy

Zilan Zhou, Apurva Badkas, and Joo-Youp Lee

Chemical Engineering Program

Department of Biomedical, Chemical, and Environmental Engineering

University of Cincinnati, Cincinnati, Ohio 45221-0012

The development of nano-scaled carriers as a drug delivery platform has made a tremendous difference in the fight against cancer. Among these nano-scaled carriers, various nanoparticles (NPs) made of biocompatible and biodegradable materials have been of particular interest. Significant progresses have been made in improving NP formulations, targeting, and drug potency. Despite the progresses, drug resistance remains a major challenge that NP-based cancer therapeutics still faces. RNA interference (RNAi) using miRNA inhibitor, a small non-coding RNA molecule, has been considered as a promising strategy to tackle the drug resistance. Recently, our group developed a novel nanoparticle system using biocompatible polymer PLGA-PHis-PEG for the co-delivery of miRNA inhibitor 221/222 and doxorubicin. In our approach, the miRNA inhibitors with negative charges were condensed with calcium phosphate (CaP) with positive charges to form solid particles via electrostatic interactions. Then an excess amount of positive charges were neutralized by coating the solid particles with anionic lipids. As a result, the entire lipid/CaP/miRNA complex (LCPR) could be encapsulated inside the polymeric nanoparticle with a chemotherapic drug, doxorubicin, via nanoprecipitation. In this talk, the efficacy of the co-delivery system will be reported.

424076 Prediction of Nanoparticles-Cell Association Based on Corona Proteins and Physicochemical Properties

Abstract:

The association of nanoparticles (NPs) with cells in biological fluids is impacted by the adsorption of proteins onto the NP surface. The multitude of proteins in the biological fluid form a protein corona which can then affect cellular bioactivity. In the present work, an extensive dataset of gold NPs protein corona was utilized to investigate the relationships between protein corona fingerprints (PCFs) and NP-cellular association. The dataset consisted of a library of 84 gold nanoparticles (NPs) of 15, 30, or 60 nm cores with cationic or anionic surface ligands, and data regarding the abundance of 129 specific corona proteins and 19 NO physicochemical properties. The specific protesins of relavance to NP-cellular association were identified via quantitative structure-activity relationships (QSARs) that were developed based on both non-linear support vector regression (SVR) and linear models. Feature selection was accomplished via sequential forward floating selection. From the initial set of 148 descriptors, the serum proteins *APOB*, *A1AT*, *ANT3*, and *PLMN* in addition to the NP zeta potential were most significant for correlating cellular association of gold nanoparticles. The non-linear model was built to a reasonable level of accuracy with only 6 serum proteins and NP zeta potential with a performance accuracy of R^2=0.895. In contrast, the linear model required 11 CFs and demonstrate performance of R^2=0.850. Model robustness was validated based on Y-randomization demonstrating that the QSARs were not “chance” correlations. The present approach demonstrates that exploration of NP-cell association data, via QSAR analysis, can be particularly useful to elucidate the role of NP corona proteins and NP physicochemical properties in affecting NP-cellular. Such information is critical for supporting the design and interpretation of toxicity studies and for the design of NPs for biomedical applications.

428378 Characterization of Detergent Residues on Surfaces

Abstract:

Characterization of detergent residues on surfaces

Sadi Gurses, Alper Uzun, Can Erkey, Seda Kizilel

Department of Chemical and Biological Engineering, Koc University, Istanbul 34450, Turkey

The problem of filmy appearance on dishwasher-washed dishes has been receiving significant attention with increased health awareness of consumers. There is a need for characterization of dishware surfaces following a complete wash cycle. Here, we focused on identifying the indicator chemical components to detect detergent residues on the dishware surfaces and their toxicity on live cells. Spectroscopic and diffraction measurements have been carried out to determine structural information about detergent components, Fourier transform infrared spectroscopy (FT-IR) and high pressure liquid chromatography (HPLC) have been used to analyze rinsing water. Surfaces of the dishware were analyzed with X-ray photoelectron spectroscopy (XPS) and inductively coupled plasma mass spectroscopy (ICP-MS). Surface analysis results showed decomposition of penta sodium tripolyphosphate and sodium carbonate, known as builders in detergent chemistry, on surfaces upon dishwashing process. We also performed toxicity assays, where the influence of these builders on human fibroblast cell viability was investigated. Cell viability results showed decreasing trend in the number of cells cultured with increasing concentrations and exposure time of penta sodium tripolyphosphate in the medium. This study is significant for characterization of unknown detergent residues on surfaces and their possible toxicity on live cells.

421609 Assessment on the Treatment of Carbon Soot for Activated Carbon Applications: Oxidative Stress and Apoptosis Induced in Human Cell Lines

Abstract:

In Singapore, approximately 30 ton/day of carbon based solid waste is produced from petroleum chemical process. Solid waste like carbon soot is generally disposed to landfill sites or re-utilized as secondary raw materials. Carbon soot has gained much attention in its re-utilization due to its high content of carbon for possible processing as absorbent materials. However, carbon soot also contains hazardous components such as heavy metals arising from either sources or petroleum chemical process. These components may be toxic to human health and environment, and pose many restrictions in its re-utilization due to the potential toxicity. Therefore, it is important to examine the possible toxicity effects of carbon soot and its mechanism on human health. In this study, we performed chemical and microbial leaching on carbon soot to remove hazardous components before further processing it as activated carbon. Subsequently, three human cell lines i.e. HepG2, MRC-5 and MDA-MB-231 were used to investigate the cytotoxicity, oxidative stress and apoptosis induction by carbon soot. Carbon soot induced dose-dependent and time-dependent cytotoxicity in HepG2, MRC-5 and MDA-MB-231 cells as demonstrated by MTS and LDH assays. Meanwhile, the flow cytometry assay and fluorescence microscope results indicated that the carbon soot induced apoptosis of HepG2, MRC-5 and MDA-MB-231 cells increased with the increasing concentration of carbon soot. Carbon soot was also found to induce oxidative stress in HepG2, MRC-5 and MDA-MB-231 cells as demonstrated by generation of reactive oxygen species (ROS) and the alteration of antioxidant superoxide dismutase (SOD) and glutathione (GSH) compared to the unexposed control cells. Further, co-treatment with N-(mercaptopropionyl)-glycine (N-MPG), a type of ROS scavenger formulations, could inhibit cells apoptosis induced by carbon soot and increase the viability of cells. These results may provide an explanation for the potential mechanism in HepG2, MRC-5 and MDA-MB-231 cytotoxicity under the effect of carbon soot. We hope that these findings could help to understand, in a microscopic scale, the threat and risk of these wastes to human.

427903 Structure-Toxicity Correlations for Ni/SiO2 Complex Engineered Nanomaterials Using High-Throughput Zebrafish Assays

Abstract:

Functional nanomaterials are developing at an ever accelerating pace and are finding application in a wide range of industrial and consumer products despite evidence that nanomaterials can show elevated toxicity. Furthermore, while bare nanoparticles (NPs) are rarely used in industrial application due to sintering and deactivation effects, embedded NPs (so-called complex engineered nanomaterials, CEN) have found little attention in nanotoxicological studies to-date. This motivates a need for the development of rapid, yet sensitive, high throughput *in vivo* nanotoxicity screening assays.

We are investigating three differently structured Ni/SiO_2 nanomaterials as model CEN with the aim to evaluate different nanotoxicity assays and develop structure-toxicity correlations for CEN that will allow derivation of predictive toxicity models. The CEN are comprised of Ni NPs embedded in or on (non-toxic) porous silica NPs, based on the hypothesis that embedding nickel nanoparticles in silica could reduce or entirely mitigate nanotoxicity while still providing accessibility and hence maintaining functionality of the embedded NP. The three CEN structures are nickel NPs (a) embedded in hollow porous silica shells (hNi@SiO_2), (b) encapsulated in non-hollow, porous silica NPs (nhNi@SiO_2), and (c) deposited on the silica nanoparticle's external surface (Ni-SiO_2). All CENs were thoroughly characterized via TEM, XRD, BET, as well as for dissolution, aggregation, and settling properties, followed by multiple 5-day zebrafish developmental toxicity assays including survival, malformation, hatching, and motility. Zebrafish (*Danio rerio*) were used as a well-suited toxicity model due to prolific breeding and rapid development time. The toxicity results were then correlated with the CEN properties to determine structure-toxicity correlations.

All three nanomaterials mitigated toxicity compared to the respective Ni^{2+} dose and showed high zebrafish survival. Zebrafish motility, which probes persistent neurotoxicity, emerged as a more sensitive assays and revealed lowest toxicity for nhNi@SiO_2, explained by lower nickel dissolution compared to the other two nanomaterials. In contrast, this material showed the highest toxicity of the three CENs in hatching assays, which can be traced back to resistance to aggregation which facilitates uptake through the chorion. Overall, the toxicity results are consistent with a 'Trojan horse' mechanism and point towards complexity and pitfalls when establishing structure-toxicity correlations and the necessity of combining multiple assays to fully assess nanomaterials toxicity.

430535 Size-Dependent Cellular Toxicity of Gold Nanoparticles on Human Embryonic Stem Cells

Abstract:

Increased demands for toxicological evaluations of engineered nanomaterials coincide with the accelerated pace of clinical translation and commercialization of nanomaterials. There is an urgent need to develop predictive and validated toxicological assessment methods for nanomaterials due to their unique size range, surface chemistry, and interactions with biological systems. Most cell-based nanotoxicity assessments are limited to long-lived cell lines or cancer cells. Human embryonic stem cells (hESCs) and their subsequent neural precursor differentiations provide a toolkit to study the effect of exposure to nanoparticles on neural specification. We employ gold nanoparticles (AuNPs) as model nanoparticles due to their versatile surface chemistry, ease of imaging, and tunability for transport across the blood-brain barrier. AuNPs are considered by most as nontoxic, which is consistent with the inertness of bulk gold, but some have found them toxic. This talk describes a toxicological study of AuNPs of different core sizes (1.5 nm, 4.0 nm, and 14 nm) as well as dendrimers (0.5 nm and 1.5 nm) using the hESCs. The various nanoparticles were coated with bovine serum albumin prior to their introduction to the cell culture medium. The nanoparticles were characterized by UV-vis spectroscopy, TEM, dynamic light scattering, and AFM. We found that colonies exposed to 1.5 nm AuNPs exhibited loss of cohesiveness, rounding up, and detachment suggesting ongoing cell death. The hESCs exposed to 1.5nm AuNPs did not form embryoid bodies but rather rapidly disintegrated into single cells within 48 hours of treatment. Cell death caused by 1.5nm AuNPs also occurred in the treatment hESC-derived neural precursor/progenitor cells. None of the other nanoparticles exhibited such toxic effect on the hSECs. We also found that 4.0 nm AuNPs caused a dramatic hypomethylation of the hESC DNA in only 24 hours. This work contributes a new toolkit for predicting neurotoxicity of nanoparticles that will ultimately impact consumers, patients, and manufacturing workers of products containing nanoparticles.

418163 Micrornas As Biomarkers for Cytotoxicity Evaluation of Multi-Wall Carbon Nanotube

Abstract:

Multi-wall carbon nanotubes (MWCNT) are one of the most promising engineered nanomaterials because of their excellent physical and chemical characteristics. However, the applications of MWCNT also raise health concerns. The small size and light weight of MWCNT allow them to enter the working environment as particulate matter of respirable sizes. Many studies have already reported that inhaled CNTs showed significant cytotoxicity both in vitro and vivo, and the toxicity potential will increase with decreasing particle sizes.

MicroRNAs, a family of short endogenous noncoding RNAs, harbor critical functions in many fundamental biological processes, such as proliferation and apoptosis. Intracellular and circulating microRNAs have been implicated as promising biomarkers in the diagnosis of lung cancer and pulmonary diseases. Here, we seek to identify both intracellular and circulating microRNAs as biomarkers for MWCNT induced pulmonary toxicity.

Total 4 types of MWCNT with different lengths (0.2-2μm and 10-50μm) and surface modifications (with and without carboxyl modification) were selected and their cytotoxicity was evaluated in human bronchial epithelial cells (BEAS-2B). The cytotoxic effects of 4 types of MWCNT were first characterized by the cell growth measurement and the reactive oxygen species (ROS) assay. BEAS-2B cells were treated with MWCNT at concentrations from 0 to 50μg/mL. At 48 h post treatment, all 4 types of MWCNT inhibited the growth of BEAS-2B cells and generated significant ROS stress. The MWCNT toxicity was dose and tube type dependent. The shorter length and carboxyl modification enhanced the MWCNT toxicity because of better dispersion in aqueous solutions and higher cellular uptake.

Next, a microRNA toxicology panel was used to identify the dysregulated microRNAs post MWCNT exposure. Profiling of 87 miRNAs related with toxicology showed that the expression of total 7 miRNAs was changed at least two folds compared with untreated controls. Among 87 miRNAs, miRNA-29b, miRNA-21, and miRNA-486 were selected for further investigation because they are closely related with lung diseases including pulmonary fibrosis and lung cancer. At 48 h post treatment, the expression of intracellular miRNA-29b was up-regulated up to ~10 folds, the expression of intracellular miR-486 was down-regulated up to ~55% and the expression of intracellular miR-21 remained unchanged compared with untreated controls. However, the expression of circulating miRNA-29b was up-regulated up to ~4 folds, the expression of circulating miR-486 was up-regulated up to ~10 folds and the expression of circulating miR-21 was up-regulated up to ~50 folds compared with untreated controls. The expression of intracellular and circulating miRNAs was also dose and tube type dependent. Among 4 types of MWCNT, short MWCNT with carboxyl modification showed the most significant impact on the expression of both intracellular and circulating miRNAs. And the expression of intracellular and circulating miRNAs were able to distinguish the toxicity effects induced by different types of MWCNT.

Our results demonstrated that intracellular and circulating microRNAs are sensitive biomarkers for MWCNT induced pulmonary cytotoxicity, and they hold great potential in the early detection of adverse health effects associated with exposure to MWCNT.

414220 Impedance Spectroscopic Determination of Fabric Thermal Sensor, Temperature Detection Mechanism

Abstract:

A temperature sensitive fabric using homemade electrospun nylon 6 (PA6) or Tecoplast[a] polyurethane (PU) nonwoven nanofiber membrane materials, commercial multiwalled carbon nanotubes (MWCNTs), and chemical vapor deposited (CVD) polypyrrole (PPy) was developed[1]. A simple sensor construction procedure facilitates reproducibly producing the fabric sensors using simple laboratory procedures. This advantageous procedure is completed without needing toxic semiconductor device construction environments. The work illustrates the viability of using polymers/carbon allotrope composites as functional materials for sensor applications. This thermal sensor provides useful data using Ohm's law and the overall equation for the temperature coefficient of resistance, or $\alpha = 1/R_0 * dR/dT$, which has been well defined previously and is common for fitting the temperature dependence of the electrical resistance of materials[2]. Sensor response is exactly linear and negative between 25C and 45C, which is a useful range for epidermal temperature measurement[3,4], and displays a high resolution, approximately equal to 0.1C to 0.5C, being that it is a mostly polymer, fabric thermal sensor. Fig. 1 illustrates the general behavior of the sensor material that classifies it as a thermistor.

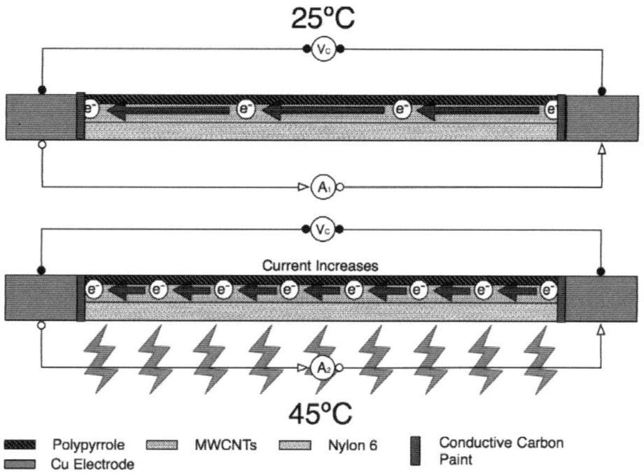

Figure 1: Schematic illustrating the thermal sensor response.

This thermal sensor was designed for biomedical issues in measuring temperature of the epidermis in load bearing situations like prosthetic sockets or shoes, and could be applied in military and sports applications. Research shows that greater than a 4C difference between the same points for each foot of a diabetic, can mean they are at a greater risk for foot ulceration, or could be an indication to more severe conditions such as neuropathy if the feet do not maintain relatively consistent temperatures regularly[5-9]. To measure the same points on each foot is cumbersome, as one imagines. Diabetics remove their shoes and socks, taking the temperature of each point on each foot using an infrared gun or even worse, a basic home use thermometer, and recording and reporting the information manually to their physician. This provides a potential for inaccuracies. A thin fabric temperature sensor could accommodate continuously monitoring the temperature between points on each foot, making it easier for diabetics to tell if they need to maintain foot temperatures. This would make it more efficient for reporting the results to their physician if the sensor is integrated to work wirelessly with a computer at home to log the data and possibly communicate directly with their physician's equipment. In addition, amputees could benefit from measuring temperature inside their prosthetic sockets to maintain a comfortable and hygienic stump socket environmen[4,10-14]. Surveys show that heat and sweat are the major contributors to whether an amputee is happy with their prosthesis, which can determine whether the amputee will even wear the prosthesis regularly[3,15,16]. Currently, temperature-measuring devices are constructed using rigid metallic and semiconductor materials[2], so it would be beneficial in some of these load-bearing situations where pressure points could be problematic, to have thermal sensors constructed from thin and soft nanocomposite materials. Even applications, where pressure points are not a main concern, like smart textiles and sports clothing could benefit from these fabric thermal sensors integrated into their increasingly more complex systems to obtain useful physiological data during periods of exercise and extreme environmental stresses and situations.

The procedure to construct these sensors starts by electrospinning PA6 or PU sheets onto a homemade rotating drum. Commercial MWCNTs are vacuum filtered onto the polymer membranes from aqueous Triton X-114 stabilized nanotube suspensions. PPy is polymerized from pyrrole in a CVD chamber at Å 100 kilopascals of vacuum and ambient temperatures for two days. The PPy overcoats on and throughout the MWCNTs on the polymer membranes, using iron(III)chloride, or ammonium persulfate as the oxidant, making a nanostructured composite material that is electrically sensitive to temperature changes. The sensors are then cut from the membrane in 1-cm^2 geometries and tested in humidity and temperature controlled environment using 2, 3, and 4 electrode configurations[17-20]. Fig. 2 depicts the 4 electrode setup.

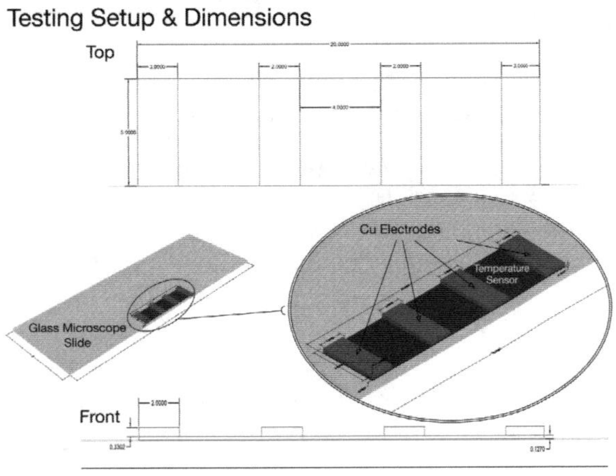

Figure 2: Design for sensor testing.

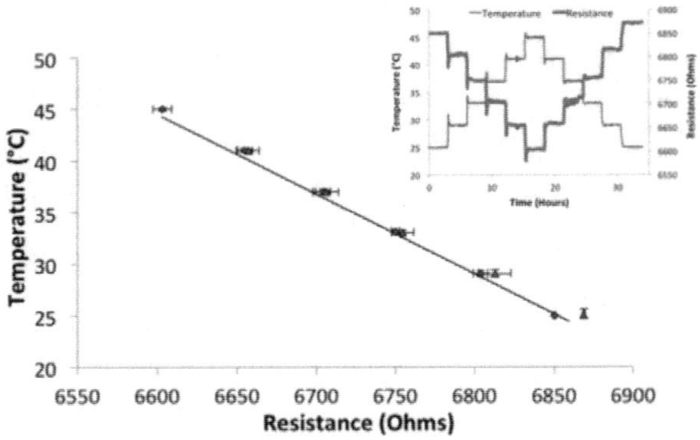

Figure 3: Steady state calibration curve showing the actual steady state temperature and resistance versus time data.

This material has unique electrical characteristics. The polymers generally have electrical surface resistivities in the 10^{10} and $10^{13 \text{ to } 15}$ Ohm-cm ranges for PA6 and PU, respectively. The application of MWCNTs and PPy to the polymer membranes has reduced those surface resistivities over seven orders of magnitude to give resistivities in the kOhm range and temperature sensitive super abilities. The material's electrical resistance responds linearly and reproducibly to temperature changes between 25C and 45C with a negative temperature coefficient of resistance ranging from -0.56%/C to -0.14%/C, a 0.1C to 0.5C thermal resolution, and sensitivities ranging from 2 Ohms/¼C to nearly 30 Ohms/¼C. Fig. 3 represents the steady state calibration.

Bulk DC resistivity, and hence bulk resistance, or $R = \rho l/a$, where R is the resistance in Ohms, r is the resistivity in Ohm/cm, l is the length travelled by the current, and a is the cross sectional area perpendicular to the flow of current, is used as a model to determine temperature. This does not really elucidate the true sensing mechanism at the nano, or atomic, scale of the sensor material. Nor does it provide any useful information regarding the nature of the interface between the sensing material, the outside environment, and the sensor's electrical connections.

Impedance spectroscopy (IS) is an important scientific tool that can provide extremely useful information about the interface between a sensor and its electrical contacts, chemi- and physi-sorption as well as any reactions and their kinetics at the material surface and within[21,22]. Ultimately IS elucidates a material's dielectric structure under electrical stress, which also translates into mechanical, stress, thereby providing a window into electronic transport through a material and its interfaces[23,24]

This presentation will focus on the value and constraints of IS in determining interfacial phenomena between this constructed temperature sensitive fabric, its electrical contacts, and the surrounding environment. An expanded equivalent circuit model for three and four electrode configurations will be proposed. In addition a simplified circuit equivalent will be shared. A short discussion on the differences in measuring techniques between solid-state devices and electrochemical cells will finish the talk, while emphasizing the use of common circuit analysis to determine and simplify "dry" and "wet" sensor models.

References:

(1) Blasdel, N. J.; Wujcik, E. K.; Carletta, J. E.; Lee, K.-S.; Monty, C. N. *IEEE Sens. J.* **2015**, *15* (1), 300–306.

(2) McGee, T. D. *Principles and methods of temperature measurement*; Wiley: New York, 1988.

(3) Huff, E. A.; Ledoux, W. R.; Berge, J. S.; Klute, G. K. *JPO J. Prosthet. Orthot.* **2008**, *20* (4), 170.

(4) Peery, J. T.; Ledoux, W. R.; Klute, G. K. *J Rehabil Res Dev* **2005**, *42* (2), 147–154.

(5) Armstrong, D. G.; Holtz-Neiderer, K.; Wendel, C.; Mohler, M. J.; Kimbriel, H. R.; Lavery, L. A. *Am. J. Med.* **2007**, *120* (12), 1042–1046.

(6) Armstrong, D. G.; Lavery, L. A.; Liswood, P. J.; Todd, W. F.; Tredwell, J. A. *Phys. Ther.* **1997**, *77* (2), 169–175.

(7) Benbow, S. J.; Chan, A. W.; Bowsher, D. R.; Williams, G.; Macfarlane, I. A. *Diabetes Care* **1994**, *17* (8), 835–839.

(8) Bharara, M.; Cobb, J. E.; Claremont, D. J. *Int. J. Low. Extrem. Wounds* **2006**, *5* (4), 250–260.

(9) Stess, R. M.; Sisney, P. C.; Moss, K. M.; Graf, P. M.; Louie, K. S.; Gooding, G. A.; Grunfeld, C. *Diabetes Care* **1986**, *9* (3), 267–272.

(10) Davidson, J. *J. Hand Ther.* **2002**, *15* (1), 62–70.

(11) Sinha, R.; van den Heuvel, W. J.; Arokiasamy, P. *Prosthet. Orthot. Int.* **2011**, *35* (1), 90–96.

(12) Da Silva, R.; Rizzo, J. G.; Gutierres Filho, P. J. B.; Ramos, V.; Deans, S. *Prosthet. Orthot. Int.* **2011**, *35* (4), 432–438.

(13) Meulenbelt, H. E.; Geertzen, J. H.; Jonkman, M. F.; Dijkstra, P. U. *Acta Derm. Venereol.* **2011**, *91* (2), 173–177.

(14) Hall, M. J.; Shurr, D. G.; VanBeek, M. J.; Zimmerman, M. B. *JPO J. Prosthet. Orthot.* **2008**, *20* (4), 134–139.

(15) Raichle, K. A.; Hanley, M. A.; Molton, I.; Kadel, N. J.; Campbell, K.; Phelps, E.; Ehde, D.; Smith, D. G. *J. Rehabil. Res. Dev.* **2008**, *45* (7), 961.

(16) Biddiss, E. A.; Chau, T. T. *Prosthet. Orthot. Int.* **2007**, *31* (3), 236–257.

(17) Castano, L. M.; Flatau, A. B. *Smart Mater. Struct.* **2014**, *23* (5), 053001.

(18) Lymberis, A. In *Engineering in Medicine and Biology Society, 2003. Proceedings of the 25th Annual International Conference of the IEEE*; IEEE, **2003**; Vol. 4, pp 3716–3719.

(19) Axisa, F.; Schmitt, P. M.; Gehin, C.; Delhomme, G.; McAdams, E.; Dittmar, A. *Inf. Technol. Biomed. IEEE Trans. On* **2005**, *9* (3), 325–336.

(20) Ciluffo, G. Therapeutic" smart" fabric garment including support hose, body garments, and athletic wear. App. 11/003,071, December **2004**.

(21) Orazem, M. E. *Electrochemical impedance spectroscopy*; The Electrochemical Society series; Wiley: Hoboken, N.J, **2008**.

(22) *Impedance spectroscopy: theory, experiment, and applications*, 2nd ed.; Barsoukov, E., Macdonald, J. R., Eds.; Wiley-Interscience: Hoboken, N.J, **2005**.

(23) Coelho, R. *Physics of Dielectrics for the Engineer*; Fundamental studies in engineering; Elsevier Scientific Publishing Company, **1979**.

(24) Jonscher, A. K. *J. Phys. Appl. Phys.* **1999**, *32* (14), R57.

(536b) Free-Standing Graphene-Based Membrane with High Density for Energy Storage Electrode

Ji, J.,
Cui, X.,
Zhu, Y.,
Chen, J.,
Liu, D.,

Abstract:

The active material selection and structure design of the electrodes is essential for high performance energy storage devices (e.g. supercpacitors, lithium ion batteries). Graphene-based electrode has been widely investigated because its high specific surface area, electrical conductivity and chemical stability. Furthermore, free-standing membrane electrodes can avoid the use of electrochemical inactive binder and metal current collector, thus highly improve the electrochemical performance of the whole electrode. However, the surface area of the directly filtering graphene membrane is limited because the large Van der Waals force between the graphene layers will cause a serious restacking; on the contrary, porous structure with relatively large surface area usually lead to a low material density, which will further influence the volumetric capacitance of the electrode. Hence, design and fabrication the free-standing graphene-based membrane electrode with high density and high ions accessible surface area may deliver a high volumetric capacitance as well as high volumetric energy/power densities.

We recently present a free-standing graphene-based membrane electrode with tunable graphene interlayer spacing, where the *in situ* polymerization of polyaniline (PANI) with tunable size as the “spacer” can adjust the graphene interlayer spacing and the membrane density. After carbonization of the PANI, this graphene-based membrane can act as the high performance supercapacitor electrode, where graphene and amorphous carbon can form the electrical conductive network, while the tunable spacing between graphene layers can balance the ions diffusion rate and the electrode density. Furthermore, when the total device volume is taken in to account, the relatively high membrane density can obtain a higher volumetric capacitance and a smaller device. These results indicate a new method to design and fabrication novel free-standing graphene-based electrode for next-generation energy storage devices.

Keywords: energy storage, graphene, polyailine, volumetric capacitance, free-standing electrode

422810 Optimization and Upscaling of Chirality-Sorting of Single-Wall Carbon Nanotubes By Aqueous Two-Phase Extraction

Abstract:

The actual implementation of single-wall carbon nanotubes (SWCNTs) in the applications for which they were put forward to be highly promising is severely hampered because current synthesis techniques generally produce low purity mixtures of many different SWCNT chiralities. Therefore, various methods to purify and separate SWCNTs according to chirality have been developed, recently including aqueous two-phase extraction (ATPE). This method makes use of two water-soluble polymers forming distinct phases with different hydrophilicity, resulting in a re-distribution and hence chirality separation of the solubilized SWCNTs. ATPE generally does not require high-end instrumentation, and has been shown to be both very robust, fast, and readily extendable. In this work, we focus on the scaling, automation, and further improvement of the ATPE method, making use of automated liquid phase separation techniques, such as countercurrent chromatography.

429626 Hyperspectral Microscopy of Near-Infrared Fluorescence Enables 17-Color Carbon Nanotube Imaging

Abstract:

The intrinsic photoluminescence of single-walled carbon nanotubes (SWCNTs)[1] exhibits unique photostability[2], narrow bandwidth[3], near-infrared penetration of biological media[4], and environmental sensitivity[5,6]. Advanced biological applications will require the spectral and spatial resolution of individual (n,m) SWCNT species' photoluminescence and its modulation within live cells and tissues. We present a wide-field hyperspectral imaging approach to resolve SWCNTs with single molecule resolution in live mammalian cells, in murine tissues ex vivo, and in live zebrafish endothelium. Taking advantage of the narrow-band emission of SWCNTs, simultaneous multicolor imaging was used to resolve 17 nanotube chiralities in the 900-1400 nm nIR spectral region, including 12 distinct fluorescent species within live cells. We used this method to identify chirality-resolved optical properties of single SWCNTs in tissues and to precisely quantify fluorescent nanotubes in live cells.

References:

1 O'Connell, M. J. *et al.* Band gap fluorescence from individual single-walled carbon nanotubes. *Science* **297**, 593-596 (2002).

2 Cherukuri, P., Bachilo, S. M., Litovsky, S. H. & Weisman, R. B. Near-Infrared Fluorescence Microscopy of Single-Walled Carbon Nanotubes in Phagocytic Cells. *J. Am. Chem. Soc.* **126**, 15638-15639 (2004).

3 Lefebvre, J., Fraser, J. M., Finnie, P. & Homma, Y. Photoluminescence from an individual single-walled carbon nanotube. *Phys. Rev. B* **69**, 075403 (2004).

4 Welsher, K., Sherlock, S. P. & Dai, H. Deep-tissue anatomical imaging of mice using carbon nanotube fluorophores in the second near-infrared window. *Proc. Natl. Acad. Sci. USA* **108**, 8943-8948 (2011).

5 Barone, P. W., Baik, S., Heller, D. A. & Strano, M. S. Near-Infrared Optical Sensors Based on Single-Walled Carbon Nanotubes. *Nature Mater.* **4**, 86-92 (2005).

6 Cognet, L. *et al.* Stepwise quenching of exciton fluorescence in carbon nanotubes by single-molecule reactions. *Science* **316**, 1465-1468 (2007).

420163 Advances in Aqueous Two-Phase Extraction Separations of Single-Wall Carbon Nanotubes

Abstract:

Aqueous two-phase extraction (ATPE) is a rapidly developing scalable separation in which a system of phase-separating polymers isutilized to spatially separate single-wall carbon nanotubes (SWCNTs) species on the basis of differential partitioning across the polymer phases. Recently advances have demonstrated that the technique is broadly applicable to the separation of a broad diameter range of nanotube species on the basis of solution composition for surfactant dispersed SWCNTs, with the typical composition needed for extraction of a specific SWCNT species strongly following a diameter dependent trendline. The basis and applicability of the method will be presented, as well as evolving understanding of the diameter dependence, enantiomeric selection, and specific SWCNT species applications.

422592 Mechanical Behavior and Properties of Graphene Nanomeshes

Abstract:

Graphene nanomeshes (GNMs) are graphene-based metamaterials consisting of a periodic arrangement of nano-scale holes or pores in the graphene lattice with neck widths less than 10 nm, mimicking dense arrays of ordered nanoribbons. Optimal design of GNMs toward enabling a broad range of technological applications requires the establishment of rigorous structure-property-function relationships in such engineered graphene nanostructures.

In this presentation, we report the results of a systematic computational study on the mechanical behavior and mechanical properties of GNMs based on molecular-statics and molecular-dynamics simulations of uniaxial tensile deformation tests using a reliable bond-order interatomic interaction potential. Both the mechanical properties and the dynamical response to mechanical loading are determined as a function of the nanomesh architecture, namely, the lattice arrangement of the pores, as well as of the pore morphology, pore size, material density, and pore edge passivation. To examine effects of pore morphology, we have studied both circular and elliptical pores, varying the ratio of the lengths of the major and minor axes in the elliptical pore shapes. To examine effects of pore edge passivation, we have studied GNMs with both unpassivated and hydrogen-terminated pore edges. For the numerous GNM structures examined, elastic moduli are computed and stress-strain curves are generated, which are used to determine the ultimate tensile strength, fracture strain, and toughness as a function of the GNM architectural, morphological, and physicochemical parameters listed above. The structural responses of the GNMs during their deformation and fracture in the simulated mechanical deformation tests are analyzed in detail and the underlying mechanisms are characterized systematically. We find that the GNM mechanical behavior undergoes a transition as the GNM density decreases, determine the critical density for the transition, and derive the dependence of this critical density on pore morphology and hydrogen pore edge termination.

416668 Fabrication of Porous Organic Framework/Graphene Oxide Hybrid Membrane with Sandwich-like Structure for Molecular Separation

Abstract:

Fabrication of Porous Organic Framework/Graphene Oxide Hybrid Membrane with Sandwich-like Structure for Molecular Separation

Yuanzhi Zhu, Yang Li, Wenchao Peng, Guoliang Zhang, Fengbao Zhang and Xiaobin Fan*

School of Chemical Engineering and Technology, State Key Laboratory of Chemical Engineering, Collaborative Innovation Center of Chemical Science and Engineering, Tianjin University, Tianjin 300072, China. E-mail: xiaobinfan@tju.edu.cn Fax: +86-22-27890090

Combining adsorption and filtration is an efficient way to realize molecular separation with low energy consumption and easy operation benefits. Porous organic frameworks (POFs), which are formed by covalent cross-linking of diverse organic moieties, can be good candidates for molecular separation due to its large surface area, rigid porous skeleton and high chemical stability. Inspired by the graphene oxide (GO) membranes, we present hereby a novel sandwich-like membrane design, which is based on the use of dispersible N-rich Schiff based POFs (SWN) as the sorbent, and the GO sheets as two-dimensional (2D) open interval layers. By alternately filtrating the SWN and GO dispersion, the sandwich-like hybrid membranes with different layered structures can be easily fabricated. Methyl orange (MO) was used as the probe to study the filtration performances of the as-prepared hybrid membranes. The breakthrough curves fit by Boltzmann model shows that the breakthrough volume (V_B) and adsorption capacity (Q) of the membranes with SWN loading of ~0.67mg quickly increases when the layer numbers increased from 1 to 4, but it becomes gradually stagnant when more GO layers are introduced. Based on the Werkhoven-Goëwie model, the theoretical plates (N) of the sandwich-like membrane with an optimized layers number can be calculated to 21.83, which is about 4.5-fold of the pure SWN membrane. The improvements of separation efficiency should be attributed to the increased effective path length and higher utilization ratios of the adsorption sites. The V_B and $Q$of the membranes can be easily scaled up by increasing the loading of SWN. Besides, the membrane can be repeatedly used for at least 7 times without decline of its performance. The strategy and methods used here may be readily changed to prepare other processable POFs and hybrid membranes with potential applications in molecular filtration.

Keywords:

Porous organic frameworks; Graphene oxide; Sandwich-Like membrane; Molecular Separation

420104 Continuous Production of Uniform Carbon-Supported Catalysts of Metal Nanocubes and Nanooctahedra Using Carbon Monoxide-Mediated Processes

Abstract:

Current technology for the production of nanoparticles often results in a broad distribution in size, shape and composition. Thus the products cannot meet the requirement when high-level control of such structural parameters are required. Although the conventional colloidal chemistry approach can generate nanoparticulates with well-defined size and morphology, the scaled-up process needs to be developed to meet the requirement for the mass manufacturing. In this talk, I will discuss our current attempt to bridge the gap between the laboratory work and industrial application. With CO as a major species to mediate the surface structures of metals and to reduce the salt precursor, highly uniform cubic Pt/C with narrow size distribution can be synthesized in a continuous fashion. Platinum alloy and other metal nanoparticles can also be produced with high-level controls over shape and composition. Our systematic studies demonstrate that capping ligands of alkylamine can play several important roles, including the dispersion of metal nanoparticles on the support. The carbon-supported Pt nanocubes are also examined as electrocatalysts. They exhibit high catalytic activity towards methanol electrooxidation when compared to commercial Pt/C catalyst.

429228 iCVD Top-Coat for Sub-10nm Patterning with Directed Self-Assembly of Block Copolymer Film

Abstract:

In technique of initiated CVD (iCVD), monomer and initiator vapors flow into a vacuum chamber. The resistively heated hot filaments in the chamber break the initiator into free-radicals. Subsequently, polymeric film growth begins on cooled substrates through free-radical polymerization of monomer. We introduce iCVD films as an effective top-coat in directed self-assembly (chemoexpitaxy) to create scalable sub-10 nm patterns. One of the benefits of vapor deposition is the ability to form smooth and pinhole-free ultrathin films. The iCVD layers, <10 nm thick, were successfully deposited as a top-coat on block copolymers (BCPs). In combination with BCPs displaying strongly segmenting domains (high χ), the iCVD top-coat enabled resolution of sub-10 nm patterns by directed self-assembly. This presentation will show and discuss on the concept of iCVD top-coat, sub-10 nm patterns driven by iCVD top-coat, and spectroscopic observations in comparison with a conventional solution-cast films applied by spin-coating.

430866 Nanomaterial Manufacturing Via Highly Parallelized Microfluidic Network

Abstract:

We have designed a network of microreactors to synthesize high-quality nanoparticles in a continuous flow process. The network design allows single inputs to be split into multiple droplet generating devices. The droplets act as ultra-low volume reactors for chemical synthesis. Traditional issues such as feedback between branches, droplet polydispersity, and low throughput are all overcome using our 3D printed network of channels. We achieve monodisperse nanoparticle products without device fouling or clogging for extended run times. Fluidic control is demonstrated for high temperature reactions, quenching, and in-line monitoring of the nanoparticle product.

432310 Precision H_2 Delivery Via a Ceramic Membrane Contactor to Continuously Manufacture Ag Nanoparticles

Abstract:

Nanoparticles have attracted substantial interest because of their novel mechanical, electrical, optical, thermal, and magnetic properties. Silver nanoparticles, in particular, are interesting as an antimicrobial, and because their color is determined by their size and shape. The production of these particles would benefit from the development of a continuous process for fabricating nanoparticles as opposed to the current batch systems. In industry, continuous processes are more attractive than batch processes because they can produce a more consistent product, and have both lower operational costs and waste. This talk will focus on the continuous manufacture of nanoparticles using a ceramic membrane as a phase contactor to induce nucleation and growth of monodispersed particles. The nanoporous membrane allows for the precise delivery of a reducing agent to the reaction zone. The model reaction chosen for this system is the reduction of aqueous Ag_2O with H_2 to Ag^+. A comparison between the particle size distribution produced by the conventional batch system and the continuous membrane system will be discussed.

433848 Novel One Step Liquid Flame Spray Pyrolysis (LFSP) Synthesis for N Doped TiO2

Abstract:

Nanosize TiO_2 and N-doped TiO_2 have been successfully prepared by one-step liquid flame spray pyrolysis (FSP). The effect of the fuel flow rate, powder production rate, precursor mixing condition, and total oxidant flow rate on product particle size, morphology, Quantity of N sources, types of N sources in the precursor and composition was also investigated. Resulting powders were characterized with X-ray diffraction (XRD), N2-physisorption, XPS, Transmission Electron Microscopy (TEM) and UV–Vis spectrophotometry. The UV-vis spectroscopy results of N incorporated TiO_2 showed enhancement of light absorption in the visible range by selected composite materials in the UV range (250-400 nm). This effect is due to the insertion of N in the TiO_2 matrix, which generated p-type impurity band. XPS study allows spotting out the role and significance of the surface species in the reaction and, in particular, the important and unique role of anionic N species in modifying the behaviour of the nanoparticles. We have found the Ti/N ratio of 13% for the materials synthesized from our XPS data. The higher surface areas of around (100-120 M^2/g) were observed for the catalysts. These interesting results will be discussed in the presentation.

434062 Scale up Production of High Performance Nanowires for Waste Heat Recovery

Abstract:

The rapid development of nanomaterials in the past decade has provided a possibility of directly converting waste heat back to electricity based on the thermoelectric Seebeck effect. In the past four years, we have developed a transformative approach to pioneered low cost and scalable solution-phase growth methods to mass produce thermoelectric nanowires and nanowire heterostructures to match the physical and economic magnitudes of energy use and economical entertainment in the manufacture/recycling. These nanostructured thermoelectric materials show a significantly enhanced performance compared to the bulk crystals based on the quantum confinement and minority carrier blocking.

413969 Facile Preparation of Nano-Sized Anisotropic Spherical Polyelectrolyte Brushes and Chemical Force Microscope Imaging for Their Regioselective Assembly

Abstract:

Here we demonstrate a facile method based on selective surface modification to prepare anisotropic spherical polyelectrolyte brushes (ASPBs) with a tailored surface property and geometry. At the first step, a strategy of Pickering emulsion was adopted to efficiently mask precursor nanoparticles, i.e. the precursor nanoparticles composed of polystyrene (PS) core and photo-initiator shell spontaneously embedded into the surface of molten wax bead, and locked by freezing the wax bead. The following step is proceeding with further modification on the unprotected surface of those trapped nanoparticles. Poly(acrylic acid) chains were chemically grafted to form robust brushes on exposed side of precursor particles under UV irradiation, which resulted in an asymmetrical geometry and property on the surface of nanoparticle. To take advantage of the anisotropism of ASPBs, we studied their alignments and orientations on solid surfaces with a different affinity. A chemical force microscope was employed to provide a high resolution imaging for their regioselective assembly. The probe was modified with mercaptan quaternary ammonium salt so as to interact with anionic brushes but inactivate the interaction with PS surface. Our work herein not only present a solution to a facile synthesis and characterization of anisotropic particles, but a way to exploit to control the geometry, the surface property, and the supraparticular assemblies of particles at nanoscale, even open the door to manipulate single nanoparticle in three dimensions.

437873 Titania Nanotube Arrays As Interfaces for Blood Contacting Interfaces

Abstract:

The constant exposure of implantable biomaterials such as titanium and titanium alloys to blood introduces serious and ongoing concerns regarding poor blood-material interactions. To date, all blood-contacting materials have been shown to initiate immunological events in the form of inflammation, thrombosis, fibrosis and infection; potentially leading to complete implant failure. Material surfaces that provide biomimetic cues such as nanoscale architectures have been shown to elicit improved cellular interaction; and thus, may provide possible solutions for enhancing blood-compatibility. However, limited information exists about the thrombogenicity of nanoscale surface architectures. In this study, we have evaluated the efficacy of titania nanotube arrays as interfaces for blood contacting devices by investigating the thrombogenic effects using whole blood plasma. This work shows the improved blood-compatibility of titania nanotube arrays, identifying this specific nanoarchitecture as a potentially optimal interface for promoting the long-term success of blood contacting biomaterials.

437871 Decreasing Medical Device Infections Using Nanotechnology, Not Drugs

Abstract:

Medical device infections are increasing not decreasing, highlighting that our current approaches to fighting bacteria are flawed. Our current approach to fighting medical device infection involves the prescription of pharmaceutical agents and/or medical device removal if infection is too aggressive. This approach is not working as there is an ever increasing presence of antibiotic-resistant bacteria and when medical device infection can not be stopped, the medical device needs to be removed. Nanotechnology, or the use of materials with at least one dimension less than 100 nm, may provide a non-drug approach to eliminating medical device infection. This is because nanoparticles can penetrate biofilms and bacteria when conventional drugs cannot. Moreover, nanoparticles of specific chemistry (such as iron, selenium, magnesium, ceria, zinc and silver) can elevate intra-bacteria levels to kill bacteria without affecting mammalian cell functions. In contrast, synthetic materials used as medical devices today are typically composed of millimeter or micron sized particles and/or fiber dimensions and have no ability to fight bacteria. Although human cells are on the micron scale, their individual components, e.g. proteins, are composed of nanometer features. By modifying only the nanofeatures on material surfaces without changing surface chemistry, it is also possible to decrease medical infection by increasing the endogenous adsorption of anti-bacterial adhesive proteins onto the medical device surface. Finally, nanomedicine has been shown to stimulate the growth and differentiation of stem cells, which may someday be used to further decrease implant infection since stem cells secrete factors known to kill bacteria. This invited talk will highlight some of these advancements and emphasize current nanomaterials currently approved by the FDA for human implantation to fight bacteria growth where our current approaches have failed.

437877 Bioactive Nanoengineered Hydrogels for Bone Tissue Engineering: A Growth-Factor-Free Approach

Abstract:

Despite bone's impressive ability to heal after traumatic injuries and fractures, a significant need still exists for developing strategies to promote healing of non-union defects. To address this issue, we developed collagen-based hydrogels containing two-dimensional (2D) nanosilicates. Nanosilicates are ultrathin nanomaterials with a high degree of anisotropy and functionality that results in enhanced surface interactions with biological entities compared to their respective three-dimensional counterparts. The addition of nanosilicates resulted in a four-fold increase in compressive modulus along with an increase in pore size compared to collagen-based hydrogels. *In vitro* evaluation indicated that the nanocomposite hydrogels are capable of promoting osteogenesis in the absence of any osteoinductive factors. A three-fold increase in alkaline phosphatase activity and a four-fold increase in the formation of the mineralized matrix was observed with the addition of the nanosilicates to the collagen-based hydrogels. Overall, these results demonstrate the multiple functions of nanosilicates conducive to the regeneration of bone in non-union defects, including increased network stiffness and porosity, injectability and enhanced mineralized matrix formation in a growth-factor-free microenvironment.

437879 An Inflammation Responsive Nano-Fiber Platform for Controlled Drug Delivery

Abstract:

We have developed an autonomous Nano-Fiber delivery system that titrates the amount of drug released in response to the level of inflammation, ensuring the drug is released only when needed at a therapeutically relevant concentration. Translation of such technology would represent a paradigm shift in the treatment of inflammatory diseases such as rheumatoid arthritis and inflammatory bowel disease which has characteristic flares followed by periods of lower disease activity. We have evaluated its ability to release drugs in response to enzymes that are abundant in human synovial fluid, ulcerated gut tissue, and in limb transplants using a series of in vitro and in vivo experiments.

437874 How Nanostructures Affect Cell Behavior?

Abstract:

A growing body of literature shows that certain nano- to micron scale surface features or surface patterns can modulate cell behaviors *in vitro and in vivo*, suggesting the potential for enhancing biomaterial and device functions *in vivo*. In recent years, a variety of techniques has been developed to create random or patterned nano-to-micron scale features on the surface of biomaterials, including metals (e.g., titanium), polymers (e.g., biodegradable polyesters), and biocomposites. Distinctive cellular behaviors on nanostructured or microstructured surfaces have been observed and reported. This talk will discuss current methods for creating these features on biomedical device surfaces and their respective effects on cell adhesion, proliferation, and differentiation, as well as their potential for clinical translation. Special attention will be given to recent results on stem cells and bacterial cells.

423773 Regulation of Human Mesenchymal Stem Cells (hMSC) Differentiation on Surface Modified 3D Printed Scaffold By Cold Atmospheric Plasma (CAP)

Abstract:

INTRODUCTION

Three-dimensional (3D) printing is a new fabrication method for tissue engineering. The significant advantage of this new scaffold fabrication method is that it can precisely control scaffold architectures, which are printed from computer-assisted software. However, a scaffold is only a small biomimic unit of integral bone tissue. It not only needs three-dimensional biocompatible structures, but it should also mimic extracellular matrix (ECM) properties, such as providing a template for cell attachment, and stimulate bone tissue formation in vitro [1]. In order to achieve the second target, we introduce cold atmospheric plasma (CAP) to modify scaffold surface roughness and chemical composition [2]. Cold atmospheric plasma is an ionized gas, which was helium in this study. The composition of cold plasma includes various electrons, positive/negative ions, radicals, excited molecules, energetic photons, and ultraviolet (UV) wavelengths. Based on those compositions, plasma treatment has been developed for the effective surface modification of tissue engineering scaffold surfaces [3]. In studying CAP treated 3D printing scaffolds, we showed that CAP surface modification can change topography. The treatment also alters the chemistry of the surface, which can significantly influence stem attachment and differentiation.

METHODS

Poly lactic acid (PLA) (Natureworks, LLC, MN) was fabricated by a 3D Makerbot printer (Makerbot, NY). After it was printed, scaffolds were directly treated for further surface modification by cold atmospheric plasma (CAP) for 1, 3, and 5 minutes. The contact angles of deionized water on control and surface modified PLA scaffolds were measured in the static mode using a Tame-hart mode 300 contract angle goniometer. Average surface roughness values were obtained by atomic force microscopy (AFM) using a Parks Systems NX-10 microscope and XEI software.

RESULTS AND DISSCUSSIONS

Under the optimized cold plasma condition, the contact angle dramatically dropped after cold plasma treated, which indicated that cold plasma treated PLA scaffolds were more hydrophilic than untreated PLA scaffolds. The contact angle on untreated PLA scaffolds treated for 1, 3 and 5 min was 70±2°, 50±2°41±2°, 24±2°, respectively. The effects of CAP treatment on the topography of the PLA scaffolds can be summarized as the roughness was controlled by exposure time. The roughness (Rq) values for the untreated scaffold and CAP scaffolds treated for 1, 3, and 5 min were 1.168 nm, 10.45 nm, 22.87 nm, and 27.60 nm, respectively. After CAP treatment, the surface had much more roughness even when exposed for 1 min. This means the nanometric roughness had a maximum 300 % increase, which is consistent with SEM results. Both hydrophilicity and nanoscale roughness changes represented a very efficient plasma treatment. Those surface changes also played a very important role in cell attachment and bone formation which will also be presented.

CONCLUSIONS

In this study, CAP treatment is introduced as a new technology for altering scaffold surface chemistry and nanoscale topography. Contact angle and AFM results demonstrated that 3D printed PLA scaffolds after CAP treated were more hydrophilic with more roughness. The above surface property modification led to an enhancement in cell attachment and bone formation. Those promising results suggest that CAP surface modification may have potential applications in bone tissue engineering.

REFERENCES

[1] Webster, T.J., Schadler, L.S., Siegel, R.W. et al. Tissue Engineering, 2001, 7, 291.

[2] Hsu S.H, Lin C.H, Tseng C.S. Biofabrication, 2012, 4, 015002.

[3] Wang M, Cheng X, Zhu W, et al. Tissue engineering: part A, 2014, 5(20): 1060-1071.

ACKNOWLEDGEMENT

This study was supported by the Northeastern University. Dr. Hilal Yazici is supported by TÜBİTAK (The Scientific and Technological Research Council of Turkey) – BİDEB (Programme Number: 2219).

417144 Preparation of Poly(L-lactic acid) Nanofibrous Scaffold with a Rough Surface in Supercritical Carbon Dioxide

Abstract:

Phase inversion using supercritical carbon dioxide (SC-CO_2) has been widely used in the development of tissue engineering scaffolds, and particular attention has been paid to obtaining desired morphology without additional post-treatments. However, the main challenge of this technique is the difficulty to generate a three-dimensional (3D) nanofiber structure with a rough surface in one step. Here, a poly($_L$-lactic acid) (PLLA) 3D nanofiber scaffold with a rough surface is obtained via phase inversion using SC-CO_2 by carefully choosing fabrication conditions and porogens. It is found that this method can effectively modulate the structure morphology, promote the crystallization process of semicrystalline polymer, and induce the formation of rough structure on the surface of nanofiber. Meanwhile, the porogen of ammonium bicarbonate (AB) can produce a 3D structure with large pores, and porogen of menthol can improve the interconnectivity between the micropores of nanofibers. A significant increase in the fiber diameter was observed as the menthol content increases. Furthermore, the menthol may affect the mutual transition between the α' and α crystals of PLLA during the phase separation process. In addition, the results of protein adsorption, cell adhesion and proliferation assays indicate that cells tend to have a higher viability on the nanofiber scaffold. This process combines the characteristic properties of SC-CO_2 and the solubility of menthol to tailor the morphology of polymeric scaffolds, which may have potential applications in tissue engineering.

Improved Cellular Functions and Reduced Bacterial Infection on MgO Nanocomposites

Daniel J. Hickey[1] and Thomas J. Webster[1,2]

[1]Department of Chemical Engineering, Northeastern University, Boston, MA 02115, USA
[2]Center of Excellence for Advanced Materials Research, King Abdulaziz University, Jeddah, Saudi Arabia

IGERT NANOMEDICINE NEU/TU/UPRM

Northeastern University

AIChE The Global Home of Chemical Engineers

Research Motivation

- More than 3 million musculoskeletal procedures are performed annually in the US
- Poor tissue integration leading to instability is the leading cause of revision surgeries

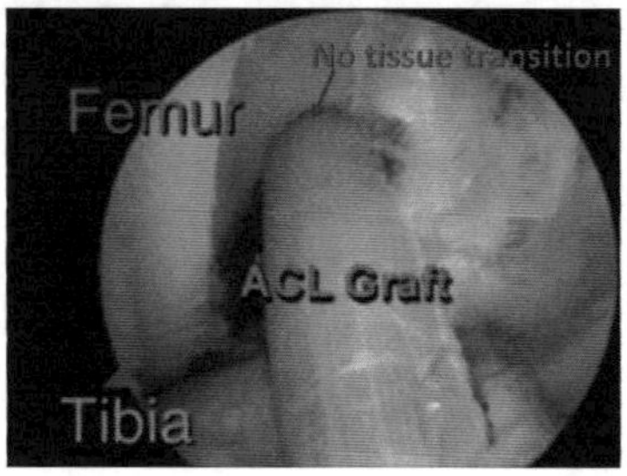

Biomaterials with improved chemical and mechanical properties are badly needed to facilitate material integration and tissue regeneration

1. Jahangir AA *et al.*, *AAOS Now*, Jan. 2008.
2. http://www.jointhealing.com/pages/knee/acl_surg.html

Research Motivation

Additionally, bacterial infection of implanted biomaterials is a recurring problem that can lead to implant failure, revision surgeries, and the spread of infection

- For total knee arthroplasty, infection is reported as the second main cause of revision (0.5-5%), after instability
- Average 5% of orthopedic fixation devices become infected (out of ~2 million implanted annually in U.S.)

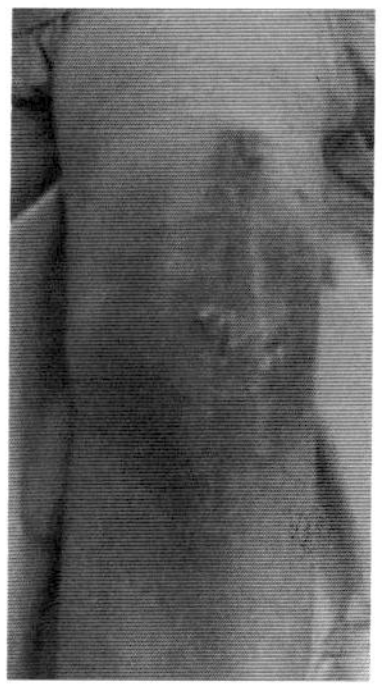

Peri-prosthetic infection: 3 *Staphylococcus* and 1 *Pseudamonas* species

1. Campoccia, D., et al., Biomaterials, 27(11), 2006.
2. Trampuz, A., and W. Zimmerli, Injury 37, (Suppl 2), 2006.
3. http://www.osteomyelitis.com/ Peri-prosthetic Infections: Case 1

3

Objective

To develop orthopedic biomaterials with:

- enhanced tissue integration abilities
- antibacterial efficacy

Specifically, using MgO nanocomposites

4

Magnesium Oxide (MgO) Nanoparticles

Biocompatible, biodegradable, low-cost,
environmentally friendly,
and present naturally in the human body

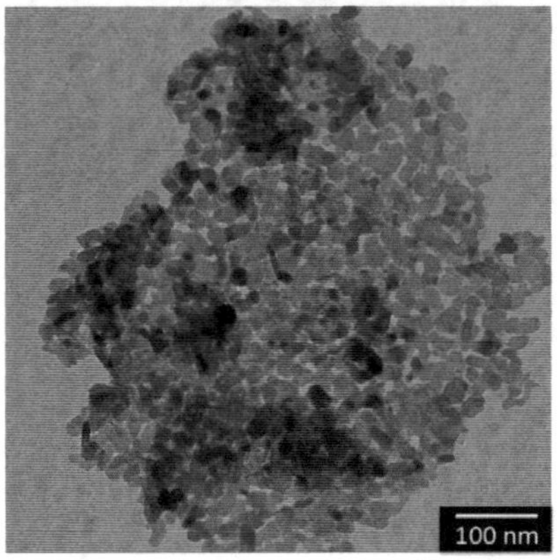

5

Magnesium for Bone

- Mg^{2+} cations coat the edges of bone apatite (HA) crystals and directly influence crystal size and density
- Indirectly influence mineral metabolism through activation of alkaline phosphatase (ALP)

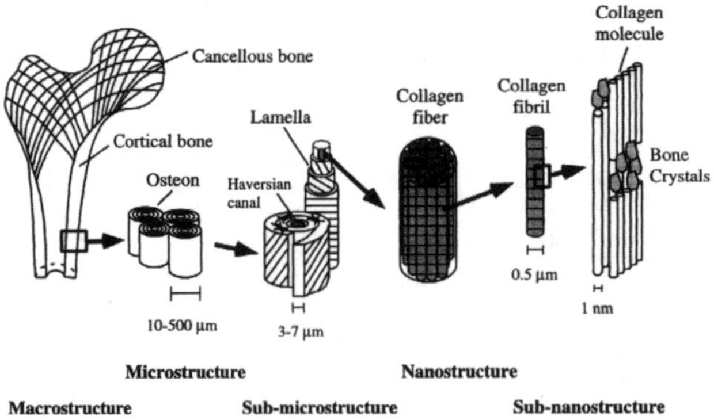

1. Serre, C.M., et al., J Biomed Mater Res., 42, 1998.
2. Rho, J.Y., et al., Medical Engineering & Physics, 20, 1998.

6

Magnesium for Cell Functions

Mg^{2+} (and Ca^{2+}) cations initiate conformational activation of integrin receptors to bind ligands

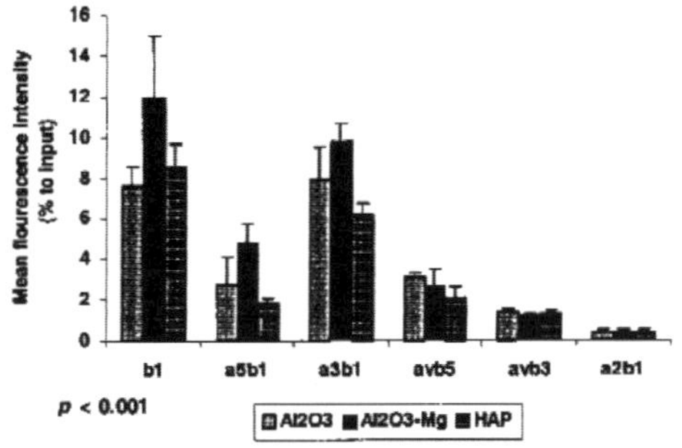

Integrin expression from human bone-derived cells (HBDC).
From Zreiqat H *et al.*, J Biomed Mat Res, 62, 2002

Integrins mediate cell-ECM interactions, which direct cell functions

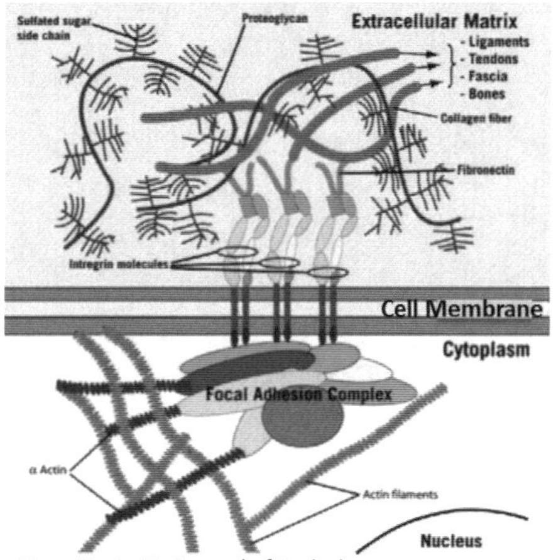

From Ravin, T., Journal of Prolotherapy, 3, 2011.

Yamniuk AP *et al.*, Protein Science, 14(6), 2005.

7

MgO NPs are Highly Antibacterial

MgO (metal oxide) nanoparticles exhibit strong broad-range antibacterial activity

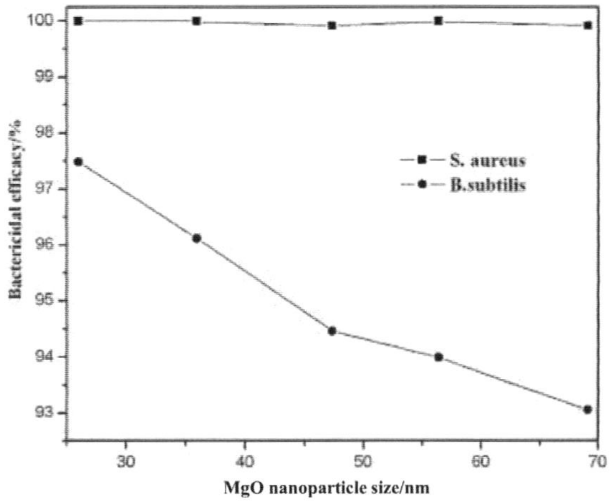

Huang, L., Journal of Inorganic Biochemistry, 99, 2005.

8

Antibacterial Efficacy of Free MgO

Free MgO NPs show outstanding toxicity towards ***S. aureus***

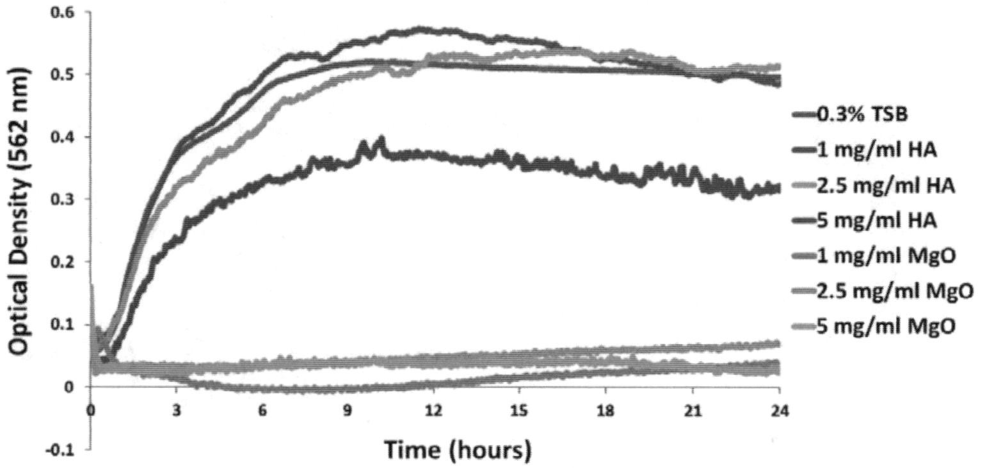

9

The Importance of Nano

Interactions at the nanoscale are more biologically relevant

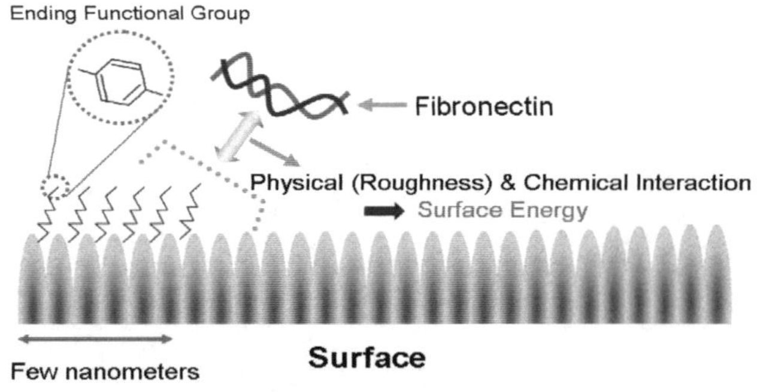

MgO only gains antibacterial efficacy in the nanoregime

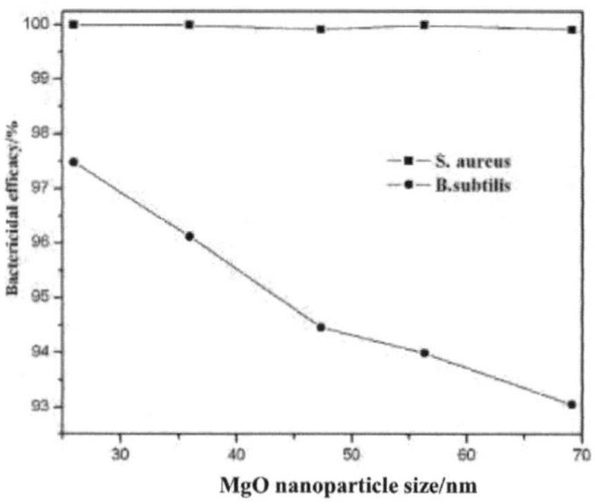

Huang, L., Journal of Inorganic Biochemistry, 99, 2005.

10

Nanocomposite Fabrication

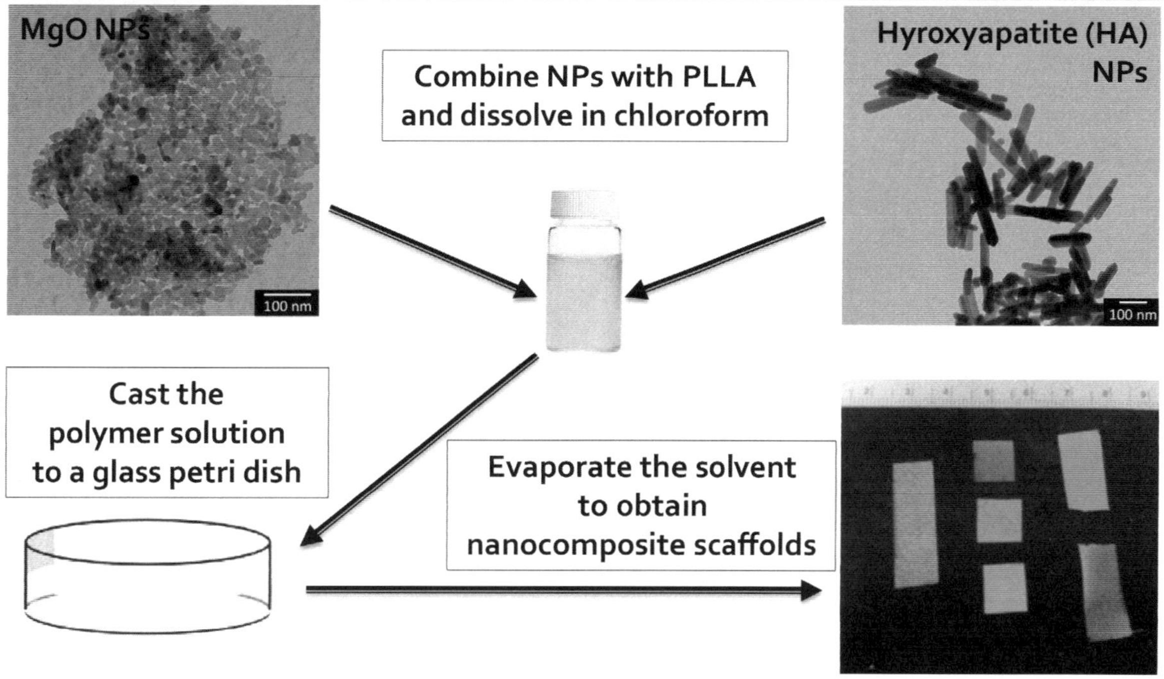

Nanocomposite Surface Topography

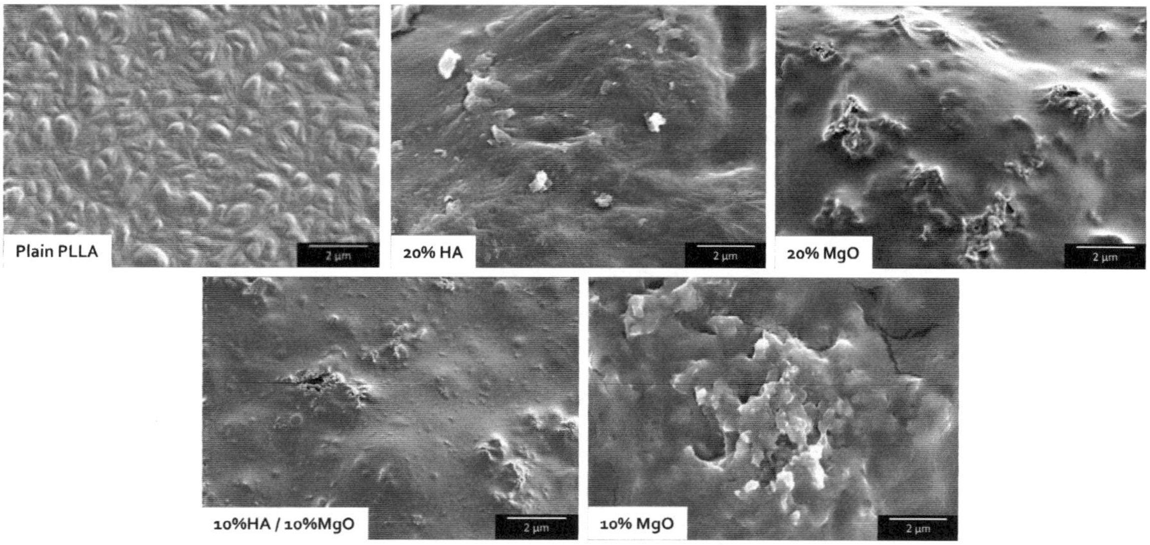

SEM images (mag = 10k) show variations in surface topography and particle exposure

12

Nanocomposite Surface Roughness

Atomic force microscopy (AFM), 2×2 μm scan size

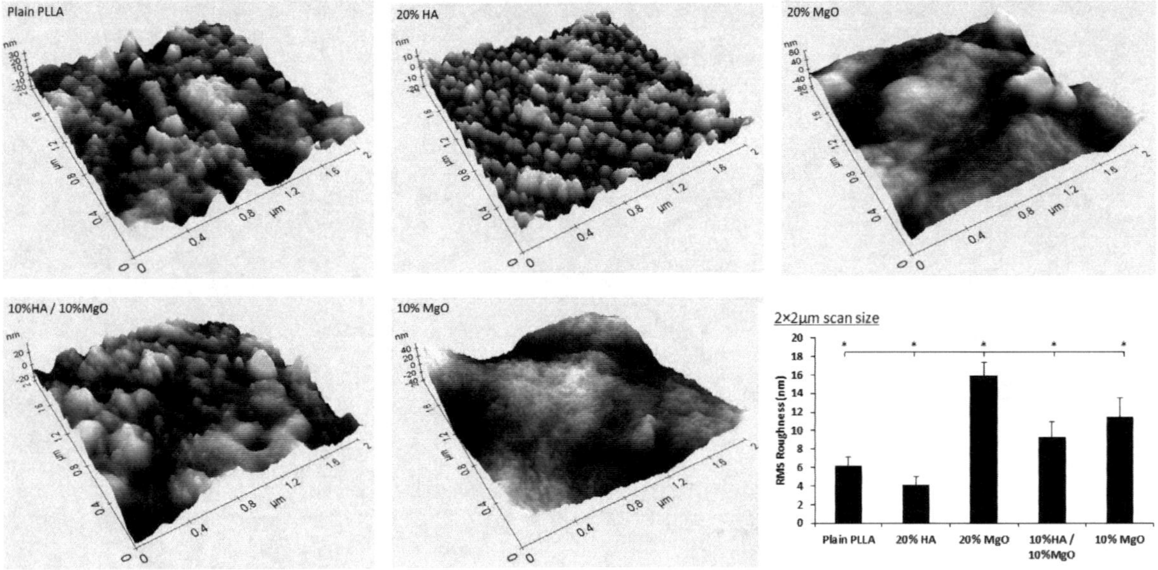

13

Mechanical Tensile Testing

- Initial gauge length = 1 cm
- Stretch rate = 0.1 mm/sec, to sample break
- Record stress vs. extension
- Calculate tensile modulus, extension, max load, load at break, max stress, stress at break

14

Mechanical Tensile Testing

Nano secondary phase increases the elastic modulus of PLLA

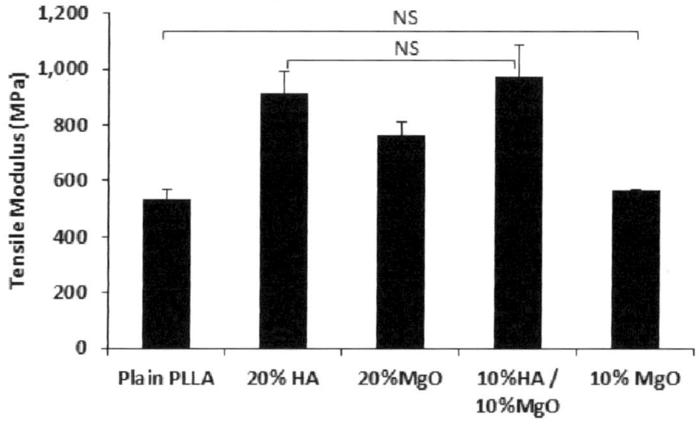

Typical Young's Moduli: [1,2]

Ligament material	~300 MPa
Tendon material	1 – 2 GPa
Cortical bone	~17 GPa
Cancellous bone	0.1 – 4.5 GPa

Data is average +/- SEM, N=3. NS signifies no statistical difference between samples

Modulus of all samples was within acceptable range for cancellous bone

1. Turner *et al.*, J. Biomech. 23, 549-561, 1990.
2. http://wings.buffalo.edu/eng/mae/courses/417-517/Orthopaedic%20Biomechanics/Lecture%203u.pdf

Material Elongation Decreases with Addition of Secondary Phase

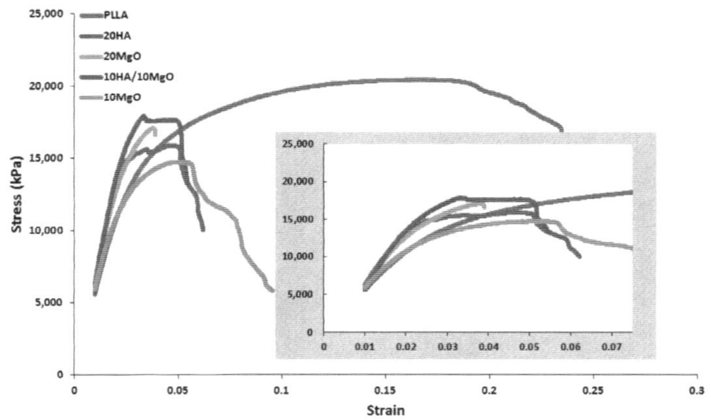

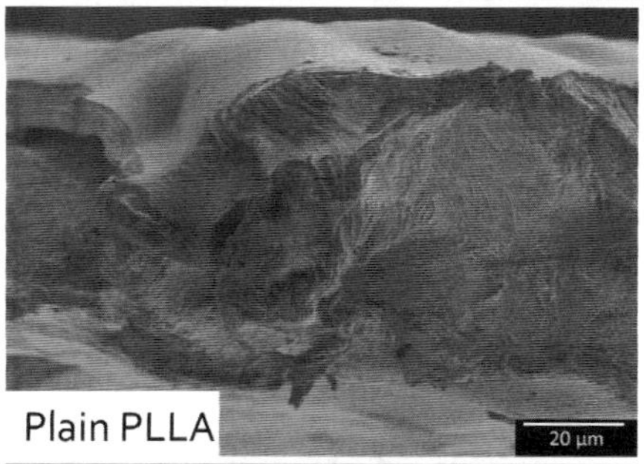

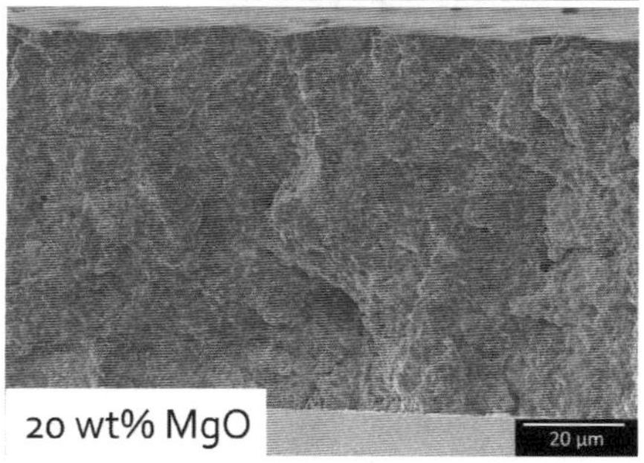

Analysis of SEM images after tensile testing gives insight into the modes of failure

Osteoblast Adhesion

MgO-NPs increase **osteoblast adhesion** on PLLA

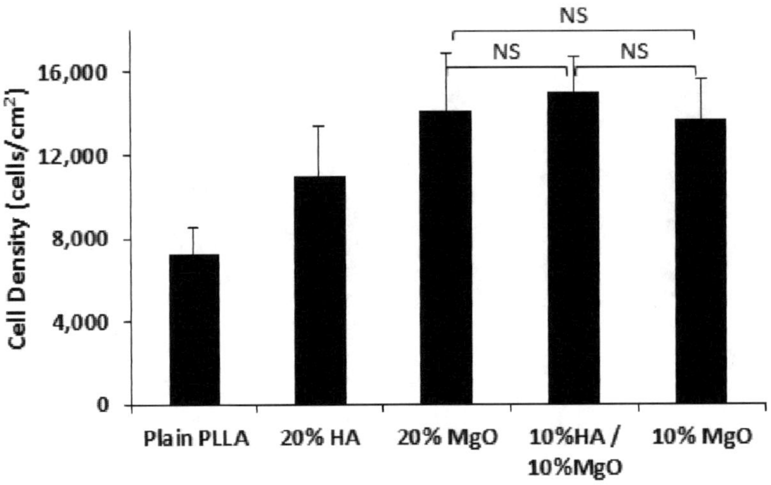

Data is average +/- SEM, N=3. NS signifies no statistical difference between samples

17

Fibroblast Adhesion

MgO-NPs increase **fibroblast adhesion** on PLLA

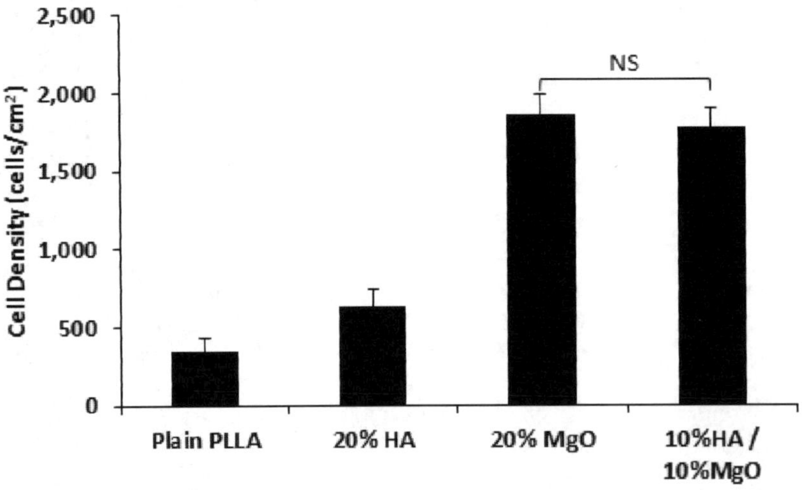

Data is average +/- SEM, N=3. NS signifies no statistical difference between samples

18

Osteoblast Cell Proliferation

MgO-NPs enhance **OB proliferation** on HA-PLLA n-comp's

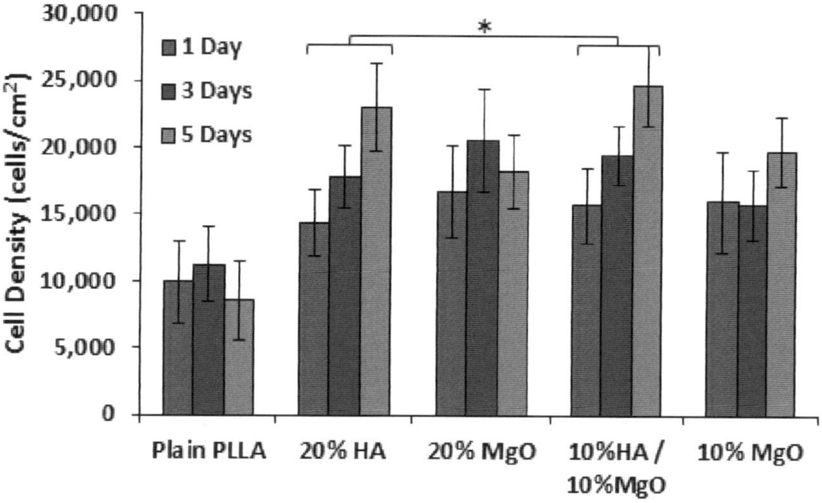

Data is average +/- SEM, N=3. *$P<0.05$.
Cell densities on all nanocomposites were statistically higher than on plain PLLA at all time points

19

Osteoblast Morphology

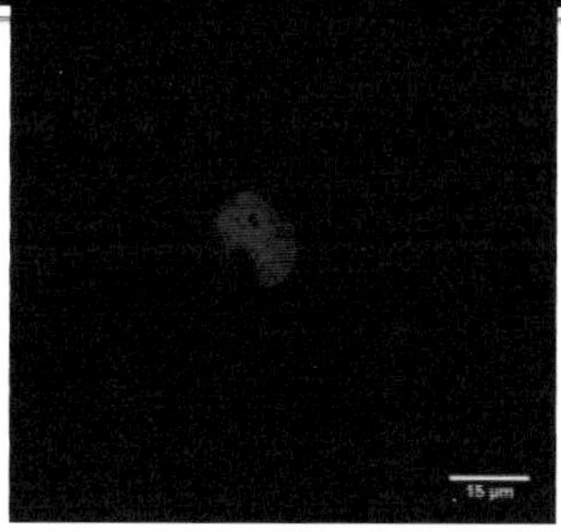

Plain PLLA

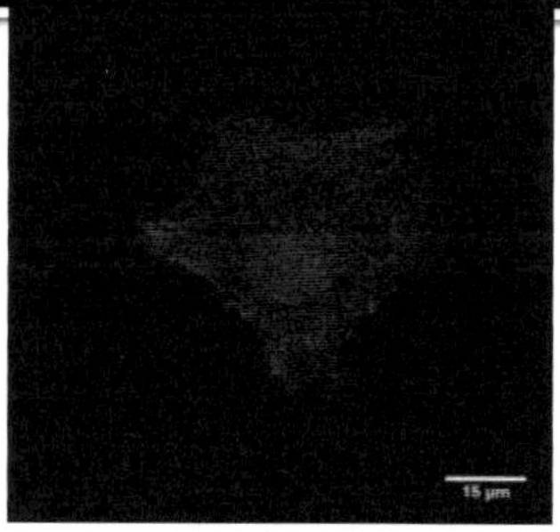

20% HA

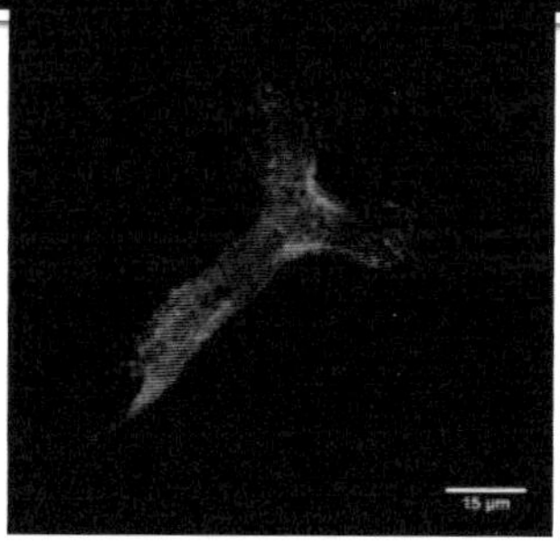

20% MgO

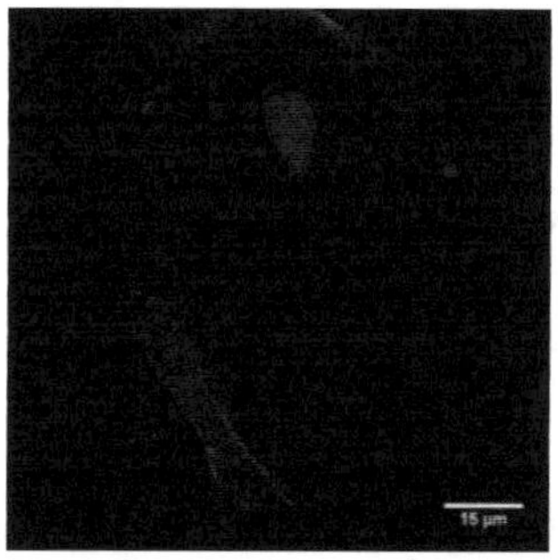

10% HA/10% MgO

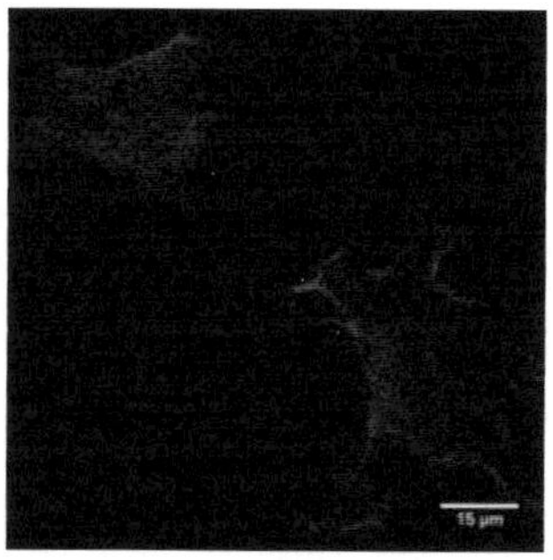

10% MgO

OBs at 3 days

- Blue: nuclei
- Red: actin
- Green: vinculin

20

Composite Degradation Study

- Record initial weight of 1-cm^2 samples
- Submerge samples in 2 ml of complete OB media and incubate at standard conditions
- Record pH values intermittently
- After 7 days of incubation (no cells), remove the samples, dry, and record final weight

21

Nanocomposite Degradation

Percent weight reduction of 1-cm^2 nanocomposite samples incubated at standard conditions for 7 days in 2 mL of complete osteoblast media

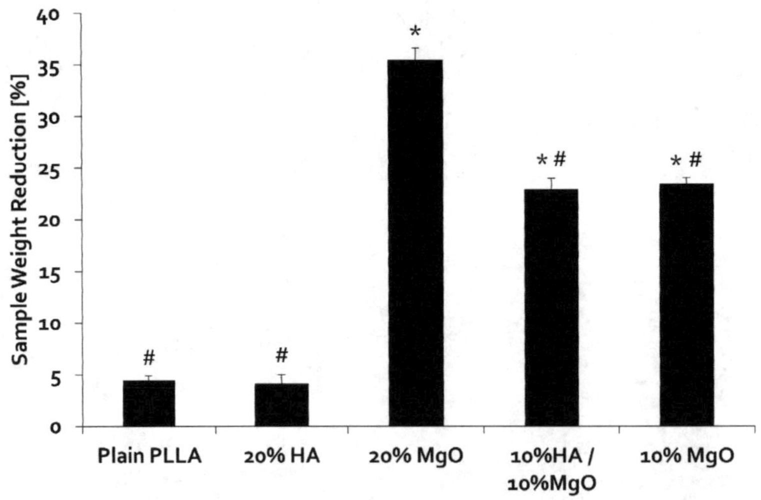

Data represents mean +/- SEM, N=3. *$p<0.05$ compared to plain PLLA, #$p<0.05$ compared to 20% MgO

22

Degraded Nanocomposites

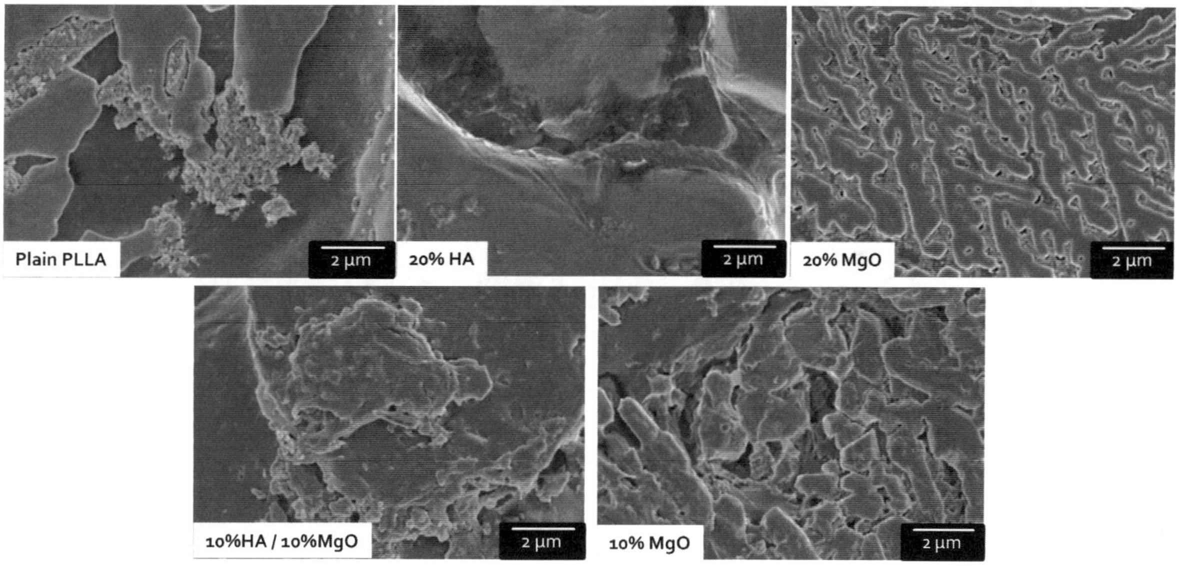

SEM images (mag = 10k) show degraded nanocomposite surfaces

Nanocomposite pH Variations

pH variations of 2 mL of complete osteoblast media containing 1-cm² nanocomposite samples incubated at standard conditions for 7 days

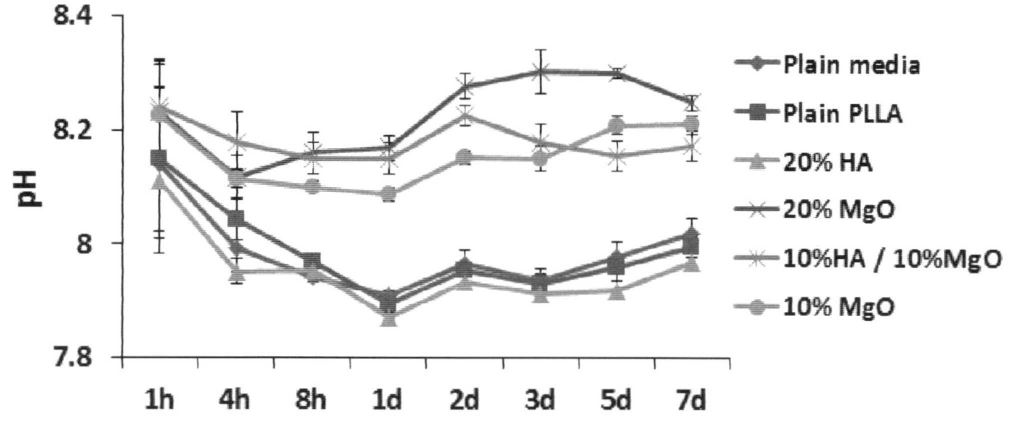

$$MgO + H_2O \rightarrow Mg(OH)_2$$
$$Mg(OH)_2 \leftrightarrow Mg^{2+} + 2\,OH^-$$

Osteoblast Proliferation in Conditioned Medium

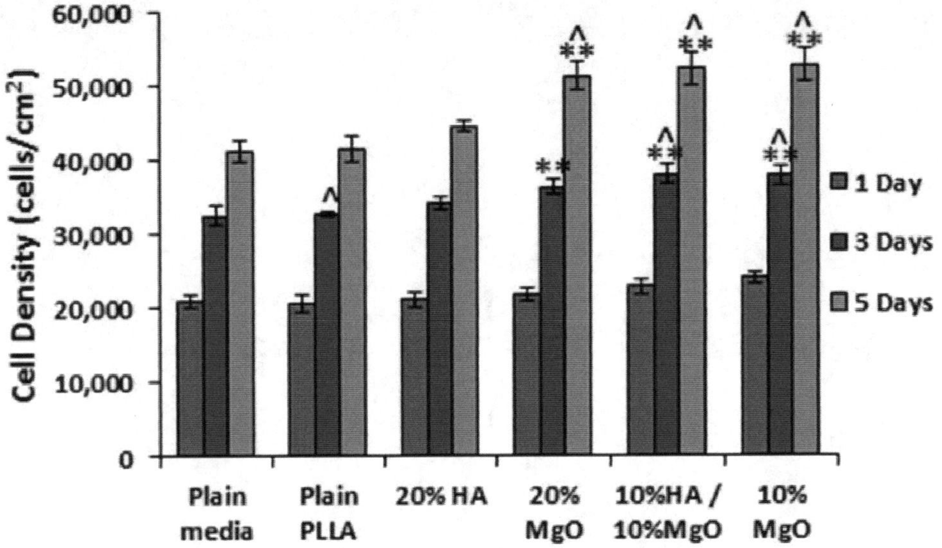

Data represents mean +/- SEM, N=3. **$p<0.05$ compared to plain media, ^$p<0.05$ compared to 20% HA
All cell densities under the same treatment were statistically greater with time

25

Antibacterial Performance: *S. aureus* adhesion

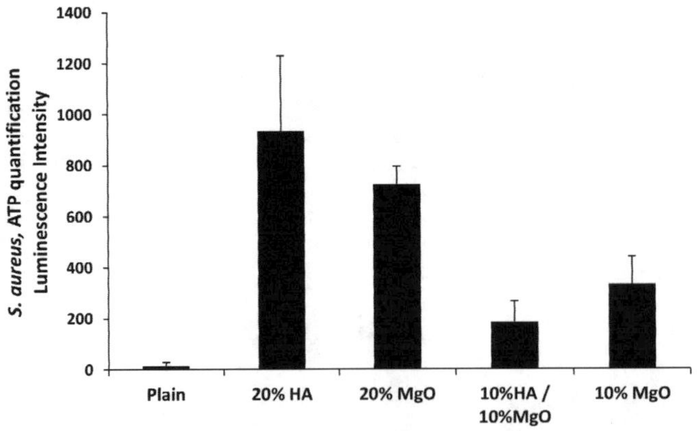

S. aureus adhesion onto scaffolds after 4 hours, measured by ATP quantification.
Data is average +/- SD, N=1.

26

S. aureus colonization

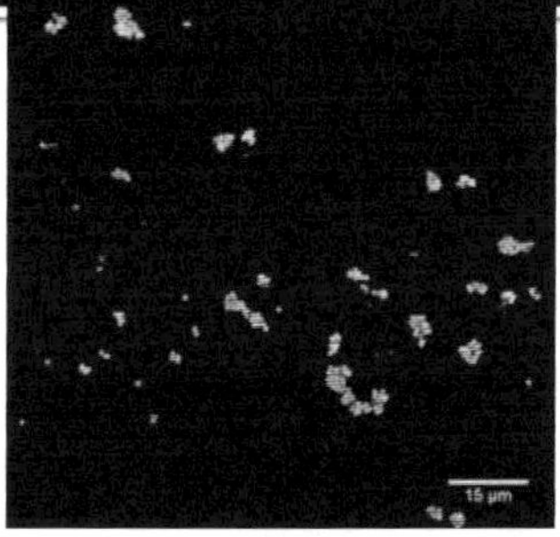

Plain PLLA

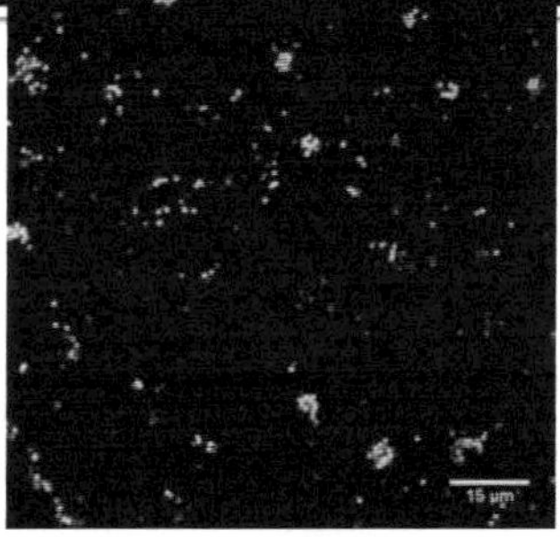

20% HA

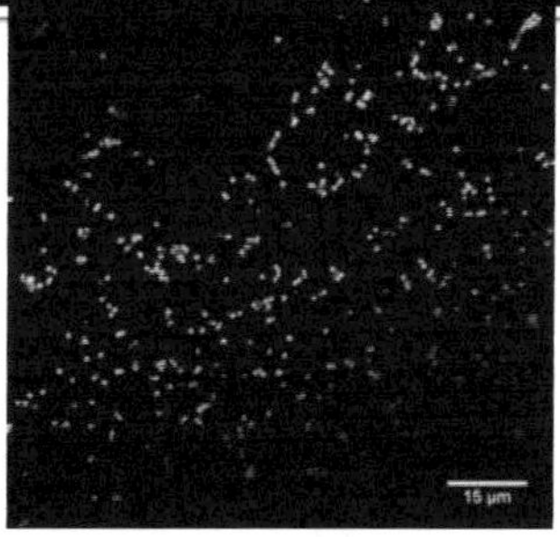

20% MgO

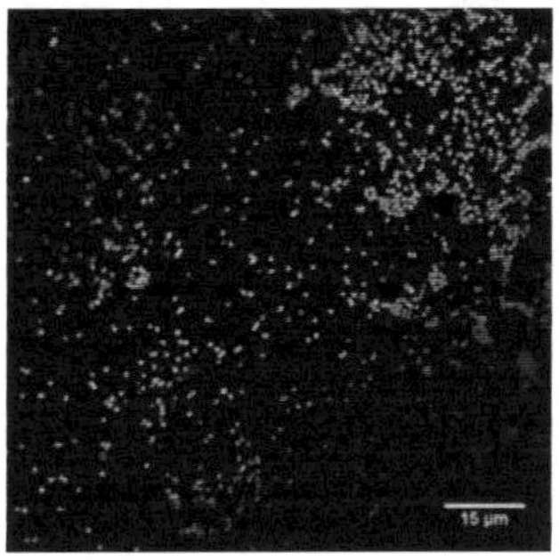

10% HA/10% MgO

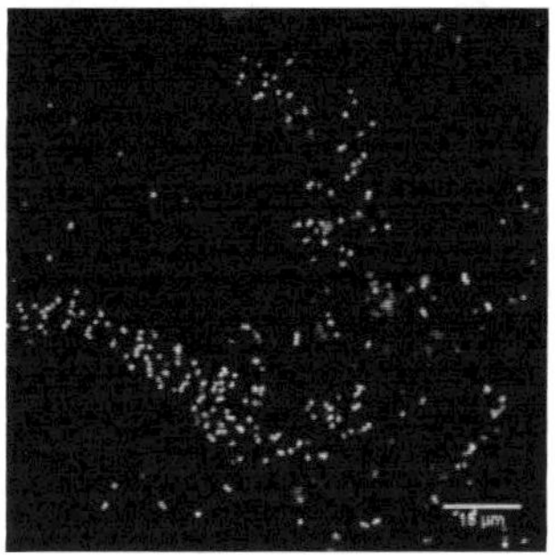

10% MgO

<u>*S. aureus at 1 day*</u>

- Green: live
- Red: Dead

27

S. aureus Biofilm Formation

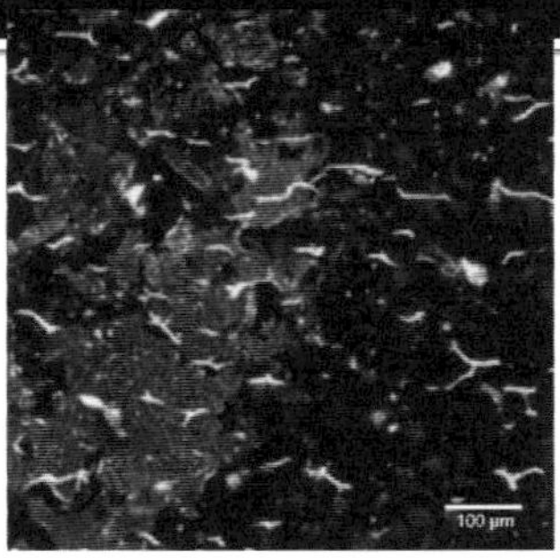

Plain PLLA

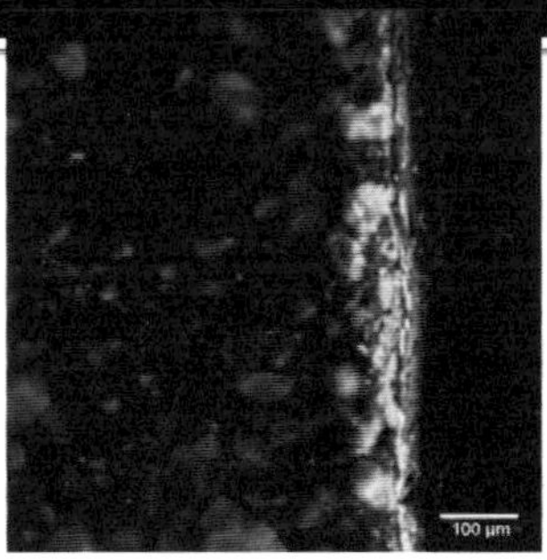

20% HA

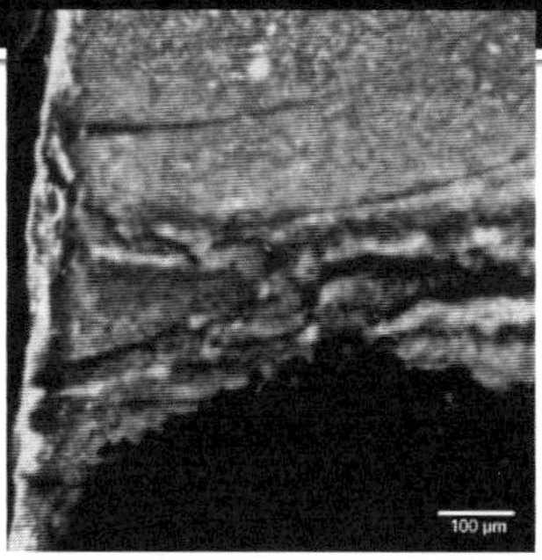

20% MgO

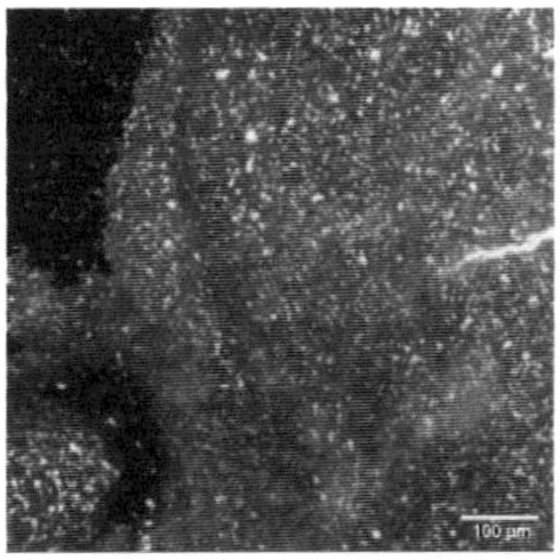

10% HA/10% MgO

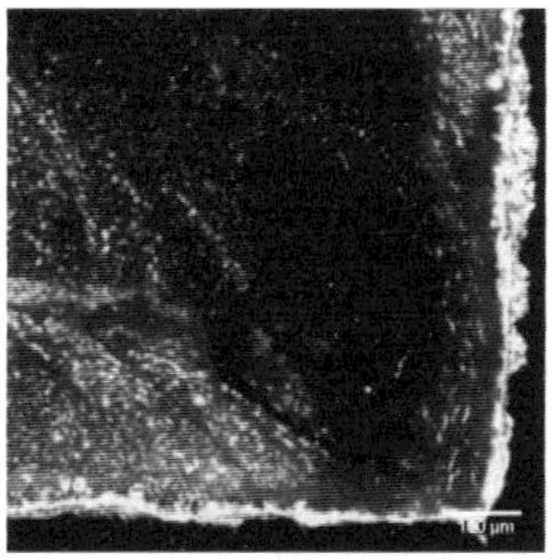

10% MgO

<u>*S. aureus at 3 days*</u>

- Green: live
- Red: Dead

28

S. Aureus growth in NP-supplemented media

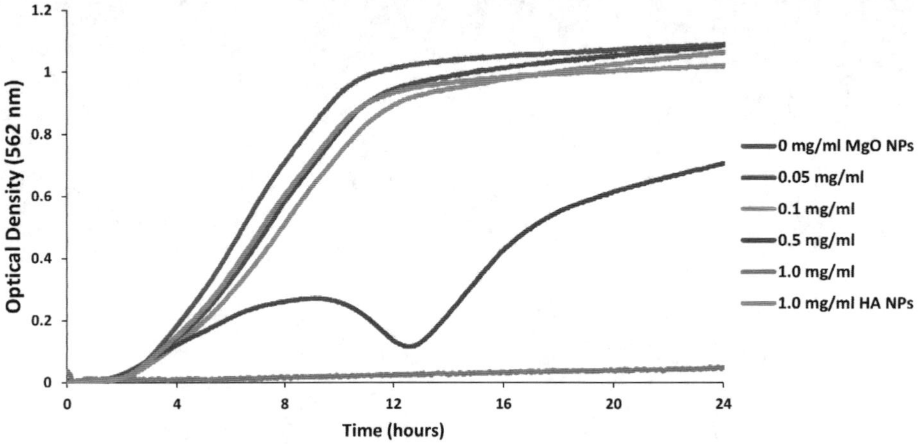

MgO NP concentrations above 500 μg/ml completely inhibit the growth of *S. aureus*, whereas high concentrations of HA NPs have little effect on bacterial growth.

29

Nanoparticle pH Variations

pH variations of 2 mL of complete osteoblast media containing indicated nanoparticle concentrations [mg/mL] incubated at standard conditions for 7 days

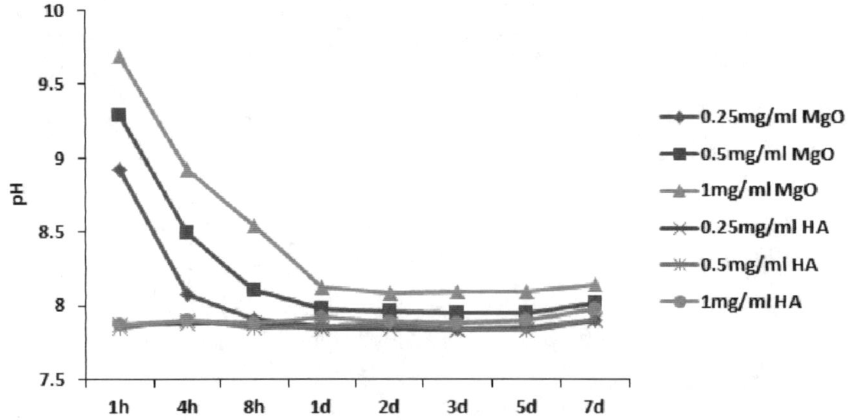

$$MgO + H_2O \rightarrow Mg(OH)_2$$
$$Mg(OH)_2 \leftrightarrow Mg^{2+} + 2\,OH^-$$

30

Next Steps

- Map the surface composition of composites
- Investigate MgO NP suspensions
- Identify bacteria-material interactions
- Try new fabrication processes

31

Conclusions

- MgO-HA-PLLA nanocomposites performed best regarding mechanical properties and OB activity
- Work is ongoing to preserve the antibacterial potential of MgO NPs when dispersed within scaffolds for TE
- MgO-HA-PLLA nanocomposites should be further studied for their promising ability to:
  - Fight infection
  - Improve orthopedic tissue regeneration

32

Acknowledgements

- Webster Nanomedicine Lab
- Rob Eagan, Bill Fowle
- NEU Department of Chemical Engineering
- NSF-IGERT Grant No. 0965843

IGERT
NANOMEDICINE
NEU/TU/UPRM

AIChE
American Institute of Chemical Engineers

413082 In-Product Anti-Counterfeiting Using Phase Change Nanoparticles

Abstract:

There is an urgent need is to develop in-product covert barcode for agrochemical anti-counterfeiting. This paper describes a new nanoparticles-based in-product barcode system, in which a panel of phase change nanoparticles with discrete and sharp melting peaks is added as a barcode into in a variety of agrochemicals. Environmentally degradable organic solids are used to make nanoparticles. The barcode can be readout by detecting melting peaks of nanoparticles using differential scanning calorimetry. This method has high labeling capacity due to small sizes of nanoparticles, sharp melting peaks, and large scan range of thermal analysis. The in-product barcode can be effectively used to protect agrochemical products from being counterfeited due to its large coding capacity, technical readiness, covertness, and robustness.

413919 A Novel Smart Microsphere with K+-Induced Shrinking and Aggregating Property Based on Responsive Host-Guest System

Abstract:

A novel type of smart microspheres with K^+-induced shrinking and aggregating properties is designed and developed based on K^+-recognition host-guest system. The microspheres are composed of crosslinked poly(*N*-isopropylacrylamide-*co*-acryloylamidobenzo-15-crown-5) (P(NIPAM-*co*-AAB15C5)) networks. Due to the formation of stable 2:1 "sandwich-type" host-guest complexes between 15-crown-5 units and K^+ ions, the P(NIPAM-*co*-AAB15C5) microspheres significantly exhibit isothermally and synchronously K^+-induced shrinking and aggregating properties at a low K^+ concentration; while other cations (*e.g.*, Na^+, H^+, NH_4^+, Mg^{2+} or Ca^{2+}) can not trigger such response behaviors. Effects of chemical compositions of microspheres on the K^+-induced shrinking and aggregating behaviors are investigated systematically. The K^+-induced aggregating sensitivity of the P(NIPAM-*co*-AAB15C5) microspheres can be enhanced by increasing the content of crown ether units in the polymeric networks; however, it is nearly not influenced by varying the monomer and crosslinker concentrations in the microsphere preparation. State diagrams of the dispersed-to-aggregated transformation of P(NIPAM-*co*-AAB15C5) microspheres in aqueous solutions as a function of temperature and K^+ concentration are constructed, which provide valuable information for tuning the dispersed/aggregated states of microspheres by varying environmental K^+ concentration and temperature. The microspheres with synchronously K^+-induced shrinking and aggregating properties proposed in this study provide a brand-new model for designing novel targeted drug delivery systems.

434379 Avidity-Driven Targeting of a Versatile Nanoscale Carrier Engineered for High Payload and Extended Release of Anticancer Drugs

Abstract:

Avidity-Driven Targeting of a Versatile Nanoscale Carrier Engineered for High Payload and Extended Release of Anticancer Drugs

Katherine A. Windham[1, 2], Ricky J. Whitener[1, 2, 3], Jacek Wower[1, 3], Mark E. Byrne[2, 3, 4]

[1] RNA Biochemistry Laboratories, Department of Animal and Dairy Sciences, College of Agriculture

[2] Biomimetic & Biohybrid Materials, Biomedical Devices, and Drug Delivery Laboratories, Department of Chemical Engineering, Auburn University

[3] US Dept of Education GAANN Graduate Fellowship Program in Biological & Pharmaceutical Engineering

[4] Biomimetic & Biohybrid Materials, Biomedical, Drug Delivery Laboratories, Department of Biomedical Engineering, Rowan University

Our work uses a nanoparticle consisting of a gold core coated with DNA strands creating a flexible and versatile nanocarrier with detection, drug carrying, and specific recognition and targeting capabilities driven to bind to cells by an avidity process. Gold nanoparticles (AuNps) were selected because of a large of number properties advantageous to our DNA based targeting and drug delivery approach. They are easy to synthesize, modify, and known to not elicit an immune response. We combine the properties of AuNps with recently developed small-cell lung carcinoma (SCLC) specific DNA aptamers in order to target and treat SCLC tumors. This allows us to construct a nanocarrier with the ability to modulate quantity of drug and specificity towards targeting of SCLC cells while resulting in an avidity driven process rather than affinity.

The first step is to functionalize 15-nm gold nanoparticles with single stranded DNA strands (anchor DNAs) modified with a thiol group on the 5' end. Slight deviations in nanoparticle size can affect their color due to surface plasmon resonance, an advantage of AuNps that can be used to determine successful functionalization of the particles. Previous work in our lab has shown that we can bind up to 101 +/- 8 of our anchor DNAs to a 15-nm AuNp using optimized experimental procedures. Due to the size of our functionalized AuNps we can target SCLC cells by a passive targeting approach. Additionally, our nanocarrier possesses an active targeting mechanism with the inclusion of SCLC-cell-specific DNA aptamers found in the literature. We have successfully engineering these aptamers to hybridize with our anchor DNAs.

Using known optimal triplet base pairs for binding of doxorubicin, one of the best drugs for treating SCLC, we can control the location and quantity of drug bound to our nanocarrier. This is done by creating a double stranded DNA region containing a series of the triplet base pair sequences, the length of which determines how many drugs can bind. Since doxorubicin is a DNA intercalator, it needs a double stranded region to bind to our nanocarrier. This property allows us to control the location of the drug on our nanoparticle platform. By extending this drug carrying region onto our DNA aptamer strands, we can introduce this ability to our nanocarrier without the need for additional components. We have been able to attach a maximum of ~1,100 molecules of drug to a 15-nm gold nanoparticle with our current design, which is much greater than the average drug loading of any 15-nm nanoparticle found in the literature.

The flexibility and versatility of this platform is evident in each of the components. The nanoparticle, the anchor strand sequence, the doxorubicin binding sequences, the aptamer sequences, and even the drug can each be individually modified depending on a patient's medical status and the heterogeneity of the SCLC tumor. By modifying the nucleic acid strands which bind the drug, we can regulate both the drug binding and control release as needed. Extending the sequence of the double stranded DNA segment of the nanocarrier increases the drug payload. Furthermore, the nanocarrier can be programmed to bind specifically to targeted cells by using either a single type of aptamer or an array of SCLC-cell-specific aptamers.

Currently, we are investigating regulation of the specificity and binding constant of our nanocarrier by modulating use of the aptamers. By attaching multiple DNA aptamers, the resulting nanoparticle is expected to act as a "molecular octopus" increasing the binding constant of our platform by avidity, and is able to target different markers on the surface of many SCLC cell lines. This means that once the SCLC cell lines present in a tumor are identified, our nanocarrier can specifically target that particular tumor as opposed to general small-cell lung cancer cases. Furthermore, the exact amount of drug needed to kill the tumor can be delivered to a patient, rather than targeting the cancer with an excess of drugs. The use of biohybrid strategies in the creation of novel controlled and targeted drug release carriers has significant potential to affect a number of cancer treatment regimes. We have synthesized, characterized, and optimized a novel gold nanoparticle platform that utilizes an avidity driven process capable of finely tuning the size of the carrier, quantity of drug, binding constant, and specificity towards particular cell lines, created as an application for a personalized treatment regime to individual cancer patients.

429777 A Diffusion Oriented NMR Investigation into the Mechanism of Phase Transfer in the Brust-Schiffin Synthesis of Alkanethiol Nanoparticles

Abstract:

The Brust-Schiffin synthesis is a widely used procedure for the production of precursor gold nanoparticles that can be further functionalized for use as sensor devices, catalytic material, or drug delivery systems. The synthesis requires the phase transfer of gold ions from an aqueous phase to an organic phase. The phase transfer is facilitated by a quaternary ammonium phase transfer catalyst, tetraoctylammonium bromide (TOABr). However, the final state of the gold ion in the organic phase is not well understood. There is an ongoing debate in the literature over whether the phase transfer results in the formation of reverse micelles or in a molecular complex.

Other research groups have suggested that reverse micelles are formed because the nuclear magnetic resonance proton shift of water deviates from that of water in an organic solvent,[1] while the formation of a molecular complex is supported by conductivity measurements and small angle x-ray measurements.[2] We utilized Diffusion Oriented Nuclear Magnetic Resonance Spectroscopy (DOSY-NMR) to quantify the hydrodynamic radius of TOABr during the phase transfer. DOSY-NMR applies a gradient magnetic pulse to spatially map the location of the chemical species through their Larmor frequency and quantifies the diffusion of the chemicals by their signal intensity attenuation following a diffusion delay. The translational self-diffusion coefficients were used to determine the radius of TOABr through a modified Stokes-Einstein relationship. DOSY-NMR couples the accuracy of Small Angle Neutron Scattering (SANS) with the accessibility of Dynamic Light Scattering (DLS) to nondestructively probe the size of chemicals at the Angstrom-scale.

Our results suggest that there is no critical micelle concentration between 2mM and 400mM of TOABr in chloroform saturated with water. Typical syntheses utilize 200 mM of TOABr. A critical micelle concentration would be evident with an increase in the hydrodynamic radius of the phase transfer catalyst. Instead, DOSY-NMR results describe the formation of a molecular complex. Our study suggests that subsequent reduction of the gold ion in the Brust-Schiffin occurs in the organic phase, and not in an aqueous phase within reverse micelles. Additionally, the study indicates that TOABr not only phase transfers gold into the organic phase, but also would necessitate that the borohydride ion be transferred into the organic phase by TOABr. Further insight into the mechanism of phase transfer and particle formation will improve the ability to synthesize monodisperse particles and ultimately lead to facile methodologies of size control.

1. Li, Y., Zaluzhna, O. & Tong, Y. J. Critical Role of Water and the Structure of Inverse Micelles in the Brust–Schiffrin Synthesis of Metal Nanoparticles. *Langmuir* **27,**7366–7370 (2011).

2. Perala, S. R. K. & Kumar, S. On the Mechanism of Phase Transfer Catalysis in Brust–Schiffrin Synthesis of Metal Nanoparticles. *Langmuir* **29,** 14756–14762 (2013).

(643f) Fabrication of Sub-Cell Size "Spiky" Nanoparticles and Their Interfaces with Biological Cells

Xie, X.,
Melosh, N., Stanford University

Originally presented on: 11/12/2015 10:05:00 - 10:24:00

Abstracts:

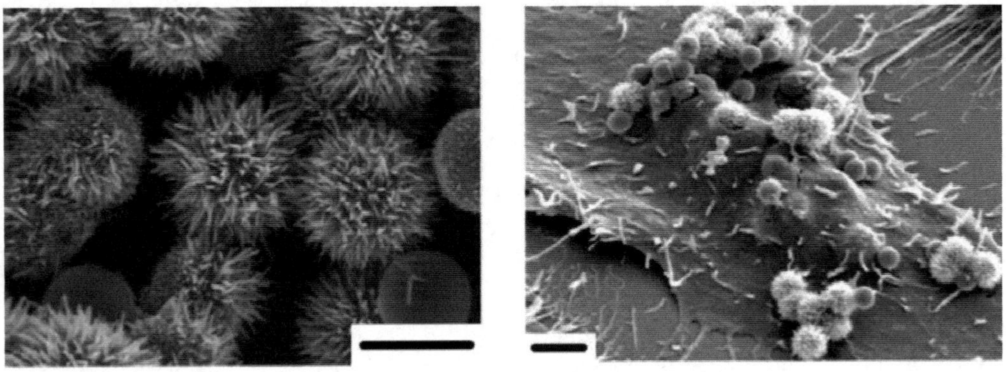

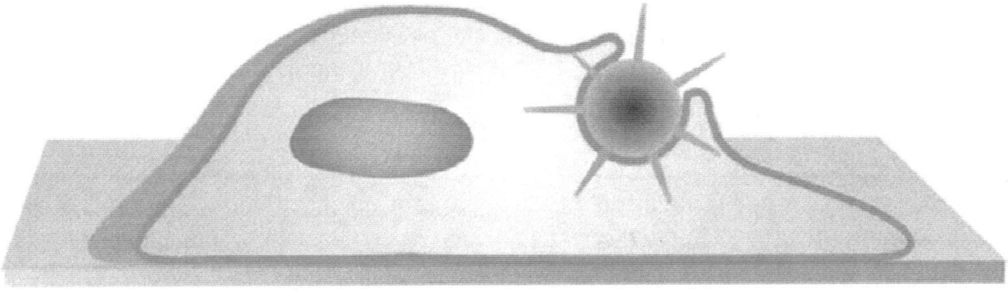

-

430095 Self Assembling Magnetic Nanostructures with Functional Polyolefin

Abstract:

Functional Polyolefin plastic additive, *i.e.*, maleic anhydride grafted polypropylene (MAPP), has been used as polymeric surfactant for synthesizing magnetic nanostructures via a one-pot in-situ bottom up wet chemistry method. The MAPP was found to serve as stabilizer to stabilize and direct the self assembly patterns/morphology of these magnetic nanostructures. Here, one dimensional nanochains including iron, cobalt, and iron-cobalt alloy will be demonstrated. In addition, the macro-assembly patterns including one-, two-, and three dimensional morphologies will also be presented. The structure manipulation mechanism were studied and the results indicated that it is primarily controlled by the surfactant tail length, capping head density, surfactant-reactant mole ratio, surfactant molecular weight, and surfactant concentration. The force balance by magnetic attraction forces (from magnetic dipolar-dipolar forces) against the repulsive forces (from the steric hindrance of surfactant long chains) were also studied and discussed in detail. The crystalline structures of iron and cobalt nanostructures were also found to be controllable by MAPP with different reaction parameters. The magnetic properties of the synthesized nanostructures will be presented and the difference will also be discussed in detail.

434991 Fabrication and Characterization of Quantum Dot-Loaded Polymeric Nanocarriers

Abstract:

The use of nanoparticles in drug delivery can address many of the difficulties faced by traditional drug delivery methods such as systemic toxicity, transport limitations, and drug solubility. By loading drugs into biocompatible polymeric nanoparticles, drugs that were previously not usable due to their solubility or toxicity can now be used. Because the properties of the polymeric carriers can be manipulated, it is possible now to improve circulatory residence time, targeting, and drug-release depending upon application needs. However, many questions about on nanoparticle biodistribution and accumulation in tissues still remain and so a method for tracking these particles is necessary. Fluorescent imaging agents are currently of high interest in biological application studies due to their high contrast and resolution potential. Quantum dots (QDs) such as CdSe nanoparticles coated with ZnS are particularly interesting due to their high quantum yield and photostability which allows for minimal use of materials while achieving good detectability. However, the presence of heavy metals in QDs pose a potential toxicity problem. Hydrophobic modification of quantum dots can be used to trap them in a hydrophobic polymer core and therefore mitigate the concern for high toxicity effects while making use of their fluorescence for particle tracking and biodistribution studies.

We report the fabrication of colloidally-stable core-shell nanoparticles comprised of hydrophobically modified CdSe/ZnS quantum dots and the poly(ethylene glycol)-*b*-poly(D,L-lactide) block copolymer [mPEG(5 kDa)-*b*-PDLLA(10 kDa)]. The CdSe/ZnS quantum dots with diameter ~ 5 nm were obtained from a commercial source. The block copolymer was synthesized by ring-opening polymerization of D,L-lactide in the presence of mPEG using stannous octoate as the catalyst. The nanoparticles were prepared by Flash Nanoprecipitation which takes advantage of low solubilities of the hydrophobically modified QDs to encapsulate them in the hydrophobic PDLLA cores of the nanoparticles. The method's scalability and efficacy in producing well-defined nanoparticles makes it a particularly attractive approach.

Nanoparticles were fabricated with hydrodynamic diameters in the range of 80-100 nm and polydispersity indices less than 0.2 as measured by dynamic light scattering. The zeta potentials of the particles were typically -20 mV in phosphate buffered saline diluted 100X. Fluorescence studies show that the nanoparticles containing the quantum dots could be excited at wavelengths of 480 nm and below and had an emission peak at 530 nm. These nanoparticles were colloidally stable in cell-growth media supplemented with fetal bovine serum (FBS) for periods up to 12 hours. Cell culture studies with macrophage cells show that they were not immunogenic. ELISA tests showed that the nanoparticles induced levels of the inflammatory cytokines interleukin-6 (IL-6) and tumor necrosis factor alpha (TNFa) in both M1 (pro-inflammatory) and M2 (anti-inflammatory) macrophage cells similar to those of mock negative control experiments. Further results will be presented on how these nanoparticles are taken up by macrophages.

437519 Quantifying Magnetic Nanoparticles Thermal Signatures at High Precision By Lock-in Thermography

Abstract:

Superparamagnetic iron oxide nanoparticles are the topic of a broad research activity in the biomedical domain because of their unique properties. One of the most investigated one is their ability to generate heat in response to an alternating magnetic field. Traditionally, the amount of heat generated by particles is measured by means of fiber optics. However, the application of this method is not optimal, and suffers from severe limitations. In this work, we present a fundamentally different approach based on lock-in thermography to probe the thermal emissions of stimuli-responsive nanoparticles. Highly monodisperse iron oxide nanoparticles, prepared from thermal decomposition of iron oleate at high temperature have been prepared in three different sizes. After ligand exchange to transfer them in water, these model nanoparticles have been exposed to an alternating magnetic field and the heat generated was measured by means of lock-in tomography. Quantitative and qualitative information at high thermal resolutions could be obtained. The amount of heat generated was modeled mathematically using Rosensweig linear response theory. Compared to conventional techniques, this approach offers a fast, sensitive and non-invasive alternative to investigate multiple and dilute specimens simultaneously, which is essential in optimizing and speeding up screening procedures for various fields of application.

418103 Enhancing Efficacy of Radiation Therapy through ROS Generation Catalyzed By Peptide-Conjugated Iron Oxide Nanoparticles

Abstract:

Non-small cell lung cancer (NSCLC) is one of the leading causes of cancer deaths in the United States. Radiation therapy is often used to treat NSCLC but can potentially result in serious side effects. To increase the effectiveness of radiation, iron oxide nanoparticles (IONPs) can be utilized for their ability to produce reactive oxygen species (ROS). Cancer cells are more susceptible to oxidative insults compared to normal cells due to fast cell proliferation and metabolism. Additional ROS stress induced by exogenous agents can overwhelm the relatively low antioxidant capacity and disrupt the redox homeostasis inside cancer cells leading to selective tumor cell toxicity. IONPs increase ROS production within cancer cells by catalyzing the Haber-Weiss reaction which generates the highly reactive hydroxyl radical through Fenton chemistry. The Fenton reaction converts hydrogen peroxide to the hydroxyl radical via a reaction with iron ions. Previous work has shown that the surface of IONPs are approximately 50 times more efficient at catalyzing the Fenton reaction than free iron ions. Radiation therapy promotes mitochondrial respiration which is often linked to higher cellular energy production. Increased respiration leads to an increase in mitochondrial production of the superoxide anion which is converted to hydrogen peroxide by superoxide dismutase. Iron oxide nanoparticles can then catalyze the reaction from hydrogen peroxide to the highly reactive hydroxyl radical. However, a significant limitation to this process is that IONPs are encapsulated within endosomes/lysosomes after internalization by a cell. The hydroxyl radicals should be generated outside of the lysosomes to increase the probability of hydroxyl radical interaction with important organelles such as the mitochondria and the nucleus. Therefore, the aim of this project was to synthesize peptide conjugated IONPs that are able to escape the lysosomes after internalization by cancer cells. The peptide of interest, TAT, has been previously shown to facilitate nanoparticle escape from endosomes likely due to destabilization of the endosome membrane by its highly positive charge. A549 lung adenocarcinoma were exposed to TAT-conjugated IONPs, and confocal and electron microscopy were used to confirm the ability of TAT-IONPs to escape endosomes/lysosomes. Furthermore, the enhancement of ROS generation via the nuclear localizing IONPs was assessed using carboxy-DCF. Viability studies were also completed for A549 cells exposed to nuclear localizing IONPs with and without radiation treatments. Additionally, the oxygen consumption rate (OCR) was monitored for A549 cells after exposure to nuclear localizing IONPs with and without radiation using the Seahorse XF96 instrument. IONPs not conjugated with the nuclear localizing signal were used as controls for these studies. When combined with radiation, TAT-IONPs have the ability to increase ROS generation to higher levels than observed with radiation alone. When radiation treatment was combined with nuclear localizing IONPs, a synergistic cancer treatment resulted, meaning that the decrease in cell viability is greater than the two individual treatments. This enhancement was not observed for the IONP systems not conjugated with the nuclear localizing signal. The presence of the TAT-IONPs also altered the OCR of A549 lung carcinoma with and without radiation treatment.

418615 Enhanced Microfluidic Immunomagnetic Separation Based on Microfabricated Patterns from Ferromagnetic Nanoparticles

Abstract:

Isolation of specific cells from complex biological samples has been of growing interest in cellular studies, disease and medicine researches. Immunomagnetic separation, which is the use of antibody functionalized magnetic beads to bind specific proteins on target cells and separate cells by external magnetic field, is a widely used method for their simple operation and high isolation specificity. Microfluidics, which allows tiny quantities of liquids to be controlled and manipulated, provides a powerful platform for sorting of small amount of cells from heterogeneous cell mixtures, especially when the sample is rare and precious. However, low magnetic force generated by limited magnetic field gradient of external magnet has been a big obstacle for enhancement of current cell sorters. Low flow rates were required for continuous microfluidic separation, which brought the problem of low throughput and non-specific cell absorption (e.g. low purity of isolated cells). Embedding a pattern of magnetic material on the bottom of a microfluidic channel is way to concentrate field lines from an external magnetic field source. However, current methods involve complicated fabrications that require advanced facilities, trained personal and high-cost materials. In our study, we proposed a simple microfluidic device containing ferromagnetic structure for cell capture. The ferromagnetic pattern was formed from magnetic nanoparticle suspension. The fabrication needs basic soft lithography instruments, cost-effective materials and simple operation only. The dimension of the pattern is controllable and reproducible. We tested the ability of our microfluidic channels in capturing magnetic beads coated RAW 264.7 cells, and found that ferromagnetic embedded channel performed up to over 4 times higher capture rate compared to no patterned channel. Mathematical models were built to simulate the magnetic field distribution inside channels, which matched the experimental data very well. We envision that our device will be useful for sorting primary cell samples from scarce sources (e.g. animals, human).

425270 HSP70 Inhibition Synergistically Enhances the Effects of Magnetic Fluid Hyperthermia

Abstract:

Introduction: Ovarian cancer is a deadly gynecological disease. Hyperthermia has been investigated as a potential treatment for cancerous tumors. However, its localized application still remains a challenge. Magnetic fluid hyperthermia (MFH) may be an alternative to surpass such challenge. Several cellular and molecular aspects of MFH have been previously elucidated, but there is still a need to further understand the implications of MFH at the cellular level and how this information could be employed to further enhance the effects of MFH. Therefore, this work focused on the examination of gene expression after MFH treatment and using such information to find target genes that when silenced or inhibited could produce an enhanced therapeutic outcome after MFH.

Materials and Methods: Ovarian cancer cell lines HeyA8, A2780 cp20 and SKOV3 were employed. Microarray analysis was performed in HeyA8 cells exposed to MFH for 30 min at 43°C. From the microarray results, PCR was performed to confirm HSPA6 and HSPA7 expression. HSPA6 gene was knocked down using Lipofectamine RNAiMax to observed MFH outcome. Cells were transfected with HSPA6 and control siRNA. MFH was applied after transfection and cells were counted after 48 hours. 2-phenylethynesulfonamide (PES), a HSP70 inhibitor, was studied with MFH. PES cytotoxicity was evaluated in 96 wells using the EUZ4 assay. For MFH, cells were incubated with the drug 2.5 hours prior to MFH for 30 min at 41 or 43 °C. Cells were tallied after 48 hours. Combination index was calculated with the model of Chou-Talay with the CompuSyn software(Chou 2006) . A subcutaneous HeyA8 tumor model was generated in athymic nude mice. HSPA6 siRNA and control siRNA were injected and mice were exposed to MFH for 30 min at a magnetic field of 62 kA/m.

Results and Discussion: Analysis revealed that heat shock protein genes, including HSPA6, were upregulated when MFH was applied at 43°C for 30 min. HSPA6 encodes the heat shock protein HSP70. HSP70 gene expression was confirmed by PCR in HeyA8 and A2780cp20 cell lines at 39, 41 and 43°C. The highest expression was observed at 1 hour after treatment. Results indicated that the effect of HSP70 inhibition in ovarian cancer cell lines during MFH could be a potential target for combined therapy. First, si-RNA inhibition of HSPA6 gene was achieved in vitro and a decrease in cell viability was observed when cells were exposed to MFH. An alternative avenue of inhibition was examined by using a small HSP70 inhibitor known as PES. An improvement in MFH efficacy was observed in various cell lines (A2780cp20, SKOV3 and HeyA8). Combination index was calculated reporting a synergistic effect. In vivo efficacy experiments were also performed in Nu/Nu mice with a HeyA8 subcutaneous tumor model. A reduction in tumor growth rate was observed in treatments with MFH and HSPA6 siRNA delivery. HSP70 inhibition appeared to be a promising target to enhance MFH therapeutic outcome.

Conclusions: HSP70 genes are upregulated when MFH is applied to ovarian cancer cells. HSP70 plays an important role in the cell and it is induced when cells are exposed to stress. HSP70 inhibition at the gene and protein level improved MFH treatment. PES showed a moderate synergistic effect with ovarian cancer cell lines. HSP70 inhibition is a promising target to improve MFH.

427241 Optimized High-Energy Dissipating Nanoparticles for Magnetic Hyperthermia in Ovarian Cancer Cells

Abstract:

Magnetic Fluid Hyperthermia (MFH) uses heat generated by magnetic nanoparticles exposed to alternating magnetic fields to cause a temperature increase in tumors to the hyperthermia range, inducing programmed cancer cell death. The development of nanoparticle synthesis strategies that maximize the rate of energy dissipation are required to reduce the amount of nanoparticles needed at the target tissue though enhanced particle uptake into cancer cells is also a requirement. To date, the aqueous co-precipitation synthesis method has only been reported to yield sufficiently high specific absorption rates (SAR) after laborious size selective fractionation. This work reports high-energy dissipating nanoparticles obtained through optimized synthesis and peptization methods. The internalization patterns of these particles in three ovarian cancer cell lines and preliminary magnetic hyperthermia results are also presented in this work. Nanoparticles were synthesized using the aqueous co-precipitation of Fe^{2+} and Fe^{3+} inorganic salts at 80-90°C and 0.08-1.0M (Fe^{2+} + Fe^{3+}). Peptization was carried out using tetramethylammonium hydroxide under low and high-energy sonication. Characterization of particles included transmission electron microscopy, dynamic light scattering, UV-Vis spectroscopy, equilibrium and dynamic magnetic measurements, and evaluation of energy dissipation rates in alternating magnetic fields. Oleic acid was adsorbed onto the surface of particles at 80°C and then ligand exchanged with PEG-Silane for 72 hours at room temperature. PEG-coated particles were washed and air-dried, and the iron oxide percentage was determined by thermo-gravimetric analysis. Nanoparticle internalization was carried out using A2780 IP-1, HeyA8 and SKOV3 SC2 ovarian cancer cell adhered to the bottom of 6-well plates, at iron oxide concentration of 0.6 mg/mL and exposure times up to 3 hours. Cells were manually counted and the iron uptake was quantified by UV-Vis spectroscopy. Cells were exposed to magnetic hyperthermia (Hf = 4.90 x 10^3 $A{\cdot}m^{-1}{\cdot}s^{-1}$) for 30 minutes at 43°C and 45°C, incubated for 48 hours at 37°C, and assayed with CellTiter Blue® for cell viability. Confocal microscopy of Alexa® 488 labeled particles was employed to investigate internalization patterns. SAR values up to 1,048 W/g_{Fe} (liquid) and 719 W/g_{Fe} (solid) were obtained under optimal conditions. High energy ultrasound applied during peptization improved particle dispersion, increasing the energy dissipation rates. Iron uptake up to 3.4 ρg/cell was quantified after 3-hours exposure, and this internalization was cell dependent. Cell viabilities of 15-40% were obtained after MFH at 43°C, whereas only 3% viability was observed at 45°C. Nanoparticle internalization in SKOV3 cells was successfully visualized through confocal microscopy, showing accumulation of particles inside the cells. We concluded that synthesis optimization yielded high energy dissipation rates, using a simple, cost-effective and scalable method. Nanoparticles were internalized in ovarian cancer cells and they were observed in the cell's cytoplasm as small clusters most likely inside lysosomes. Cell viability reduction occurred after MFH treatment for 30 minutes. Experiments are underway to increase the *in vitro* nanoparticle internalization using various methods.

418905 Synthesis and Characterization of Environmentally-Benign Silver Ion/Lignin Antimicrobial Nanoparticles

Abstract:

The introduction of antimicrobial silver nanoparticles (AgNPs) in industrial applications may lead to persistent and potentially toxic nanowaste in the environment. We will present a new class of antimicrobial environmentally-benign nanoparticles (EbNPs) designed with green chemistry principles, which can serve as highly efficient microbicidal substitutes of the AgNPs. The EbNP core is made of biodegradable lignin, and is infused with an optimal amount of silver in the form of adsorbed Ag^+ ions. We show that Ag^+ ion functionalized EbNPs exhibit higher antimicrobial activity towards Gram-negative human pathogens *Escherichia coli* and *Pseudomonas aeruginosa* in direct comparison with silver nanoparticles and silver nitrate solution. The EbNps are also effective against quaternary ammonium resistant *Ralstonia* species. The enhanced antimicrobial action is due to highly biocidal Ag^+ ions released from the EbNP adhered to the cell. The functionalized EbNPs exhibit broad spectrum biocide action, while using 10× less silver when compared with conventional AgNPs and $AgNO_3$ aqueous solution. These particles are fabricated by environmentally friendly water-based acid precipitation methods. We obtained EbNPs with diameters ranging from 40 to 200 nm and with pH stability ranging from pH 4.0 to 9.0. High-throughput bioactivity screening using mammalian cell and zebra fish embryo assays performed in collaboration with the U.S. Environmental Protection Agency (US EPA) did not reveal increased toxicity of the EbNPs, when compared to equivalent amount of AgNPs or $AgNO_3$ solution. The EbNP synthesis illustrates how green chemistry principles including atom economy, use of renewable feedstocks, and design for degradation can be applied to design more sustainable nanomaterials with increased functionality and decreased environmental footprint. This abstract does not necessarily reflect US EPA policy.

Potential Analysis for Biosynthesis of Copper Nanoparticles Using Fungus *Hypocrea Lixii* Cell Extract

Mariana Marangoni, Mariana de P. Eduardo, Meriellen Dias, Maria Anita Mendes, Cláuio A. Oller do Nascimento - Polytechnic School of USP - Brazil

AGENDA

- Introduction and Context
- Precedents
- Materials and Methods
- Results and Discussion
- Conclusions
- Acknowledgments

AGENDA

Introduction and Context

Precedents

Materials and Methods

Results and Discussion

Conclusions

Acknowledgments

AIChE 2015 - Salt Lake City, UT

Nanoparticles production processes: what are we looking for?

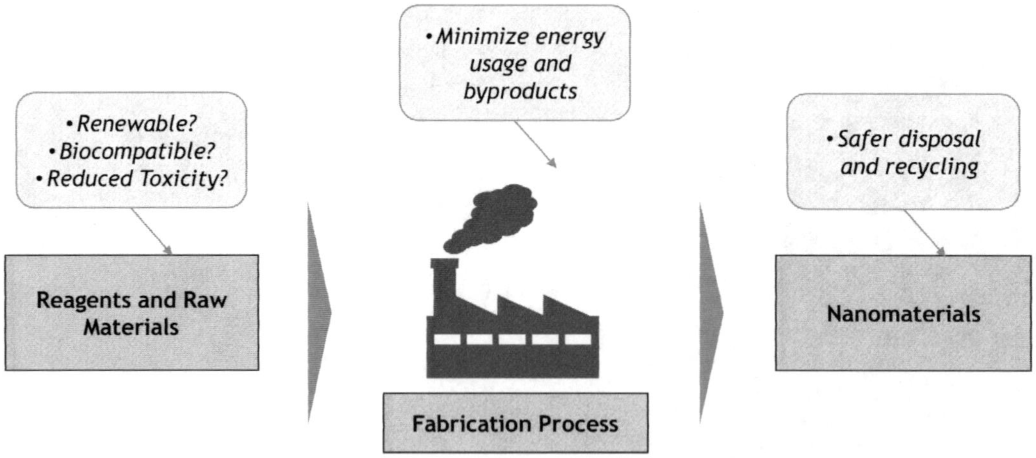

AIChE 2015 - Salt Lake City, UT

Introduction and Context

Why copper nanoparticles?

- Nanomaterials have innumerous applications of high economical interest. **In this context, metal nanoparticles are in the spotlight**. (Zhang et al., 2010)
- There is a **high growth perspective in the market** of metallic oxides nanoparticles in the near future. (Research and Markets, 2013)
- Copper nanoparticles can be used in **sensors, catalytic processes, solar cells, wood preservatives, etc**. (Evans et al., 2008; Kathiresan et al., 2009; Norashikin et al. 2011)
- The current average value of copper nanoparticles is roughly **100x higher** than metal plates or ingots.

Introduction and Context

Why fungal biomass?

- Possibility to use **dead biomass** and **existing processes' wastes as raw material**
- **Easy to handle and scale-up** (Varshney; Bhadauria; Gaur, 2012a)
- **Bioprocesses are ususally able to synthesize smaller particles with higher purity** (Ramanathan et al., 2010; Zhang et al., 2010)
- **The use of microorganisms is an alternative to traditional processes which involve high energy demands and/or toxic solvents**

AGENDA

- Introduction and Context
- Precedents
- Materials and Methods
- Results and Discussion
- Conclusions
- Acknowledgments

Precedents in Scientific Literature

Biosynthesis of copper nanoparticles using *Hypocrea lixii*:

Hypocrea lixii
=
Trichoderma harzianum

OPEN ACCESS Freely available online

PLOS | ONE

Biosynthesis and Uptake of Copper ... cles by Dead Biomass of *Hypocrea lixii* Isolated from the Metal Mine in the Brazilian Amazon Region

Marcia R. Salvadori[1]*, Luiz F. Lepre[2], Rômulo A. Ando[2], Cláudio A. Oller do Nascimento[3], Benedito Corrêa[1]

1 Departamento de Microbiologia, Instituto de Ciências Biomédicas II, Universidade de São Paulo, São Paulo, São Paulo, Brazil, 2 Departamento de Química Fundamental, Instituto de Química, Universidade de São Paulo, São Paulo, São Paulo, Brazil, 3 Departamento de Engenharia Química, Politécnica, Universidade de São Paulo, São Paulo, São Paulo, Brazil

Synthesis with *H. lixii* cells

Transmission Electron Microscopy: dead cells control samples

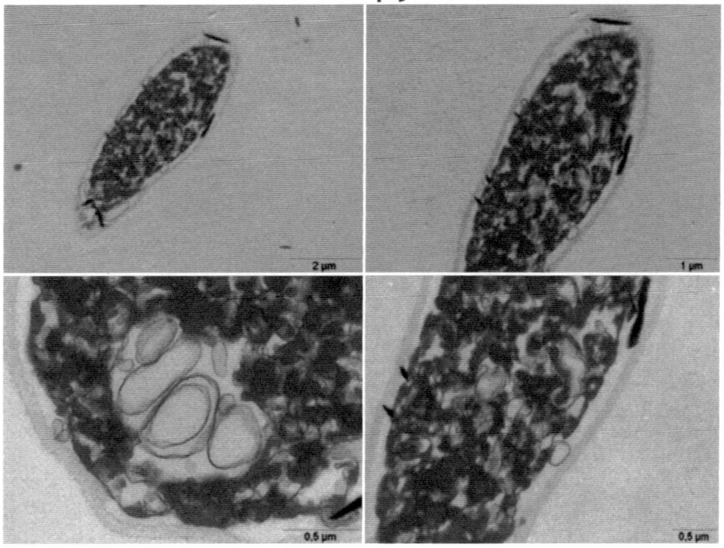

Synthesis with *H. lixii* cells

Transmission Electron Microscopy: dead cells after synthesis

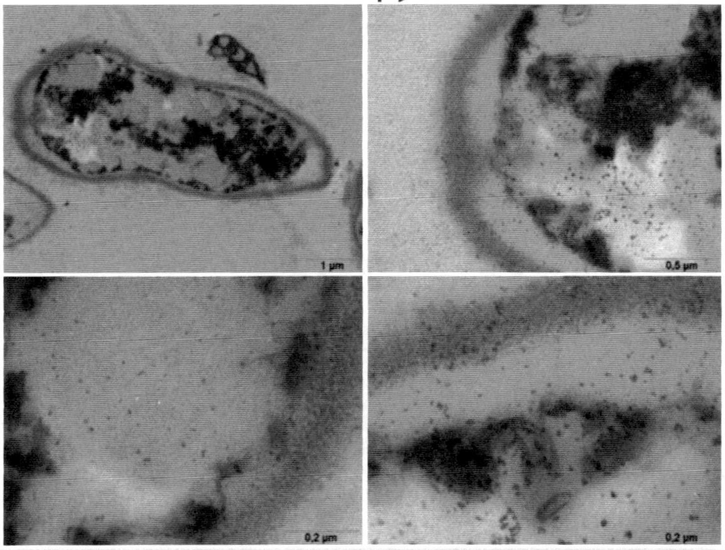

Synthesis with *H. lixii* cells

Transmission Electron Microscopy: live cells control samples

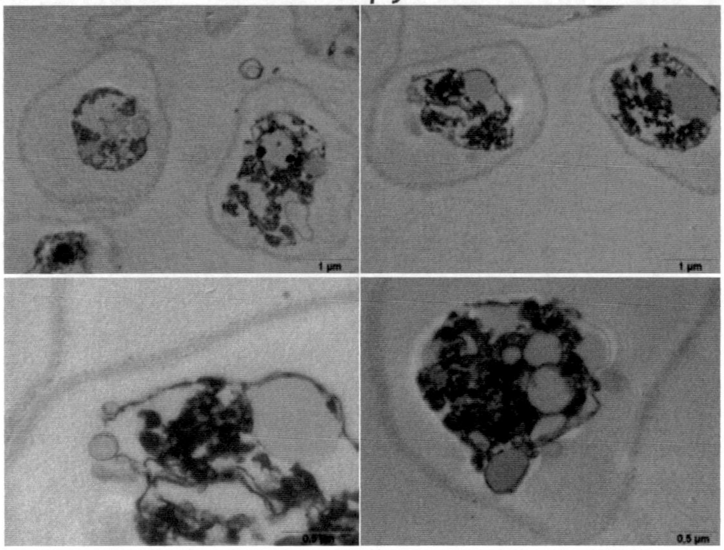

AIChE 2015 - Salt Lake City, UT

Synthesis with *H. lixii* cells

Transmission Electron Microscopy: live cells after synthesis

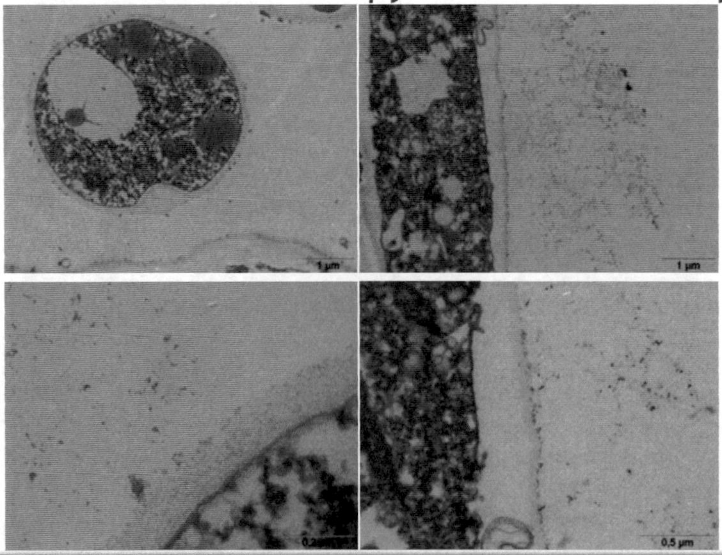

AIChE 2015 - Salt Lake City, UT

Precedents in Scientific Literature

Fungal cell extract to synthesize copper/copper oxide NPs:

Extracellular Biosynthesis of Copper and Copper Oxide Nanoparticles by *Stereum hirsutum*, a Native White-Rot Fungus from Chilean Forests

R. Cuevas,[1,2] N. Durán,[3] M. C. Diez,[2,4] G. R. Tortella,[2,4] and O. Rubilar[2,4]

[1]*Doctoral Program of Science of Natural Resources, Universidad de La Frontera, Chile*
[2]*Scientific and Technological Bioresource Nucleus, Instituto Química, Universidad de La Frontera, Chile*
[3]*Biological Chemistry Laboratory, Instituto Química, Universidade Estadual de Campinas, SP, Brazil*
[4]*Department of Chemical Engineering, Universidad de La Frontera, Chile*

Correspondence should be addressed to O. Rubilar; olga.rubilar@ufrontera.cl

Received 29 July 2014; Revised 16 December 2014; Accepted 30 December 2014

AGENDA

Materials and Methods

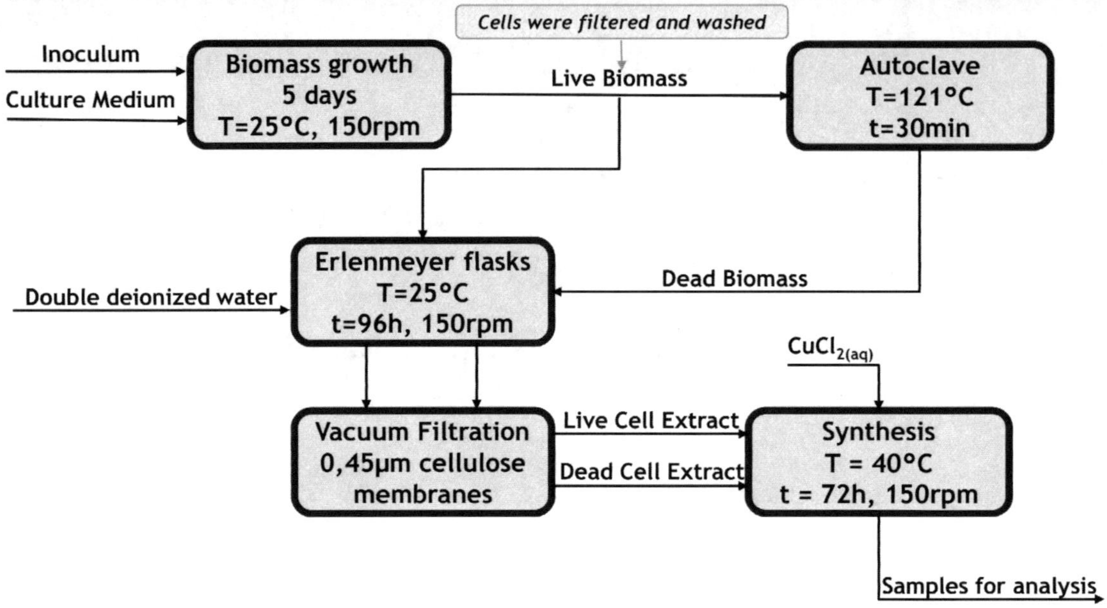

AGENDA

Introduction and Context

Precedents

Materials and Methods

Results and Discussion

Conclusions

Acknowledgments

Results and Discussion

Transmission Electron Microscopy

Live cell extract after synthesis

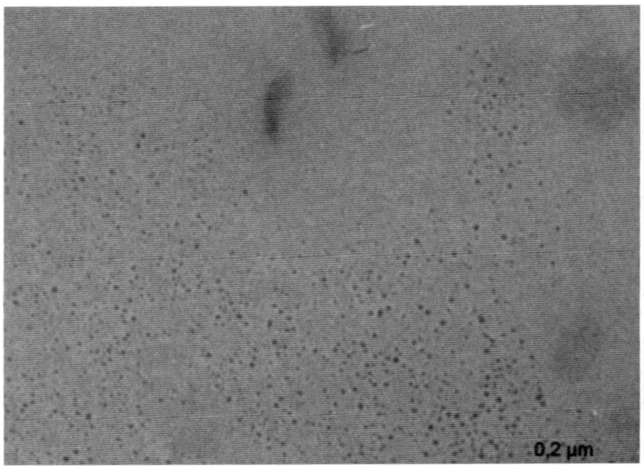

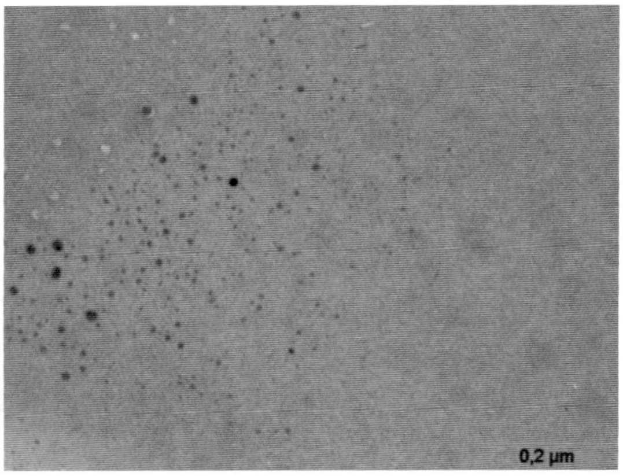

Results and Discussion

Transmission Electron Microscopy

Dead cell extract after synthesis

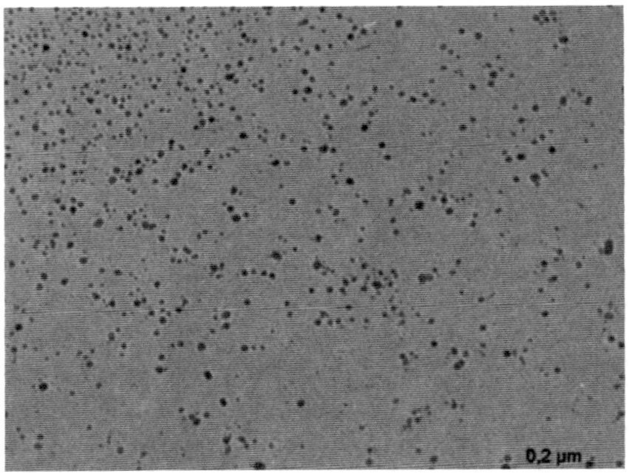

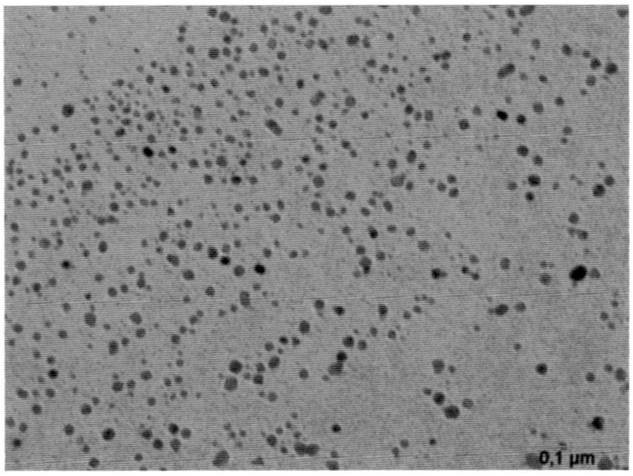

Results and Discussion

Characterization of particles synthesized:

- **Spectrophotometry**: maximum absorption around 570-600nm
- **Infrared Spectroscopy (FT-IR):** indications of changes/shifts in chemical bonds
- **X-ray Diffraction (XRD):** patterns corresponding to crystalline copper/copper oxide
- **Mass spectrometry (MALDI-TOF):** indications of proteins or high molecular weight compounds

AIChE 2015 - Salt Lake City, UT

Results and Discussion

Gas Chromatography - GC-MS (DB1 column)

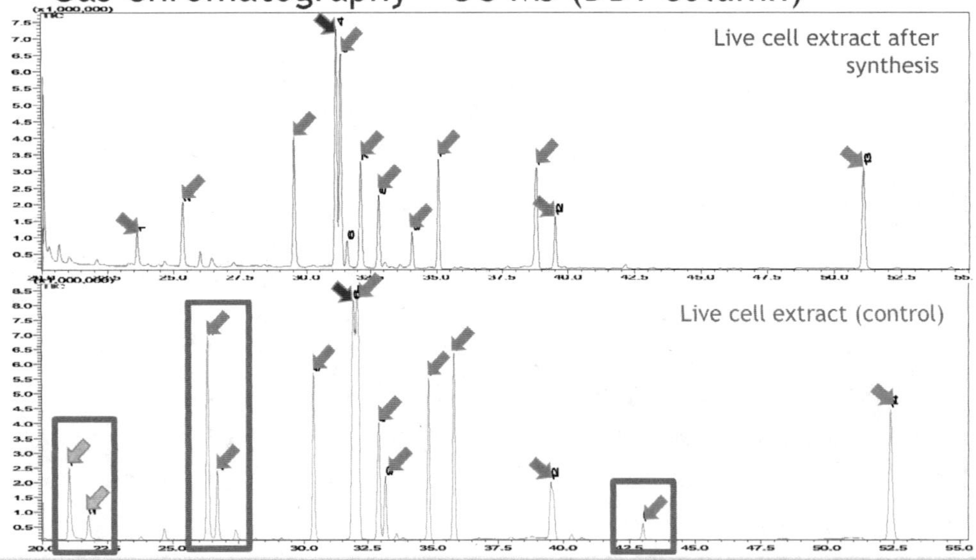

Amino acids

Phosphoric acid

Organic acids

Sugar alcohol

Carbodiimide and amine

AIChE 2015 - Salt Lake City, UT

Results and Discussion

Gas Chromatography – GC-MS (DB1 column)

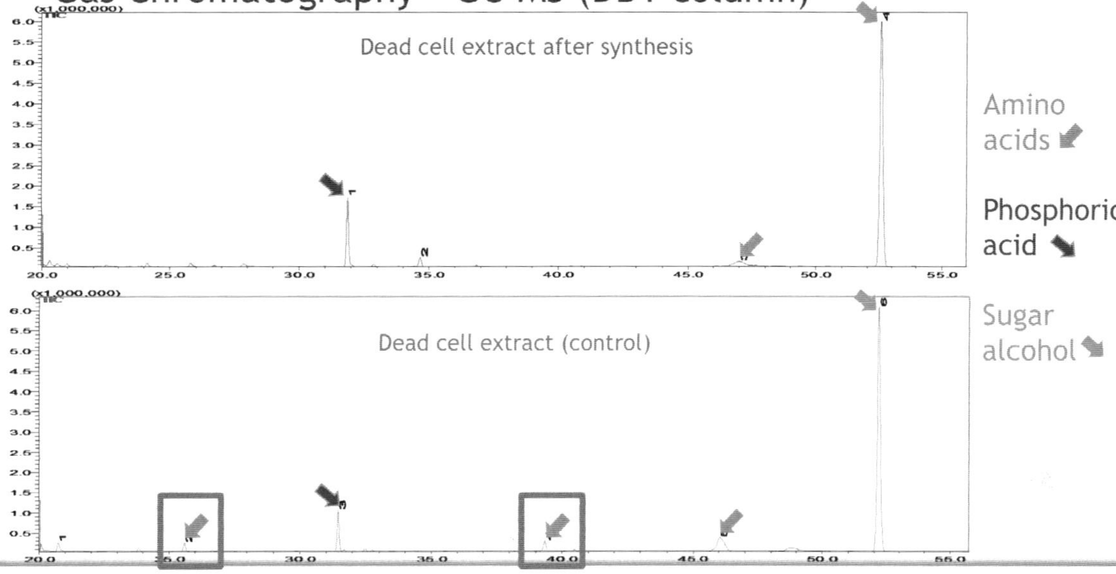

AIChE 2015 - Salt Lake City, UT

Results and Discussion

Gas Chromatography – GC-MS (WAX column)

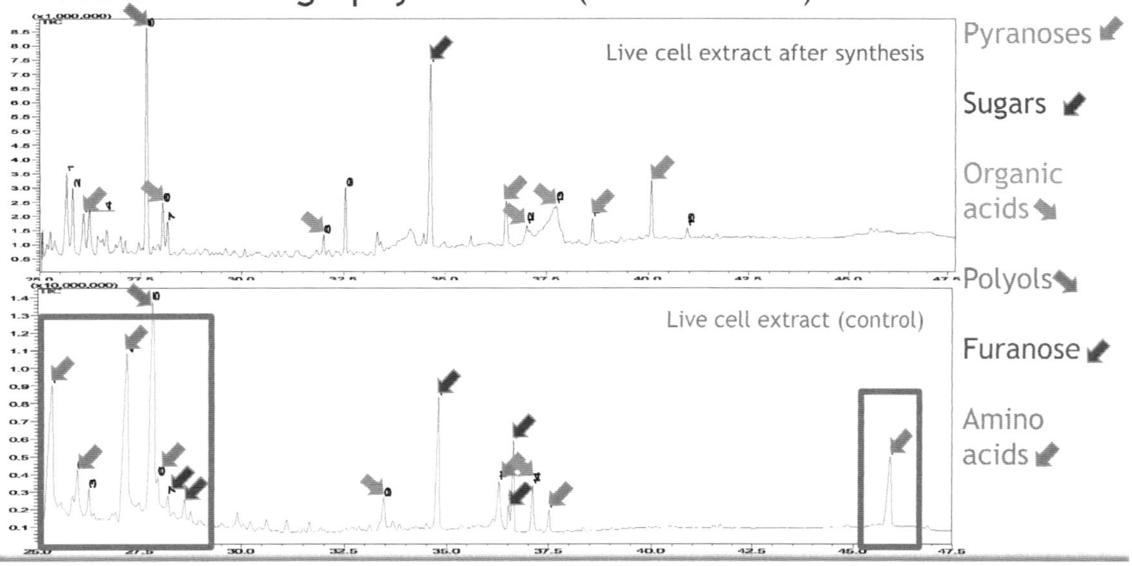

AIChE 2015 - Salt Lake City, UT

Results and Discussion

Gas Chromatography – GC-MS (WAX column)

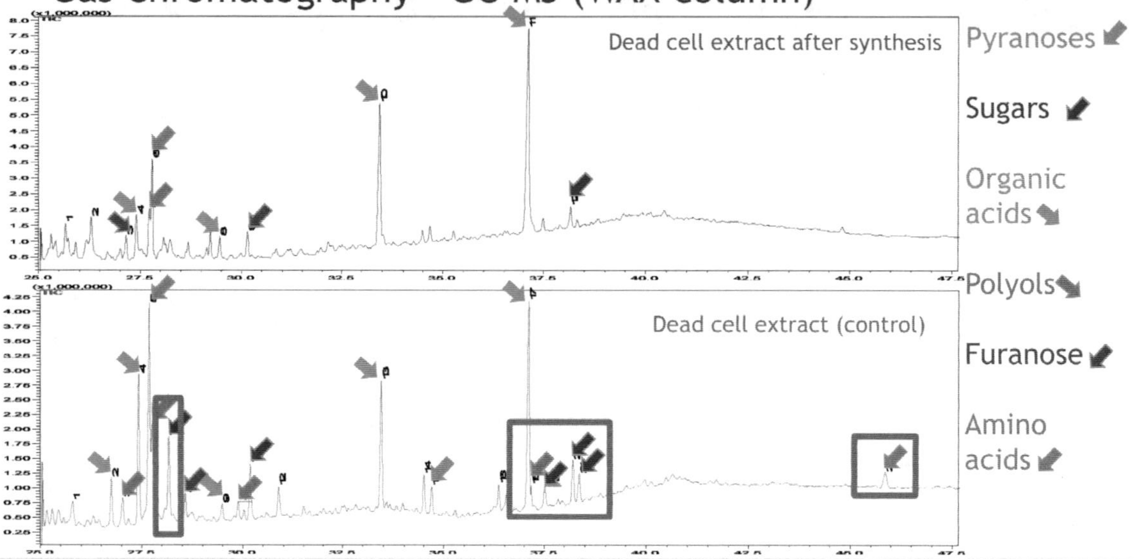

AIChE 2015 - Salt Lake City, UT

AGENDA

Introduction and Context

Precedents in literature

Materials and Methods

Results and Discussion

Conclusions

Acknowledgments

AIChE 2015 - Salt Lake City, UT

Conclusions

1 Composition of cell extracts:

- Main components are sugars, amino-acids and other low molecular weight organic compounds
- Proteins and primary metabolites were not identified

2 Nanoparticles characterization:

- Analyses commonly used in literature where inconclusive for the nanoparticles synthesized

3 Reducing agent and synthesis mechanism:

- The compounds that showed higher interaction with copper were sugars and amino acids

4 Future development:

- It is necessary to further characterize the particles formed
- The next step is to perform quantitative studies to determine the yield

AGENDA

Introduction and Context

Precedents in literature

Materials and Methods

Results and Discussion

Conclusions

Acknowledgments

AIChE 2015 - Salt Lake City, UT

Acknowledgments

FAPESP - São Paulo Research Foundation

BNDES - Brazilian Development Bank

- Dr. Lidiane Andrade

Any Questions?

428331 Life-Cycle-Based Sustainability Assessment and Decision Making for Nanocoating Material Development

Abstract:

The remarkable development on nanocoating materials brings a wide range of potential applications in the automotive, aerospace, and pharmaceutical industries. Despite the obvious technical benefits of nanocoating such as anti-scratch and corrosion prevention, the unintended health and environmental risks as well as the economic and social benefit associated with the use of nanoproducts are not yet fully understood. The proactive and deep understanding of nanocoating materials requires a comprehensive assessment over each stage of its life cycle in order to develop nanocoating systems with improved product performance and reduced impact on environment and society. It becomes urgent to develop systems approaches for comprehensive evaluation of performance of nanocoating products and assurance of sustainability performance over their life cycle.

The life cycle of nanocoating materials consists of the stages ranging from (nano)material selection and processing, through nanopaint/nanocoating manufacturing, to product use and disposal. In this work, a model-based sustainability assessment and decision-making methodology is introduced. It can be used to assess the economic, environmental, and social aspects in every life cycle stage. In addition, the optimal sustainability improvement strategy can be derived at individual stages as well as lifespan. To perform a comprehensive assessment, different sets of sustainability metrics have been identified for use in different life cycle stages. These metrics are analyzed to ensure the consistency of the assessment. The methodology has been used to study the sustainability performance of nanopaint and its application to automotive coatings. A comprehensive case study will highlight critical issues concerning the material's development and nanoparticles emission to the environment and health impact, economic incentive and social satisfaction.

437681 Life Cycle Environmental Implications of Silver Nanoparticle Applications

Abstract:

The benefits of nanomaterials in environmental and medical applications must be judged alongside potential negative impacts from their production or uncontrolled release. In a series of studies, we have examined the life cycle of silver nanoparticles (AgNPs) in consumer products. These studies include comparison of chemical and physical synthesis routes, determination of appropriate functional unit based on antimicrobial activity, leaching during use, and potential ecological impacts of releases. Modeling and experimental work has been assembled into nanomaterial life cycle assessments, used for risk analysis and product design. The relative contribution of AgNPs to life cycle benefits or impacts of consumer products depends on many factors, including method of particle incorporation, silver concentration, material composition, and the role of AgNPs in the function of the product, and was found to range from 10-90% across 15 product types and multiple impact categories. For products used for antimicrobial purposes, results were rescaled based on efficacy using empirical zone of inhibition tests. These analyses can be used to guide AgNP and nano-enabled product design as well as evaluate loading levels of AgNPs in products that indicate either direct or indirect risk to the environment and public health.

(657a) Multifucntional Assemblies of Amphiphilic Nanocrystals

Duan, H.,

Originally presented on: 11/12/2015 8:30:00 - 8:55:00

Metallic nanostructures with localized surface plasmon resonance arising from the collective excitation of conduction electrons have found widespread use in biosensing, bioimaging, surface-enhanced spectroscopy, drug delivery, and catalysis. LSPR spectral shifts induced by interparticle plasmonic coupling have attracted considerable research interest in controlled assembly of plasmonic nanoparticles, which, in conjugation with responsive "smart" coatings, has been exploited to detect a wide range of molecular targets and environmental factors. We have developed versatile approaches to engineer nanoparticle surface with polymer ligands and assemble the nanoparticles into well-defined structures in aqueous dispersion and at oil-water interfaces. This talk will cover synthesis and assembly of the nanoparticles and their applications in traceable drug delivery, electrochemical biosensors, and chemical catalysis.

428065 Self-Assembled Nanoparticle-in-Nanoparticle Metal/Lipid Complex

Abstract:

Metal and metal oxide nanoparticles (NPs) such as gold (Au), silver, iron oxide NPs, quantum dots (QDs) have shown many unique properties including plasmonic effect, magnetism, bring emission, *etc*., which have potential applications for *in vivo* imaging as well as local, on-demand heating. It is desirable to improve their biocompatibility prior to these applications. The experimental results from dynamic light scattering, small angle X-ray scattering, cryogenic transmission electron microscopy and UV-vis absorption have demonstrated that nanometer-sized AuNPs and QDs, after their surface being modified with hydrophobic hydrocarbons, can be successfully incorporated into well-defined lipid NPs, e.g., discoidal bicelles or vesicles, forming NP-in-NP complexes. The morphology of such complex varies with the chain length of the conjugated hydrocarbons associated with AuNPs. The result of a coarse-grained molecular dynamics also indicates that 1-nm AuNPs can be entrapped and aggregate in the bilayer, further confirming the experimental observation. I will discuss over possible effects of AuNPs on the morphological variation of the bicelles, in order to provide insight to the metal/lipid self-assembled nanocomplex.

431944 Analysis and Large-Scale Synthesis of Bimetallic Nanocube Substrate for Surface Enhanced Raman Scattering

Abstract:

Analysis and Large-Scale Synthesis of Bimetallic Nanocube Substrate for Surface Enhanced Raman Scattering

Joshua Weatherston[1], Nolan Worstell[1], Hung-Jen Wu[1]

[1]Dept. Chemical Engineering, Texas A&M University, USA

Surface Enhanced Raman Scattering (SERS) is a highly sensitive spectroscopic technique capable of identifying surface-adsorbed chemical species at extremely low concentrations, down to single molecule detections. The low detection threshold and the "fingerprint" nature of this vibrational spectroscopic method, as well as its compatibility with equipment found in many research and diagnostic labs, make SERS an ideal candidate for multiplexed biomolecule detection. SERS is a consequence of the large electric field enhancement present near nanostructured surfaces; as such, its signal is dependent on the morphology and dielectric properties of the substrate and the analyte's proximity to the substrate. To create a suitable substrate for high-throughput SERS measurements, highly monodisperse plasmonic silver nanocubes were synthesized and coated with several atomic layers of gold in a large-scale batch synthesis. The theoretical SERS enhancement of Ag@Au core-shell nanocubes, pure Ag nanocubes, hollow Au cages, and other similar structures were calculated using finite element method (FEM), which predicted large enhancement factors. The FEM computation was confirmed by acquiring Raman spectra of the nanocubes functionalized with Raman-active organic dyes. It was found that the Ag@Au nanocubes showed large enhancement factors. Compared with pure silver nanocubes, the gold coated substrate showed greater Raman enhancement, increased resistance to particle degradation, and better compatibility with biological analytes. The multiplexing capabilities of this system were successfully demonstrated by detecting the co-adsorption of multiple dye species on the nanocube surface, including labeled and unlabeled peptide sequences. Furthermore, this platform was used to analyze the enzymatic digestion kinetics of peptide-functionalized nanocubes.

Submitting author: Joshua Weatherston, Dept. Chemical Engineering, Texas A&M University, College Station, TX 77843, USA, Tel: +1-361-232-3982; Email: jdub@tamu.edu

435195 Scaleable Production of Long Wavelength Fluorescent Nanoparticles to Enable Targeting and Multiplexed Imaging

Abstract:

Long wavelength (NIR) fluorescent dyes enable deep tissue penetration and avoidance of tissue autofluorescence. In diagnostic applications and therapeutic drug delivery, targeting is often required. Nanoparticles offer advantages in targeted delivery because avidity of binding can be enhanced by presenting multiple copies of the targeting ligand on the nanoparticle surface. The combination of long wavelength fluorescence and targeting presents challenges in the production and scaleup of nanoparticles with repeatable and quantifiable characteristics. We present a novel technology --Flash NanoPrecipitation – a controlled precipitation process that produces stable nanoparticles at high concentrations of encapsulated components using amphiphilic block copolymers to direct self-assembly. Uniform particles with tunable sizes from 50 – 500 nm can be prepared in an economical, scalable, and reliable manner.[2] The key to the process is the control of time scales for micromixing, self-assembly, and nucleation and growth. Stoichiometric encapsulation of components enables the assembly of complex nanoparticles with tailored optical and targeting properties.

Most bio-imaging, even for long wavelength dyes, has employed aqueous dyes that are conjugated onto the surfaces of nanoparticles. The long wavelength dyes are intrinsically large, conjugated structures and significant surface attachment may interfere with targeting. Instead we employ extremely hydrophobic dyes that remain sequestered in the cores. This enables higher loadings than are achievable with surface-attached dyes. Also, we show that several dyes which the community considers "hydrophobic" such as Nile red or ICG, partition out of nanoparticle cores to other lipid compartments and do not act as true reporters of the nanoparticle concentration or location. We report on two novel classes of NIR dyes: a hexacene based dye that has been first used in photo-voltaic applications, and a series of chlorin and bacteriochlorin dyes from Nirvana Sciences. The later dyes are particularly interesting because the wavelengths can be tuned at ~10nm intervals over the NIR window as shown in Fig. 1. The native form of these dyes is hydrophobic, so they are ideally suited for encapsulation in our nanoparticle constructs. The narrow emission spectra will allow multiplexing of these dyes so that multiple populations of dyes with different targeting ligands can be injected simultaneously and their location/fate in vivo can be ascertained.

We present targeting studies with these longwavelength
dyes to show ligand density can
be easily varied to determine optimum ligand
concentration on the nanoparticle surface.[3]

1. Taniguchi, M., et al., Accessing the near-infrared spectral region with stable, synthetic, wavelength-tunable bacteriochlorins. New Journal of Chemistry, 2008. 32(6): p. 947-958.

2. Akbulut, M., et al., Generic Method of Preparing Multifunctional Fluorescent Nanoparticles Using Flash NanoPrecipitation. Advanced Functional Materials, 2009. 19(5): p. 718-725.

3. D'Addio, S.M., et al., Optimization of cell receptor-specific targeting through multivalent surface decoration of polymeric nanocarriers. Journal of Controlled Release, 2013. 168(1): p. 41-49.Long wavelength (NIR) fluorescent dyes enable deep tissue penetration and avoidance of tissue autofluorescence. In diagnostic applications and therapeutic drug delivery, targeting is often required. Nanoparticles offer advantages in targeted delivery because avidity of binding can be enhanced by presenting multiple copies of the targeting ligand on the nanoparticle surface. The combination of long wavelength fluorescence and targeting presents challenges in the production and scaleup of nanoparticles with repeatable and quantifiable characteristics. We present a novel technology --Flash NanoPrecipitation – a controlled precipitation process that produces stable nanoparticles at high concentrations of encapsulated components using amphiphilic block copolymers to direct self-assembly. Uniform particles with tunable sizes from 50 – 500 nm can be prepared in an economical, scalable, and reliable manner.[2] The key to the process is the control of time scales for micromixing, self-assembly, and nucleation and growth. Stoichiometric encapsulation of components enables the assembly of complex nanoparticles with tailored optical and targeting properties.

Most bio-imaging, even for long wavelength dyes, has employed aqueous dyes that are conjugated onto the surfaces of nanoparticles. The long wavelength dyes are intrinsically large, conjugated structures and significant surface attachment may interfere with targeting. Instead we employ extremely hydrophobic dyes that remain sequestered in the cores. This enables higher loadings than are achievable with surface-attached dyes. Also, we show that several dyes which the community considers "hydrophobic" such as Nile red or ICG, partition out of nanoparticle cores to other lipid compartments and do not act as true reporters of the nanoparticle concentration or location. We report on two novel classes of NIR dyes: a hexacene based dye that has been first used in photo-voltaic applications, and a series of chlorin and bacteriochlorin dyes from Nirvana Sciences. The later dyes are particularly interesting because the wavelengths can be tuned at ~10nm intervals over the NIR window as shown in Fig. 1. The native form of these dyes is hydrophobic, so they are ideally suited for encapsulation in our nanoparticle constructs. The narrow emission spectra will allow multiplexing of these dyes so that multiple populations of dyes with different targeting ligands can be injected simultaneously and their location/fate in vivo can be ascertained.

We present targeting studies with these longwavelength dyes to show ligand density can be easily varied to determine optimum ligand concentration on the nanoparticle surface.[3]

1. Taniguchi, M., et al., Accessing the near-infrared spectral region with stable, synthetic, wavelength-tunable bacteriochlorins. New Journal of Chemistry, 2008. 32(6): p. 947-958.

2. Akbulut, M., et al., Generic Method of Preparing Multifunctional Fluorescent Nanoparticles Using Flash NanoPrecipitation. Advanced Functional Materials, 2009. 19(5): p. 718-725.

3. D'Addio, S.M., et al., Optimization of cell receptor-specific targeting through multivalent surface decoration of polymeric nanocarriers. Journal of Controlled Release, 2013. 168(1): p. 41-49.

Figure

434910 Layer-By-Layer Nanoparticles for Delivery of Second Window Near-Infrared Probes in Vivo

Abstract:

Second window near-infrared (NIR-II) imaging provides fluorescence-based anatomical information deep to skin as NIR-II light (1000 nm to 1700 nm) features deep tissue penetration, reduced tissue scattering, and negligible tissue autofluorescence. NIR-II fluorescence imaging probes; including down-conversion particles, quantum dots, single-walled nanotubes, and organic dyes, are formulated in biocompatible layer-by-layer (LbL) nanoparticles. 150 - 200 nm size range is obtained and hyaluronic acid is used as outer layer for extended blood circulation. Systemic delivery of LbL nanoparticles in Balb/c mice demonstrates that down-conversion particles have least autofluorescence due to its relative longer emission wavelength (>1500 nm). Organic dyes with emission at 1100 nm show most autofluorescence and elevated tissue scattering. Both quantums dots and single-walled nanotubes give moderate autofluorescence and tissue scattering. Pharmacokinetics studies of the LbL probes indicate that layer-by-layer formulation extends the blood circulation with half-life from 14 ~ 23 hrs. Tissue histology suggests none of the LbL probes present major toxicity. A further imaging involving orthotopic ovarian tumor is conducted using down-conversion particles due to its relative low autofluorescence. At 96 hrs, tumor is visualized in deep intraparatoneal cavity with high resolution. In this work, we present a first example of head-to-head comparison of all current NIR-II probes using LbL formulation. In vivo imaging performance is evaluated and suggests NIR-II imaging is versatile and robust technique for biomedical applications.

432494 Design and Construction of Supramolecular Imaging Agents

Abstract:

Intracellular sensing of pathologically relevant biomolecules could provide essential information for accurate evaluation of disease staging and progression, yet the poor cellular uptake of water soluble molecular probes limits their use as effective protease sensors. Rational design of molecular probes capable of self-assembling into supramolecular nanoprobes presents a potential strategy to alter their interaction mechanisms with cells thus promoting their effective cellular entry and intracellular accumulation. Herein we report on the design and synthesis of peptide-based self-assembling molecular beacons that could spontaneously associate into supramolecular protease sensors of either spherical or cylindrical shapes depending upon both their growth kinetics and assembly conditions. Our confocal imaging and flow cytometry results reveal that cancer cells experience difficulties internalizing high-aspect-ratio cylindrical nanobeacons regardless of their charge status, but could rapidly internalize positively charged spherical nanobeacons. These studies corroborate the shape-dependent cellular uptake of filamentous nanostructures and the charge-dependent internalization of spherical nanoparticles, and more importantly provide guiding principles for rational design of supramolecular nanoprobes with tunable cellular uptake characteristics.

435074 Tunable Composite Nanocarriers for Magnetic Resonance Imaging, Multi-Modal Imaging and Theranostic Applications

Abstract:

MRI and NIR-active, multi-modal and theranostic Composite NanoCarriers (CNCs) have been developed using a simple one-step process, Flash NanoPrecipitation (FNP). The CNCs have higher MRI T2 contrast and drug loading than have been reported for carriers produced by more complex assembly processes. In the controlled precipitation process, 10 nm iron oxide-based nanocrystals (IONCs) were encapsulated into poly(ethylene glycol) stabilized CNCs to make biocompatible T2 contrast agents. By adjusting the formulation, the CNC size was tuned between 80 and 360 nm. Holding the CNC size constant at 99 nm ± 3 nm, the particle relaxivity varied linearly with encapsulated IONC content ranging from 66 to 533 mM-1s-1 for CNCs with 4 to 16 wt% IONCs. By functionalizing the CNC surface with hydroxyl groups, the CNCs targeted the liver. The CNCs darkened healthy liver tissue, which enabled the detection of 0.8 mm3 non-small cell lung cancer metastases in vivo. Incorporation of the hydrophobic NIR dye, PZn3, into CNCs enabled visualization with long wavelength fluorescence at 800 nm as well as MRI imaging. Both the dye and IONC were encapsulated into the core of the CNCs. In vivo imaging demonstrated the ability of these particles to act both as MRI and fluorescent imaging agents and to circulate in the blood over 24 hours. To demonstrate the high drug-loading capabilities of CNCs formed via FNP, a theranostic CNC was created by the single step, block copolymer-directed assembly of IONCs and itraconazole. The CNC had an ITZ drug loading of 47%.

428996 Magnetic Nanoparticles for Multispectral Optoacoustic Imaging of Proteolytic Activity in the Tumor Microenvironment

Abstract:

Prostate Cancer (PCa) is the most diagnosed non-skin malignancy and the second leading cause of death from cancer among men in the United States. In order to achieve more effective therapy, early and accurate detection of prostate cancer is necessary to identify patients at high risk of disease progression. This allows for aggressive treatment to patients with tumors that carry certain metastatic potential. Unfortunately, current screening technologies for PCa, including the prostate-specific antigen (PSA) test, are limited by false-negatives and false-positives. A recent study showed that there is no PSA threshold that accurately predicts the presence or absence of PCa. Additionally, even when PCa is successfully detected, these screens provide no indication of the aggressiveness of the disease. Improved molecular imaging techniques, that are both quantitative and sensitive, and that can non-invasively detect processes deep within the human body are required to monitor changes which may be predictors of treatment outcomes.

The expression of matrix metalloproteinase-2 (MMP-2), an important tumor-associated protease, has been correlated with increasing metastatic potential in PCa and is therefore an attractive target for PCa targeting and diagnosis. We developed an MMP-2 sensitive, dual fluorochrome probe based on superparamagnetic iron oxide nanoparticles (SPIONs), with one fluorochrome (AF 700) that resisted proteolytic activation and thus served as an internal reference, while another optically distinct fluorochrome (AF 750) was conjugated to SPIONs through an MMP-2 sensitive peptide linker. We were able to quantify the presence of both fluorochromes using multispectral optoacoustic tomography (MSOT). The probe displayed sensitivity to MMP-2 activity in an MMP-2 overexpressing cell line *in vitro* and *in vivo*; slowly releasing AF 750 in low activity microenvironments and more rapidly in the presence of high MMP-2 activity. The MSOT ratio of AF 750 signal over AF 700 signal, which corrects for differences in the lesion size and depth, reflected the level of MMP-2 activity in the surrounding tissue; decreasing in areas of high MMP-2 activity, while remaining relatively constant in areas of low MMP-2 activity.

431347 Continuous Separation of Magnetic Nanoparticles to Enhance Size and Surface Homogeneity

Abstract:

Magnetic nanoparticle has foreseeable potential in biomedical applications. Due to the size and magnetic property, nanoparticle is able to access anywhere in the human body. This creates a great interest in pharmaceutical and engineering researches, including drug delivery, hyperthermia, and other in vivo applications. However, because of the small size of the magnetic nanoparticle, it limits the control the particle properties, such as size, surface property, composition and shape. Unfortunately, most of the known separation techniques are not applicable at the nanoscale. This research suggests a separation process to separation magnetic nanoparticle with different size and surface properties. A mixture of poly-disperse and PEGylated superparamagnetic iron oxide nanoparticles (SPION) is immobilized inside an external magnetic field. The field is provided by a series of computer controlled electromagnetic coils. SPION can be separated into multiple narrow distributions by size or surface property in accordance with the control in field strength and frequency. The size of the SPION particle is proportional to its iron content, which also determines its response to magnetic fields. By manipulating the field strength, different SPION size distributions can be separated. The size distribution is quantified by dynamic light scattering, and surface homogeneity is reported by its zeta potential.

431880 Encapsulation of Iron Oxide Nanoparticles into Poly(lactic acid) Micro/Nanoparticle System and Poly(lactic acid)/Polystyrene Bio-Blend Microparticle System

Abstract:

Magnetic iron oxide nanoparticles (IO NPs) have a wide application in biomedical engineering, catalysis, magnetic resonance imaging and so on. However, free IO NPs have several challenges such as aggregation and chemical activity, causing loss of magnetism and dispersibility. In order to protect naked IO NPs, a controllable process to encapsulate either hydrophobic or hydrophilic IO NPs into the hydrophobic biodegradable polymer micro/nanoparticles has been developed to meet various applications of magnetic nanoparticles. By applying the same experimental setup, we can generate polymer particles with different shapes and sizes, encapsulating either hydrophobic or hydrophilic NPs, by simply adjusting the water phase viscosity and the emulsion temperature. In this study, both a pure poly (lactic acid) (PLA) system and a bio-blend of poly(lactic acid)/polystyrene (PS) were employed to encapsulate hydrophobic (~5 nm) or hydrophilic (~10 nm) iron oxide nanoparticles (IO NPs). In order to encapsulate hydrophilic IO NPs, a water-in-oil-in-water (W/O/W) emulsion was formed via only one emulsion step. The transformation between hollow microparticles and solid nanospheres was realized by repeating the emulsion process between high temperature (HT) and room temperature (RT). Hydrophobic IO NPs were encapsulated into nanospheres via an oil-in-water (O/W) emulsion or microparticles via W/O/W emulsion.

434501 Injectable, Magnetic "Plum Pudding" Hydrogel Nanocomposites: Improving Externally-Mediated, Enhanced Pulsatile Drug Release

Thursday, November 12, 2015: 2:25 PM
253B (Salt Palace Convention Center)

Scott B. Campbell, Danielle Maitland and Todd R. Hoare, Chemical Engineering, McMaster University, Hamilton, ON, Canada

INTRODUCTION: "Smart", polymer-based biomaterials that respond to specific stimuli have been extensively studied in recent years, particularly as potential therapeutic release agents. The temperature-responsive polymer poly(*N*-isopropylacrylamide) (PNIPAM) has been used to develop several hydrogels and microgels for drug delivery applications due to the thermosensitive volume phase transition temperature (VPTT) that this polymer experiences upon heating.[1] However, such thermosensitive materials alone are unable to repeatedly release a drug of interest in the body, as safe application of a thermal stimulus *in vivo* is not straight-forward. Composite materials that combine thermosensitive materials with nanomaterials that generate heat in response to external systems (such as gold nanorods, carbon nanotubes, or superparamagnetic iron oxide nanoparticles (SPIONs)) can overcome this issue, facilitating an advanced drug releasing mechanism whereby temperature-dependent swelling/deswelling responses can be non-invasively controlled *in vivo*.[2,3]

We recently used this approach to produce injectable, externally-responsive PNIPAM hydrogels in which SPIONs were covalently bound into the hydrogel network.[4] An alternating magnetic field (AMF) was used to heat the SPIONs and induce deswelling of the matrix to produce pulsatile release of drug over multiple AMF applications. However, the observed increase in the drug release was relatively small and too short-lived for practical use (most of the drug was released after one day).[4]

Building on this result, we have fabricated nanocomposite materials with SPIONs and thermosensitive microgels entrapped in an injectable thermosensitive hydrogel matrix, resulting in "plum pudding"-like nanocomposite hydrogel materials. Upon AMF application, the SPIONs will generate heat that is transferred to the thermosensitive microgels within the nanocomposite, causing them to deswell and open up free volume that encourages drug release (Figure 1). This proposed release mechanism should also be entirely reversible, so that when the AMF application is halted, the nanocomposite (and the drug release rate), will return to its initial state. This study not only involves the elucidation of the true mechanism of externally-induced enhanced drug release, but also the degree to which various parameters (i.e., the microgel content, the microgel VPTT, the hydrogel swelling characteristics, etc.) effect the efficacy of enhanced release due to an AMF exposure over multiple days.

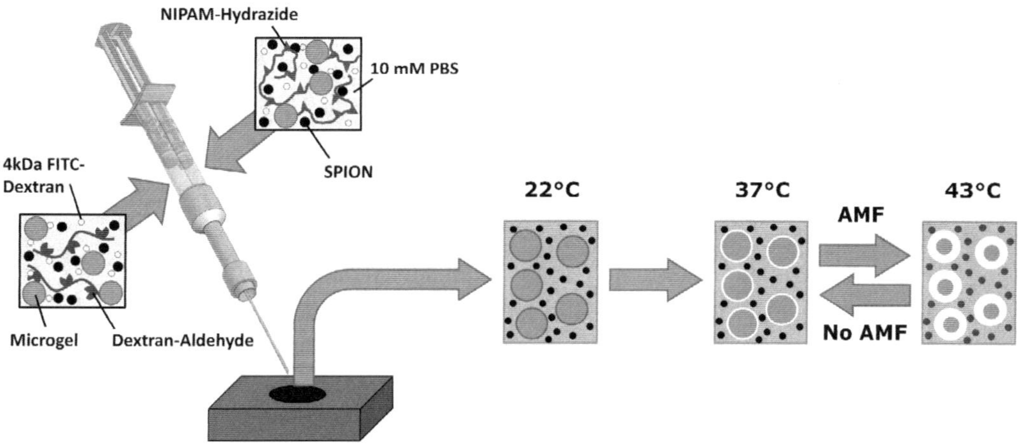

Figure 1: Fabrication of the injectable "plum pudding" nanocomposite hydrogels and their proposed AMF-mediated release mechanism.[5]

EXPERIMENTAL: Hydrazide-functionalized PNIPAM was prepared by carbodiimide-mediated conjugation of adipic acid dihydrazide to a p(NIPAM-AA) copolymer.[6] Aldehyde-functionalized dextran was generated by oxidizing dextran with sodium periodate.[6] The thermosensitive microgels were prepared via precipitation polymerization of NIPAM (36 mol%), *N*-isopropylmethacrylamide (NIPMAM, 58 mol%), and acrylamide (6 mol%) with *N,N*-methylenebisacrylamide (crosslinker) and ammonium persulfate (initiator). SPIONs were fabricated via coprecipitation and peptized with 8 kDa PEG.[7]

The nanocomposites were generated using a double barrel syringe in which each barrel contains 8 wt% of the hydrogel precursor polymer (with hydrazide-functionalized PNIPAM and aldehyde-functionalized dextran in opposite barrels), 8 wt% thermosensitive microgels, 1 wt% 4 kDa FITC-dextran, and 5 wt% SPIONs, in 10 mM PBS. The hydrazide- and aldehyde-functionalized polymers rapidly react upon injection to form the hydrolytically-degradable hydrazone crosslinks of the hydrogel while physically entrapping the microgels, SPIONs, and drug within the nanocomposite matrix.

Drug release experiments were performed using an AMF setup that allows for multiple composites ($n = 4$) to be held in equivalent positions in the magnetic field while maintaining a constant baseline temperature of 37°C (or 22°C or 43°C for the varying temperature experiments). Drug release experiments involved immersing the nanocomposites in 4 mL of PBS and applying the AMF for certain intervals. For pulsed release experiments, samples were collected every 10 minutes, including directly before and after 10 minute AMF pulses (4-6/day) that were applied every 40 minutes. The FITC-dextran concentration in each release sample was measured with a fluorescent plate reader. The increase in the rate of release due to an AMF application was reported as the percent increase between the measured release rate following an AMF pulse and the baseline release rate, which was calculated via linear interpolation between the 2 sample points directly before and the 2 points immediately after the AMF pulse.

RESULTS: The nanocomposites were confirmed by TGA to possess ~5 wt% SPIONs and by SQUID analysis to be superparamagnetic in nature. Their degradability was confirmed in accelerated hydrolysis conditions, with full dissolution of the gel observed after ~240 hours of incubation in pH 1 buffer. The nanocomposites were quite elastic, with the storage modulus increasing with microgel content until a certain critical point (between 12 and 16 wt% microgel) above which the microgel inhibited the formation of hydrazone crosslinks upon injection.

To determine whether or not drug release was taking place via the free volume mechanism proposed in Figure 1, several different AMF-mediated drug release experiments were performed. Drug release throughout a 2 hour AMF application showed that the composites with microgel and SPIONs released significantly more drug than composites without either microgel or SPION content, with the slope of the release versus time curve roughly double for the AMF-exposed composites. Following, short 10 minute pulses were performed that showed that composites with both microgel and SPION content could achieve much greater enhancements in release due to a pulse compared to composites without either microgel or SPION content; furthermore, after one day, only the microgel+SPION composites could achieve pulsatile AMF-mediated release, with the rate of release after a pulse increasing as high as fourfold relative to the baseline release.

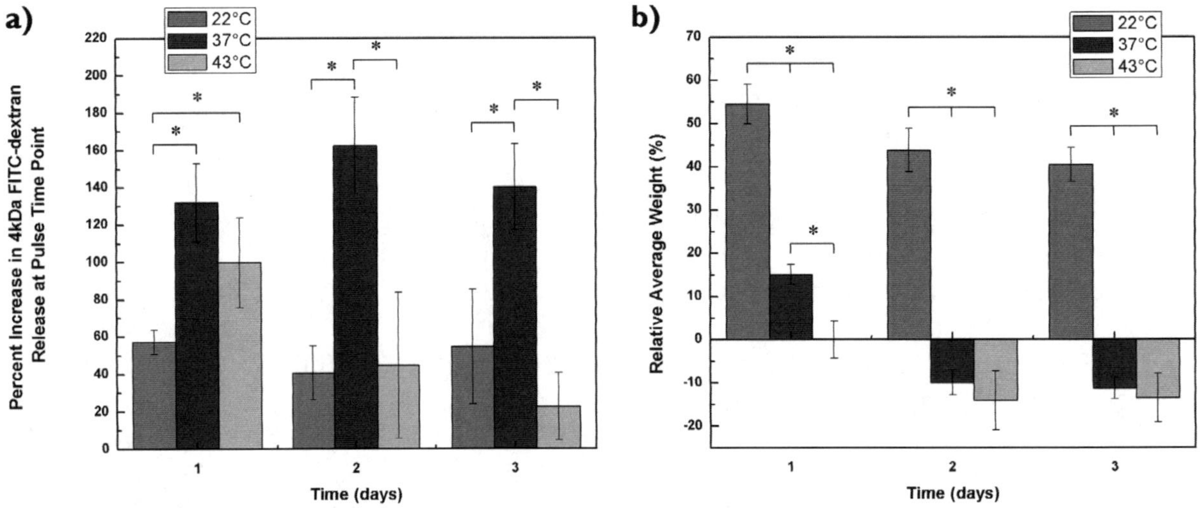

Figure 2: Percentage increases in FITC-dextran release due to an AMF pulse (a) and relative swelling (b) of the nanocomposites at different baseline temperatures. *$p < 0.05$ in a pairwise comparison.[5]

Now that the importance of microgels to improving the control over release over multiple days was confirmed, experiments were performed at baseline temperatures well below (22°C), slightly below (37°C), and slightly above (43°C) the VPTT of the microgels (~39°C) to confirm that the phase transition of the microgels is the origin of the enhanced release (Figure 2). Figure 2 shows that when the AMF-controlled system operates from a baseline slightly below the VPTT of the microgels there is a much greater enhancement of release due to short AMF pulses, particularly after the first day. Overall, these results confirmed our proposed mechanism of AMF-mediated release.

The influence of various factors on release was then observed. Studies using microgels with differing PNIPAM:PNIPMAM ratios, and thus different VPTTs, indicated that microgels with VPTT values slightly above the baseline physiological temperature facilitated higher pulsatile release, owing to the increased volume change those microgels can undergo when heated in an AMF. Composites with higher microgel contents (up to 10 wt%) lead to greater external control over release for the entire 5 day period studied. In addition, the swelling of the hydrogel scaffold, and thus the swelling of the nanocomposites, was altered by replacing a certain percentage of the aldehyde-functionalized dextran with identically functionalized CMC, a more hydrophilic polymer that enhances the bulk gel swelling (Figure 3).

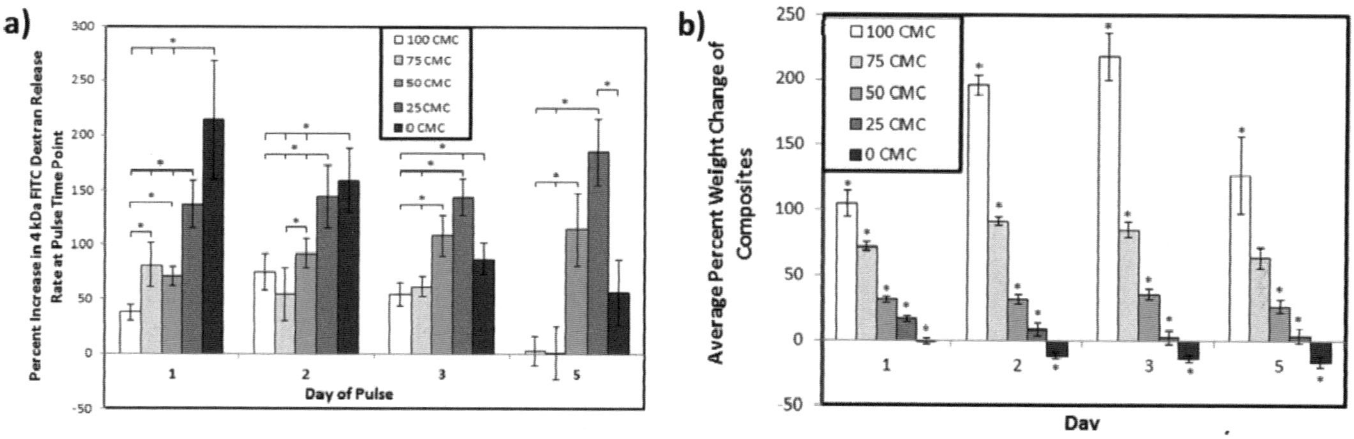

Figure 3: Effect of hydrogel scaffold swelling on the percent increase in 4 kDa FITC-dextran release due to a 10 minute AMF application (a) and relative swelling (b) of the nanocomposites. *$p < 0.05$ in a pairwise comparison.

Figure 3 shows that as the amount of CMC relative dextran in the bulk hydrogel increased, the degree of swelling dramatically increased and the magnitude and duration of pulsatile drug release both decreased. This result is attributable to the swelling bulk hydrogel phase consuming the free volume generated by the deswelling of the microgels following AMF application.

All of the components of the nanocomposites, and the nanocomposites themselves, also displayed no significant cytotoxicity via an MTT assay with *3T3* mouse fibroblast cells.

This further suggests that these materials may be promising candidates as "smart", on-demand drug delivery platforms that could be used to deliver a variety of drugs for therapies that would benefit from repeated, pulsatile release.

CONCLUSIONS: The addition of thermosensitive microgels inside injectable, magnetic hydrogel composite materials significantly improved the externally AMF-controlled enhanced release. This enhanced release occurs due to microgels deswelling during the AMF-mediated heating of the nanocomposite to create free volume that enhances release. Such enhancements in pulsatile release can be tuned by adjusting the microgel VPTT, microgel content, and the hydrogel swelling characteristics and have shown to persist over multiple days rather than hours.

REFERENCES: (1) Pelton, R. *Adv. Colloid Interface Sci.* **2000**, *85*, 1-33. (2) Merino, S. et al. *ACS Nano* **2015** [*In Press*]. (3) Campbell, S.B.; Hoare, T. *Curr. Opin. Chem. Eng.* **2014**, *4*, 1-10. (4) Campbell, S.B.; Patenaude, M.; Hoare, T. *Biomacromolecules* **2013**, *14*, 644-653. (5) Campbell, S.; Maitland, D.; Hoare, T. *ACS Macro Lett.* **2015**, *4*, 312-316. (6) Patenaude, M.; Hoare, T. *Biomacromolecules*, **2012**, *13*, 369-378. (7) Mahmoudi, M. et al. *Adv. Drug Deliv. Rev.* **2011**, *63*, 24-46.

ACKNOWLEDGEMENTS: This research is funded by the J.P. Bickell Foundation (Medical Research Grants), and the Natural Sciences and Engineering Research Council of Canada (Discovery Grant and Vanier Scholarship programs).

424939 ALD-Grown SiO2 Protective Layer on TiO2 Nanoparticles Under Mild Conditions for the Suppression of Photocatalytic Activity

Abstract:

ALD-grown SiO_2 Protective Layer on TiO_2 Nanoparticles under Mild Conditions for the Suppression of Photocatalytic Activity

Jing Guo, Shaojun Yuan, Hairong Yue, Siyang Tang, Changjun Liu, Bin Liang*

Multi-phase Mass Transfer & Reaction Engineering Lab, College of Chemical Engineering, Sichuan University, Chengdu 610065, China

*corresponding author: liangbin@scu.edu.cn

Titanium dioxide (TiO_2), an inert and nontoxic chemical, is one of the best white pigment due to its excellent optical properties. It is widely used in the paint, plastic, and paper industries. However, the excellent photocatalytic activity of TiO_2 particles facilitates the oxidation and decomposition of organic substrates in the paint layer, which severely restrict the performance of the optical properties and the color of TiO_2 products. Therefore, it is necessary to suppress the photocatalytic activity of the TiO_2 with inert protective layers, such as SiO_2, Al_2O_3. In view of the rich surface hydroxyl group on TiO_2 particles, the gas phase atomic layer deposition (ALD)-grown method was used as an efficient approach to coat much thin and uniform protective layers on the TiO_2particles. The ALD-grown method also possesses inherent advantages, such as no solvent waste, the feasibility of continuous processing, and the potential for scaling-up.

The experimental results indicated that $SiCl_4$ could easily reacts with the hydroxyl group of TiO_2 via gasification and hydrolyzation in the ALD-grown method at room temperature. The TiO_2 particles coated with SiO_2 protective layer exhibited a significant decrease of the photocatalytic activity (K_{app} from 0.158 to 0.017 min^{-1}). The shielding effect and the coating thickness were strengthened in accordance with the increase of the coating number. This simple gas phase coating process with less reaction time, no use of acid and alkali solutions and dense and uniform coating layers has the potential application in the material synthesis and catalysis.

Keywords: TiO_2coating, ALD reaction, nanoparticles

References:

[1] Lin Y L, Wang T J, Jin Y. Surface characteristics of hydrous silica-coated TiO_2 particles. *Powder Technology,*2002, 123(2): 194-198.

[2] Rossi E M, Pylkkänen L, Koivisto A J, *et al*. Airway exposure to silica coated TiO_2 nanoparticles induces pulmonary neutrophilia in mice. *Toxicological Sciences*, 2009: kfp254 .

[3] Feng X, Zhang S, Lou X. Controlling silica coating thickness on TiO_2 nanoparticles for effective photodynamic therapy. *Colloids and Surfaces B: Biointerfaces*, 2013, 107: 220-226.

[4] Van Ommen J R, Yurteri C U, Ellis N, *et al*. Scalable gas-phase processes to create nanostructured particles. *Particuology*, 2010, 8(6): 572-577.

427858 Uio-66-NH_2 Assembly on ALD Oxide Surfaces Using b-Cyclodextrin and Cetyltrimethylammonium Bromide (CTAB)

Abstract:

A wide range of metal-organic frameworks (MOFs) have been synthesized through isoreticular synthesis using organic linkers and metal ions or metal clusters, to create a new class of porous materials with large surface area for more efficient gas storage, catalysis, separation, and gas sensing.

However, MOFs are typically formed as crystalline powders making them difficult to handle, manipulate and position for desired application. One way to address this constraint is to integrate or couple MOFs onto fibrous scaffolds. Our previous work has demonstrated atomic layer deposition (ALD) can impart uniform and conformal MOF nucleation sites onto the fiber mats, thus leading to nucleation and growth of highly absorbent MOFs that are covalently attached to the mats. We previously attained high mass loading (up to 300 wt.%) for Cu(BTC) MOFs on fibers, but mass loading for UiO-66-NH_2 is typically smaller (30 wt.%).

Here we report a novel synthesis route to assemble UiO-66-NH_2 onto polymeric fibrous scaffolds to achieve additional MOF mass loading and mesoporosity in the MOF-fiber structures. Especially UiO-66-NH_2 has been well known to reveal high thermal stability over 400 °C, decent chemical stability to atmospheric moisture, and photocatalytic effect in visible range at high efficiency. Using solvothermal synthesis at 85 °C, we prepared a UiO-66-NH_2 seed layer on ALD-coated polypropylene fiber mats. Subsequently, we assembled UiO-66-NH_2 onto the seeded fiber surface using cetyltrimethylammonium bromide (CTAB) and β-cyclodextrin (β-CD). Owing to host-guest interaction, we were able to obtain a supramolecular complex, thereby further increasing the mass loading of UiO-66-NH_2. X-ray diffraction confirms good crystallinity of the UiO-66-NH_2 coating. Furthermore, SEM images show UiO-66-NH_2 agglomerates fill the void space inside the MOF-seeded fiber mesh. Moreover, N_2 adsorption analysis shows the creation of extra mesoporosity and reveals the Brunauer-Emmett-Teller (BET) surface area of the MOF-functionalized fiber mats, exceeding 90 $m^2/g_{MOF+Fiber}$. We will discuss how synthesis conditions affect the assembly of UiO-66-NH_2 and show how robust these MOF coatings are on the fiber substrates. Similar assembly strategy can also be applied to other MOFs such as CuBTC, and is promising for new MOF-based composites for a wide variety of applications.

428646 Anomalous Dispersion of 'hedgehog' Particles

Abstract:

Dispersion of colloids in ‘phobic’ solvents is unstable and leads to irreversible aggregation. A stable dispersion in ‘phobic’ environment requires chemical camouflages that provides polarity matching with the environment, thereby increasing the particle-solvent affinity, and that provides steric and electrostatic effects, thereby increasing the particle-particle repulsion. Tuning such colloidal behavior by physical canopy would overcome the limitations imposed by interfacial chemical modification and also preserve the pristine properties and functionalities. In this research, we show that the ‘hedgehog’ particles with arrays of high aspect ratio ZnO nanowires coated on a surface of microsphere exhibit stable dispersion in both ‘philic’ and ‘phobic’ media. From the experimental findings and calculations, high surface corrugation leading to limited contact area and increased minimum interaction distance drastically reduces the attractive potential. Solvent-ionization at the air-water interfaces of trapped air-pockets in the nano-topography provides extra repulsive potential in the case of aqueous dispersion of hydrophobic HPs. Low ionic strength provides longer-range repulsive potential in the case of hydrophilic HPs in non-polar organic media. The findings could lead to reduction in the use of the volatile organic compound in a wide spectrum of industries ranging from pharmaceutical to paint, enable new strategies for exotic colloidal behavior and self-assembly, and exhibit unexplored or enhanced physicochemcial properties ranging from photophysics to catalysis.

428007 Chromogenic Vapor Sensors Enabled By Novel Shape Memory Polymers

Abstract:

Here we report a new type of chromogenic vapor sensor that exhibits an abrupt change from a transparent color to a striking iridescent color when exposed to a large variety of organic vapors, such as acetone, ethanol, and toluene. Importantly, these chromogenic vapor sensors enable the specific detection of different vapors – i.e., different vapors can trigger different colors which can be distinguished by our naked eyes or by using a conventional smart phone. They can also be reused for hundreds of times with no apparent degradation in their chromogenic performance. These smart sensors integrate scientific principles drawn from two disparate fields that do not typically intersect - the fast-growing shape memory polymer (SMP) and photonic crystal technologies. Shape memory polymers can memorize and recover their permanent shape in response to an external stimulus, such as heat, light, and solvent, while photonic crystals can control the flow of light in miniature volume. The striking chrmogenic effect (from transparent to iridescent) is induced by a disorder-order transition of the macroporous SMP photonic crystals. These reusable and inexpensive chromogenic vapor sensors that show easily perceived color changes are useful for environmental monitoring (e.g., monitoring toxic vapors in buildings), homeland security (such as detecting explosives), and healthcare screening (e.g., breath testing the aount of acetone for diabetes screening).

429030 Scalable Assembly of Nanoparticle Antireflection Coatings

Abstract:

Antireflection coatings on transparent substrates (e.g., glass) are important components for a large number of optical and optoelectronic devices, such as displays, lenses, and photovoltaic panels. Traditional quarter-wavelength antireflection coatings produced by vacuum deposition suffer from high operating and equipment costs, limited material selection, low throughput, and small coating areas. Although some solution processing technologies have been developed to reduce the manufacturing cost and improve the production throughput, many of these techniques involve multiple steps, are limited to single-sided coatings on planar substrates, are not very reproducible over large areas, and/or are not inherently parallel for industry-scale manufacturing. Here, we report a simple, inherently parallel, and scalable bottom-up approach for fabricating nanoparticle anti-glare coatings on large glass substrates. Negatively charged silica nanoparticles are electrostatically adsorbed onto a surface-functionalized glass substrate with positive surface charges to form a monolayer nanoparticle coating. This innovative technology enables the simultaneous coating of multiple 5-in.-sized glass substrates with high and reproducible qualities. Fundamentally, we combine experiments with theoretical simulations to address a basic question for monolayer nanoparticle-based anti-glare coatings - what is the optimal nanoparticle surface area coverage for best anti-glare performance? Surprisingly, the theoretical optimum nanoparticle surface area coverage is found to be ~60% instead of 74% for a close-packed colloidal array. We have demonstrated that this optimal nanoparticle coverage is readily achievable by the current electrostatics-assisted bottom-up technology.

418977 Solvent Influence on Dodecanethiol Interactions Measured By Atomic Force Microscopy

Abstract:

Introduction:

The control of nanoparticle dispersability is important for many applications, such as size-selective nanoparticle precipitation, thin film deposition, and nanoparticle composite formation. Current methods for the preparation of monodisperse nanoparticles are costly, require a large quantity of solvent, or have a low throughput [1-3]. Surfaces of nanoparticles are often passivated with stabilizing ligands which prevent agglomeration by introducing thermodynamic repulsive forces. Understanding the forces between such ligands can provide insight into a better way to control nanoparticle dispersion [3]. Recently, modelling efforts have shown that traditional steric repulsion and van der Waals attractions are not sufficient to characterize the interactions between nanoparticles in poor solvents [4]. In this study, the importance of solvation forces on regulating the interactions of dodecanethiol ligands were quantified *via* atomic force microscopy (AFM). We chose toluene, hexane, and ethanol to represent good, mild and bad solvents, respectively to elucidate the forces involved in the interactions of dodeceanthiol ($C_{12}SH$) ligands.

Materials and Methods:

Silicon nitride (Si_3N_4) AFM probes were coated with a 5 nm chromium adhesive layer followed by a 40 nm gold layer. The gold-coated cantilevers were then cleaned in absolute methanol and ethanol solution for 15 minutes each, respectively. Cantilevers were then immersed for 18 hours at room temperature in 1 mM ethanolic solution of 1-dodecanethiol for self-assembled monolayer (SAM) formation. Functionalized AFM cantilevers and substrates were then loaded into the AFM and imaging and force measurements were conducted in ethanol, hexane, and toluene. Areas of 2×2 μm^2 were scanned at a rate of 1 Hz. Force-distance curves were collected with a ramp size of 1 μm at 1 Hz and with a 5 nN threshold force. Jump into contact and adhesion forces in the approach and retraction curves were analyzed, respectively.

Results and Discussion:

As the ligand-decorated AFM cantilever approaches or retracts from the opposing ligand surface, the ligands interact with each other. Representative approach and retraction force-distance curves are shown in Figure 1. As the ligand-decorated cantilever moves away from the opposing ligand surface, adhesion forces can be measured. The mean adhesion force was highest in ethanol (0.172 ± 0.073 nN) followed by toluene then hexane with adhesion forces of 0.165 ± 0.100 nN and 0.147 ± 0.063 nN, respectively. The high adhesion force between the hydrophobic $C_{12}SH$ layers in ethanol is largely due to the unwillingness of hydrophobic molecules to interact with hydrophilic solvents. Because ethanol is hydrophilic, the $C_{12}SH$ will want to interact with other $C_{12}SH$ molecules and will therefore not want to be separated. This increases the adhesion force or force required to separate the two $C_{12}SH$ SAM layers. In comparison to hexane, hydrophobic SAMs interacted with a higher force in toluene. This may be due to the π bonds that form between toluene and the extended $C_{12}SH$ polymers. In hexane, SAMs had the lowest retraction adhesion forces. This is mainly because hexane is hydrophobic and $C_{12}SH$ will likely interact with the solvent just as much as it will interact with the opposing $C_{12}SH$ layer.

In the approach curve or as the SAM layers come together, there is a jump into contact adhesion force that was analyzed with respect to its magnitude, frequency, and jump distance. The magnitude of these adhesion forces were 0.05 ± 0.02 nN for hexane, 0.08 ± 0.03 nN for ethanol, and 0.10 ± 0.03 nN for toluene. Although toluene seems to have the strongest adhesion force, these events only made up 11% of the curves whereas in hexane and ethanol, they were observed in 37 % and 60% of the registered approach curves, respectively. Since toluene is expected to extend the ligands, steric repulsion between the ligands will increase. Such increase in the repulsive forces hinder the attractive forces between ligands and thus reduce the probability of the occurrences of attractive forces. Ligands in hexane interacted with each other intermediately with the lowest strength. This is because the ligands are able to interact with the solvent as both are hydrophobic. However, in the presence of ethanol (hydrophilic solvent), the hydrophobic interactions will dominate the affinity of ligands and thus minimize the steric repulsion between the ligands.

The molecules interacted together in toluene at the longest distance (5.04 ± 3.58 nm). In comparison, molecules needed to be closer to each other for the jump into contact interactions to occur in ethanol and hexane with distances of 2.78 ± 1.11 nm and 2.90 ± 1.48 nm, respectively. These later distances were not statistically significantly different. The longer distance registered in toluene suggests that ligands adopt an extended conformation. It is important to note that jump distance does not represent the actual length of the ligands but the distance where the ligands start to attract each other. Currently, approach data is being fit to a model of steric repulsion to quantify the conformational properties of the SAM chains. All adhesion force and jump into contact data were statistically different among solvents tested. With respect to jump distance, toluene was statistically significantly different from the other treatment groups.

Conclusions:

When in hydrophobic or good solvents, the $C_{12}SH$ ligands were extended whereas hydrophilic or bad solvents caused the ligands to collapse. Ligands interacted more favorably with each other when in hydrophilic solvents due to solvation excluding effects and vice versa when they were in good hydrophobic solvents. As a result, the use of hydrophilic solvents may be detrimental to dispersing nanoparticles.

Acknowledgements: This work was supported by the, NIH Protein Biotechnology Training Program 24280305, a

NASA Space Grant, a WSU DRADS fellowship, and a Harold P. Curtis Scholarship.

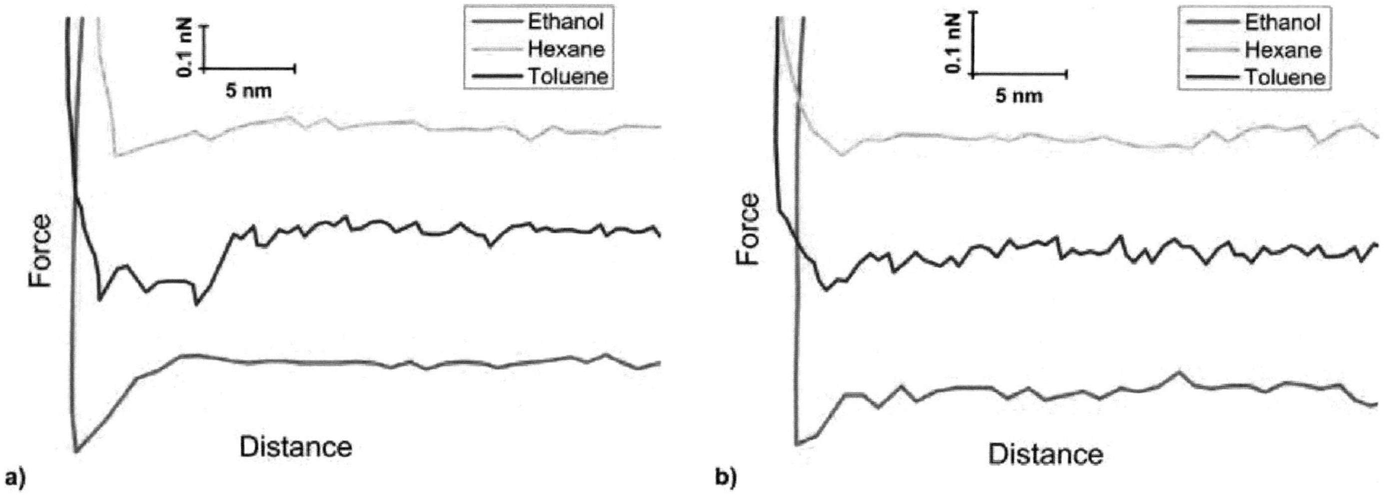

Figure 1. Representative force-distance curves of DDT SAMs in various solvents. a) Retraction curves with the adhesion forces as quantified form the adhesion peaks b) Approach curves with jump into contact peaks. The jump distances were quantified as the difference between the distance of maximum peak value and the distance where there is no more attractive forces between ligands.

References

1. Murray, C.B., C.R. Kagan, and M.G. Bawendi, *Synthesis and characterization of monodisperse nanocrystals and close-packed nanocrystal assemblies.* Annual Review of Materials Science, 2000. **30**: p. 545-610.

2. Murray, C.B., D.J. Norris, and M.G. Bawendi, *Synthesis and characterization of nearly monodisperse CdE (E = sulfur, selenium, tellurium) semiconductor nanocrystallites.* Journal of the American Chemical Society, 1993. **115**(19): p. 8706-8715.

3. Saunders, S.R. and C.B. Roberts, *Nanoparticle separation and deposition processing using gas expanded liquid technology.* Current Opinion in Chemical Engineering, 2012. **1**(2): p. 91-101.

4. Saunders, S.R., M.R. Eden, and C.B. Roberts, *Modeling the Precipitation of Polydisperse Nanoparticles Using a Total Interaction Energy Model.* The Journal of Physical Chemistry C, 2011. **115**(11): p. 4603-4610.

430908 Single-Molecule Visualization of Corona Phase Molecular Recognition

Abstract:

Molecular recognition is central to the design of therapeutics, chemical catalysis, and sensor platforms, with the most common mechanisms involving biological structures such as protein antibodies. However, identifying and isolating an antibody for a particular biological molecule is often costly and time-consuming, and their use is limited to physiological conditions in which protein antibodies remain stable. This work reports the development of a synthesis platform for the design of synthetic sensors, based on the principle of carbon nanotube Corona Phase Molecular Recognition (CoPhMoRe) [1]. The design of sensors with CoPhMoRe is based on the adsorptive properties of engineered heteropolymers to carbon nanotubes, which create a unique corona with molecular recognition properties for a target analyte. This phenomenon is shown to be generic, with new recognition complexes demonstrated for riboflavin, l-thyroxine, and estradiol, and confirmed with single-molecule imaging of the nanotube and the corona phase [2]. This platform has recently been expanded for optical detection of neurotransmitters [3], and signaling molecules in living plant tissues [4]. The complexes can be used as new types of sensors based on modulation of carbon nanotube optical fluorescence, as demonstrated using a synthetic antibody for real time spatio-temporal detection of riboflavin in live murine macrophages. The analyte library for these synthetic sensors could be amplified to identify synthetic sensors for a variety of different molecules for which there currently exists no detection method.

1. Landry, M.P.*, Zhang, J.Q.*, et al., *Molecular recognition using corona phase complexes made of synthetic polymers adsorbed on carbon nanotubes.* Nature Nanotechnology, 2013. **8**(12): p. 959-968.

2. Landry, M.P., et al., *Experimental tools to study molecular recognition within the nanoparticle corona.* Sensors (Basel), 2014. **14**(9): p. 16196-211.

3. Landry, M.P.*, Kruss, S.*, et al., *Neurotransmitter Detection Using Corona Phase Molecular Recognition on Fluorescent Single-Walled Carbon Nanotube Sensors.* Journal of the American Chemical Society, 2014. **136**(2): p. 713-724.

4. Giraldo, J.P., et al., *Plant nanobionics approach to augment photosynthesis and biochemical sensing (vol 13, pg 400, 2014).* Nature Materials, 2014. **13**(5): p. 530-530.

425796 A Colorimetric Plasmonic Nanosensor for Dosimetry of Therapeutic Levels of Ionizing Radiation

Abstract:

Of the several types of cancer prevalent in men today in the US, prostate cancer is the second most commonly occurring with an estimated 220,800 new cases by the American Cancer Society in 2015. Hypofractionated radiotherapy as a treatment procedure for the ablation of cancerous tumors, delivers the total dose planned by the therapists over several sittings. However, owing to repeated radiation there is a high possibility of accidental damage to the nearby tissues and organs during treatment. It is extremely difficult to reduce the associated risk in this procedure, taking into consideration the absence of an in vivo dosimeter which could quantify the exact dose received by the rectum and the tissues. Moreover, the commercial dosimeters available for use are often bulky and require extensive post processing for data readout. Here, we present a simple technique to detect therapeutically relevant doses of ionizing radiation based on nanoparticle formation. We were successful in achieving this by holding gold in its univalent state and stabilizing it using a cationic surfactant. This precursor solution presents potential to be employed within an inflatable endorectal balloon which is often times used during radiotherapy to support the prostate and detect the dose received by the rectal wall. Upon irradiation of the precursor solution, free radicals are generated which subsequently reduce the univalent gold to Au(0), forming nanoparticles. The shape and size of the resultant nanoparticles give rise to characteristic colors while the concentration results in differential intensities of color which is a pseudo measure of the incident radiation. Characterization of the resultant nanoparticles was carried out using UV-Vis absorbance spectroscopy, Dynamic Light Scattering and Transmission Electron Microscopy. We also investigated the role of different cationic surfactants in assisting nanoparticle formation by maintaining the same head group, counter ion and varying the chain lengths. The nanosensor we synthesized is sensitive to radiation levels as low as 1 Gy and being therapeutically relevant, could potentially be used to validate the doses during radiotherapy.

NANOELECTRODE BASED BIOSENSORS FOR RAPID PATHOGEN DETECTION AND IDENTIFICATION

Foram Madiyar, PhD
Visiting Assistant Professor

Embry Riddle Aeronautical University, Department of Physical Science, Daytona Beach, FL 32114

Kansas State University, Department of Chemistry, Kansas State University, Manhattan, KS 66502

madiyarf@erau.edu

Nanoscale Science and Engineering Forum
Nanotechnology and Nanobiotechnology for Sensors and Imaging II

AIChE Annual Meeting, Salt Lake City, Utah, 2015
November 12, 2015

KANSAS STATE UNIVERSITY

1

Contents

- **Importance of Pathogen Detection**
  - Traditional methods of detection
- **Introduction of Biosensors**
- **Principle of Dielectrophoresis**
  - Fabrication of vertically aligned carbon nanofibers nanoelectrode arrays (VACNF NEAs)
  - Fabrication of microfluidic device
  - Simulation of particles on VACNF
- **Assessment of Pathogens**
  - Fluorescence Technique
  - Surface Enhanced Raman Scattering
  - Impedance Technique
- **Acknowledgements**

KANSAS STATE UNIVERSITY

2

Importance of Pathogen Detection

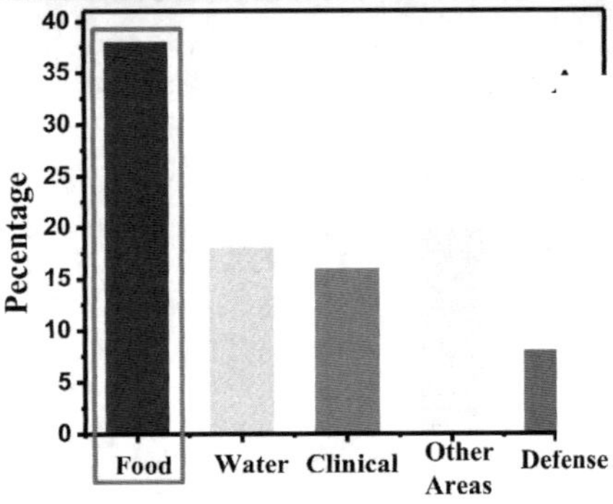

Source: CDC (Center for Disease Control and Prevention)

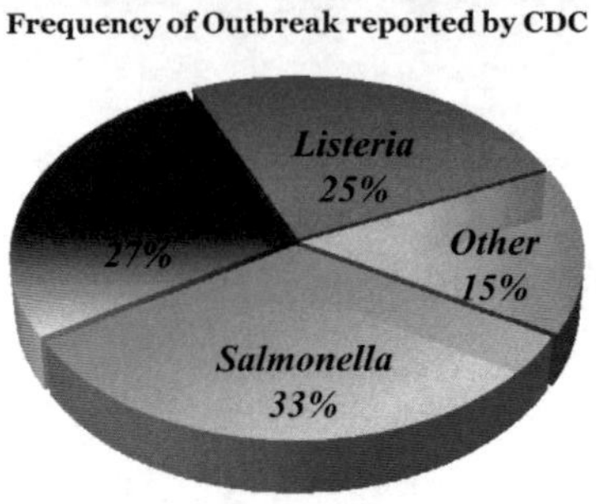

http://www.cdc.gov/listeria/outbreaks/ http://www.cdc.gov/listeria/salmonella/ecoli/ http://www.cdc.gov/listeria/salmonella/ecoli/

2015 2014 2013

Importance of Pathogen Detection

2014 FOOD SAFETY PROGRESS REPORT

Pathogen	Change in number of food borne illnesses compared with 2006-2008	
Campylobacter	13 % increase	☹
E. Coli O157	23 % increase	☹
Listeria	No change	😐
Salmonella	No change	😐
Vibrio	52 % increase	☹

Traditional Methods of Detection

Culture and Colony Method

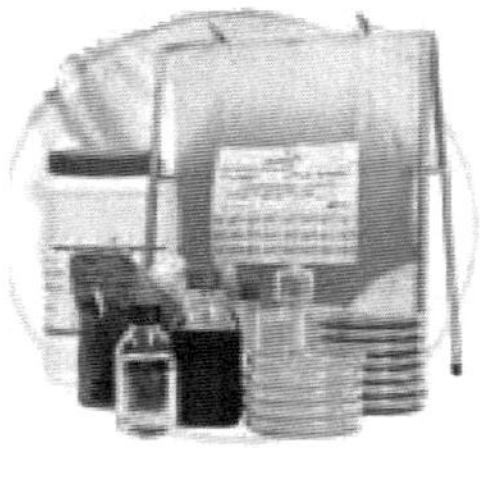

Amplification is observed by division of bacteria ~ 20min

Aseptic conditions
Sterilized culture media
Temperature 37 ^{0} C and
Results in 3-5 Days

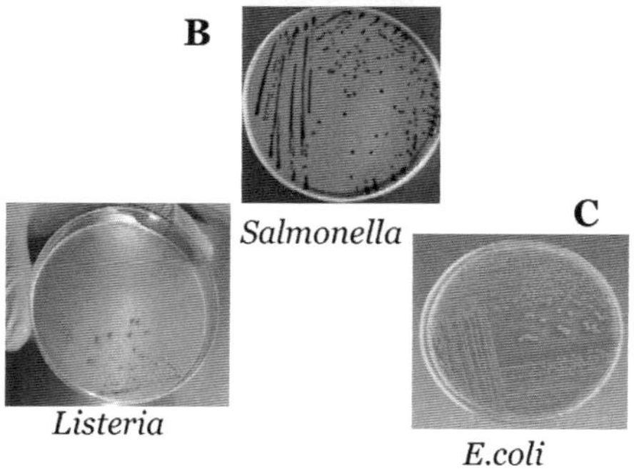

Microbiology Research Assocs. Inc.

Bacteriological Analytical Manual, 8th Edition, Revision A, 1998 Chapter 7.

ELISA

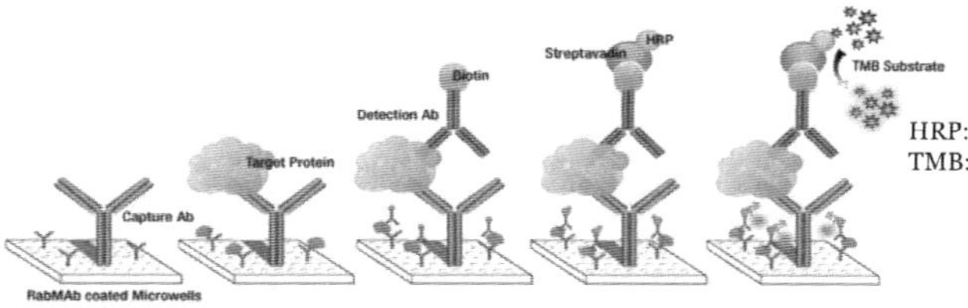

HRP: horseradish peroxidase
TMB: 3,3',5,5'-tetramethylbenzidine

Abcam Ltd., Total ELISA Kit antibody

Analyte is amplified by enzymatic conversion. Rate of reaction depends on the k_{cat} of enzyme.

KANSAS STATE UNIVERSITY

1. Lazcka, O.; Campo, F. J. D.; Muñoz, F. X., *Biosens. Bioelectron.* **2007**, *22* (7), 1205-1217
2.Bacterial Enteric Infections Detected by Culture-Independent Diagnostic Tests — FoodNet, United States, 2012–2014, March 13, 2015 / 64(09);252-257
3. Fung DYC., J. Rapid.Methods Autom. Microbiol. **1992**, 1,:1-14

5

Biosensor

A **biosensor** is an analytical device which converts a biological response into a quantifiable and processable signal."

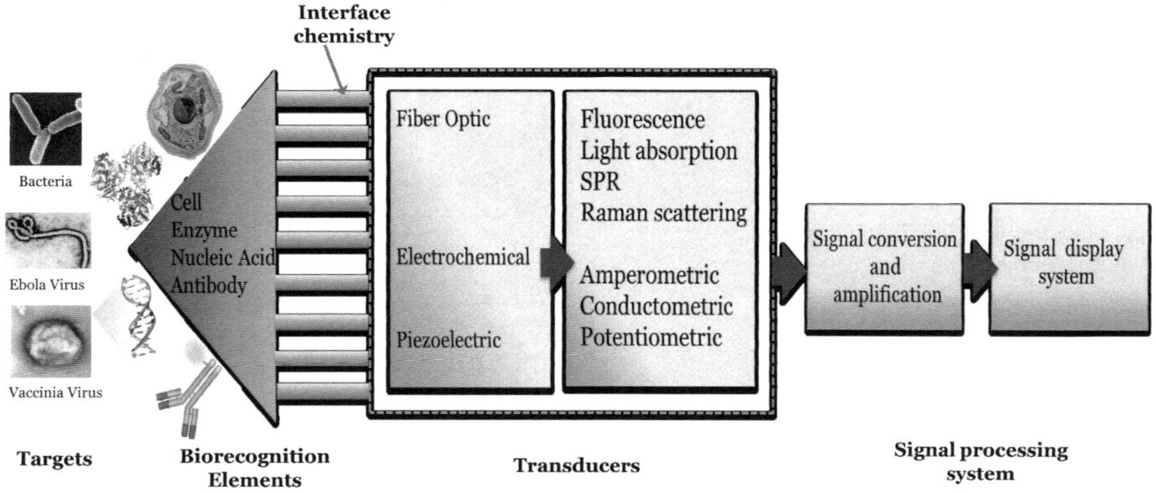

Long, F., et al. *Sensors* **2013**, *13*(10), 13928-13948

KANSAS STATE UNIVERSITY

1.Clark, L.C., *Journal of Applied Physiology*, **1953**,6,189-193.
2. Long, F., et al. *Sensors* **2013**, *13*(10), 13928-13948
3. www.biosensors-congress.elsevier.com/about.htm

6

Principle of Dielectrophoresis

DEP – Governing Equation:

$$\langle \vec{F}_{DEP} \rangle = 2\pi\, r^3 \varepsilon_m \,\mathrm{Re}[CM(\omega)]\, \nabla \left| \vec{E}_{rms} \right|^2$$

$$CM(\omega) = \frac{\varepsilon_p^* - \varepsilon_m^*}{\varepsilon_p^* + 2\varepsilon_m^*} \qquad \varepsilon^* = \varepsilon - j\frac{\sigma}{2\pi f}$$

Pohl, H. A., *Cambridge University Press*, Cambridge. **1978**

Clausius-Mossotti (*CM*) factor

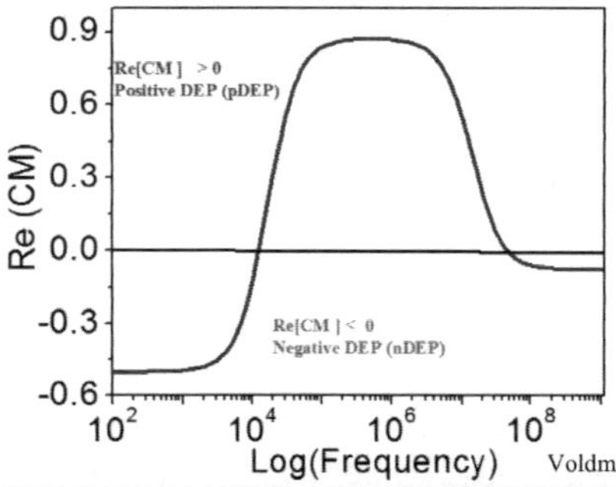

Voldman et.al., *J. Biophysical*, **2001**, 80:531–541

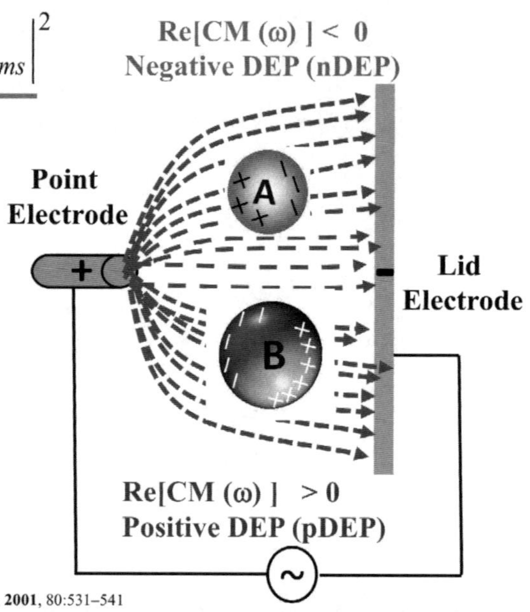

How Do We Make Point Electrode?

Steps involved

A. Ion beam coating (IBC)

B. Plasma enhanced chemical vapor deposition (PECVD) growth of VACNF

C. Encapsulation of CNF by tetra ethylene ortho silicate (TEOS), by chemical vapor deposition (CVD)

D. Chemical mechanical polishing (CMP) and reactive ion etching (RIE)

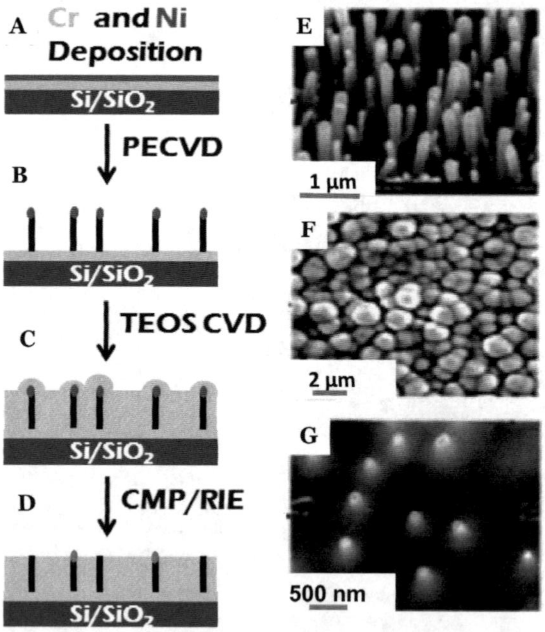

J. Li, et al, *Appl. Phys. Lett.*, **2003**, (82), 15, 2491.

Fabrication of DEP Device

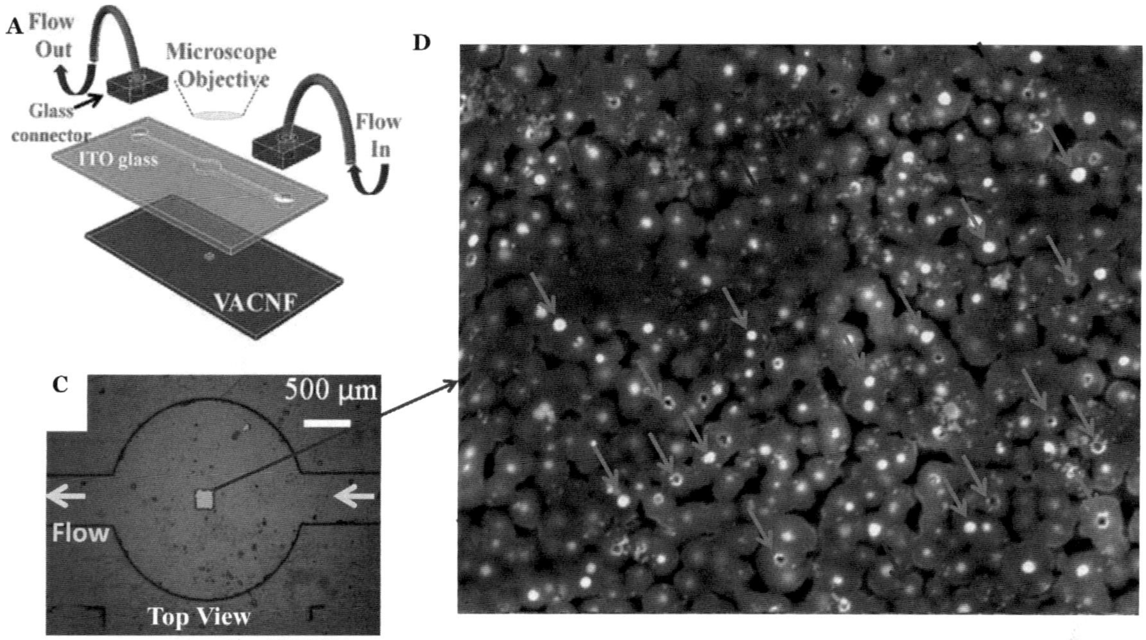

200 mm x 200 mm active CNF NEA area defined by SU-8 lithography

Top view of VACNF tips in region 200 µm x 200 µm

1. L. Syed, et al., *Electrophoresis*, **2011**, 32 (17), 2358-2365
2. Madiyar, F. R., Syed, L. U., Culbertson, C. T., Li, J., *Electrophoresis* **2013**, *34* (7), 1123-1130

In collaboration with Dr. Christopher T. Culbertson.

Simulation and Forces Acting on the Particles

Channel dimension:
20 mm (L) x 1 mm (W) x 20 mm (H)

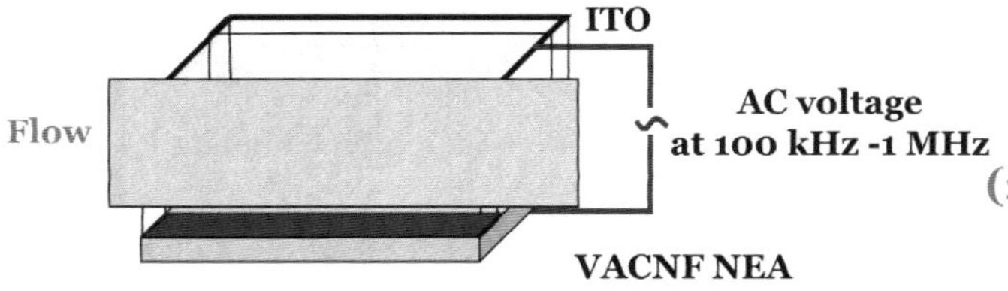

P. Arumugam, et al, *J. Phys. Chem. A*, **2007**, 111 (49), 12772-12777

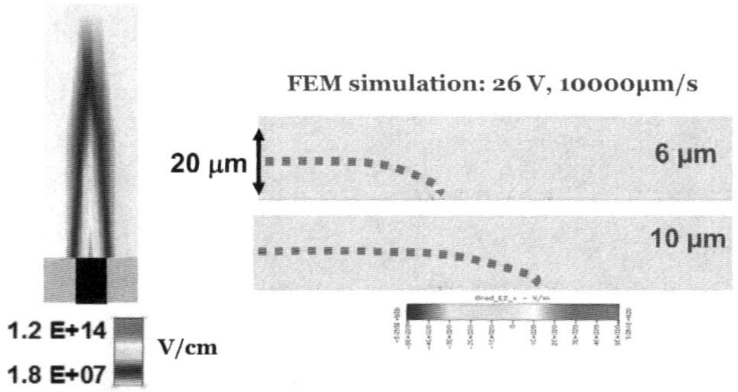

P. Arumugam, et al, *J. Phys. Chem. A*, **2007**, 111 (49), 12772-12777

(1) **Drag force:**

$$\vec{F}_{drag} = 6\pi k \mu r \vec{v}$$

(2) **Dielectrophoresis (DEP) force:**

$$\langle \vec{F}_{DEP} \rangle = 2\pi r^3 \varepsilon_m \mathrm{Re}[CM(\omega)] \nabla |E_{rms}|^2$$

(3) **Sedimentation force** (negligible)

(4) **Brownian force** (negligible)

P. Arumugam, et al, *J. Phys. Chem. A*, **2007**, 111 (49), 12772-12777

Microfluidic Dielectrophoretic Device

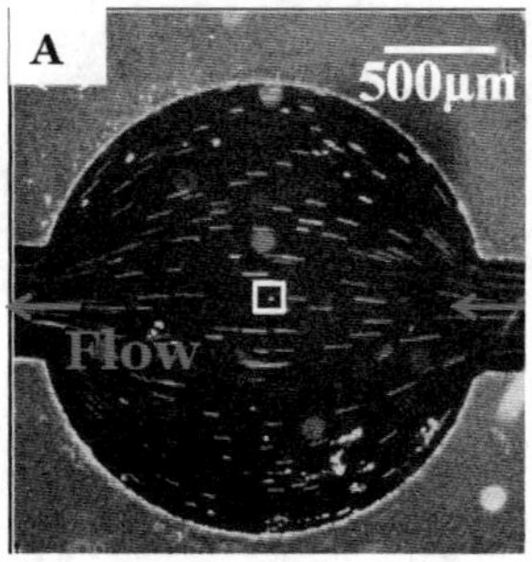

SYBR Green –I intercalates with the DNA to give out green fluorescence

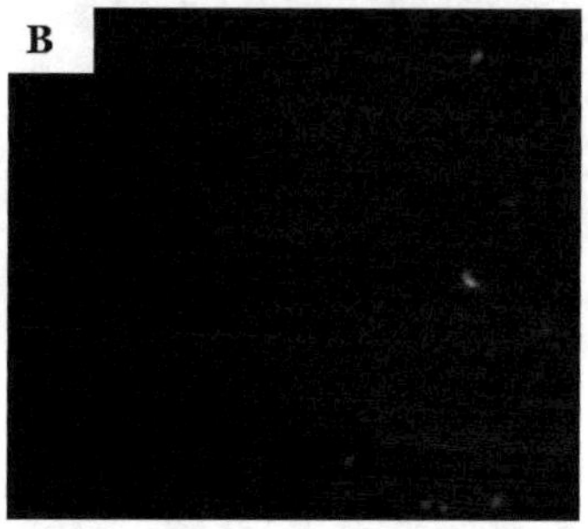

10 kHz, 330 µm/sec, 10 V_{pp}

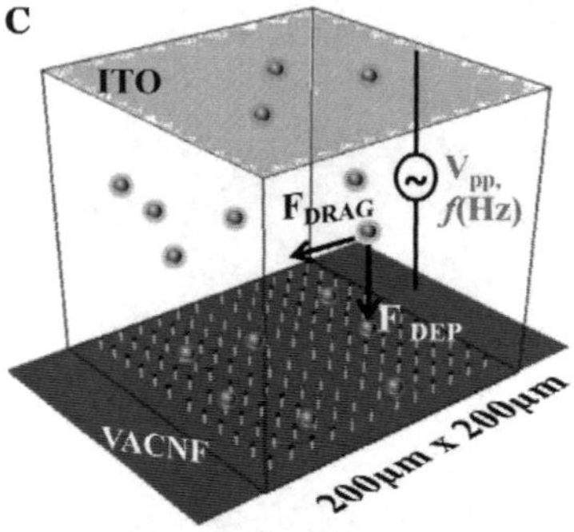

F_{DEP} **is orthogonal to** F_{DRAG}

Drag force

$$\vec{F}_{drag} = 6\pi k \mu r \vec{v}$$

Dielectrophoresis (DEP) force

$$\left\langle \vec{F}_{DEP} \right\rangle = 2\pi r^3 \varepsilon_m \, \mathrm{Re}[CM(\omega)] \, \nabla \left| E_{rms} \right|^2$$

Microfluidic Dielectrophoretic Device in Action

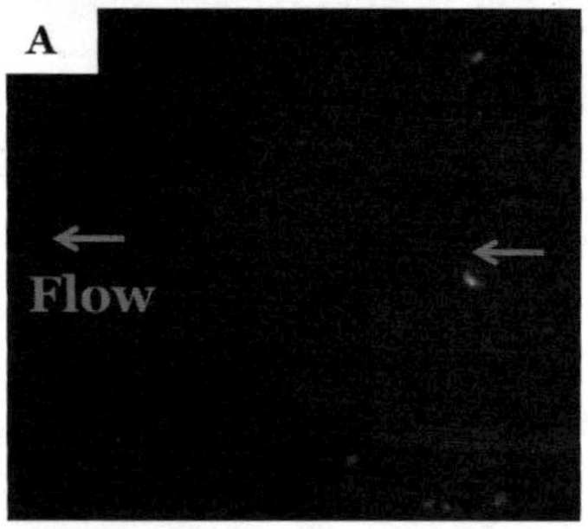

Linear velocity of 330 µm/sec

Vir[redacted]

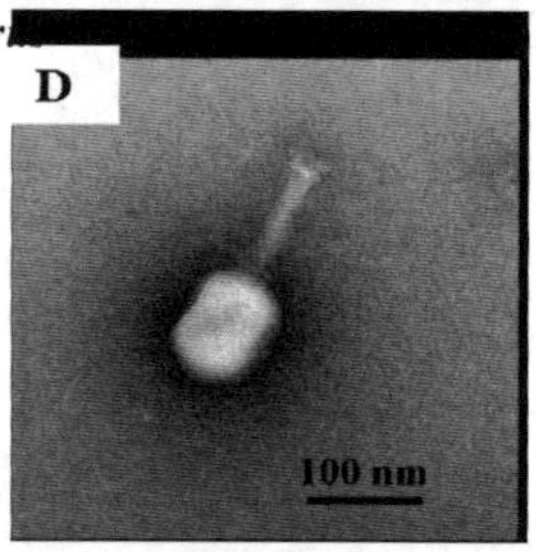

Linear velocity of 730 µm/sec

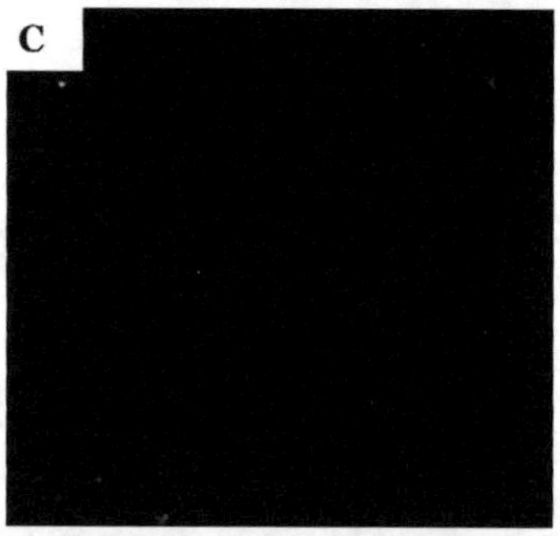

Linear velocity of 2720.0 µm/sec

Quantification of Dielectrophoretic Capture

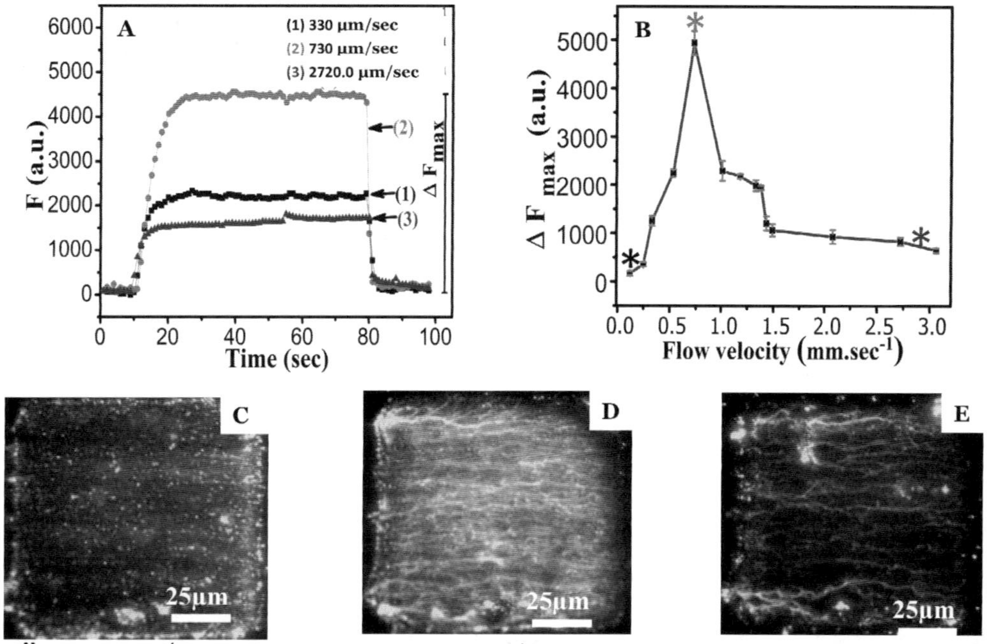

* 330 μm/sec
* 730 μm/sec
* 2720.0 μm/sec

Quantification of Dielectrophoretic Capture

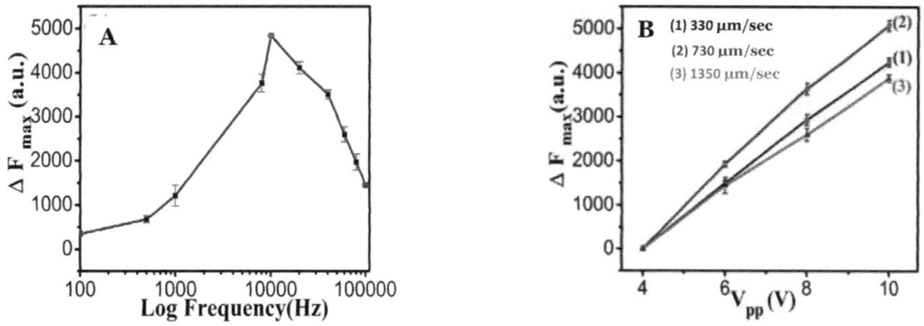

Linear flow velocity was kept constant at 730.0 μm/sec, Voltage at 10.0 V_{pp}

Frequency at 10 .0 kHz was kept constant

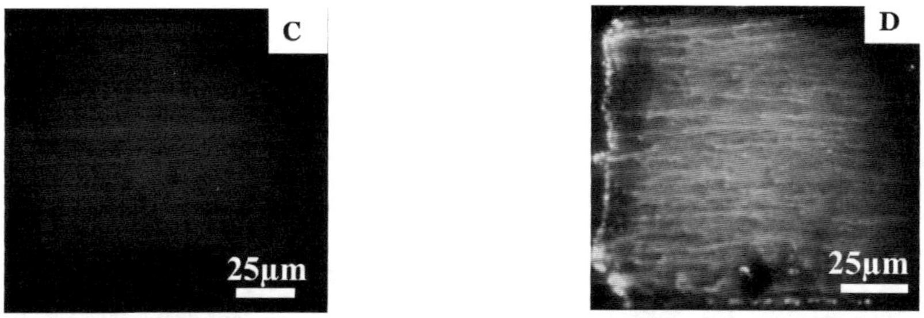

Frequency : 10 kHz and Voltage : 4.0 V_{pp}

Frequency : 10 kHz and Voltage : 8.0 V_{pp}

Quantification of Dielectrophoretic Capture

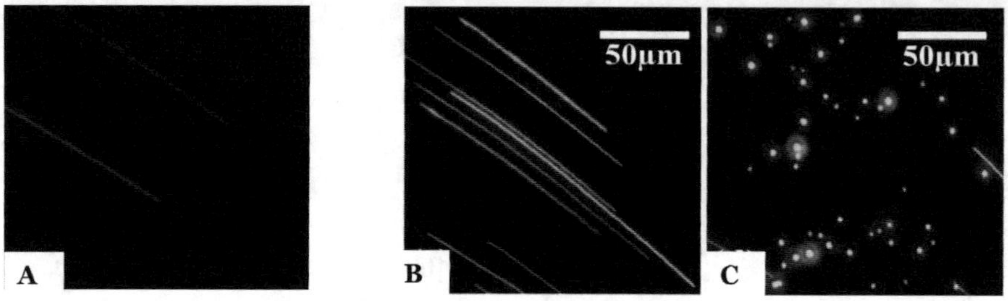

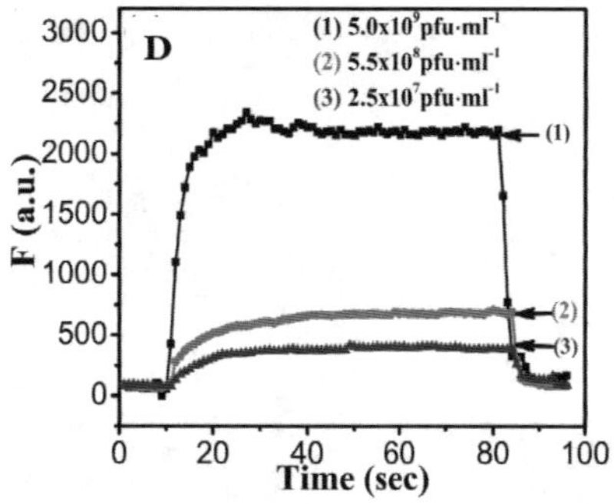

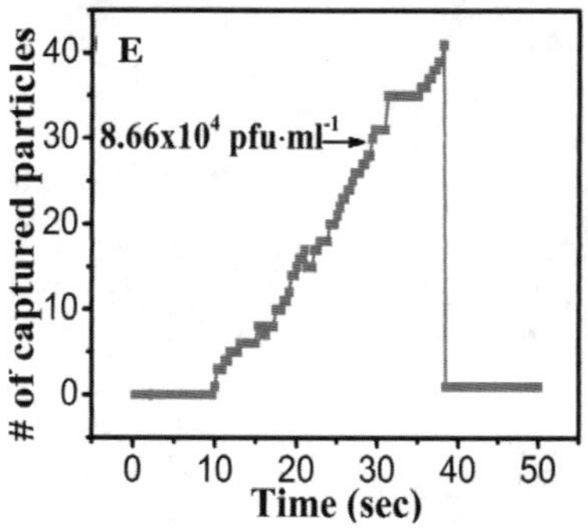

Linear flow velocity : 870 μm/sec **Capture effiency = 60%**

Madiyar, F. R., Syed, L. U., Culbertson, C. T., Li, J., *Electrophoresis*, **2013**, *34* (7), 1123-1130

Capture Profiles for Bacteria and Virus on VACNF

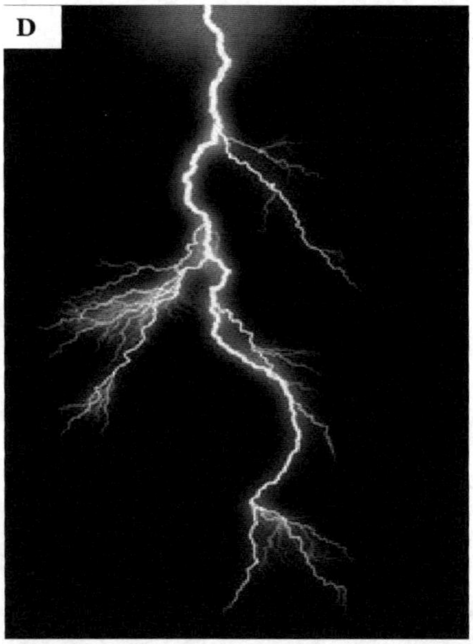

Madiyar, F. R., Syed, L. U., Arumugam, P., Li, J., Electrical Capture and Detection of Microbes Using Dielectrophoresis at Nanoelectrode Arrays. American Chemical Society Books, **2013**, Vol. 1143, pp 109-124

en.wikipedia.org/wiki/Lichtenberg

Capture of Bacterial Particles on VACNF

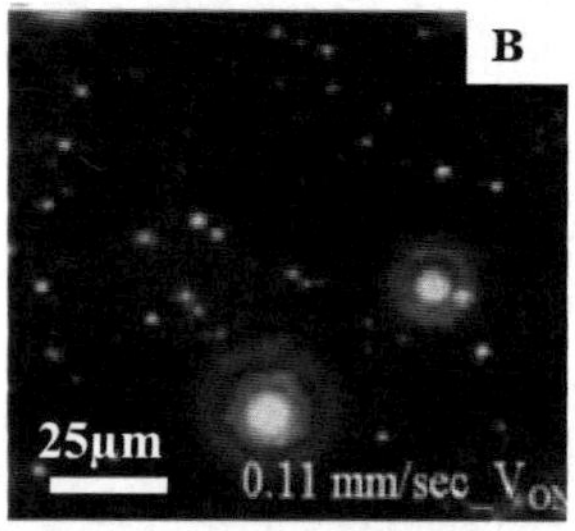

Syed, L. U., Li, J., *Electrophoresis* **2011**, *32* (17), 2358-2365

Capture of Virus Particles on VACNF

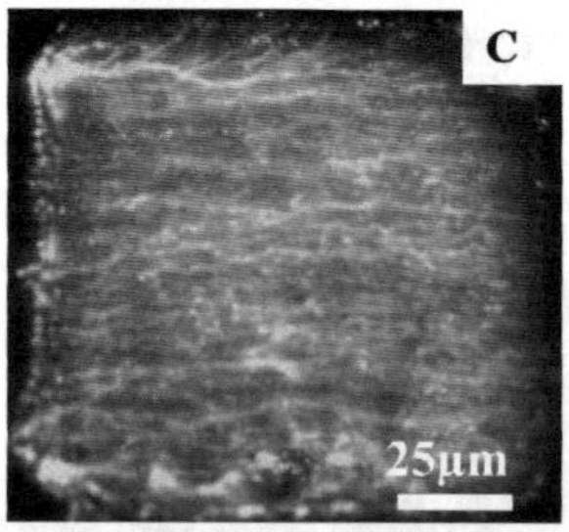

Madiyar, F. R., Li, J., *Electrophoresis* **2013**, *34* (7), 1123-1130.

1. Madiyar, F. R., Syed, L. U., Arumugam, P., Li, J., Electrical Capture and Detection of Microbes Using Dielectrophoresis at Nanoelectrode Arrays. American Chemical Society Books, **2013**, Vol. 1143, pp 109-124
2. Madiyar, F. R., Syed, L. U., Culbertson, C. T., Li, J., *Electrophoresis* **2013**, *34* (7), 1123-1130
3. Syed, L. U., Liu, J., Price, A. K., Li, Y.-f., Culbertson, C. T., Li, J., *Electrophoresis* **2011**, *32* (17), 2358-2365

Surface Enhanced Raman Spectroscopy

Sir Chandrasekhara Venkata Raman

C. V. Raman and K. S. Krishnan, Nature, (1928) 121, 501–502

1. Chemical Enhancement Effect:

- Molecules comes onto close physical contact with the rough metal surface (1-2 nm)
  - a. Adsorbed
  - b. Chemically bonded
- Charge transfer between molecules and the metal.

2. Electromagnetic Enhancement Effect:

- The molecules experience greater electromagnetic field intensity near the nanoparticle surface.

Raman enhancement calculation

$$EF = \frac{I_{SERS}}{I_{Bulk}} \times \frac{N_{Bulk}}{N_{SERS}}$$

1. Willets, A., and Van Duyne, R.P., Annual Reviews 2007, 58,267-297.
2. Orendorff, C. J.., Sau, T. K.; Murphy, C. J. , Small 2006, 2, 636-639

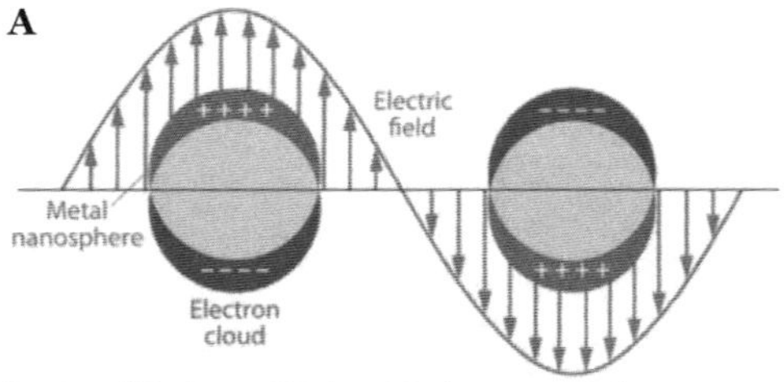

Willets, A., and Van Duyne, R.P., Annual Reviews 2007, 58,267-297.

1. www.nobelprize.org
2. C. V. Raman and K. S. Krishnan, Nature, (1928) 121, 501–502
3. Van Duyne R P and Haynes C L, Raman spectroscopy Encyclopedia of Physical Science and Technology 2001
4. Orendorff, C. J.., Sau, T. K.; Murphy, C. J. , Small 2006, 2, 636-639

Comparison of Raman Detection Techniques

Fingerprint Detection

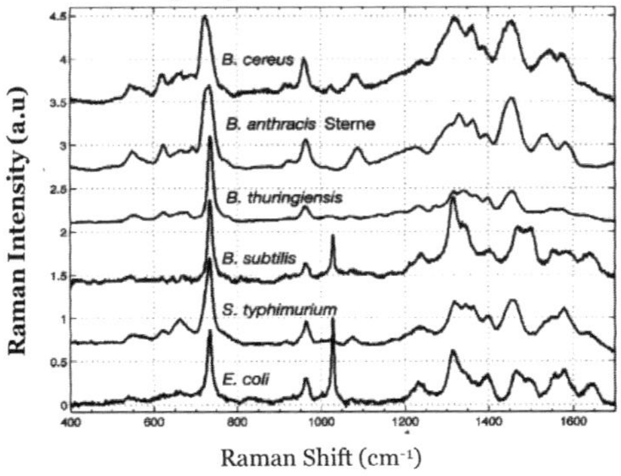

Premasiri, W. R., et al., *J. Phys. Chem. B*, **2005**, *109* (1), 312-320.

- **Pros:**
  - Simple sample preparation and quick method
  - No external labels required
- **Cons:**
  - Data analysis is complicated; usually requires statistical method like Principle Component Analysis (PCA)

Raman Tag Detection

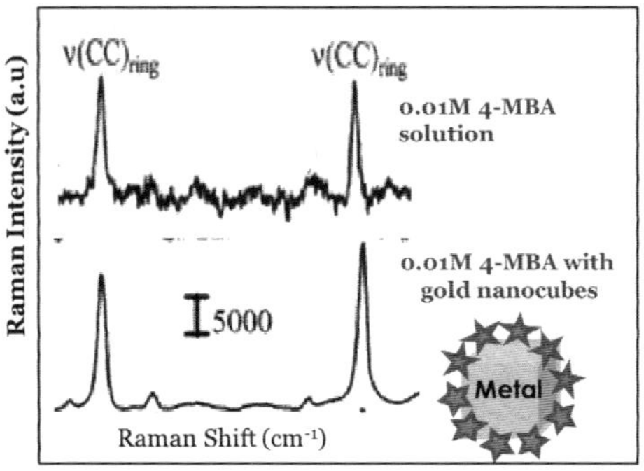

Orendorff, C. J., Murphy, C. J., *Anal. Chem.* **2005**, *77* (10), 3261-3266

4-MBA (4-Mercapto Benzoic Acid)

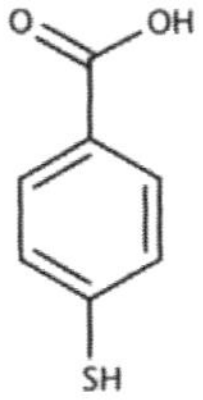

- **Pros:**
  - Enhanced Raman signal
  - Detection possible at low concentration
  - Data analysis is simple
- **Cons:**
  - Extra Fabrication steps
    - (a) Attach Raman Tags to nanoparticle
    - (b) Make nanoparticle biocompatible

1. Premasiri, W. R., et al., *J. Phys. Chem. B*, **2005**, *109* (1), 312-320.
2. Orendorff, C. J.; Gole, A.; Sau, T. K.; Murphy, C. J., *Anal. Chem.* **2005**, *77* (10), 3261-3266

Attaching Metal Nanoparticles to Bacteria

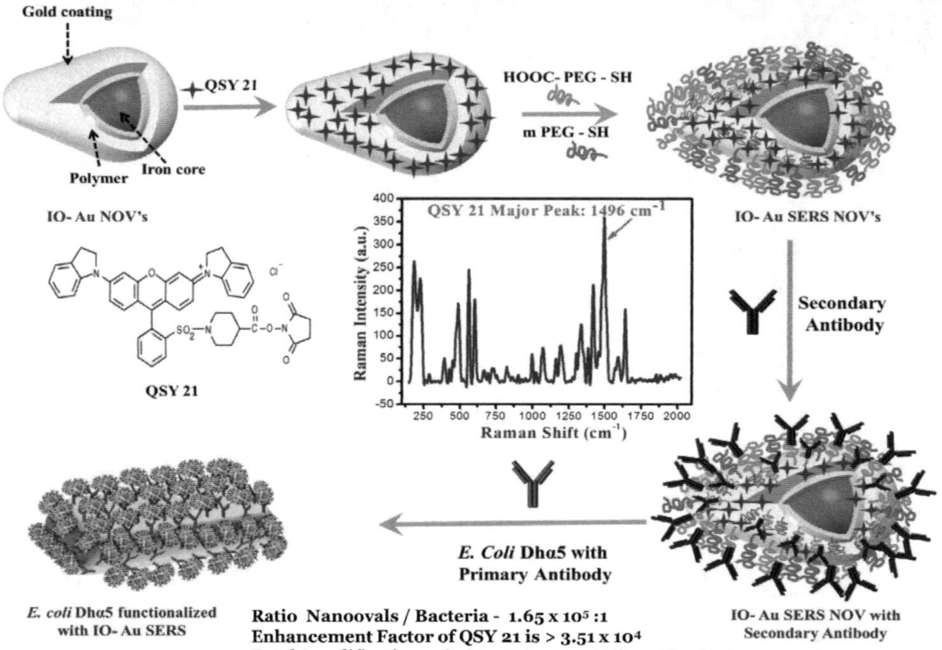

Ratio Nanoovals / Bacteria - $1.65 \times 10^5 : 1$
Enhancement Factor of QSY 21 is > 3.51×10^4
Total Amplification = $(1.65 \times 10^5 \times 3.5 \times 10^4) = 5.78 \times 10^9$

KANSAS STATE UNIVERSITY

Madiyar, F. R., Li, J., et.al., *Nanoscale*, **2015**, (7), 3726-3736
In collaboration with Dr. Huang's Lab , University of Memphis

19

Transmission Electron Microscope and Confocal Images

TEM of Nanoovals | **TEM of Bacteria coated with Nanoovals** | **Confocal Fluorescence Image Bacteria coated with Nanoovals**

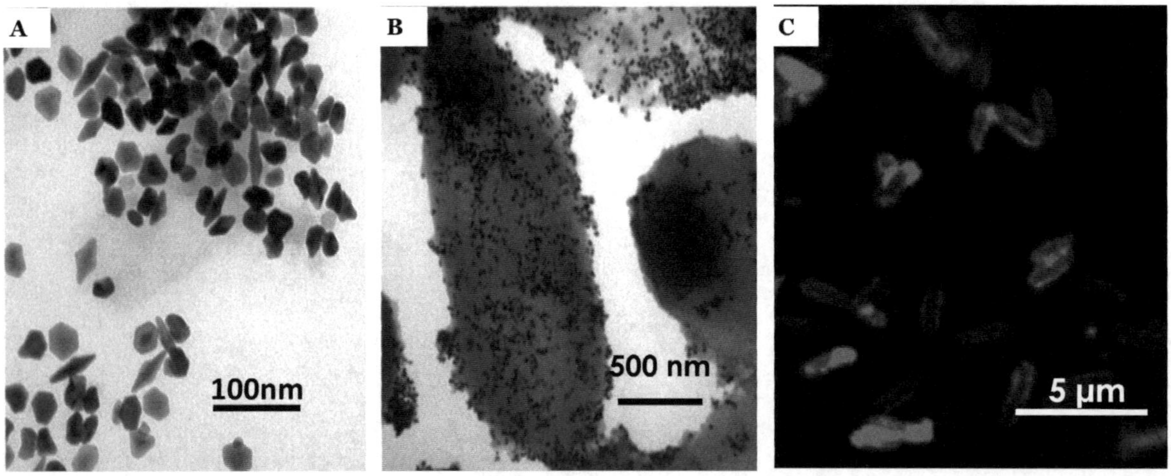

KANSAS STATE UNIVERSITY

Madiyar, F. R., Li, J., et.al., *Nanoscale*, **2015**, (7), 3726-3736

20

Device in Action to Capture *E. coli*

A
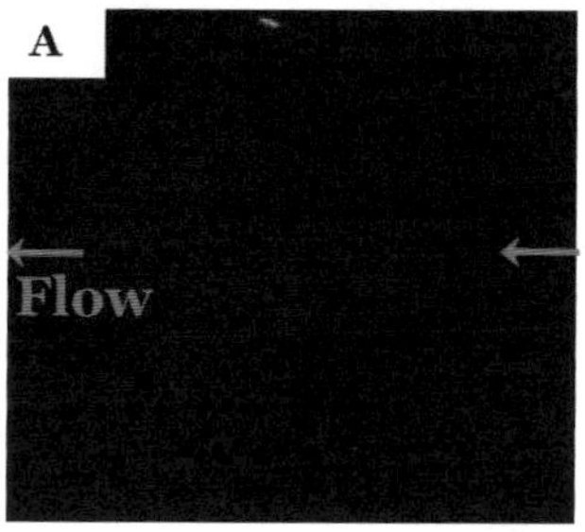

Linear Velocity 210 μm/sec

B
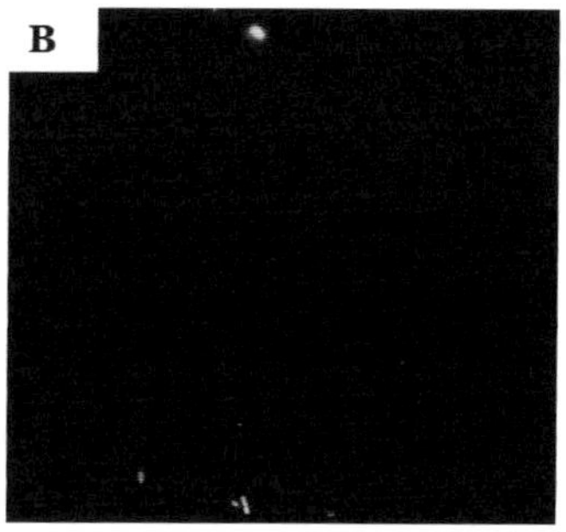

Linear Velocity 330 μm/sec

C
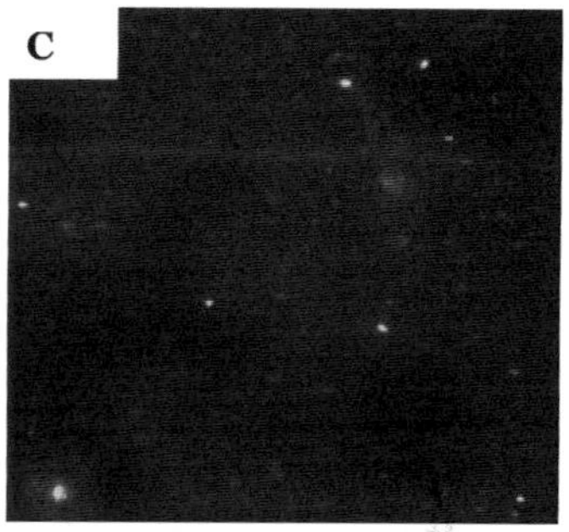

Linear Velocity 2430 μm/sec

Real - Time Detection and Portability

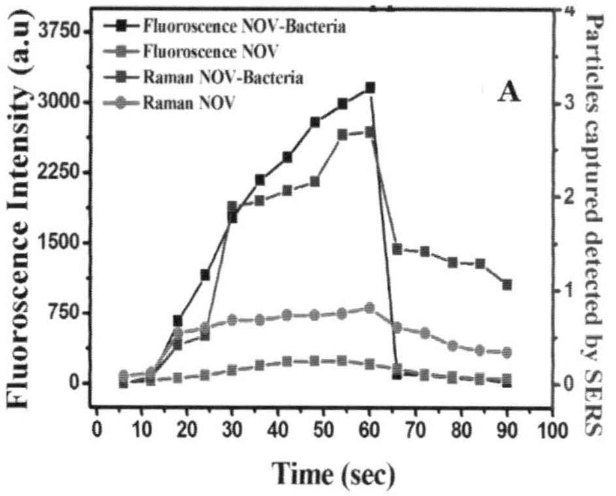

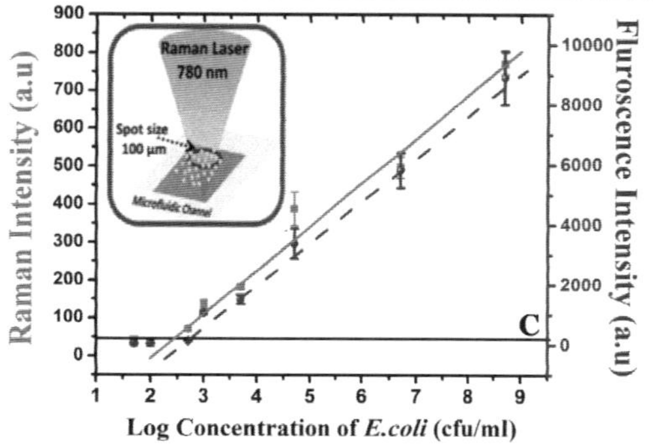

Probe Size: 100 μm (Diameter at the focused spot)
Power/Area (Power Density): 0.0414 mW/μm²

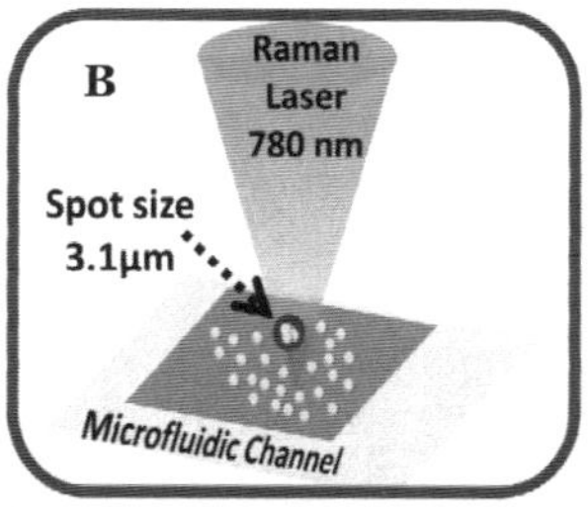

Probe Size: 3.10 μm (Diameter at the focused spot)
Power/Area (Power Density): 0.66 mW/μm²

$(\text{Raman Intensity})_{\text{portable}}$ = 108.8 x log C −214.7
Fluorescence Intensity = 1224 x log C − 2650

$\log C_{dl} = \log C_o + 3\sigma_{blank}/m$
$(RI)_{blank}$ = ~36 a.u.
σ_{blank} = ~11.7 a.u.

The detection limit SERS measurement was ~210 CFU/mL
The detection limit with fluorescence measurements was **~827 CFU/mL**

Capture of Bacteria in Complex Matrix Samples

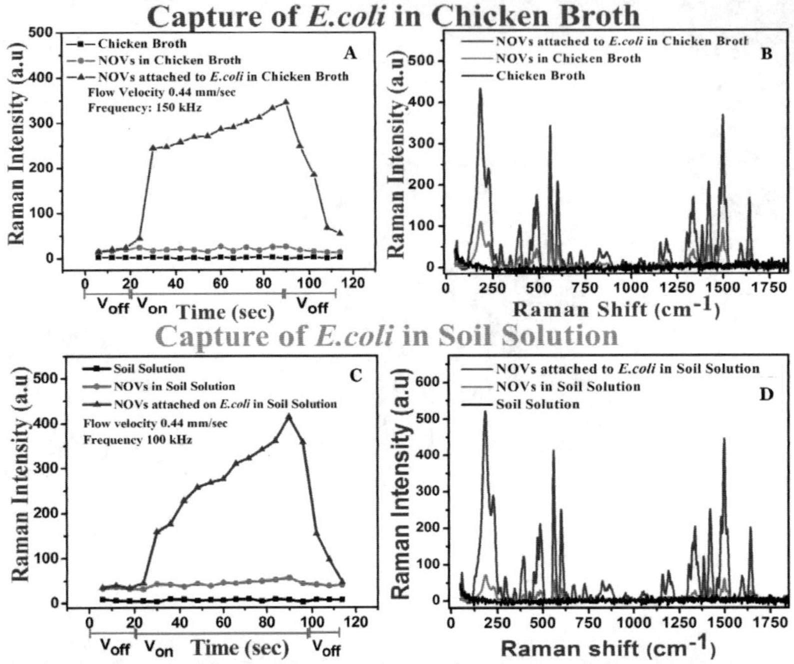

Madiyar, F. R., Li, J., et.al., *Nanoscale*, **2015**, (7), 3726-3736

Introduction to Impedance Biosensor

Electrical impedance is defined as the ratio of an incremental change in sinusoidal voltage to the resulting change in current.

$$Z^* = \frac{V(t)}{I(t)} = \frac{V_o\sin(\omega t)}{I_o\sin(\omega t + \varphi)} = \frac{1}{Y^*}$$

Z^* is described as a complex number

$Z^* = Z' + jZ''$, where $j = \sqrt{-1}$

A. J. Bard, L. R. Faulkner, Electrochemical Methods: Fundamentals and Applications, Wiley, **2001**

Impedance Microbiology

Bulk property measurement

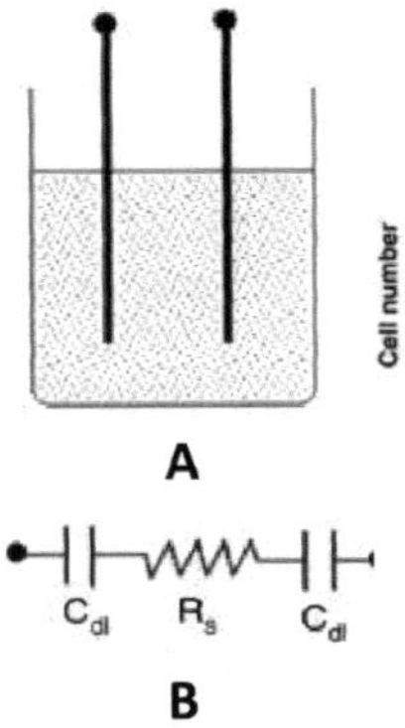

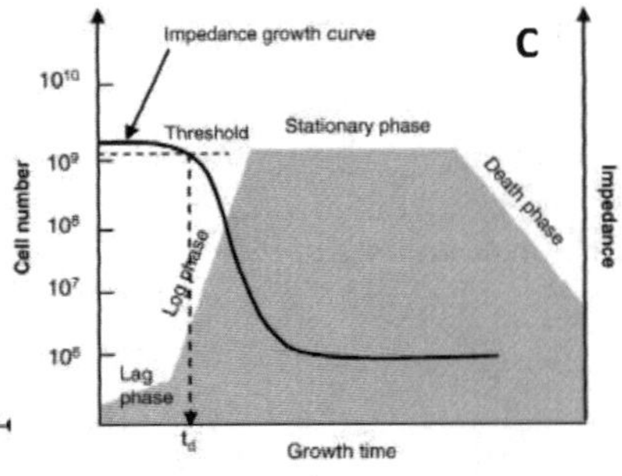

Talaro KP. Foundations in microbiology. 5th edition. New York, NY: McGraw Hill; **2005**. p. 209-10.

Advantages of Impedance Biosensors

1. Low cost
2. Small instrument size
3. Speed of analysis
4. Cutting edge accuracy
5. Point of care diagnosis
6. Label free

Bueno, P. R., J. Anal. Bioanal. Tech., **2015**, S7:016

Daniels, J. S.; Pourmand, N., *Electroanalysis*, **2007**, *19* (12), 1239-1257

1. Bueno, P. R., J. Anal. Bioanal. Tech., 2015, S7:016
2. Liju Yang, R. B., *Biotechnology Advances*, **2008**, *26*, 135-150.
3. Daniels, J. S.; Pourmand, N., *Electroanalysis*, **2007**, *19* (12), 1239-1257
4. A. J. Bard, L. R. Faulkner, Electrochemical Methods: Fundamentals and Applications, Wiley, **2001**

Impedance Biosensor Mechanisms

(1) Electrochemical impedance biosensors using redox probes

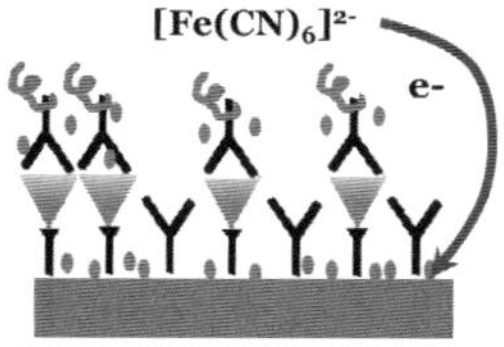

Liju Yang, R. B., *Biotechnology Advances* **2008**, *26*, 135-150.

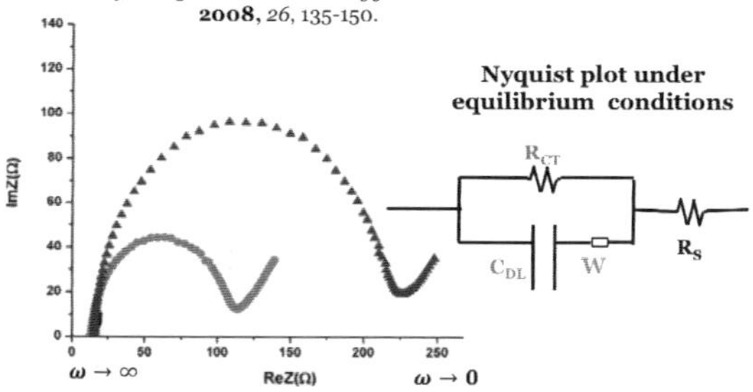

The bacteria *E. coli* detection limit of 6.0×10^2 CFU/mL
Ruan, C., Yang, L., Li, Y., Anal Chem, **2002**,74:4814–20

(2) Impedance biosensors for bacterial detection without redox probes

Radke, S. M., Alocilja, E. C., *Sensors J.* **2005**, *5* (4), 744-750

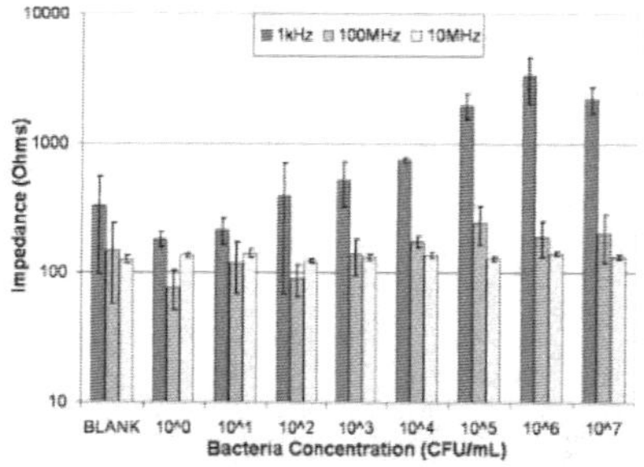

Detection of bacterial cells in food samples 10^4 CFU/mL
Radke S.M., Alocilja E.C., Biosens Bioelectron , **2005**, 20:1662–7

KANSAS STATE UNIVERSITY

1. Gómez-Sjöberg, R.; Morisette, D. T.; Bashir, R., *J. Microelectromech. Syst.* **2005**, *14* (4), 829-838
2. Park, K.; Akin, D.; Bashir, R., *Biomed Microdevices* **2007**, 9 (6), 877-883
3. Yang, L.; Li, Y.; Erf, G. F., *Anal. Chem.* **2004**, *76* (4), 1107-1113
4. Varshney, M.; Li, Y., *Biosens. Bioelectron.* **2009**, *24* (10), 2951-2960

25

Vaccinia Virus Detection by Impedance Method

Frequency study of *Vaccinia* Virus

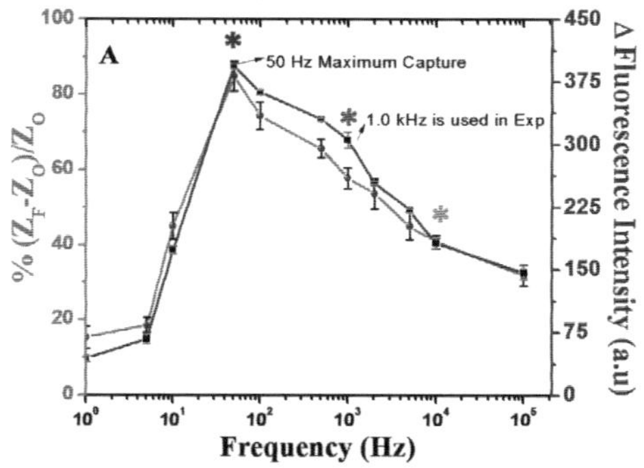

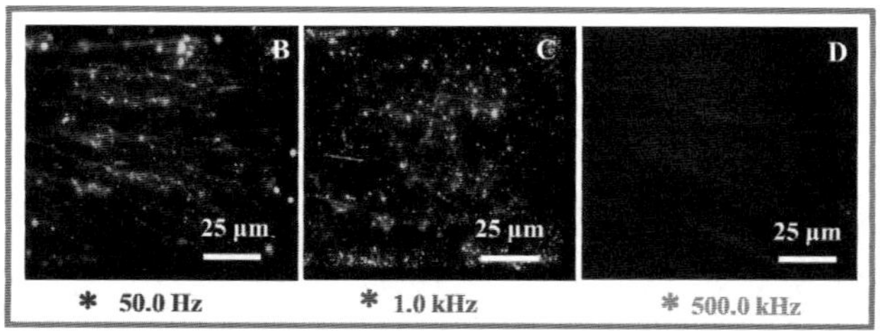

Vaccinia Virus

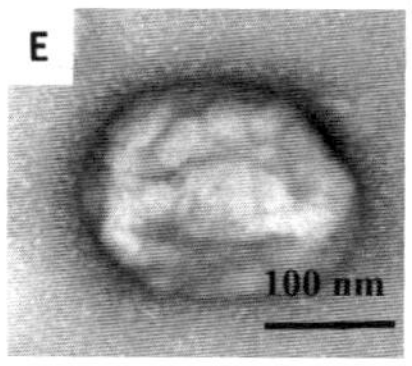

Real-time impedance measurement at various concentrations

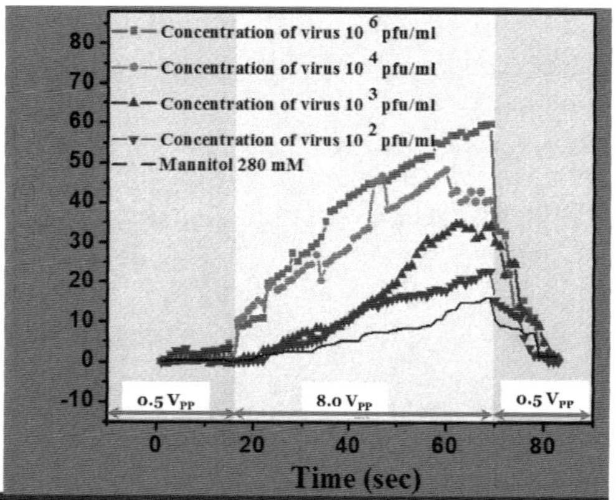

KANSAS STATE UNIVERSITY

Madiyar, F. R., Basset, S., Swisher, L., Culbertson, C. T., Rothenburg, S. , Li, J. (In preparation)

26

Vaccinia Virus Detection by Impedance Biosensor

Electroporation of Vaccinia Virus

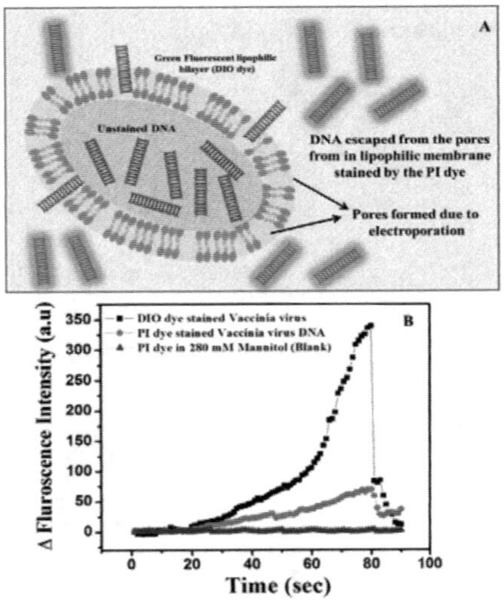

DiO dye stains the lipophilic membrane
Exi/Emi: 484/501 nm
Propidium Iodide (PI) stain the DNA in solution
Exi/Emi: 493 / 636 nm

Detection Limit using Impedance Method

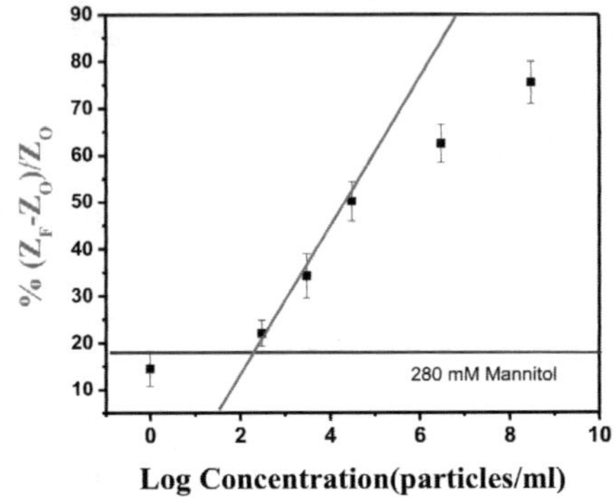

$$\frac{\Delta Z}{Z_0} = 12.1 \times \log C - 8.09$$

$\log C_{dl} = \log C_o + 3\sigma_{blank}/m$
I_{blank} = ~ 14.53 a.u.
σ_{blank} = ~3.77 a.u.

The detection limit with Impedance measurement was ~ **2.58×10^3 particles/ mL**

KANSAS STATE UNIVERSITY

1. Madiyar, F. R., Basset, S., Swisher, L., Culbertson, C. T., Rothenburg, S., Li, J. (In preparation)
2. Molecular probes, Invitrogen detection technologies, Lipophilic Tracers—DiI, DiO, DiD, DiA, and DiR handbook.

Summary

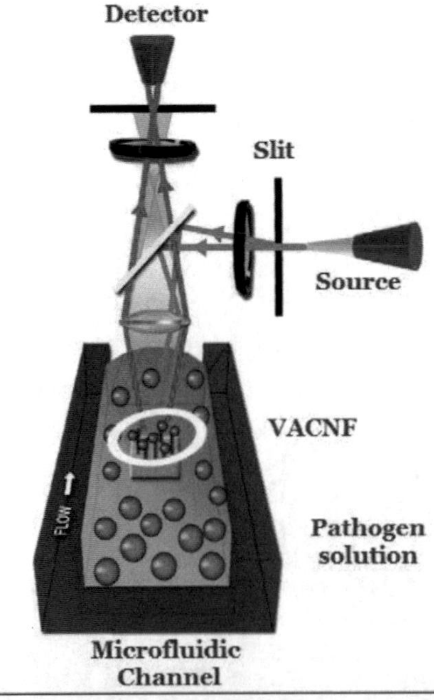

KANSAS STATE UNIVERSITY

1. Thiel, A.; Scheffold, A.; Radbruch, A. *Immunotechnology* **1998**, 4, 89-96
2. Davey, H. M.; Kell, D. B. *Microbiological Reviews* **1996**, 60, 641-696
3. Deering, A. J. et al. In *Micro/Nano Symposium (UGIM)*, **2010**, 18th Biennial University/Government/Industry, p 1-6
4. Zourob, Mohammed, *Principles of Bacterial Detection: Biosensors, Recognition Receptors and Microsystems*, Springer **2008**
5. Culbertson, C. T., et al., *Anal. Chem.*,**2014**, *86* (1), 95-118

Summary

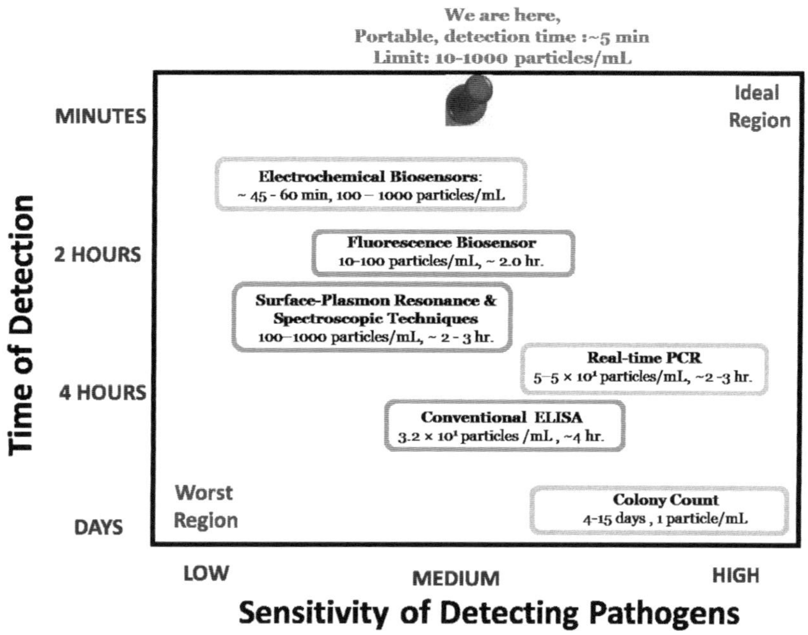

KANSAS STATE UNIVERSITY

1. Thiel, A.; Scheffold, A.; Radbruch, A. *Immunotechnology* **1998**, 4, 89-96
2. Davey, H. M.; Kell, D. B. *Microbiological Reviews* **1996**, 60, 641-696
3. Deering, A. J. et al. In *Micro/Nano Symposium (UGIM)*, **2010**, 18th Biennial University/Government/Industry, p 1-6
4. Zourob, Mohammed, *Principles of Bacterial Detection: Biosensors, Recognition Receptors and Microsystems*, Springer **2008**
5. Culbertson, C. T., et al., *Anal. Chem.*,**2014**, *86* (1), 95-118

29

Conclusions

- The DEP based on a VACNF NEA acts as an effective and reversible electronic manipulation technique to rapidly concentrate virus and bacteria into a micro area from the solution flowing through the microfluidic channel.
- The exposed tips carbon nanofibers (diameter ~100 nm) generate high electric fields at relative low AC potential that is sufficient to capture the *Bacteriophage* T4r.
- Lowest concentration that we could capture was 1×10^4 pfu/mL at the optimum frequency of the AC voltage of 10 kHz and the flow rate of 0.73 mm/sec.
- A highly sensitive SERS nanotag based on QSY21 adsorbed on IO-Au NOVs provides greatly enhanced Raman signals and specific recognition of *E. coli* DHα5 with a concentration detection limit of 210 CFU/mL using the portable Raman system, and 827 CFU/mL by fluorescence microscope.
- Lastly, label free impedance detection of *Vaccinia* Virus was accomplished with a concentration detection limit of 2.58×10^3 particles/ mL

KANSAS STATE UNIVERSITY

30

Acknowledgements

Thanks to my major advisor

- Dr. Jun Li

Thanks Chemistry and Biology Department, Kansas State University.

Special Thanks to:

- Luxi Swisher
- Steven Klankowski
- Yiqun Yang
- Emery Brown
- Dr. Gaind Pandey
- Dr. Lateef Syed (former member)
- Tao Liu
- Dr. Yifen Li
- Dr. Dan Boyle (Dept. of Biology, KSU)
- Sherry L. Haller (Dept. of Biology, KSU)
- Dr. Stefan Rotherburg (Dept. of Biology, KSU)
- Saheel Bhana (Dept. of Chemistry, University of Memphis)
- Dr. Xiaohua Huang (Dept. of Chemistry, University of Memphis)

Funding from:

- Sponsored by Emery Riddle Aeronautical Univeristy, Physical Science Department, Daytona Beach, FL 32114
- Kansas Bioscience Authority through the Centre of Excellence for Emerging Zoonotic Animal Disease (CEEZAD)
- National Cancer Institute grant number R15CA159250
- US Department of Homeland Security Scientific Leadership Award grant # 2012-ST-062-000055
- Alpha Upsilon Honorary Chemical Society, KSU.
- Graduate Student Council and College of Arts & Sciences Graduate Student Research Travel Awards, 2014-2015

C · E · E · Z · A · D
Center of Excellence
for Emerging and Zoonotic Animal Diseases

Johnson Center for
Basic Cancer Research
KANSAS STATE UNIVERSITY

EARLY WARNING
Detect Pathogens...
...Prevent Outbreaks

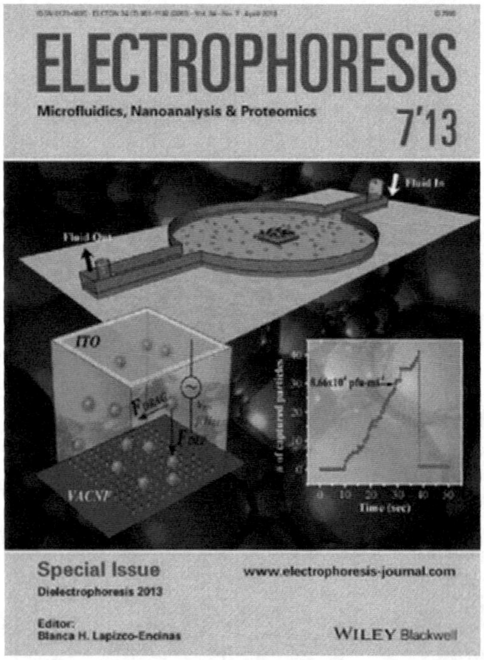

Thank you
Welcome for questions

429148 Protein Targeted Corona Phase Molecular Recognition

Abstract:

Corona Phase Molecular Recognition (CoPhMoRe) is a method whereby a heteropolymer is adsorbed onto a nanoparticle surface, templating it for the recognition of a specific target analyte. Recently demonstrated for small molecules [1,2], CoPhMoRe applicability has not been extended yet to macromolecules such as proteins. In this work a CoPhMoRe screen of single walled carbon nanotubes (SWCNT) wrapped in a library of phospholipid-poly ethylene glycol derivatives and oligonucleotides against a panel of human blood proteins reveals a phase highly selective to fibrinogen. This corona phase has one of the highest relative coverage of the nanotube surface, of more than 85%, and upon the interaction with fibrinogen, an 80% decrease in fluorescent emission intensity is observed. The three nodules of fibrinogen are shown to bind sequentially in a three step mechanism, starting with the outer D-regions followed by the central E-region, with dissociation constants that increase with the SWCNT diameter, ranging from 3.5 nM for the (6,5) chirality to 31.4 nM for the (11,3) chirality. Our results open a new direction in the fundamental search of novel synthetic non-biological antibodies for the recognition of bio-macromolecules, proteins, and peptides, for biological applications.

References

[1] Zhang J. *et. al.*, *Nature Nanotechnology*, 8, 959–968, 2013

[2] Kruss S. *et. al.*, *J. Am. Chem. Soc.*, 136 (2), 713–724, 2014

430111 Sensing Applications of Single-Walled Carbon Nanotube Imaging Spectroscopy in Live Cells

Abstract:

The intrinsic near-infrared photoluminescence from an individual single-walled carbon nanotube (SWCNT) can be detected within a live cell with epifluorescence emission. As the emission spectrum of a SWCNT is exquisitely modulated by its immediate microenvironment, information contained in the peak emission wavelength, intensity, and full width half maximum can reflect physicochemical properties of the microenvironment. We have developed a hyperspectral imaging platform for spatially localizing the spectra from individual SWCNTs within live cells. Thus, by spatially localizing a polymer-functionalized SWCNT to a subcellular organelle, and measuring the spectral response to its environment, we demonstrate the use of single SWCNT spectroscopy to quantify specific metabolites in live cells.

416331 Nanoenzyme Catalysis Factory on DNA Origami Platform

Abstract:

DNA, not only widely studied as the secret of life but also have been one of the widely used candidate to fabricate structures in nano-meter domain for past 30 years or so using bottom-up methodology. Current trend in DNA nanotechnology is to add functional aspects to these well defined DNA nano-structures. Lately a lot of minute structural details have been successfully programmed and demonstrated; especially since the beginning of DNA origami era in last decade. Much work has also been done to functionalize DNA origami for irreversible and single step functions. This talk will narrate the design and demonstration of first ever completely reversible enzyme catalysis nano-factory on DNA origami surface which is able to modify DNA and the modified products can be specifically removed such that one can modify new target DNA with the same or a different chemical group in the next nano-machine cycle. Visualization of this nano-factory is done through atomic force microscope (AFM). These kind of nano-enzyme factory can be a well studied stepping stone to the new generation nano-bio-chip for parallel detection in nano-dimension.

431404 Cationic Nanogels for the Co-Delivery of siRNA and Chemotherapeutics

Abstract:

Cancer cells can be intrinsically drug resistant or can become drug resistant when exposed to chemotherapeutics. Specifically, cancer cells can develop resistance through up-regulation of efflux pumps or anti-apoptotic pathways. In addition, chemotherapeutic regimens can essentially select for a cancer cell subpopulation resistant to the chosen therapy. Currently, the preferred method of treatment for the majority of solid tumors is surgical resection followed by chemotherapy and radiation. As such, methods to improve chemotherapeutic regimens to combat drug resistant cancer are needed. Small interfering RNAs (siRNAs) are 21-23 nucleotide double stranded RNAs that have the ability to specifically silence the expression of a target gene. However, siRNAs are rapidly degraded by nucleases in the body. As such, delivery vehicles including cationic polymers and liposomes have been studied to enhance chemotherapeutic efficacy because of their ability to load and release chemotherapeutic agents and siRNA. Here, cationic nanogels were synthesized with either randomly incorporated or localized pH responsive and hydrophobic functional groups. The presence of localized functional groups are hypothesized to enhance pH responsive properties as well as drug and siRNA loading to reduce off target affects associated with premature release of therapeutic agents.

In this work, activators regenerated by electron transfer (ARGET) atom transfer radical polymerization (ATRP) was utilized to synthesize cationic nanogels with either localized or randomly distributed functional groups. For random polymerizations, tertiary amino ethyl methacrylate monomers, poly ethylene glycol methyl ether methacrylate, surfactants, crosslinker and ATRP reagents were combined in water and sonicated into an emulsion. The emulsion was nitrogen purged and polymerized for 24 hours upon addition of a reducing agent. Cationic nanogels with localized functional groups were similarly synthesized through sequential addition of monomers to a PEG macroinitiator. Cationic nanogels were purified though five cycles of swelling and ionmer collapse in hydrochloric acid and acetone followed by dialysis against water for one week and lyophilized. Incorporation of functional groups for nanogels with localized functional groups was tracked with proton nuclear magnetic resonance and gel permeation chromatography on linear analogs of nanogels.

The swelling properties of cationic nanogels were observed with dynamic light scattering from pH 4-9. The onset of the pH dependent swelling was modulated by varying monomer feed ratios. Typical nanogels exhibited hydrodynamic diameters of 80-100nm in collapsed state to 120-140nm swollen in 1x PBS at 25C. Similarly, zeta potential was used to track apparent surface charge on the nanogels through the same pH range. Positive zeta potential on the order of 15mV was observed for swollen gels with almost neutral zeta potential observed at pH 7.5.

Initial screening of the randomly polymerized nanogels exhibited minimal cytotoxicity when incubated with RAW 264.7 murine macrophages at concentrations up to 0.1mg/mL for 24 hours. Results were calculated using a lactate dehydrogenase membrane integrity assay. Further, randomly polymerized cationic nanogels exhibited the ability to electrostatically bind siRNA with high efficiency post synthesis. In a typical experiment, 95% of siRNA incubated with the nanogels was electrostatically bound as determined by Quant-iT™ RiboGreen® RNA Assay.

Cationic nanogels with randomly incorporated and localized functional group were synthesized. Swelling and surface potential tracked as a function of pH demonstrated desirable pH responsive behavior, concentration dependent cytotoxicity demonstrated limited toxicity in the anticipated dosing range, and high siRNA loading demonstrated the promise of these cationic nanogels for the co-delivery of siRNA and chemotherapeutics. Further studies will compare the co-delivery potential of nanogels with localized functional group to those with randomly distributed functional groups with respect to swelling properties, drug and siRNA loading and release, and in vitro toxicity and transfection.

This material is based upon work supported by the National Science Foundation Graduate Research Fellowship Program.

420128 Electronic Platform for Real-Time Multi-Parametric Analysis of Cellular Behavior Post Exposure to Single-Walled Carbon Nanotubes

Abstract:

Single-walled carbon nanotubes (SWCNTs) implementation in a variety of biomedical applications from bioimaging, disease targeting to controlled drug and gene delivery, has raised awareness of their potential toxicity. Nanotubes structural aspects which resemble asbestos, as well as their ability to induce cyto and genotoxicity upon interaction with biological systems by generating reactive oxygen species or inducing membrane damage, just to name a few, have led to focused efforts aimed to assess associated exposure risks prior their user implementation. In this study, we employed a non-invasive and real-time electric cell impedance sensing (ECIS) platform to monitor the cellular behavior of human lung epithelial cells upon exposure to a library of SWCNTs with user-defined physico-chemical properties. Using the natural sensitivity of the cells as primary transducers, we evaluated SWCNT-induced cellular changes in relation to cell attachment, cell-cell interactions and cell viability respectively. In particular, our results showed functional differences in the cellular behavior upon exposure to pristine and acid treated SWCNTs according to nanotube's length, surface chemistry, purity and agglomeration state. Our methods have the potential to lead to the development of standardized assays for risk assessment of other nanomaterials as well as risk differentiation based on the nanomaterials physico-chemical properties.

425561 Silver Nanoparticle-Embedded Polymersome Nanocarriers for the Treatment of Antibiotic-Resistant Infections

Thursday, November 12, 2015: 3:45 PM
253A (Salt Palace Convention Center)
Benjamin M Geilich, Bioengineering, Northeastern University, Boston, MA and Thomas J. Webster, Chemical Engineering, Northeastern University, Boston, MA

Introduction

Antibiotics have been extensively used since their commercialization in the 1930s to treat patients suffering from a wide variety of infectious diseases. Unfortunately, however, these drugs have been used so prevalently over the last 80 years that the bacteria they were designed to kill have begun to evolve and adapt, rendering them ineffective. According to the Center for Disease Control, at least 2 million people acquire serious infections from antibiotic resistant bacteria each year, and over 23,000 die as a direct result. Even when alternative treatments exist, patients with antibiotic-resistant infections have significantly higher mortality rates, and survivors often have increased hospital stays and long-term complications. These infections cost an estimated $20 billion in excess direct healthcare expenses.1 Infections caused by Gram-negative bacteria are particularly difficult to treat because these bacteria have a robust and hydrophobic outer lipopolysaccharide membrane that can impede the influx of drugs into the cell.2 Unfortunately, the number of new antibiotic drugs in the pipeline has also been rapidly decreasing, largely due to the fact that new drugs are extremely expensive to bring to market, and antibiotics are less financially lucrative to develop when compared to treatments for chronic conditions. Thus, the need to develop alternative strategies to treat such antibiotic-resistant bacteria, while still utilizing existing drugs, has never been more urgent than today.

Over the past decade, interest in using nanomedicine-based approaches to combat difficult infections has rapidly grown due to the many advantages offered over conventional treatment with free antibiotics. This study explored encapsulating the drug inside nano-sized structures called polymersomes (that is, artificial vesicles made from biodegradable, high molecular weight, amphiphilic block co-polymers). These vesicles typically display a spherical morphology and are composed of hydrated hydrophilic coronas both at the inside and outside of a hydrophobic polymer membrane. This allows for hydrophilic bioactive materials to be loaded into the particle's aqueous core, and hydrophobic bioactive materials to be loaded into the particle's membrane bilayer. Along these lines, metallic nanoparticles have long been investigated as potential antibacterial agents due to their many unique physiochemical properties which are not present at the macro scale. Among these metals, silver is perhaps the most well-known for its antimicrobial effects. Recent studies have shown that there may even be a synergistic effect when silver nanoparticles and antibiotics are used simultaneously to treat a Gram-negative infection.3 However, there is little information regarding whether a combined treatment is sufficient to overcome bacteria which display genetic antibiotic resistance. Additionally, there has been almost no investigation into the effect of dually encapsulated antibiotics and nanoparticles.

Thus, for all of the above reasons, the objective of the present in vitro study was to design, characterize, and optimize a polymersome nanocarrier to co-localize and deliver both antibiotics and hydrophobic silver nanoparticles, while protecting the drug from hydrolysis by β-lactamase enzymes, into a single "nanoformulation" in order to kill antibiotic- resistant bacteria. Specifically, these particles were then tested for efficacy against a strain of *Escherichia coli* (*E. coli*) which had been genetically modified to be antibiotic- resistant.

Results

Silver nanoparticle-embedded polymersomes (AgPs) were synthesized using a modified stirred-injection technique. Monodispersed hydrophobic silver nanoparticles 5 nm in diameter were suspended in an organic solvent containing dissolved mPEG-PDLLA. This mixture was injected through a syringe atomizer at high speed into actively stirring phosphate buffered saline (PBS, pH 7.4) containing the antibiotic ampicillin. The resulting suspension was allowed to dialyze against PBS to remove the organic solvent and non-encapsulated drug.

Physicochemical characterization was performed to assess AgPs size, surface charge, and loading. Transmission electron microscopy (TEM) revealed polymersomes of highly uniform size and shape with clusters of silver nanoparticles embedded inside (Figure 1A). These silver clusters frequently appeared as a single layer of nanoparticles that was off-center from the nanoparticle core, suggesting that they may be intercalated into the membrane bilayer. Dynamic light scattering (DLS) indicated that the average hydrodynamic diameter was 104.3 nm ± 15.6 nm (Figure 1B). The AgPs surface was found to have a near neutral zeta potential of 0.315 mV ± 1.13 mV at pH 7.4. The number of silver nanoparticles embedded per polymersome was quantified from TEM images. The nanoparticles were shown to load in a normal tailed distribution with an average of 9.29 ± 6.07 silver nanoparticles per polymersome (Figure 1C) . The mass of silver loaded was estimated using the density of silver and the volume of a 5 nm sphere.

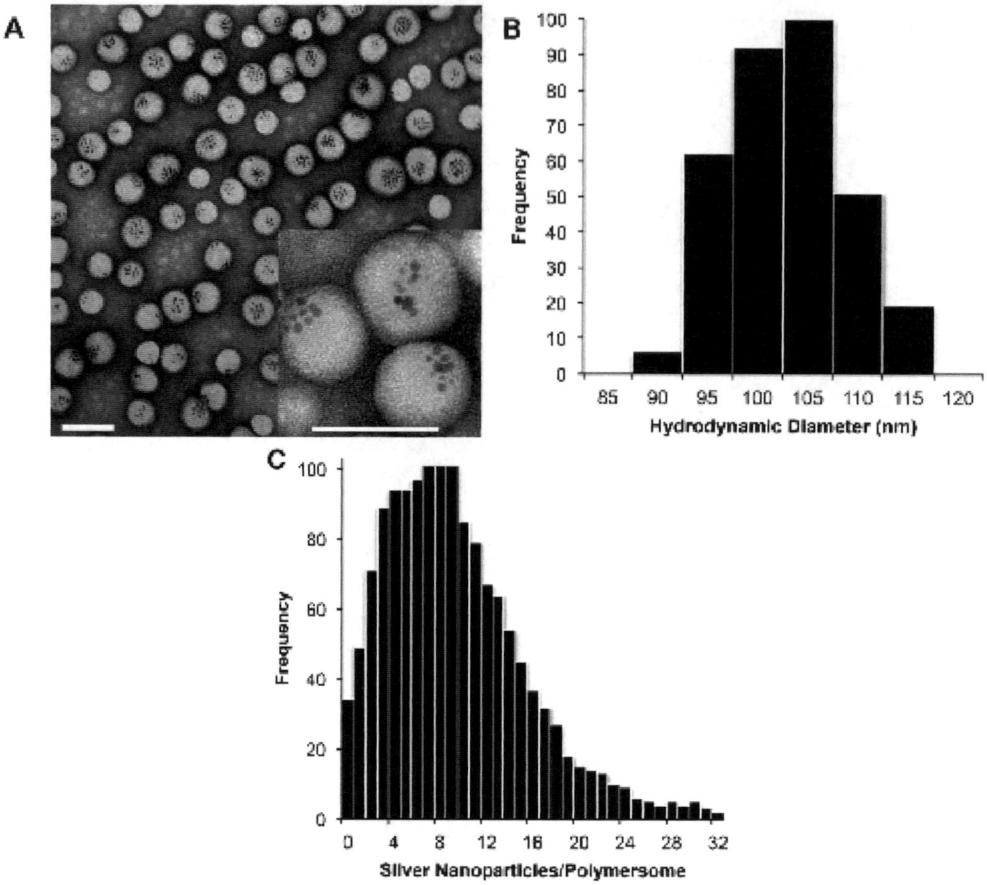

Figure 1: Physiochemical characterization of the AgPs

The growth and proliferation of a 106 colony forming units mL-1 (CFU mL-1) suspension of ampicillin-resistant *E. coli* was examined by measuring the optical density at 600 nm (OD600) for 24 hours following treatment with volumes of AgPs containing a silver : ampicillin (Ag : Amp) ratio of 1:0.28 (Figure 2A), 1:0.44 (Figure 2B), or 1:0.64 (Figure 2C). Ampicillin- loaded AgPs displayed significant bacteriostatic action against the *E. coli*, manifesting as a delay in the time taken to reach exponential growth phase. This response was dose-dependent, with higher concentrations of ampicillin producing a longer delay in bacterial growth. Bacteria treated with ampicillin concentrations above 55 µg mL-1 failed to proliferate within 48 hours. In the absence of silver nanoparticles, no bacteriostatic effect was observed for all ampicillin concentrations tested. This suggests that the presence of silver potentiates the therapeutic efficacy of ampicillin. AgPs without ampicillin like- wise produced no therapeutic benefit. Additionally, no significant differences were observed between bacteria treated with free ampicillin (200 µg mL-1), PBS, AgPs without ampicillin, and ampicillin loaded polymersomes (200 µg mL-1) without silver nanoparticles. When bacteria were treated with sub-optimal concentrations of AgPs, bacterial growth was always observed within 17 hours. The time to exponential phase was found to vary with both silver concentration and ampicillin loading.

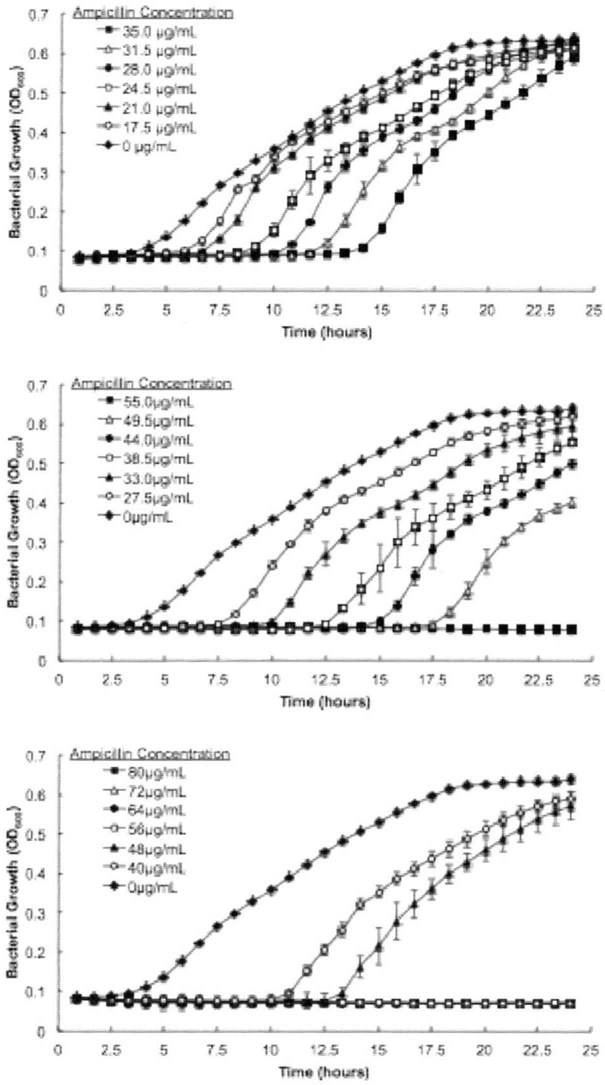

Figure 2: Bacterial growth inhibition from the AgPs

A Bliss Model was utilized to determine the degree of synergy for different silver and ampicillin combinations.4 Drug interactions were found be synergistic (S > 0) in all cases where ampicillin was supplied at concentrations of 24 µg mL-1 and above. At lower concentrations, no synergism was observed (S = 0). The degree of synergy was dose-dependent and increased with both silver and ampicillin concentrations. The therapeutic benefit of ampicillin reached a plateau at 50 µg mL-1 over a range of silver concentrations due to complete inhibition of bacterial growth. When the silver concentration was held constant, the degree of synergism was directly determined by the amount of ampicillin loaded.

Interactions between *E. coli* and AgPs were visualized using TEM (Figure 3). Indentation of the bacterial cell membrane was observed in regions of AgPs contact. Silver nanoparticles inside AgPs appeared to be polarized in an orientation perpendicular to the bacterial cell membrane, suggestive of hydrophobic interactions with the outer cell membrane. In order to assess physical intracellular changes caused by AgPs, cells treated with an intermediate particle concentration (44 μg mL-1 Amp, 1 Ag:0.44 Amp) for 24 hours were sectioned. Bacteria in contact with AgPs displayed significant protein aggregation and diffuse widening of the cell envelope. This phenomenon has been observed by other researchers following silver ion treatment, and has been shown to correlate with increased membrane permeability and protein misfolding due to disulfide bond disruption.4 Regions of the cell envelope with little to no AgPs contact appeared morphologically normal.

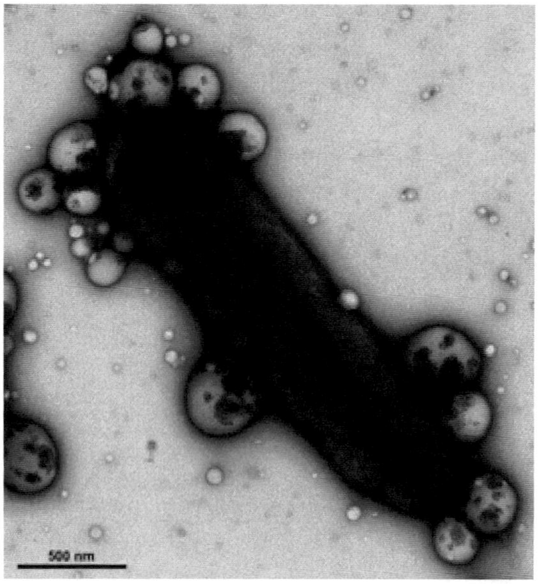

Figure 3: Transmission electron micrograph displaying the AgPs aggregating around an *E. coli* bacterium.

Discussion

This work has many important implications. A long-lasting, single-particle treatment capable of overcoming antibiotic resistance would be extremely beneficial in the clinic. This potential is compounded by the fact that the initial bacteria density found in vivo is rarely as high as the CFU studied here. It has been shown that silver nanoparticles are also effective at treating bacterial persister cells, and therefore AgPs may show promise for biofilm-forming infections.5 One of the most commonly occurring bacterial infections, urinary tract infection, is frequently caused and perpetuated by biofilm-forming (and often antibiotic-resistant) strains of *E. coli*.6 Additionally, the absence of a significant cytotoxic effect from AgPs towards human fibroblasts is a promising sign for toleration by mammalian cells. Reports of toxicity associated with nanoscale silver in vivo have been varied, and toxicity is generally considered to depend on nanoparticle size, concentration, and surface coating. As designed, the AgPs particle is easily loaded with a variety of aqueous drugs, and combinations thereof, opening avenues for the creation of a library of therapeutic particles.

References

[1] Antibiotic Resistance Threats in the United States, Centers for Disease Control and Prevention, 2013.

[2] J. M. Pages et al, The porin and the permeating antibiotic: a selective diffusion barrier in Gram-negative bacteria, *Nat. Rev. Microbiol.*, 2008, 6, 893.

[3] P. Li et al, Synergistic antibacterial effects of β-lactam antibiotic combined with silver nano- particles, *Nano.,* 2005, 16(9), 1912.

[4] J. R. Morones-Ramirez et al, Silver Enhances Antibiotic Activity Against Gram-Negative Bacteria, *Sci. Transl. Med.*, 2013, 5(190), 190ra81.

[5] D. Roe et al, Antimicrobial Surface Functionalization of Plastic Catheters by Silver Nanoparticles, *J. Antimicrob. Chemother.*, 2008, 61(4), 869–876.

[6] T. Mah, Biofilm-Specific Antibiotic Resistance, *Future Microbiology*, 2012, 7(9), 1061.

430782 Photo-Activation of Near Infra-Red Light Absorbing Nanoparticles : Continuous Wave and Pulse Irradiation

Abstract:

Currently, a popular area in nanomedicine is the implementation of plasmonic gold nanoparticles for cancer treatment. The wide-ranging implementation of gold nanoparticles for medical applications is based largely on the unique and highly tunable optical properties that gold nanomaterials provide. Hollow gold nanoparticles (HGNs) have been appealing to medical applications, since the HGN-absorbance maximum can be easily tuned to 700 – 900 nm, near infra-red (NIR) light, by adjusting the ratio of the shell thickness to the nanoparticle diameter. The advantage of NIR is that it can penetrate several centimeters of soft tissues without showing any significance harmful effect on tissues. In previous works, the HGN synthesis, which involves the galvanic exchange of gold on a silver nanoparticle template, provides a relatively polydisperse population, which broadens the absorption maximum. We have developed new methods of creating monodispersed hollow gold nanoparticles in different sizes (25 – 40 nm) and shapes (nanospheres and nanocubes). Moreover, we optimized the reaction to synthesize nanoparticles as small as 10 nm, by adjusting the reaction time. We use NIR light in two different ways to irradiate the HGNs, continuous wave and pico-second pulsed laser light. We show two different mechanism that the HGNs are photo-activated. We present the effect of size, shape, and concentration of HGNs at a specific power density with both continuous wave and pico-second pulsed laser irradiation. Nanoparticles were continuously irradiated with NIR for certain amount of time (e.g., 5 min) at a specific power density; substantial increments in temperature were observed. We compared the size, shape, and concentration effect of nanoparticles on heating. On the other way, nanoparticles were irradiated with picosecond NIR pulses. The generation of nanobubbles around nanoparticles were detected as evidence of fast (picoseconds) NIR absorption. We compared the threshold energy for nanobubble generation in different sizes of nanoparticles.

430793 Protein Confinement in Mesoporous Silica – Effects of Surface Curvature Investigated By Neutron Scattering and Catalysis

Abstract:

Confinement of biomolecules in structured mesoporous materials offers advantages in both biological and synthetic systems[1]. The application of these systems spans a diverse range of purposes, from industrial production of biofuels to laboratory-on-a-chip diagnosis and controlled drug delivery. Therefore, it is necessary to develop a fundamental understanding of biomolecular confinement. Our previous studies on confinement effects in the ordered mesoporous silica, SBA-15, which contains a hexagonal array of cylindrical pores with tunable diameter, have shown that geometric properties, such as pore surface curvature, and physicochemical properties, such as electric charge and hydrophobicity, play a significant role in the stability of the confined protein[2]and that these effects differ from proteins immobilized on the external surface of nanoparticles. High surface curvature, in particular, was shown to have a dramatic stabilization effect for the model proteins. This stabilization was most evident when comparing the secondary structure and extraordinary enzymatic activity of the confined protein samples. Expanding on these results, we have studied the structural and packing effects of curvature using the SBA-15 immobilized myoglobin and lysozyme model proteins with small angle neutron scattering. Additionally, we investigate these model proteins while confined in mesoporous materials with different pore network topologies and pore morphology, with a focus on enzymatic activity, loading capacity, and structural changes.

References:

1. Siefker J, Karande P, Coppens M-O. Packaging Biological Cargoes in Mesoporous Materials: Opportunities for Drug Delivery. Expert Opin Drug Deliv. 2014 Nov;11(11):1781-93
2. Sang L-C, Coppens M-O. Effects of surface curvature and surface chemistry on the structure and activity of proteins adsorbed in nanopores. Phys Chem Chem Phys. 2011;13(14):6689

432519 Focused Ultrasound Triggered Temperature Sensitive Liposomes for Treating Prostate Cancer

Abstract:

High-intensity focused ultrasound (HIFU) emerges as a powerful technology for noninvasive or minimally invasive non-ionizing treatment of cancer. HIFU deposits a large amount of acoustic energy at the focal region within the target tissue (i.e., tumor), causing localized heating and necrosis. HIFU treatment also causes pores in the cellular membrane, an effect known as sonoporation. The objective of this study is to combine the HIFU treatment with the triggered release of a chemotherapeutic agent (Sorafenib). The rise in temperature caused by HIFU was used to trigger the release of Sorafenib from temperature sensitive liposomes (TSLs). TSLs are self-assembled lipid based nano-carriers which can be triggered to rapidly release their contents above a certain temperature. HIFU alone can be used to effectively kill tumor cells at high enough intensities but even if few cells survive, they can grow back into a tumor. Thus, to effectively treat the tumor, a chemotherapeutic agent is used in synergy with the ablative effect of the HIFU treatment. The HIFU waves were administered at moderate intensities, enough to raise the temperature above the transition temperature of the TSLs to quickly release the drug. The HIFU treatment of cells also causes sonoporation, which increases the cell permeability to the chemotherapeutic drug allowing for increased treatment efficacy.

Liposomes were characterized for particle size (DLS), morphology (Cryo-TEM) and encapsulation efficiency (UV-VIS assay). The sensitivity of the liposomes to temperature was confirmed by monitoring the release of a dye at elevated temperatures. Cryogenic Scanning Electron Microscopy (Cryo-SEM) was used to visualize sonoporation in the cancer cell membranes. Pellets of DU145 human prostate cancer cells were placed in thin-wall PCR tubes to mimic dense tumor aggregation. Confocal microscopy confirmed the uptake of liposomes by the cells. The cell viability was found to be significantly lower for the cells exposed to both HIFU and Sorafenib encapsulating TSLs (SFTSLs) as compared to the cells exposed to Sorafenib only or SFTSLs only. The DU145 cell viability when treated with both HIFU and SFTSLs was greatly reduced post treatment. The results from a cell proliferation assay showed no significant regrowth of the tumor after the synergistic treatment. This implies that the sonoporation which increased the permeability of the cells, resulted in more bioavailability of the drug to the cells, effectively containing the tumor.

Acknowledgments: Authors acknowledge funding from National Science Foundation, Department of Defense, Louisiana Board of Regents, Newcomb-Tulane College, and Tulane Center for Engaged Learning and Teaching.

434305 Pnipmaam Based Core/Shell Systems for Improved Retention of Hydrophobic Chemotherapeutics

Abstract:

Chemotherapeutic drug delivery demands localization to the tumor site do to deleterious side effects that occur when they are allowed to spread throughout the body. The issue with delivering these therapeutics is twofold, a lack of solubility and systemic circulation. One solution to this is the development of externally triggerable theranostic vehicles that are capable of delivering these devastating agents to the site of disease, preventing off site effects. These systems, composed of a temperature sensitive coating around a stimuli responsive core, are questioned for a number of reasons. One major issue is the premature release of their payload, leaving little improvement over systemic delivery. This is particularly problematic for hydrogel based carriers, whose hydrated neutral state leaves little compatibility with, what are commonly, hydrophobic drugs, such as chemotherapeutics.

The premature release of the chemotherapeutics can be addressed by taking advantage of the low solubility of the chemotherapeutics in physiological conditions. Carrier design that focusses on the polymer-drug interactions can take advantage of partitioning resulting in retention of the chemotherapeutic payload. In the past, the use of N-Isopropyl acrylamide has demanded the incorporation of hydrophilic comonomers in order to raise the lower critical solution temperature (LCST) above physiological temperature. This can be avoided by the use of other N-Alkyl substituted acrylamides such as N-Isopropylmethacrylamide (NIPMAAm); which has an LCST of about 45°C. This allows for copolymerization with hydrophobic comonomers for the depression of the LCST increasing the partitioning of the hydrophobic drug in the polymer. This can again be improved upon by developing a carrier that has a coating of a drug compatible shell. This provides a hydrophobic sink that provides a high partitioning coefficient of the drug in the carrier.

Nanogels were synthesized via an emulsion precipitation polymerization. This involves the dissolution of NIPMAAm, a cross linker (N'N'-methylene bisacrylamide (MBAM)), and surfactants in water heated to 70°C and nitrogen purged. The reaction is then initiated by the injection of ammonium persulfate. For core/shell systems, after 6 hours the shell monomer; phenyl methacrylate (PhMA), tert-butyl methacrylate (TBMA), and ethylene glycol or phenyl acrylate (EGPhA), is injected and the reaction is allowed to proceed overnight. The resulting mixture is then dialyzed against 18.2 MΩ water and freeze dried.

The hydrogel systems are then characterized by dynamic scanning calorimetry (DSC), nuclear magnetic resonance (NMR) spectroscopy, and dynamic light scanning (DLS) to confirm the development of core/shell systems. Loading was performed by mixing nanogels with fluorescein or Doxorubacin hydrochloride (DOX) and allowed to equilibrate overnight. The free drug is then separated by centrifugation filtration and washed several times with water. Drug loaded is measured by fluorescence and thermogravimetric analysis.

The nature of the drug in the polymer is further analyzed by comparing DSC curves of free drug to that of drug encapsulated in the nanogels. The drug loaded nanogels were then also assessed for their biocompatibility *in vitro* with RAW 264.7 macrophage cells.

Release was characterized by dissolving drug loaded nanoparticles in phosphate buffered saline (PBS). The particles were initially allowed to reach equilibrium at 37°C, at 6 hours. After the 6 hour timepoint the temperature is raised to 45°C and the drug released is measured for another 2 hours.

The NMR spectrographs show the characteristic peaks of the shell material alone, this is due to the fact that NMR of crosslinked systems in aqueous environments only display the peaks of the solubilized polymer chains on the surface. DSC theromograms of core/shell materials show multiple glass transition temperatures (T_g) compared to that of PNIPMAAm. This demonstrates multiple distinct regimes of different polymer networks. These two facts lend to the conclusion that we successfully synthesized core/shell structures with a polymer compatible shell and a temperature sensitive core. DLS swelling curves and DSC curves in water demonstrate that these gels still maintain the characteristic swelling response of PNIPMAAm, without depression due to copolymerization. Showing that the shell has little impact on the responsiveness of the nanogel.

Loading studies show an increase in the drug loaded into the core/shell particles. These studies reveal a significant increase in the partitioning of the drug into the nanogels. This is further observed in the release studies. While PNIPMAAm nanogels release a large portion of their payload at 37°C. EGPhA showed mild, but no significant increase in partitioning. The hydrophobic systems, based on PhMA and TBMA, show undetectable release at 37°C. After heating to 45°C burst release is achieved, and a large portion of the drug payload is released. Higher levels of drug release is attained by pulsatile release achieved by cooling back down to 37°C and heating back up to 45°C.

The synthesis of core/shell nanogels is possible through the sequential addition of comonomers. These novel structures utilize a core and a shell that are completely separate domains. These systems have been shown to improve both the loading and release due to increased partitioning of common model chemotherapeutics. These results lend themselves to application *in vivio* as it provides extra control over these composite systems. This increased control leads to increased localization of the chemotherapeutic pay load to the site of the disease. This should allow the avoidance of the deleterious side effects that are common with standard chemotherapy.

434333 Selectively Permeable Microbial Nanoculture System

Abstract:

The advances in the field nanotechnology provide new means for isolating microbes, and replicating their growth conditions in microenvironments to answer new biological questions. For instance, the spatial confinement of bacteria in micropatterns or microcompartments has provided new insights on the inherent advantages of cell clustering in survival mechanisms and drug tolerance. As the worldwide antibiotic crisis persists, novel technologies are emerging to screen for new compounds that could potentially eradicate recalcitrant and multidrug-resistant pathogens. In this study, Polydimethylsiloxane (PDMS) is used to develop a selective microbial nanoculture system allowing the diffusion of only low-molecular weight molecules such as water, antibiotics, signaling molecules and fluorescent dyes. This nanoculture system enables the encapsulation and growth of microbes, the testing of antibiotic susceptibility, the labeling and sorting of the encapsulated bacteria and fungi based on their physiologic properties. The PDMS nanocultures selectivity of small molecules establishes a new tool for investigating interspecies or cross-kingdom interactions in isolated and co-cultures models. In addition, it facilitates the identification of potential drug candidates among bacteriocins, farnesols, and enzymes secreted by interesting microbes. Thus, the development of such selective and functionalized nanoculture system using PDMS has the potential for studying microbiome dynamics and antagonistic interactions relevant to drug discovery.

434741 Degrading the Toxic Chemicals Deposited in the Lungs

Abstract:

Lung cancer is a significant public health and economic burden in the United States. This is the most diagnosed cancer in both men and women. Lung cancer is also the leading cause of cancer death in the country. It is estimated that approximately $9.6 billion are spent in the country on lung cancer treatment alone. Lung cancer is caused due to the harmful chemicals deposited in the lungs. These chemicals are metabolically activated in the body by the enzymes present which cause DNA adducts further causing mutations forming a tumor. This tumor metastasis leading to cancer. Lung cancer is usually diagnosed after this tumor is metastasized; but we know that lung cancer is caused due to the toxic chemicals deposited. Toxic chemicals like asbestos, silica, benzene, aromatics amines, polycyclic aromatic hydrocarbons etc. are deposited in the lungs. These chemicals are deposited due to smoking (active or passive), levels of air pollution, and inhalation of harmful substances at work place. The strongest carcinogens known are benzene, PAHs like benzo[a]pyrene which lead to lung cancer in human beings. Toxicological studies on the absorption of these chemicals on the cell lines and degrading these chemicals is being done. Novel nano-composites possessing muco adhesive properties will be developed to deliver them to the alveoli in the lungs. These nano composites can trap these chemicals and eventually be removed from the body reducing the concentration of the toxic chemicals in the lungs reducing the probability of a person getting cancer.

409164 Cytosolic Delivery of Doxorubicin to Overcome Multidrug Resistance

Thursday, November 12, 2015: 5:30 PM
253A (Salt Palace Convention Center)

Jacob B Williams, Chemical Engineering, Brigham Young University, Provo, UT and William G. Pitt, Department of Chemical Enginering, Brigham Young University, Provo, UT

Cytosolic Delivery of Doxorubicin to Overcome Multidrug Resistance

Jacob B. Williams[1], William G. Pitt[1]

[1] Chemical Engineering Department, Brigham Young University

We have developed a method to deliver a payload directly to the cytosol of multidrug resistant cancer cells. The payload is encapsulated inside of a liposome along with a liquid nanodroplet of immiscible perfluorocarbon having a high vapor pressure (eLiposome) [1]. Since folate receptors are overexpressed on over 1/3 of cancer cells [2], folate is attached to the surface of the liposome as a targeting ligand to induce folate-mediated endocytosis. Insonation causes the nanodroplet to expand to gas and rupture the liposome, thereby rapidly releasing the payload [3]. If the payload is released after the liposome has been endocytosed, then there is a sudden high concentration of drug released directly to the cytosol of the cell.

The multidrug resistance of cancer cells is often a result of an increased the number of permeability glycoprotein (P-gp) pumps on the surface of the cell that export undesired compounds out of the cytosol [4]. Normally the resistant cancer cells are able to keep the cytosol at sub-lethal levels of anti-cancer drugs, but the sudden high concentration of drug released directly to the cytosol allows the drug to have a cytotoxic response before all of it can be pumped out of the cell. Furthermore, the combination of folate-induced endocytsosis and triggered release (via ultrasound) localizes the drug to the inside of the cell and will therefore minimize negative side-effects of cancer drugs commonly observed when delivered systemically.

Our liposomes were synthesized with DPPC (dipalmitoylphosphatidylcholine) and cholesterol and phospholipid-folate ligand was inserted in the surface of the liposome to bind the liposome to the cell and induce endocytosis. The liposomes were loaded with a liquid nanodroplet of perfluorohexane and the anti-cancer drug doxorubicin (Dox). Resistant breast cancer cells (MCF-7/ADR) were grown in a 24-well plate and kept in folate-free media (RPMI) for 48 hours prior to drug delivery. Cells were then exposed to Dox at 10 μM as 1) free Dox; 2) Dox in eLiposomes (eLipoDox); 3) Dox in eLiposomes with folate ligands incorporated in the surface (folated eLipoDox). The media was aspirated 2 hours after the drug was administered and replaced with fresh media.

Ultrasound (1W/cm^2) was applied to some of the cells. Cell viability was measured with a trypan blue exclusion after 24 hours.

As shown in Figure 1, ultrasound applied to folated eLipoDox (red solid bar) is more effective at killing MCF-7/ADR cells than the other conditions tested. If ultrasound is not applied to folated eLipoDox (red hatched bar) or folate is not incorporated into the eLipoDox (yellow bars) there is no significant killing of MCF-7/ADR cells. This suggests that both ultrasound and folate ligands are necessary to achieve significant killing of resistant breast cancer cells.

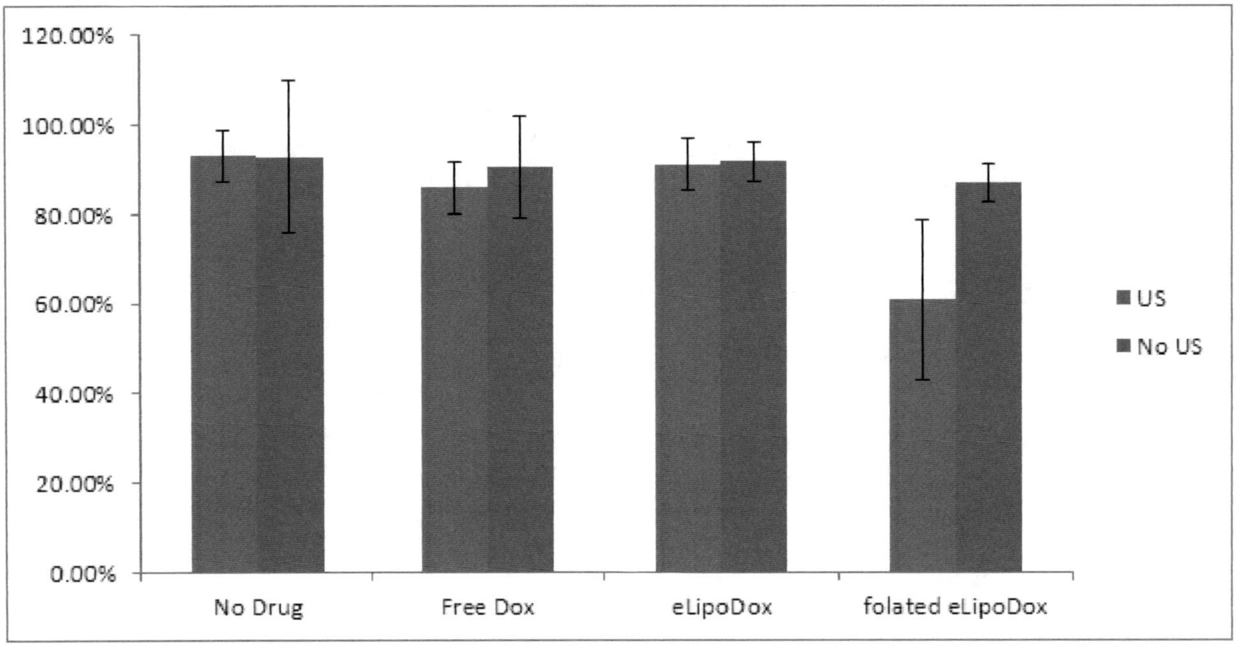

Figure 1. Viability of MCF-7 resistant cells to treatment with Dox, eLipodox, and folated eLipodox. Each drug formulation had a 10 μM Dox concentration. Ultrasound was applied at 20-kHZ, 1 W/cm^2. Error bars indicate the standard deviation (n = 3). Folated eLipoDox with US cell viability is statistically different from all other conditions except for No Drug without US.

References

[1] M. Javadi, W.G. Pitt, D.M. Belnap, N.H. Tsosie, J.M. Hartley, Encapsulating Nanoemulsions Inside eLiposomes for Ultrasonic Drug Delivery, Langmuir, 28 (2012) 14720-14729.

[2] L.A. Bareford, P.W. Swaan, Endocytic mechanisms for targeted drug delivery, Adv Drug Deliver Rev, 59 (2007) 748-758.

[3] M. Javadi, W.G. Pitt, C.M. Tracy, J.R. Barrow, B.M. Willardson, J.M. Hartley, N.H. Tsosie, Ultrasonic gene and drug delivery using eLiposomes, J Control Release, 167 (2013) 92-100.

[4] M.M. Gottesman, T. Fojo, S.E. Bates, Multidrug resistance in cancer: role of ATP-dependent transporters, Nature reviews. Cancer, 2 (2002) 48-58.

AUTHOR INDEX

AUTHOR INDEX

AUTHOR INDEX

AUTHOR INDEX

AUTHOR INDEX

AUTHOR INDEX

AUTHOR INDEX